U0412682

# Java Web实操

## 基于IntelliJ IDEA、JDBC、Servlet、Ajax、Nexus、Maven

高洪岩 / 编著

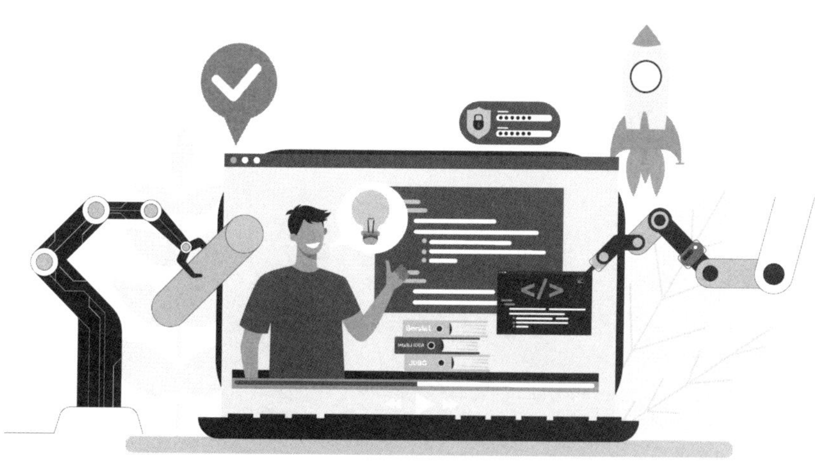

電子工業出版社
Publishing House of Electronics Industry
北京•BEIJING

## 内 容 简 介

本书根据实战项目的技术需求，垂直讲解技术要点，内容简洁，不绕弯，不拖沓，实用性强。通过对本书 16 章内容的学习，读者可以独立完成 IntelliJ IDEA 软件开发项目，使用 JDBC 操作 Oracle/MySQL 数据库，使用 Servlet 技术开发基于 B/S 架构的 Java Web 项目，还有 Cookie，HttpSession，ServletContext，Filter，Listener，JSTL/EL 等 Java Web 的核心技术，以及基于 Web 环境的 AJAX 异步编程，还有在 IntelliJ IDEA 中使用 Maven 搭建实战的软件开发环境，包括 Nexus 私服的搭建、父子模块、依赖、聚合、继承等常用的方式。

在章节安排上，本书遵循学习的连贯性，前面的知识点都是后面章节要使用的技术，以提升读者学习的效率。



**图书在版编目（CIP）数据**

Java Web 实操：基于 IntelliJ IDEA、JDBC、Servlet、Ajax、Nexus、Maven / 高洪岩编著. —北京：电子工业出版社，2021.11
ISBN 978-7-121-42166-2

Ⅰ. ①J… Ⅱ. ①高… Ⅲ. ①JAVA 语言-程序设计 Ⅳ. ①TP312.8

中国版本图书馆 CIP 数据核字（2021）第 202183 号

责任编辑：李淑丽
印　　刷：三河市良远印务有限公司
装　　订：三河市良远印务有限公司
出版发行：电子工业出版社
北京市海淀区万寿路 173 信箱　　　邮编：100036
开　　本：787×980　1/16　印张：40.75　　字数：796 千字
版　　次：2021 年 11 月第 1 版
印　　次：2021 年 11 月第 1 次印刷
定　　价：138.00 元

凡所购买电子工业出版社图书有缺损问题，请向购买书店调换。若书店售缺，请与本社发行部联系，联系及邮购电话：（010）88254888，88258888。

质量投诉请发邮件至 zlts@phei.com.cn，盗版侵权举报请发邮件至 dbqq@phei.com.cn。

本书咨询联系方式：（010）51260888-819，faq@phei.com.cn。

# 前　言

每当有 Java 学习者问我，怎么样才能学好 SSM（Spring-SpringMVC-MyBatis）时，我都会陷入沉思，原因并不是不能立即给予答案，而是内心非常愧疚，又有学习者遇到了迷茫点，而我能为他们做些什么呢？这就是本书出版的主要原因。

有 SSM 开发经验的人都知道，SSM 的基础是 Java SE 和 Java Web。如果在学习 SSM 之前没有打好 Java Web 的基础，则往往会遇到前面学习者同样的问题。因为 SSM 内部就是把 Java Web 相关的技术进行封装，如果不知道 Java Web 基础的技术原理，又怎么能学好和写好 SSM 呢？

笔者认为，在从 Java SE 到 SSM 的过程中，最关键的技术是 Java Web 开发，它能起到承上启下的作用。Java Web 能把 Java SE 传统的控制台开发方式转移到基于 B/S 架构的 Web 开发，能把从 Java SE 中学习的技术点应用到实际的 Web 场景。学习 Java Web 后，读者能立即做一些小项目，如记事本、博客、留言板、企业网站等，增加了他们继续学习的信心。当看到自己写的程序成功运行在互联网上时，那种学会之后的成就感和自信心，只有经历过的人才能体会。

本书的写作风格是以案例的形式结合手把手式的教学，每一个小节都是一个技术点，每一章就是一个学习目标和学习结果。本书由浅入深地介绍知识点，中间还穿插一些浏览器开发者工具的使用，还深入 Tomcat 源代码探究技术的实现原理、剖析 JDBC 驱动源代码内部的细节等，力求尽可能多地介绍更有实战价值的知识点，让读者有所收获。

全书共分为 16 章：

第 1 章主要介绍 IntelliJ IDEA 开发工具的 50 多个常用的使用技巧，这些都是笔者在开发中高频率使用的，有些技巧能大幅提升开发效率。学习后，也许你会感叹

IDE 具有如此强大的功能，操作方便，插件丰富，可以对提升开发效率产生事半功倍的效果。

第 2 章主要介绍 IntelliJ IDEA 开发工具的核心技能，如调试程序代码、创建非 Maven 和 Maven 的 Java Web 项目、导出 jar 和 war 包文件、创建非 Maven 和 Maven 的多模块项目环境，还介绍了非 Maven 环境下的多 Web 模块的项目搭建，这些都是 Java 开发要创建的项目类型。本书基本上把常用的项目搭建场景都做了介绍，涵盖面较广。

第 3 章主要介绍 JDBC 的核心技术，包含 JDBC 接口的介绍、不同注册驱动写法的区别、使用 JDBC 实现基本的 CURD 操作、如何避免 SQL 注入、预编译的原理、JDBC 使用 finally 的必要性、多条件查询、封装 DAO 类等。

第 4 章主要介绍 JDBC 的实战技术，本章也是学习 JDBC 的重点内容。可以说，MyBatis 框架中提供的核心功能的内部实现在本章都有介绍，如 JDBC 结合高性能的连接池框架 HikariCP、在 JDBC 中如何处理事务、转账操作发生异常时事务为什么不回滚及解决办法、使用 JDBC 操作 CLOB 和 BLOB 类型的数据、Batch 批处理的使用、操作 Date 数据类型。另外，还介绍了 ACID 特性及对脏读、不可重复读和幻读的理论解释。最为重要的是，本章用大量篇幅介绍了事务隔离性的代码实现，以及与脏读、不可重复读和幻读的关系，并配有完整的代码进行论证，遵守“少废话，上代码”的写作方式。

第 5 章主要介绍 Servlet 的核心技术，属于 Java Web 最核心的技术，详细介绍了 Servlet 接口、ServletConfig 接口和 Servlet 接口的关系，以及接口中全部 API 的应用实现，还详细介绍了 HttpServletRequest 和 HttpServletResponse 接口中常用 API 的使用，以及两种请求提交方式在传输内容及格式上的区别，并使用 Servlet+JDBC 实现基于 Web 的 CURD。

第 6 章主要介绍如何使用 Cookie 对象，在此基础上结合浏览器开发者工具，把 Cookie 在请求和响应中的传输过程进行可视化介绍，这样有助于读者学习 HttpSession，因为 HttpSession 的技术原理就是 Cookie。另外，还介绍了对 Cookie

进行 CURD 操作、在 Cookie 中读写中文等常用案例。

第 7 章主要介绍如何使用 HttpSession 接口，同时结合浏览器开发者工具，从内部执行流程上查看 Cookie 和 HttpSession 的关系和完整的处理过程，还介绍了 HttpSession 中的 URL 重写技术、使用 HttpSession 实现简易版购物车等案例。可以说，通过对本章内容的学习，读者会对 HttpSession 的原理有更加深入的了解。

第 8 章主要介绍如何使用 ServletContext 接口，介绍了 ServletContext 接口的使用场景、与初始化参数的结合使用、实现 charset 编码可配置等常用案例。

第 9 章主要介绍如何使用 Filter 接口，该接口是 Java Web 开发中比较常用的技术点，主要介绍了 Filter 的生命周期、私有/公共初始化参数的使用、使用注解声明 Filter、过滤链、使用 Filter 实现编码处理、过滤转发和请求、实现权限验证、结合 ThreadLocal 对 CookieTools 进行解耦等常用的使用案例。

第 10 章主要介绍如何使用 Listener 接口，将常用的 Listener 接口结合代码进行详细介绍。

第 11 章主要介绍 JSP，JSTL 和 EL 必备技术，包含 JSP 和 Servlet 的关系、常用 JSP 指令的使用、JSP 内部对象的使用、JSTL 和 EL 的使用，以及设计纯正的基于 MVC 模式的分层应用。

第 12 章主要介绍异步处理 Ajax 技术，包含无参和有参、无返回值和有返回值等常见场景的使用、get()和 post()两种方法的使用、异步和同步在执行上的区别、formdata 和 payload 在提交格式上的区别、实现 form 有刷新和 Ajax 无刷新文件上传等实战案例。

第 13 章主要介绍搭建 Maven Nexus 私服环境，包含下载 Nexus、安装 Nexus、配置 Nexus，还介绍了 hosted，proxy 和 group 仓库的区别及在使用上的联系，并结合 IDEA 实现 RELEASE 和 SNAPSHOT 版本的管理。

第 14 章主要介绍在 IDEA 中处理 Maven 项目的生命周期，并把 IDEA 中与 Maven 生命周期有关的菜单以案例的方式进行详细讲解。

第 15 章主要介绍 Maven 依赖的应用，包含依赖范围、依赖调解、可选依赖、排除依赖、集中处理版本、显示依赖结构、源代码打包、跳过测试等常用使用方式。

第 16 章主要介绍 Maven 的聚合与继承，包含使用聚合和继承的目的及优势，并在 IDEA 环境下创建这两种环境，目的是在实际的软件项目中更好地管理 Maven 项目。

虽然在此不能全部罗列本书讲解的技术点，但相信读者用心阅读本书后一定会有所收获。笔者对本书的整理花费大量精力，力求使 Java Web 技术最核心的内容得以体现，因为不管是现阶段的 SSM 框架还是未来更高级 Java Web 框架的出现，它们的核心原理还是这些内容，因此，学好这些内容，再学习任何框架都不会怕。

本书的出版离不开背后辛勤工作的朋友，非常感谢李淑丽编辑对本书出版的付出，李姐，辛苦了！还要感谢那些与我并没有直接交集的编辑们，虽然与他们素不相识，但内心依然有很真诚的感谢。最后，也要感谢我的家人，我的爸爸、妈妈和老婆，还有我最可爱的儿子高晟京，看到你们为家庭默默地付出，我该做些什么予以报答呢？只有好好学习，好好工作！祝所有人身体健康。

高洪岩　于天津工业大学

2021-07-06

# 目　　录

# 第 1 章
# IntelliJ IDEA 常用技巧

本章采用功能列举的方式，将 IntelliJ IDEA 比较常用的知识点进行展示和说明。

JetBrains 是捷克的一家软件公司，成立于 2000 年，总部位于捷克的首都布拉格，该公司最为著名的产品是 IDE 集成开发工具 IntelliJ IDEA。在 Java 开发中常说的 IDEA 就是指 IntelliJ IDEA。

IntelliJ IDEA 的 1.0 版本发布于 2001 年 1 月。

IntelliJ IDEA 包含针对 Java 语言的 IDE（Integrated Development Environment，集成开发环境），它的市场占有率遥遥领先，是现在最流行的 IDE，这也是我们学习 IntelliJ IDEA 的主要原因。

IntelliJ IDEA 通过安装插件可以支持 PHP，Python，Ruby，Scala，Kotlin 等主流的编程语言。

JetBrains 公司的 Logo 如图 1-1 所示，下载 IntelliJ IDEA 的页面如图 1-2 所示。

图 1-1

图 1-2

单击“Download”按钮进入下载页面，效果如图 1-3 所示。

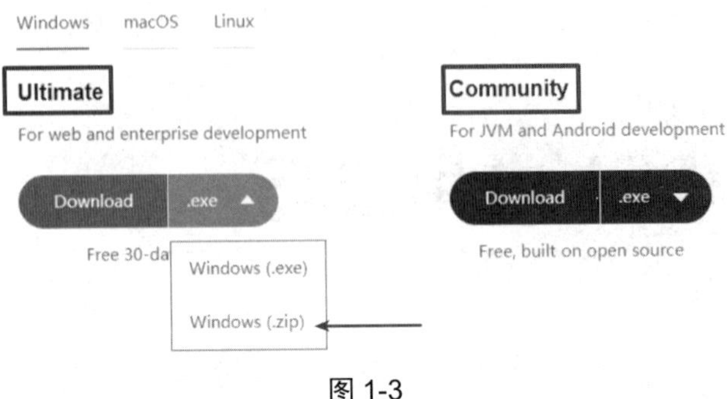


图 1-3

IntelliJ IDEA 官方提供了两个版本：一个是 Ultimate 旗舰商业版本，功能最多，需要付费，本章使用其进行介绍；另外一个是功能较少的 Community 社区版本，免费使用。Android Studio 是 Google 基于 IntelliJ IDEA 的 Community 社区版再结合 Android 插件进行升级开发的。

## 1.1 初次配置IntelliJ IDEA

将下载的“idea.zip”文件解压，效果如图 1-4 所示。

进入 bin 文件夹，双击“idea64.exe”文件可执行安装程序，如图 1-5 所示。

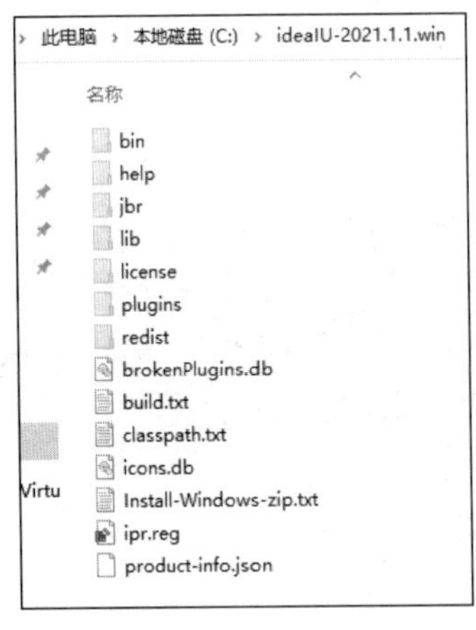


图 1-4

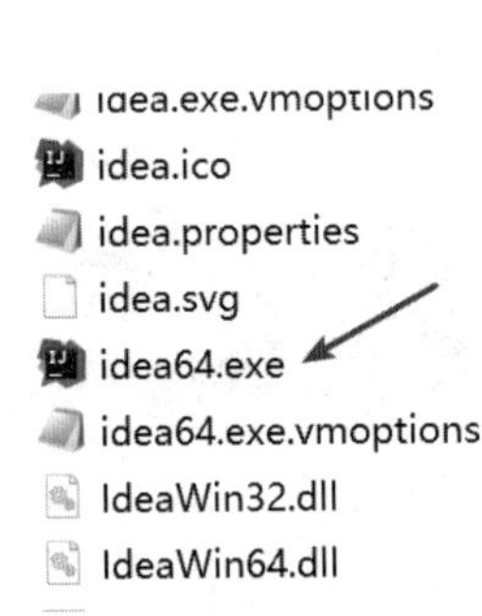


图 1-5

进入如图 1-6 所示的界面，接受许可，单击“Continue”按钮；单击“Don't Send”按钮，如图 1-7 所示。

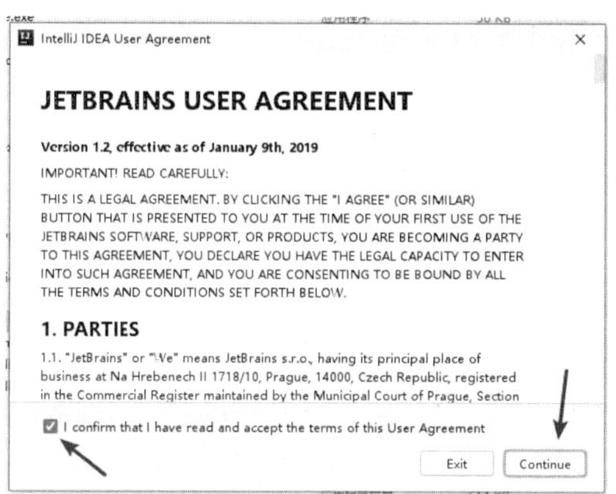


图 1-6

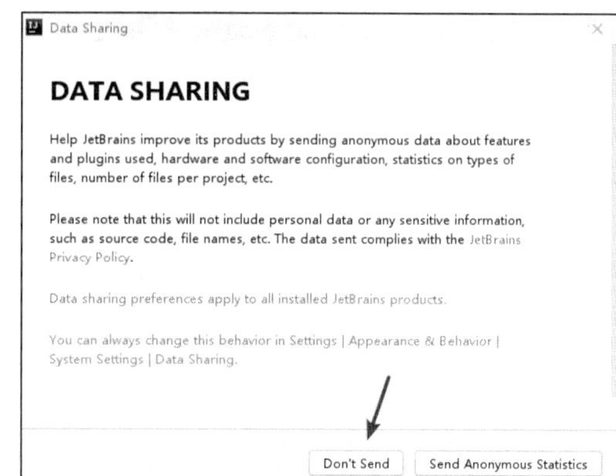


图 1-7

使用 30 天试用版本，如图 1-8 所示；单击“Continue”按钮，如图 1-9 所示。

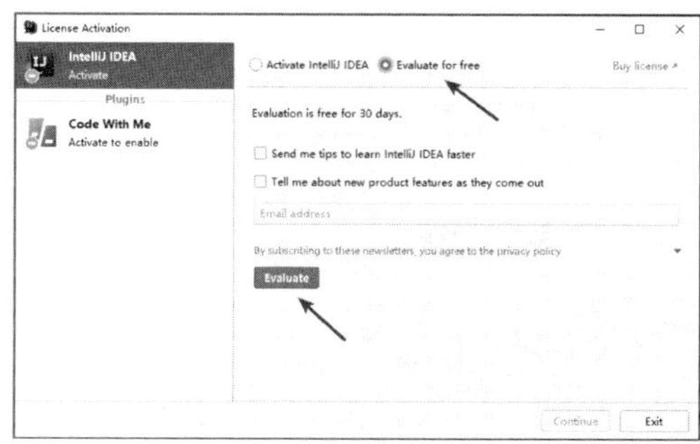


图 1-8

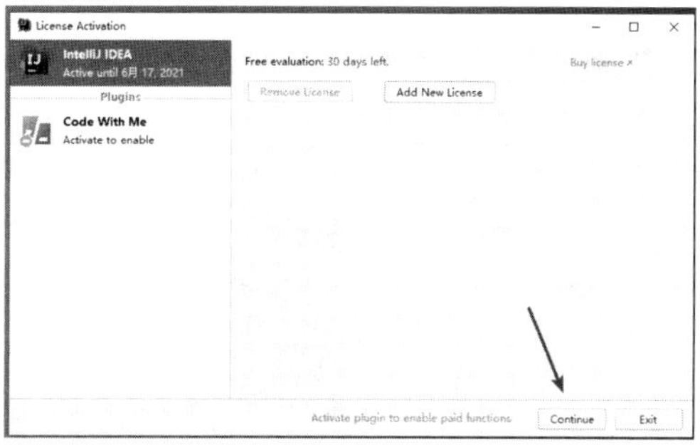

图 1-9

显示欢迎界面，如图 1-10 所示。在此界面中可以进行如下 3 项操作。

（1）New Project：创建项目。

（2）Open：打开项目。

（3）Get from VCS：从版本控制系统中获得项目。

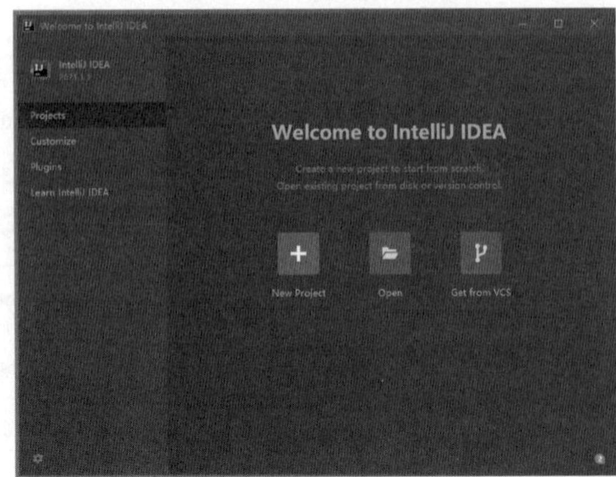


图 1-10

## 1.2 创建Java项目

本节介绍在 IDEA 中创建 Java 项目的方法。

### 1.2.1 关联 JDK 并创建 Java 项目

单击“New Project”按钮创建新的项目，弹出界面如图 1-11 所示。

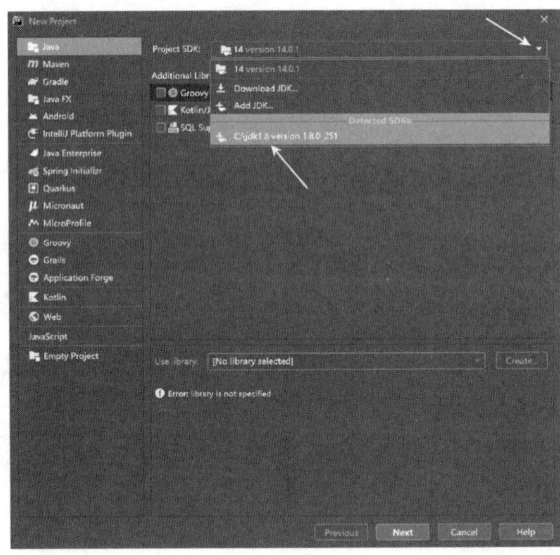


图 1-11

如果在当前 IntelliJ IDEA 环境中没有关联任何版本的 JDK，则可以单击“Add JDK...”菜单进行关联。界面中的其他选项选择默认即可，如图 1-12 所示，单击“Next”按钮。

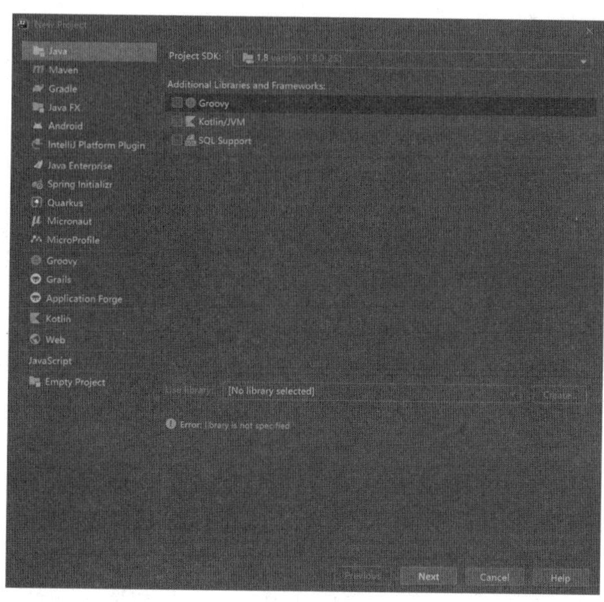


图 1-12

弹出界面如图 1-13 所示，因为创建的项目不需要依赖于某个模板，所以此界面使用默认配置即可，单击“Next”按钮。

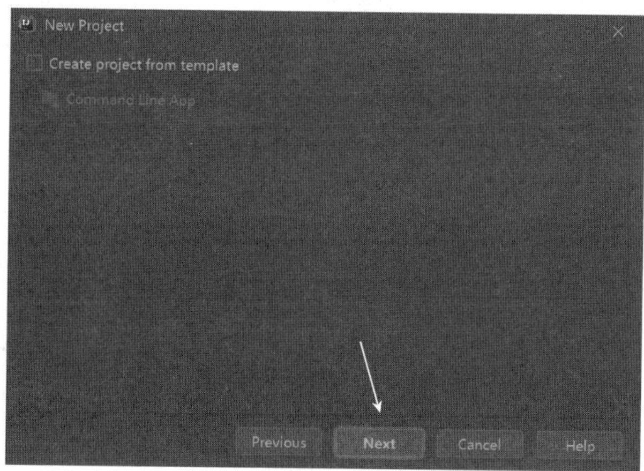


图 1-13

弹出界面如图 1-14 所示，设置项目名称为 test1，在此界面中还可以设置项目保存的路径。单击“Finish”按钮完成 Java 项目的创建。

### 1.2.2 查看项目结构

进入 IDEA 主界面，单击“Project”标签，生成项目结构如图 1-15 所示，其中节点 src 就是存放 Java 类的位置。

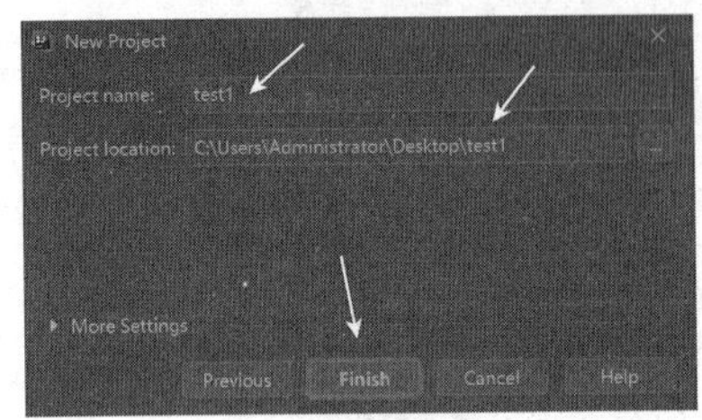


图 1-14

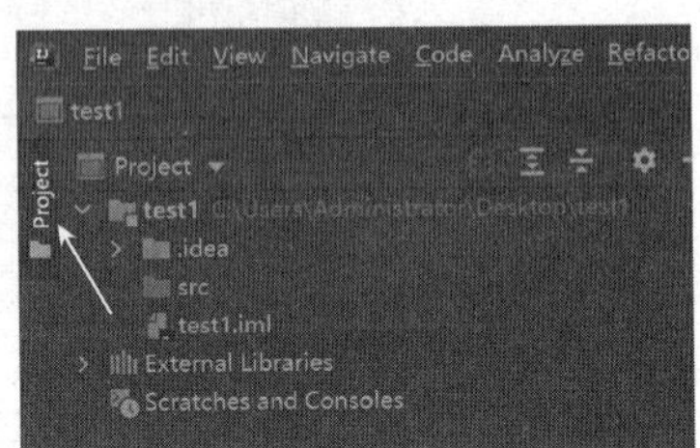


图 1-15

### 1.2.3 创建 Java 类

在 src 节点上右击，创建 Java 类，如图 1-16 所示。

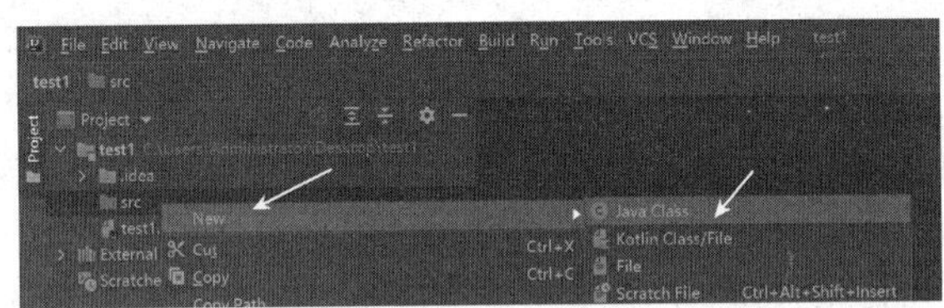


图 1-16

界面如图 1-17 所示，输入“test1.Test1”代表在 test1 包中创建名称为 Test1 的 Java 类，按“Enter”键开始创建 Java 类，显示 Test1.java 类的代码如图 1-18 所示。

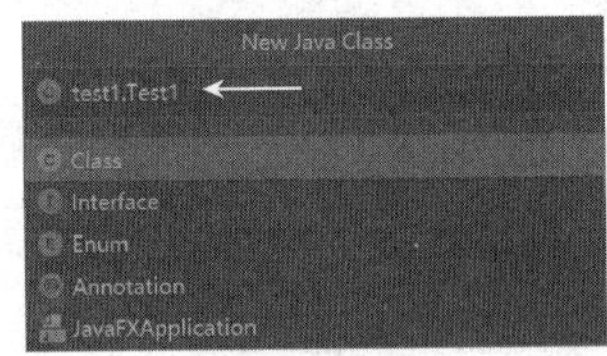


图 1-17

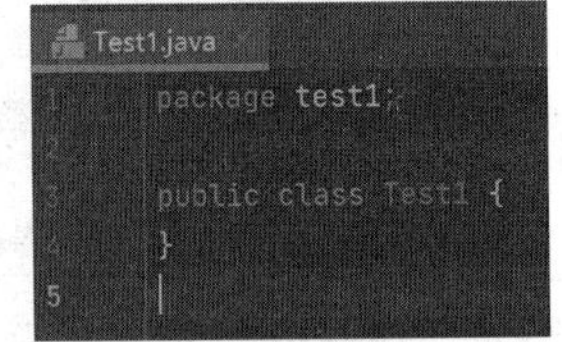


图 1-18

由于 Test1.java 类中没有 main()方法，因此在输入“main”后会自动出现创建 main()方法的提示，如图 1-19 所示。成功生成 main()方法，代码如图 1-20 所示。

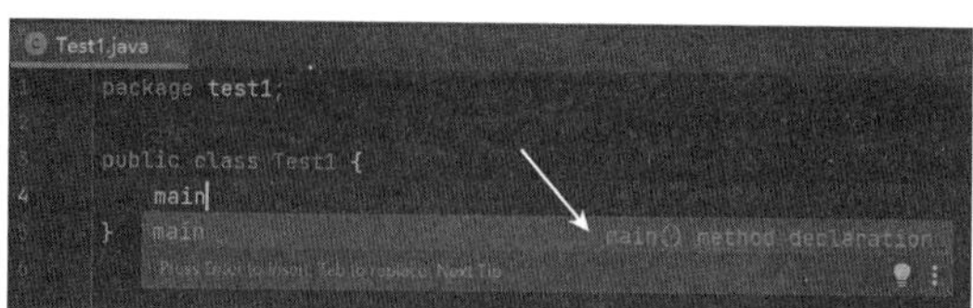

图 1-19

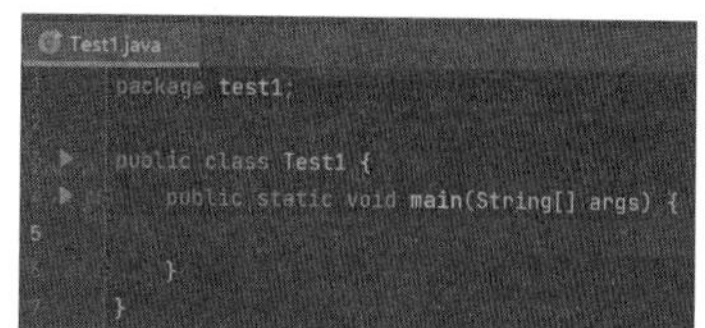

图 1-20

更改代码如下：

```
package test1;

public class Test1 {
    public static void main(String[] args) {
        System.out.println("Hello IntelliJ IDEA ~");
    }
}
```

### 1.2.4　运行 Java 类

单击“Run'Test1.main()'”选项，如图 1-21 所示，成功运行的效果如图 1-22 所示。

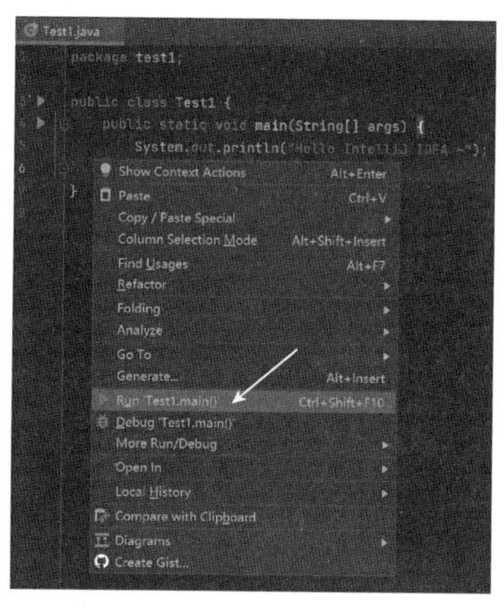

图 1-21

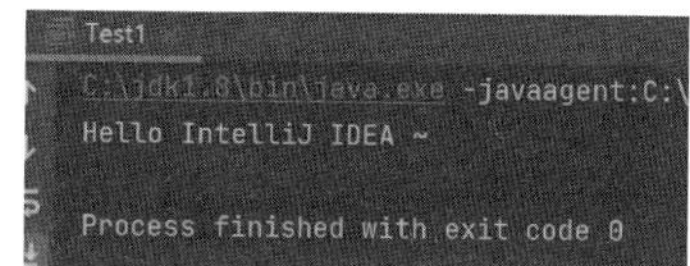

图 1-22

## 1.3　IntelliJ IDEA相关配置与使用技巧

IntelliJ IDEA 版本的差异会造成界面的不同，我们仅以当前使用的 IntelliJ IDEA 版本的界面作为参考。

### 1.3.1 配置界面皮肤

单击“Settings...”菜单，如图 1-23 所示，配置“IntelliJ Light”选项如图 1-24 所示。

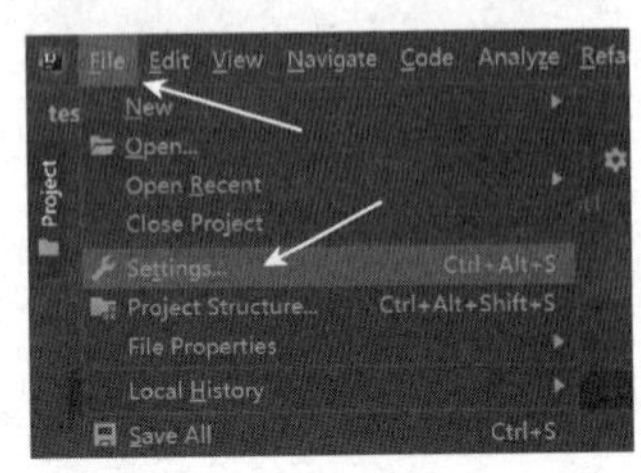


图 1-23

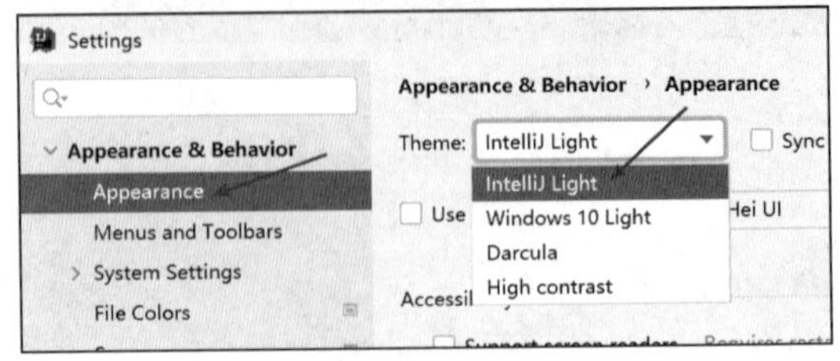


图 1-24

### 1.3.2 设置 Eclipse 风格的快捷键

在默认情况下，IntelliJ IDEA 的快捷键和 Eclipse 的不一样，但可以将 IntelliJ IDEA 的快捷键设置成 Eclipse 风格的。本实验使用 Eclipse 风格的快捷键进行讲解。

设置 Eclipse 风格快捷键的方法如图 1-25 所示，设置成功后，立即生效。

### 1.3.3 更改代码编辑器文字大小

如果代码编辑器中的程序代码文字过小，则不利于查看代码。下一步增加代码编辑器显示的字号，设置字号为 20，如图 1-26 所示。

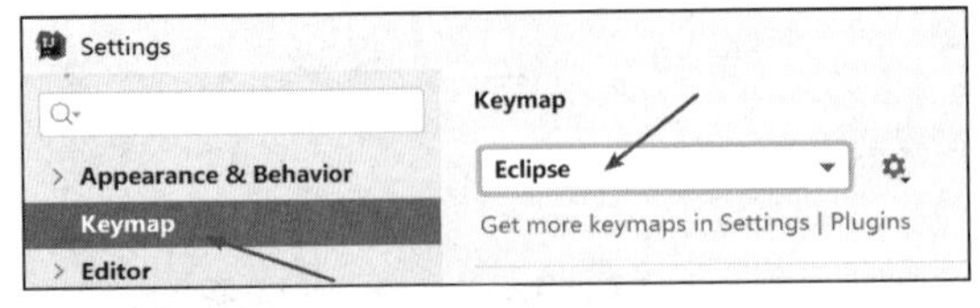


图 1-25

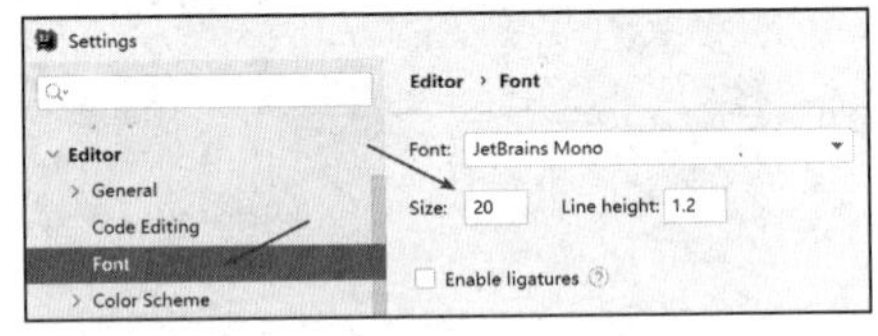


图 1-26

设置成功后，字的大小发生改变，此设置也会改变控制台输出信息的文字大小。

还有一种更加便捷的方法：按住“Ctrl”键同时滚动鼠标滚轮，设置界面字号的大小，如图 1-27 所示，代码编辑区和控制台输出信息的文字大小都可以通过此方式设置。

### 1.3.4 快速生成代码

使用 sout+Enter/Tab 键可以快速生成代码，效果如图 1-28 所示。

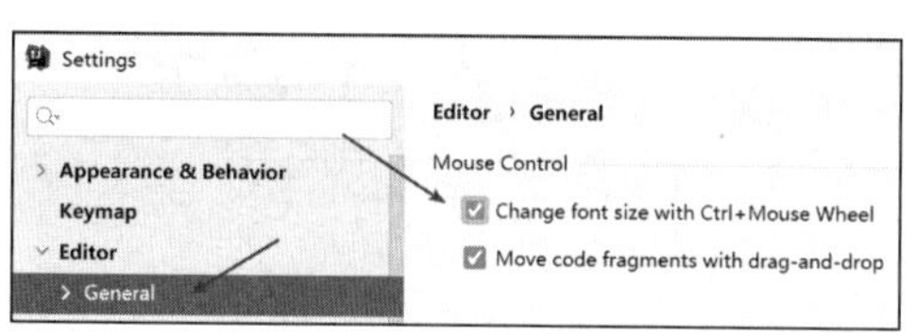


图 1-27

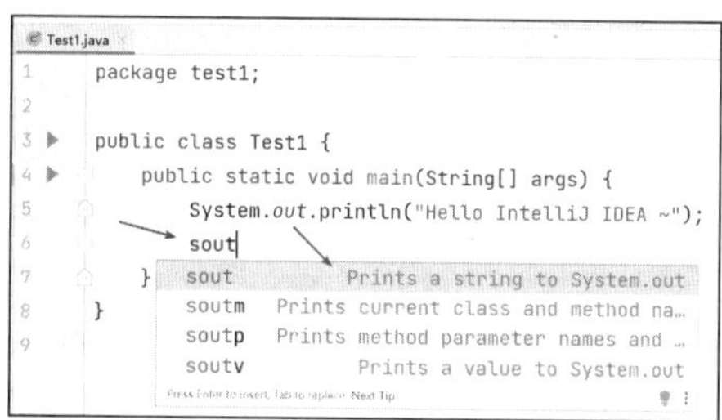


图 1-28

## 1.3.5　设置快捷键“Alt+/”

当在 Eclipse 中按快捷键“Alt+/”时，会自动弹出代码提示，但在 IntelliJ IDEA 中却没有弹出提示，如图 1-29 所示。

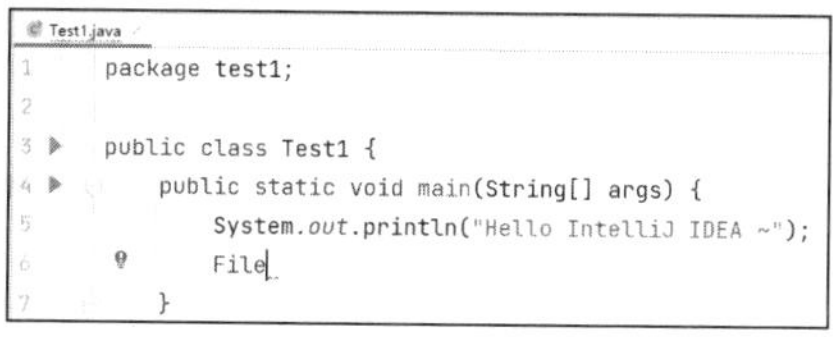

```
Test1.java
package test1;

public class Test1 {
    public static void main(String[] args) {
        System.out.println("Hello IntelliJ IDEA ~");
        File
    }
```

图 1-29

这时，我们可以进入“Keymap”界面进行设置，依次选择“Main menu”→“Code”→“Code Completion”选项，右击“Cyclic Expand Word”的快捷键，如图 1-30 所示。

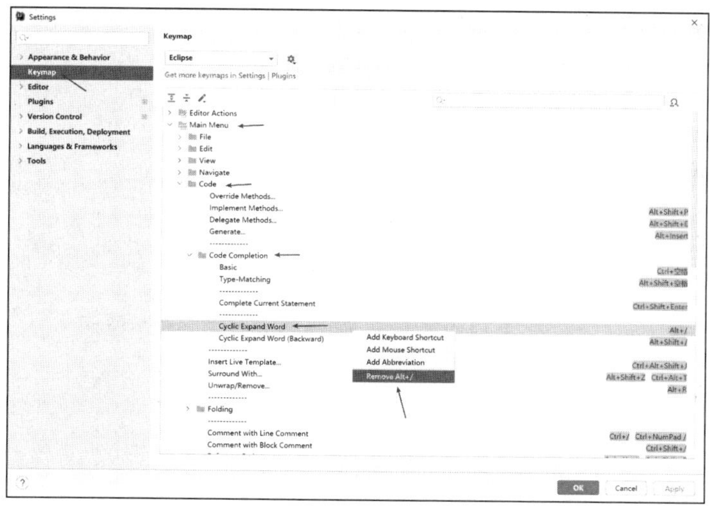


图 1-30

然后对 Basic 执行右击，在弹出的菜单中单击“Remove Ctrl+空格”选项，如图 1-31 所示。

再对 Basic 执行右击，选择“Add Keyboard Shortcut”选项，如图 1-32 所示。

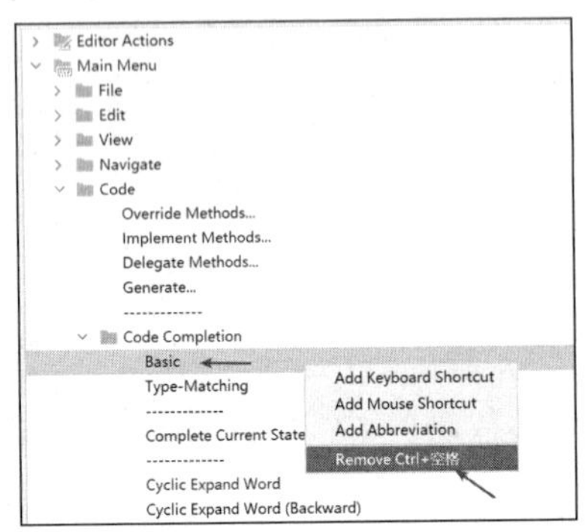


图 1-31

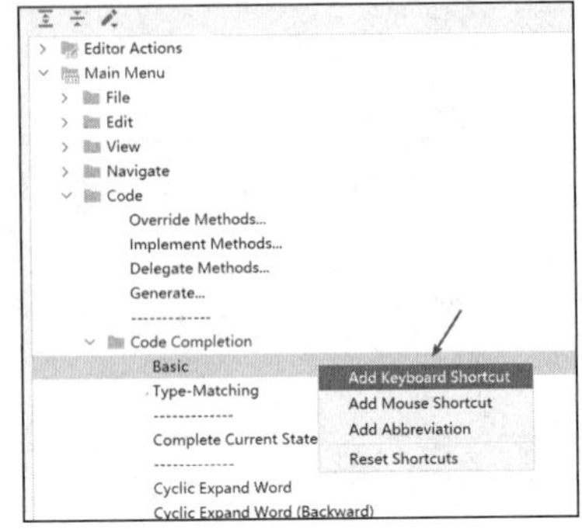


图 1-32

设置快捷键“Alt+/”，如图 1-33 所示，设置成功的效果如图 1-34 所示。

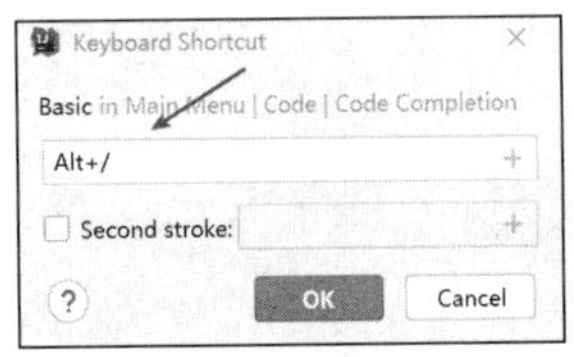


图 1-33

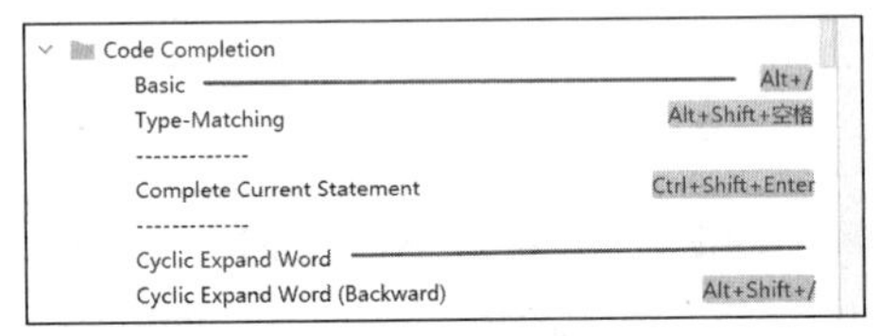


图 1-34

这时，在代码编辑器中就可以使用“Alt+/”快捷键实现代码自动完成的功能了，效果如图 1-35 所示。

## 1.3.6 设置代码完成对大小写不敏感

虽然成功设置了“Alt+/”快捷键代码自动完成功能，但其对大小写是敏感的，如图 1-36 所示。

下面，设置代码完成对大小写不敏感，配置界面如图 1-37 所示。

这时，输入小写的“file”也可以出现提示，如图 1-38 所示。

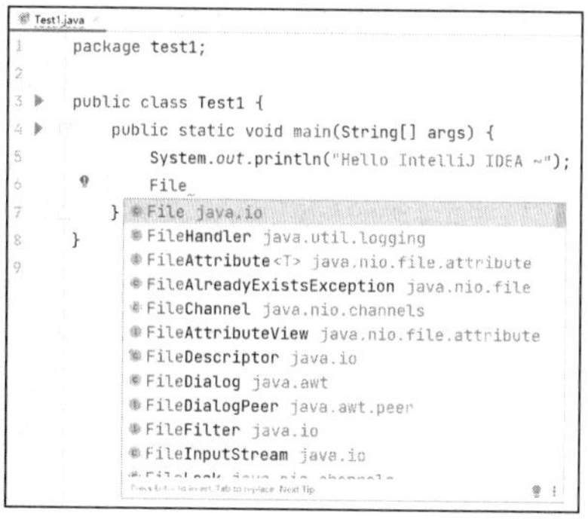

图 1-35

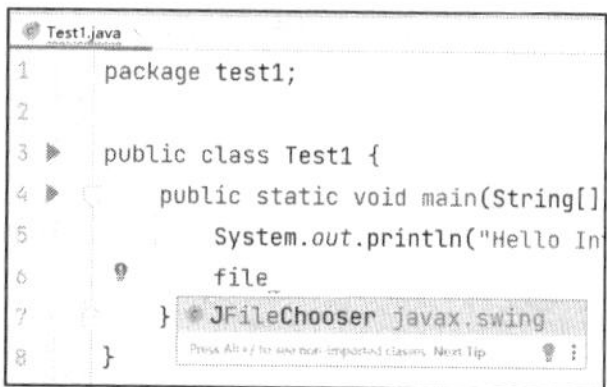

图 1-36

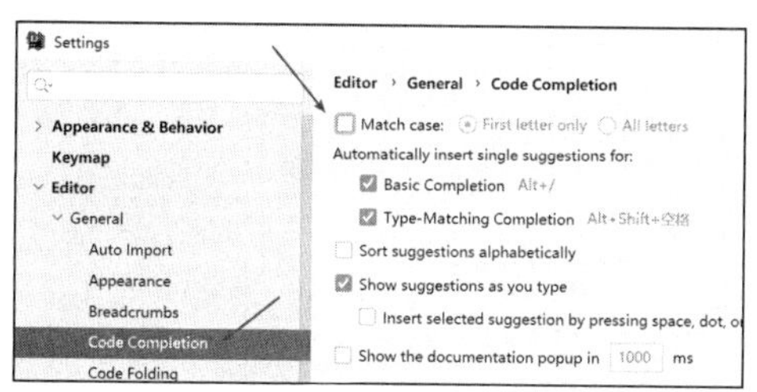


图 1-37

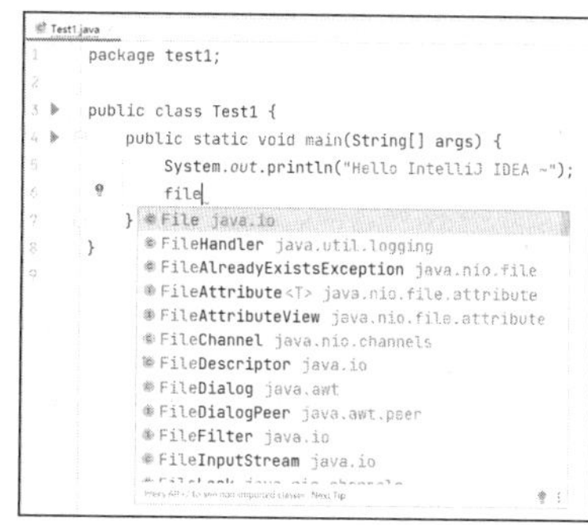


图 1-38

## 1.3.7　配置 JDK

要在 IntelliJ IDEA 中配置其他版本的 JDK，需要先单击“Project Structure...”选项，如图 1-39 所示，然后进入如图 1-40 所示界面。

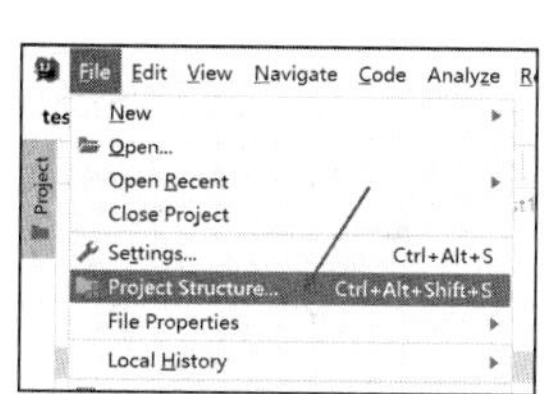


图 1-39

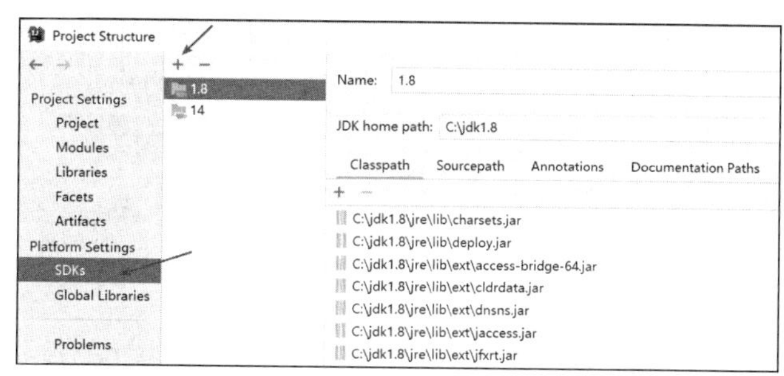


图 1-40

你可以在线下载及手动添加 JDK，如图 1-41 所示。JDK 配置完成后，可以对当前项目设置使用指定版本的 JDK，如图 1-42 所示，也可以对某一个 Module 模块设置指定版本的 JDK，如图 1-43 所示。

Project，Modules，Global Libraries 三者的区别如图 1-44 所示。

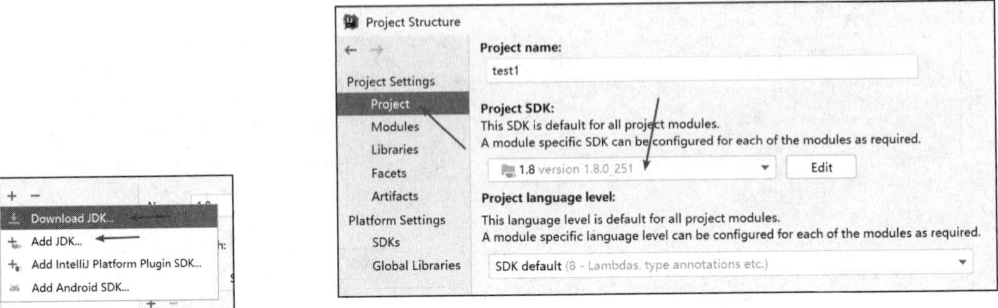

图 1-41　　　　图 1-42

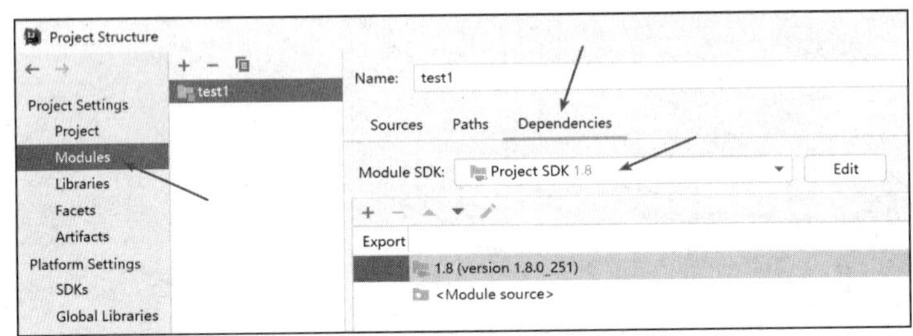

图 1-43

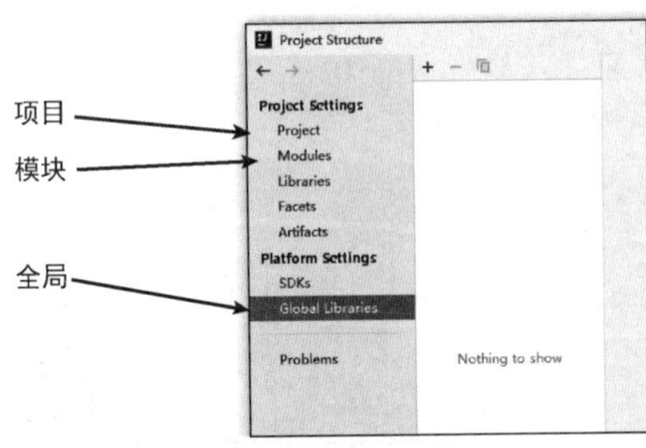


图 1-44

（1）Global Libraries（全局）：可选的 JDK 列表。

（2）Project（项目）：整体项目使用的 JDK。

（3）Modules（模块）：指定模块使用的 JDK。

### 1.3.8　设置自动编译

在默认情况下，IntelliJ IDEA 不会对保存后的.java 文件进行自动编译形成最新版的.class 文件，而是在运行项目的时候才进行编译，想实现自动编译的功能需要进行如图 1-45 所示的设置。

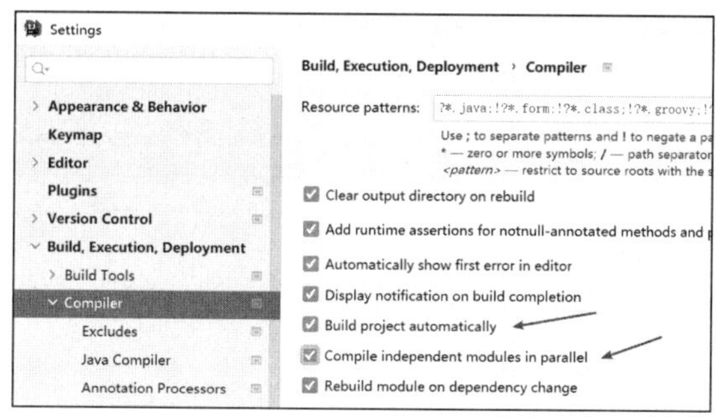


图 1-45

这时，保存一个.java 文件就会自动编译成最新版本的.class 文件了。

### 1.3.9　抛出异常

单击如图 1-46 所示的链接可以抛出异常，抛出异常的代码如图 1-47 所示。

```
package test1;

public class Test1 {
    public static void main(String[] args) {
        Class.forName("java.util.Date");
    }
}
```

图 1-46

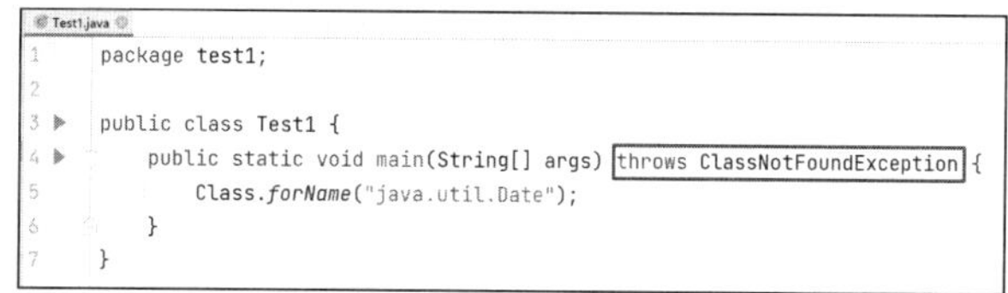


图 1-47

### 1.3.10　使用“Ctrl+Alt+T”快捷键生成代码块

当想对某一段代码使用 try/catch/finally 进行包围时，可以先选中待处理的代码，再使用“Ctrl+Alt+T”快捷键进行生成，如图 1-48 所示，生成的代码块如图 1-49 所示。

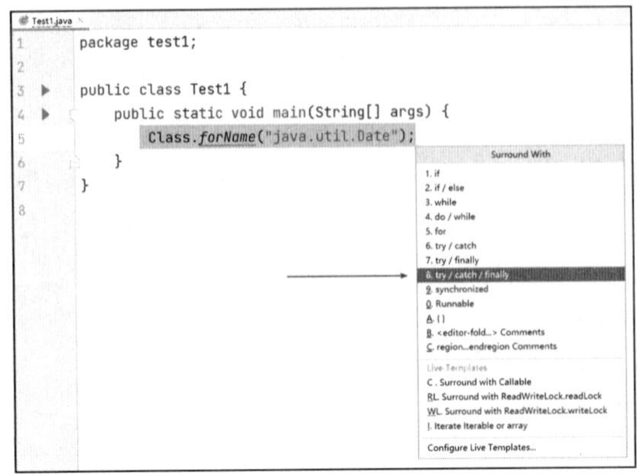


图 1-48

```
Test1.java
package test1;

public class Test1 {
    public static void main(String[] args) {
        try {
            Class.forName("java.util.Date");
        } catch (ClassNotFoundException e) {
            e.printStackTrace();
        } finally {
        }
    }
}
```

图 1-49

当然，也可以使用“Ctrl+Alt+T”快捷键生成其他代码块，代码块的种类如图 1-50 所示。

### 1.3.11 使用 Generate 菜单生成方法

首先创建新的类 Test2，并添加 3 个属性，然后在代码编辑器中对其右击，选择“Generate...”菜单，如图 1-51 所示，弹出如图 1-52 所示的菜单。

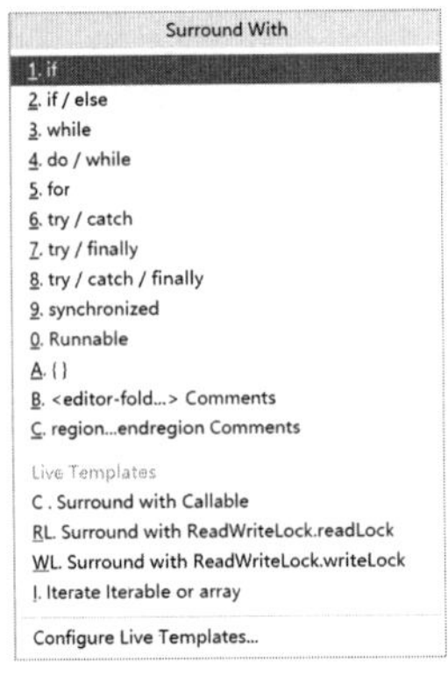


图 1-50

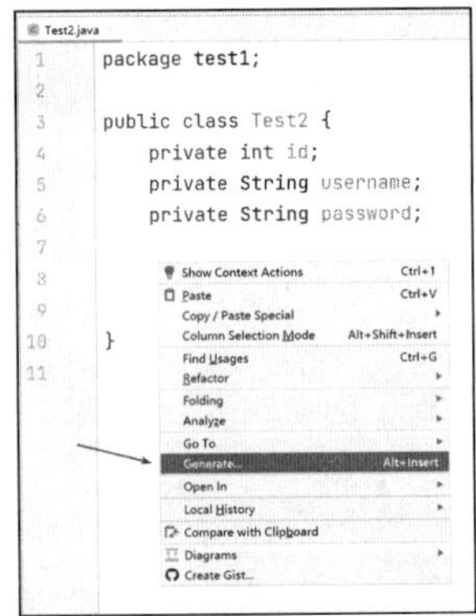


图 1-51

在该菜单中可以生成 Constructor（构造），Getter/Setter，equals() and hashCode()，Override（重写）等方法。

请注意，一定要对属性生成对应的 Getter and Setter 方法才可以生成 equals() and hashCode()方法。在生成 equals() and hashCode()方法时，要对两个选项进行打钩，如

图 1-53 所示。

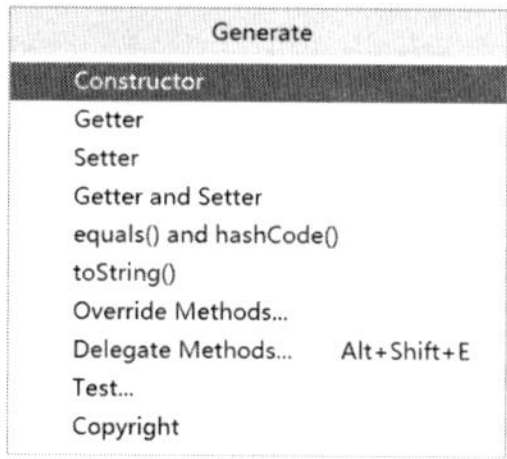


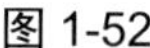
图 1-52

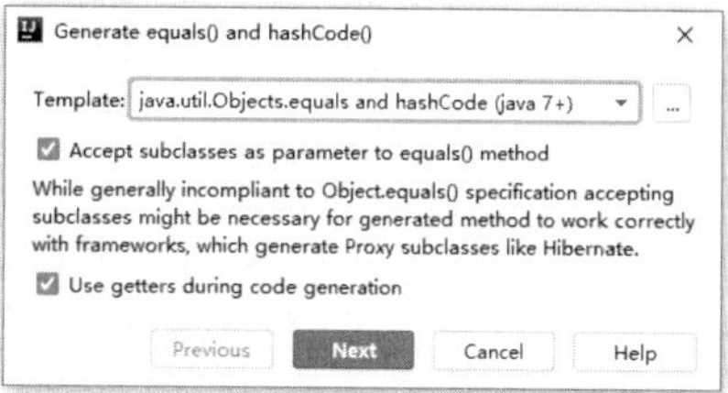


图 1-53

（1）选择“Accept subclasses as parameter to equals() method”代表可以对具有继承关系的父子类进行 equals 比较，其生成效果如图 1-54 所示。

（2）选择“Use getters during code generation”代表在获得属性值时使用 get()方法，而不是直接通过属性名访问属性值，其生成效果如图 1-55 所示。

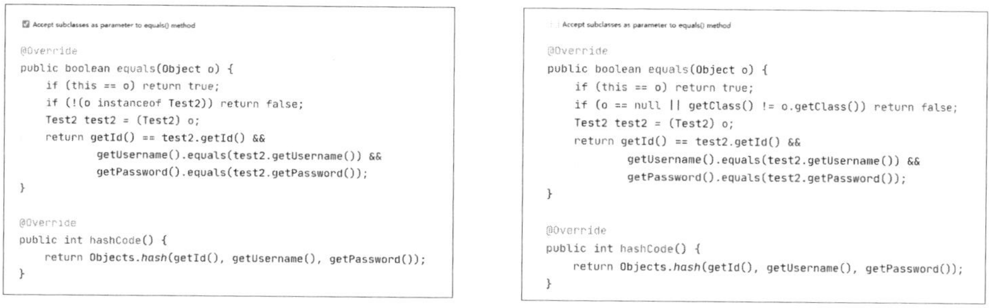

Accept subclasses as parameter to equals() method

```
@Override
public boolean equals(Object o) {
    if (this == o) return true;
    if (!(o instanceof Test2)) return false;
    Test2 test2 = (Test2) o;
    return getId() == test2.getId() &&
            getUsername().equals(test2.getUsername()) &&
            getPassword().equals(test2.getPassword());
}

@Override
public int hashCode() {
    return Objects.hash(getId(), getUsername(), getPassword());
}
```

（a）

Accept subclasses as parameter to equals() method

```
@Override
public boolean equals(Object o) {
    if (this == o) return true;
    if (o == null || getClass() != o.getClass()) return false;
    Test2 test2 = (Test2) o;
    return getId() == test2.getId() &&
            getUsername().equals(test2.getUsername()) &&
            getPassword().equals(test2.getPassword());
}

@Override
public int hashCode() {
    return Objects.hash(getId(), getUsername(), getPassword());
}
```

（b）

图 1-54

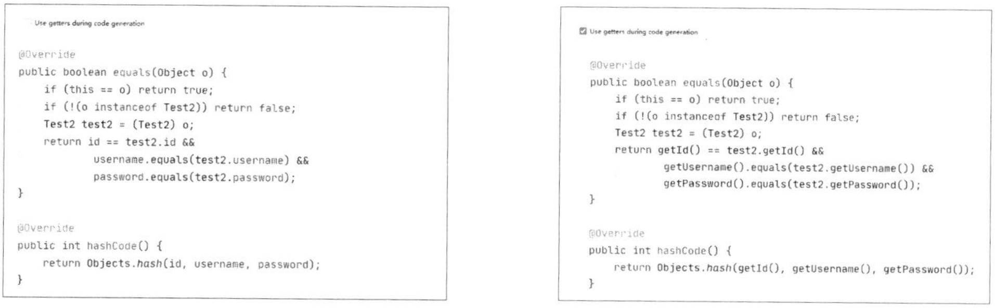

Use getters during code generation

```
@Override
public boolean equals(Object o) {
    if (this == o) return true;
    if (!(o instanceof Test2)) return false;
    Test2 test2 = (Test2) o;
    return id == test2.id &&
            username.equals(test2.username) &&
            password.equals(test2.password);
}

@Override
public int hashCode() {
    return Objects.hash(id, username, password);
}
```

（a）

Use getters during code generation

```
@Override
public boolean equals(Object o) {
    if (this == o) return true;
    if (!(o instanceof Test2)) return false;
    Test2 test2 = (Test2) o;
    return getId() == test2.getId() &&
            getUsername().equals(test2.getUsername()) &&
            getPassword().equals(test2.getPassword());
}

@Override
public int hashCode() {
    return Objects.hash(getId(), getUsername(), getPassword());
}
```

（b）

图 1-55

### 1.3.12 使用 fori 生成 for 语句

先在编辑器中输入“fori”，如图 1-56 所示；再按“Enter”键生成 for 语句，如图 1-57 所示。

```
package test1;

public class Test1 {
    public static void main(String[] args) {
        fori
    }       fori                    Create iteration loop
}           ForwardPositionIterator com.sun.org.a…
            FontRunIterator sun.font
            FormatData_ia sun.text.resources.cldr…
            FormatData_id sun.text.resources.cldr…
            FormatData_ig sun.text.resources.cldr…
            FormatData_ii sun.text.resources.cldr…
            FormatData_in sun.text.resources.in
            FormatData_in_ID sun.text.resources.in
            FormatData_is sun.text.resources.is
            FormatData_is sun.text.resources.cldr…
```

图 1-56

```
package test1;

public class Test1 {
    public static void main(String[] args) {
        for (int i = 0; i <_; i++) {

        }
    }
}
```

图 1-57

### 1.3.13 实现自动导入的功能

当把如下代码：

```
public static void main(String[] args) {
    SimpleDateFormat format;
}
```

粘贴到 Java 类中时，不会自动添加导入语句，如果想要实现自动导入的功能，则需要对图 1-58 所示的两个选项打钩。

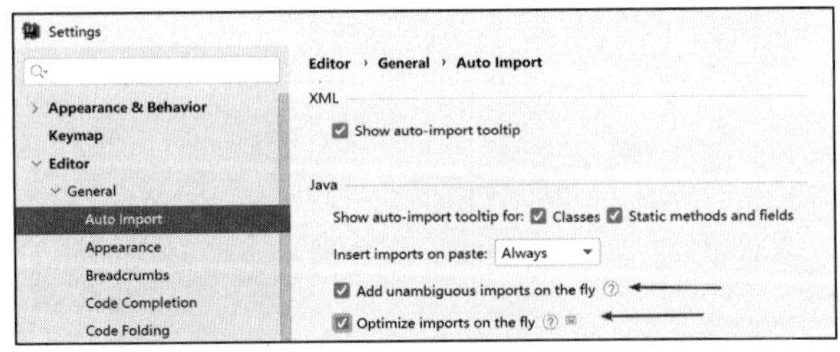


图 1-58

这时，再粘贴代码就会自动添加导入语句。

### 1.3.14 实现水平或垂直分屏的功能

在当前文件标签上右击，实现左右、上下分屏，如图 1-59 所示。

打开 5 个类，实现如图 1-60 所示的布局效果，成功的布局效果如图 1-61 所示。

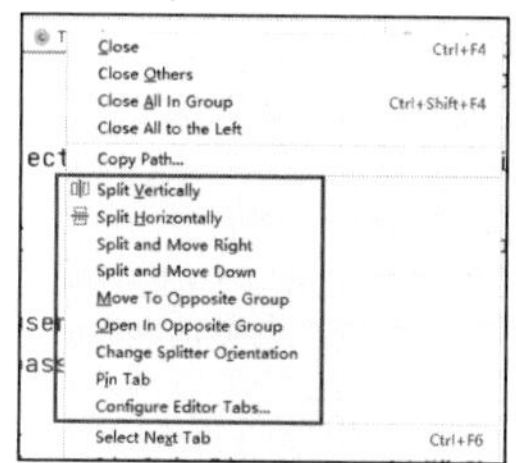


图 1-59

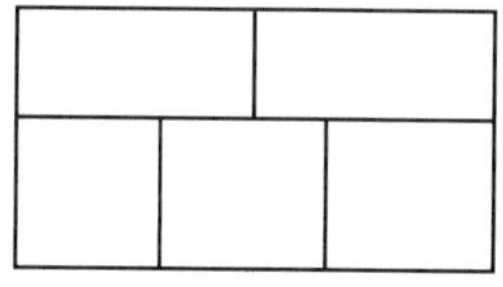

图 1-60

请自行实现如图 1-62 所示的代码编辑窗口布局效果。

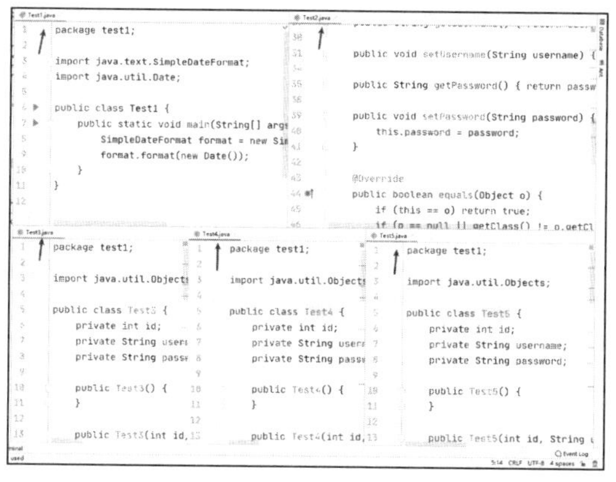

图 1-61

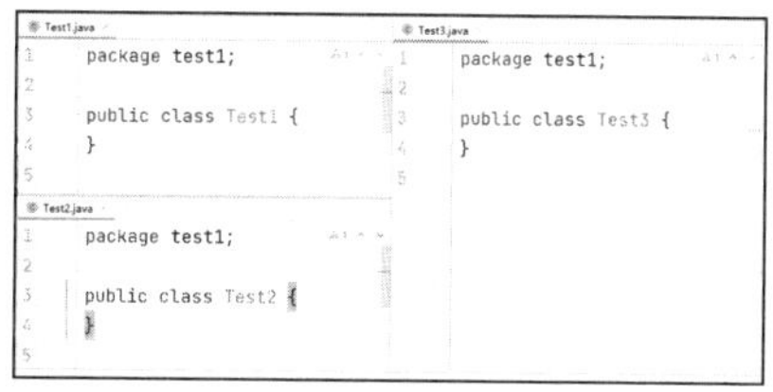


图 1-62

## 1.3.15 树形显示包结构

IntelliJ IDEA 默认以树形显示包结构，如图 1-63 所示。

如果使用平级的方式显示包结构，则需要使“Flatten Packages”选项呈选中状态，如图 1-64 所示。这时，包结构以平级方式显示，如图 1-65 所示。

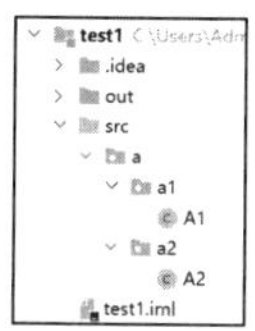


图 1-63

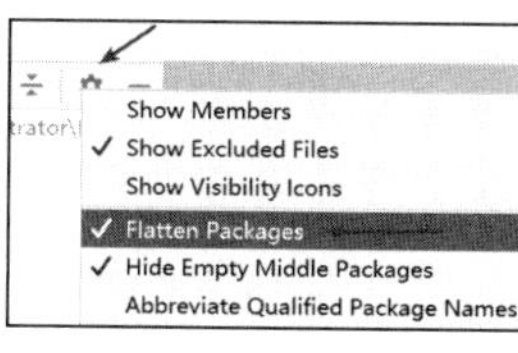


图 1-64

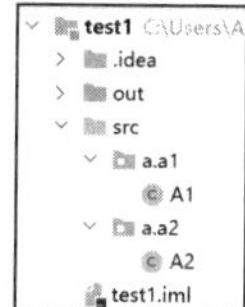


图 1-65

### 1.3.16 设置“F2”键用于改名

设置“F2”键用于改名的方法如图 1-66 所示。重启 IDEA，这时选中类并按“F2”键，会弹出改名界面，如图 1-67 所示。

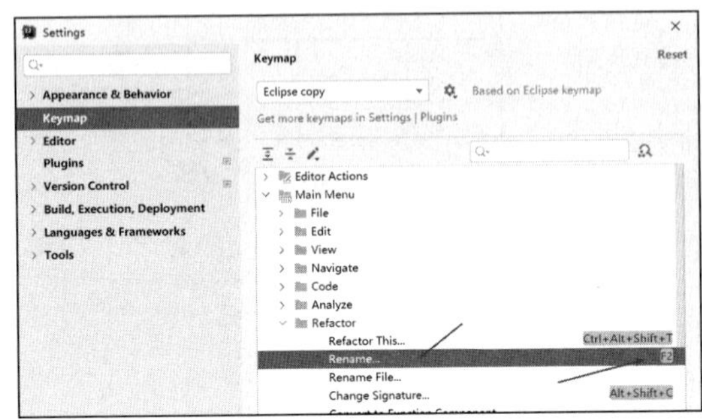


图 1-66

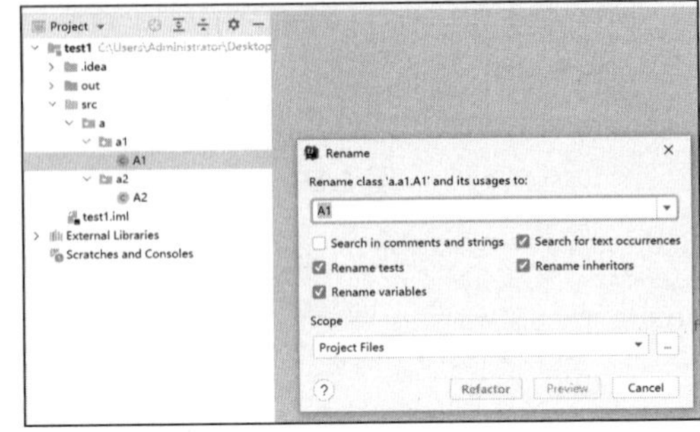


图 1-67

### 1.3.17 将类的方法抽取成接口

将类的方法抽取成接口的设置方法如图 1-68 所示。

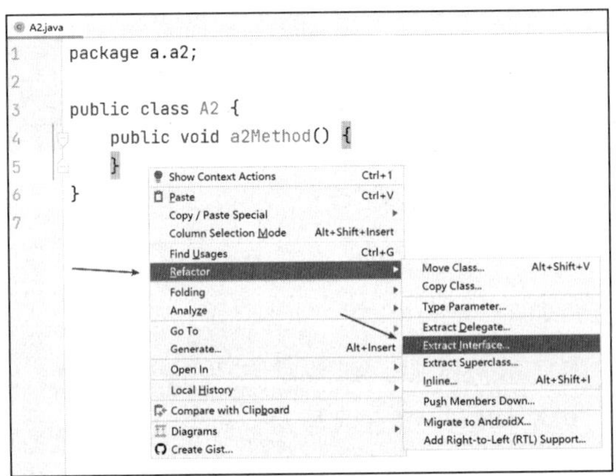


图 1-68

配置接口信息的方法如图 1-69 所示，生成的接口与实现类的信息如图 1-70 所示。

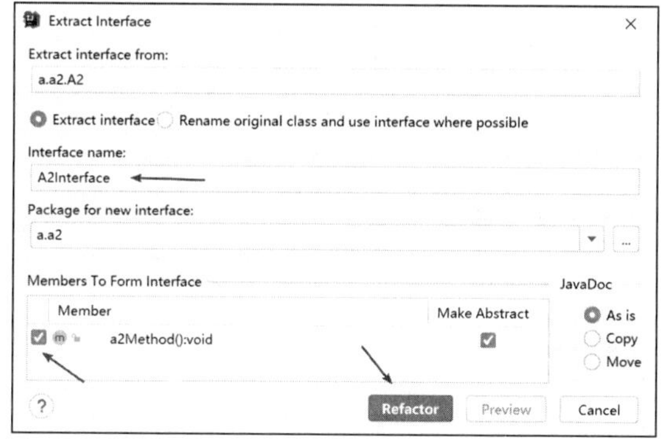


图 1-69

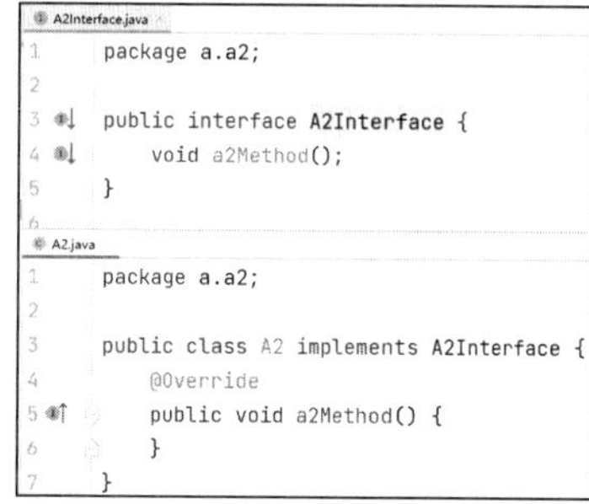


图 1-70

## 1.3.18　Maven 项目下载源代码和帮助文档

Maven 项目下载源代码和帮助文档的配置界面如图 1-71 所示。

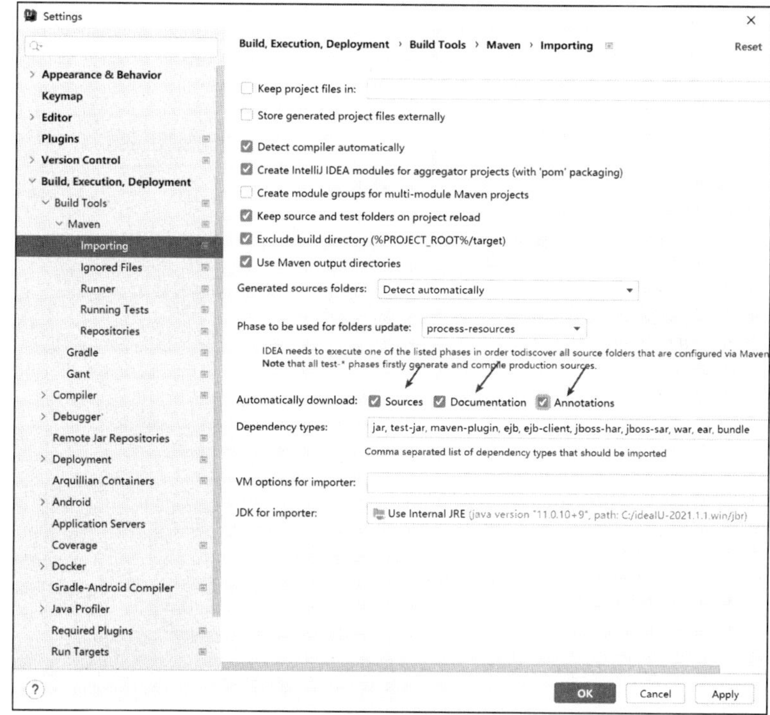


图 1-71

## 1.3.19 Find Usages 菜单查看方法调用

在方法名上右击，并单击“Find Usages”选项，如图 1-72 所示，此时可以查看哪些方法调用了 test1Method()方法，如图 1-73 所示。

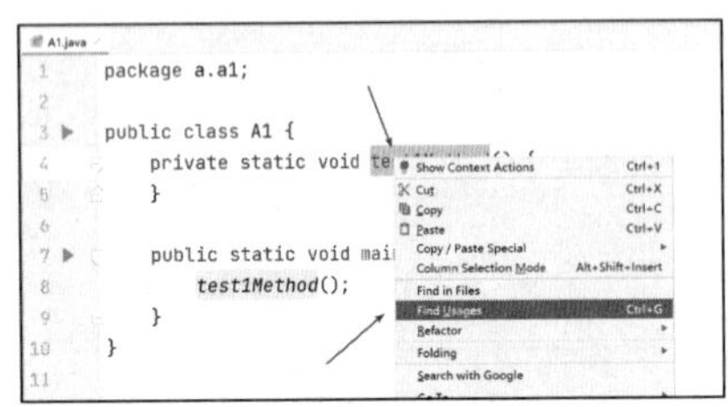


图 1-72

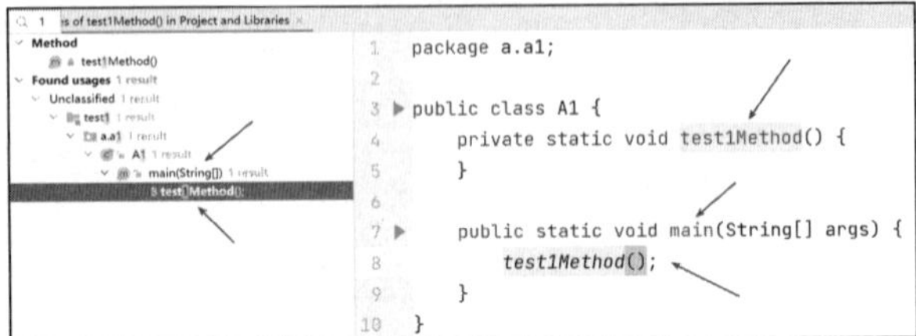


图 1-73

你也可以使用“Ctrl+Alt+H”快捷键，显示方法被调用的层级过程，如图 1-74 所示。

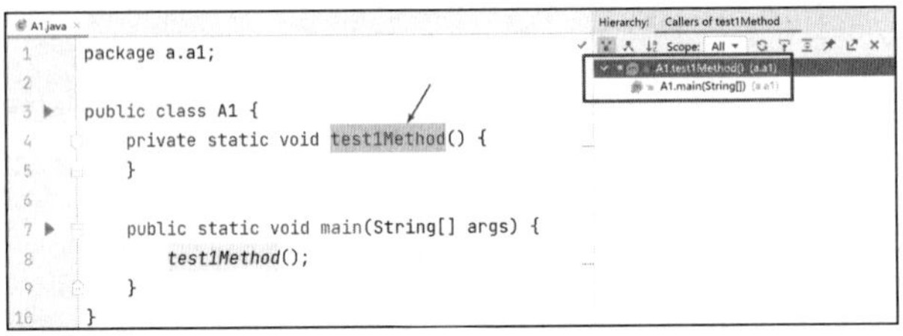

图 1-74

## 1.3.20 增加控制台保存输出信息的缓冲区大小

增加控制台保存输出信息的缓冲区大小配置如图 1-75 所示。

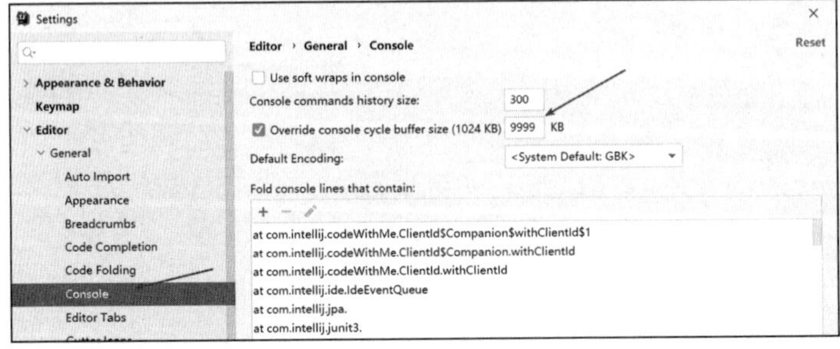


图 1-75

## 1.3.21　增加可用内存

你可以增加 IntelliJ IDEA 的可用内存，以提高 IntelliJ IDEA 的运行速度。打开安装文件夹中的 bin 子文件夹，先编辑并保存“idea64.exe.vmoptions”文件的内容，如图 1-76 所示，然后重启 IntelliJ IDEA 就会使用最新版的内存占用策略。

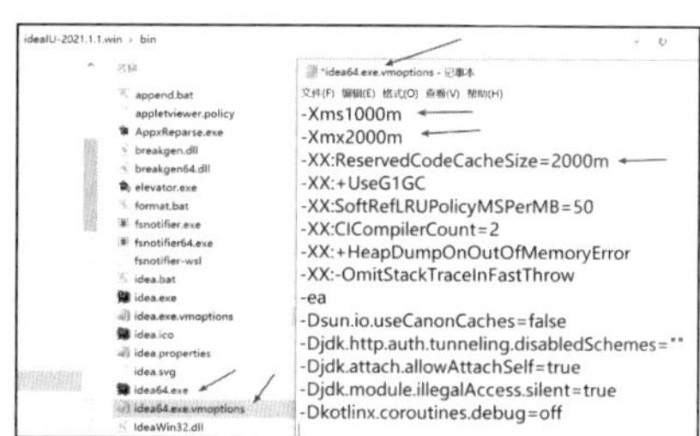


图 1-76

## 1.3.22　启用 Toolbar

启用 Toolbar 可以方便单击某些功能按钮，如图 1-77 所示。

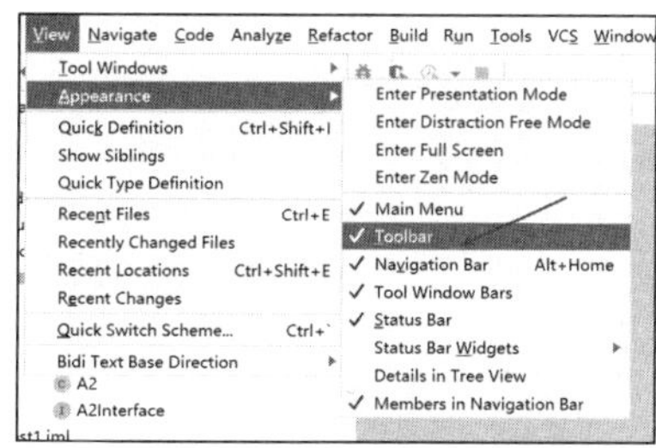


图 1-77

显示工具栏，如图 1-78 所示。

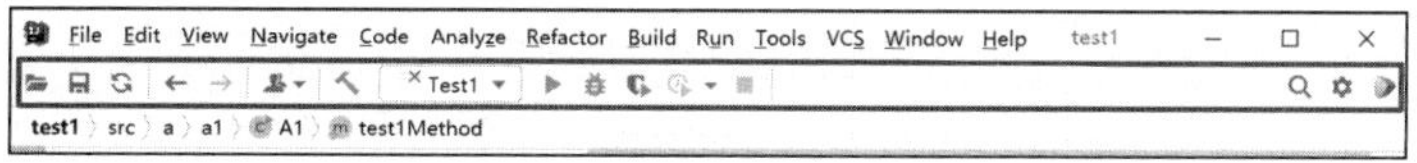


图 1-78

单击工具栏右边齿轮按钮可以进行快速配置，如图 1-79 所示。

## 1.3.23　显示方法分割符

默认情况下，方法之间没有分割符，如图 1-80 所示。

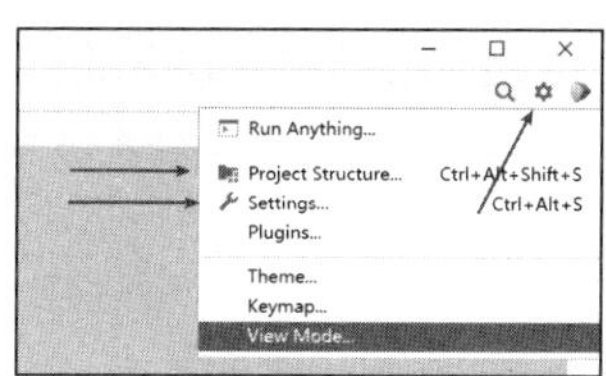


图 1-79

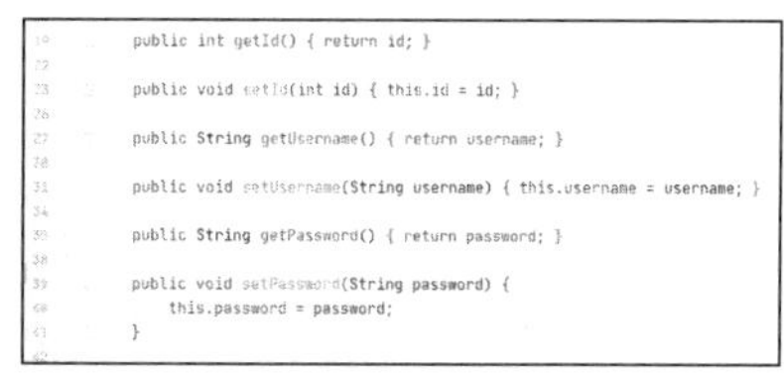


图 1-80

启用分割符的方法如图 1-81 所示；分割符显示效果如图 1-82 所示。

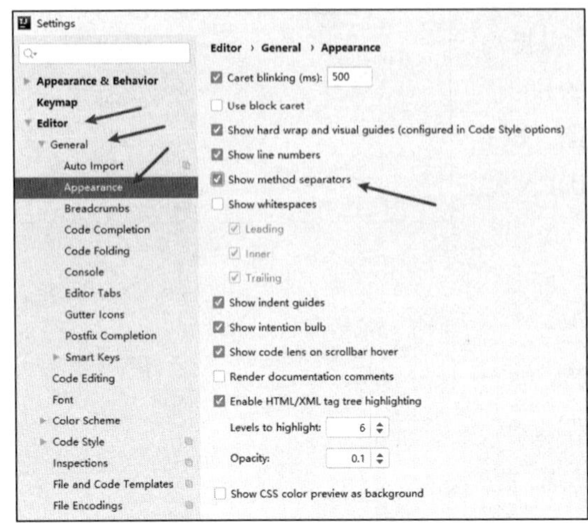


图 1-81

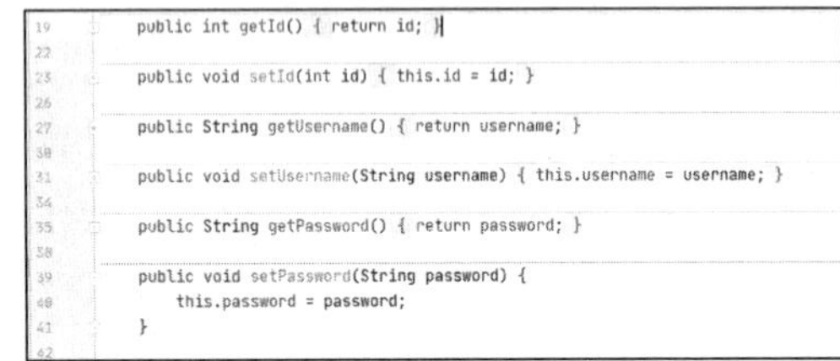


图 1-82

## 1.3.24 设置文件编码

设置文件编码，如图 1-83 所示。

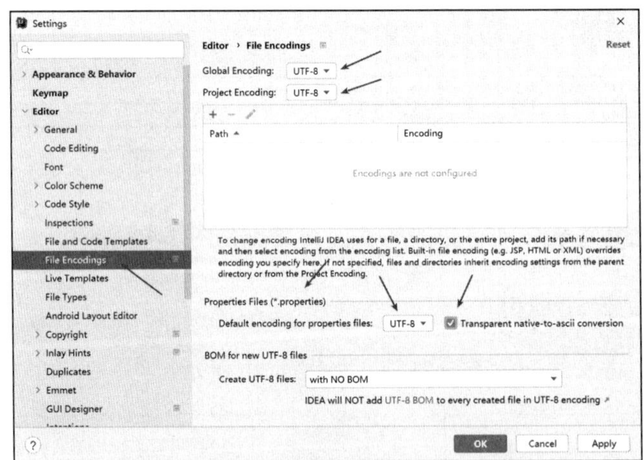


图 1-83

选项“Transparent native-to-ascii conversion”会影响.properties 属性文件中中文的显示，选中的效果如图 1-84 所示，未选中的效果如图 1-85 所示。

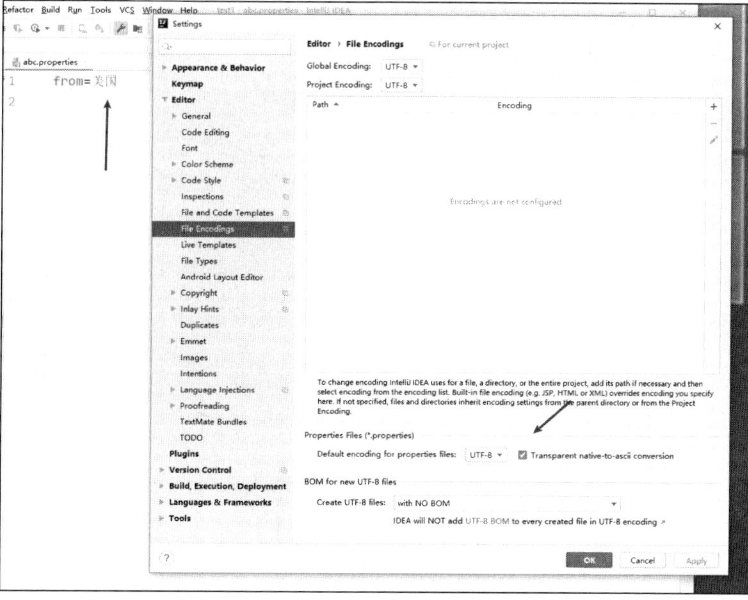
Refactor Build Run Tools VCS Window Help
Settings
abc.properties
from=美国
Appearance & Behavior
Keymap
Editor
General
Code Editing
Font
Color Scheme
Code Style
Inspections
File and Code Templates
File Encodings
Live Templates
File Types
Android Layout Editor
Copyright
Inlay Hints
Duplicates
Emmet
Images
Intentions
Language Injections
Proofreading
TextMate Bundles
TODO
Plugins
Version Control
Build, Execution, Deployment
Languages & Frameworks
Tools
Editor › File Encodings
For current project
Global Encoding: UTF-8
Project Encoding: UTF-8
Path
Encoding
Encodings are not configured
To change encoding IntelliJ IDEA uses for a file, a directory, or the entire project, add its path if necessary and then select encoding from the encoding list. Built-in file encoding (e.g. JSP, HTML or XML) overrides encoding you specify here. If not specified, files and directories inherit encoding settings from the parent directory or from the Project Encoding.
Properties Files (*.properties)
Default encoding for properties files: UTF-8
Transparent native-to-ascii conversion
BOM for new UTF-8 files
Create UTF-8 files: with NO BOM
IDEA will NOT add UTF-8 BOM to every created file in UTF-8 encoding
OK
Cancel
Apply

图 1-84

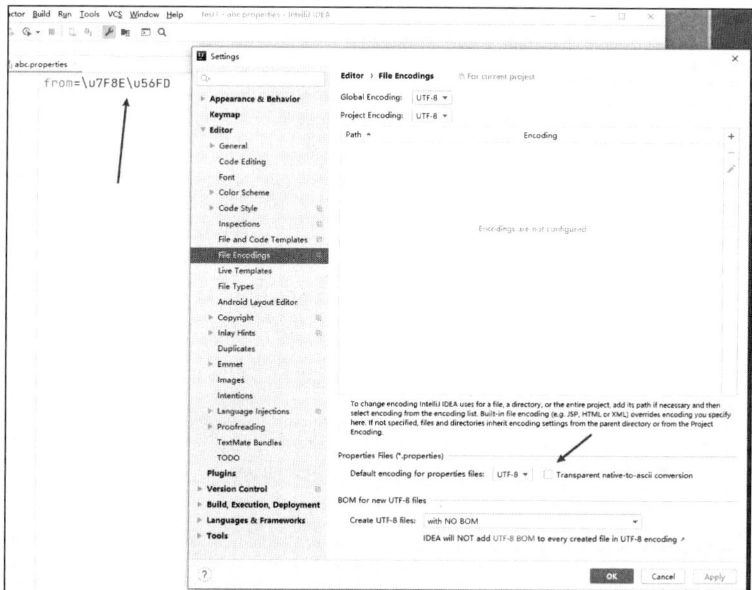
Build Run Tools VCS Window Help
Settings
abc.properties
from=\u7F8E\u56FD
Appearance & Behavior
Keymap
Editor
General
Code Editing
Font
Color Scheme
Code Style
Inspections
File and Code Templates
File Encodings
Live Templates
File Types
Android Layout Editor
Copyright
Inlay Hints
Duplicates
Emmet
Images
Intentions
Language Injections
Proofreading
TextMate Bundles
TODO
Plugins
Version Control
Build, Execution, Deployment
Languages & Frameworks
Tools
Editor › File Encodings
For current project
Global Encoding: UTF-8
Project Encoding: UTF-8
Path
Encoding
Encodings are not configured
To change encoding IntelliJ IDEA uses for a file, a directory, or the entire project, add its path if necessary and then select encoding from the encoding list. Built-in file encoding (e.g. JSP, HTML or XML) overrides encoding you specify here. If not specified, files and directories inherit encoding settings from the parent directory or from the Project Encoding.
Properties Files (*.properties)
Default encoding for properties files: UTF-8
Transparent native-to-ascii conversion
BOM for new UTF-8 files
Create UTF-8 files: with NO BOM
IDEA will NOT add UTF-8 BOM to every created file in UTF-8 encoding
OK
Cancel
Apply

图 1-85

### 1.3.25 使用快捷键找到对应的功能名称

使用快捷键对功能名称的查询操作如图 1-86 所示。

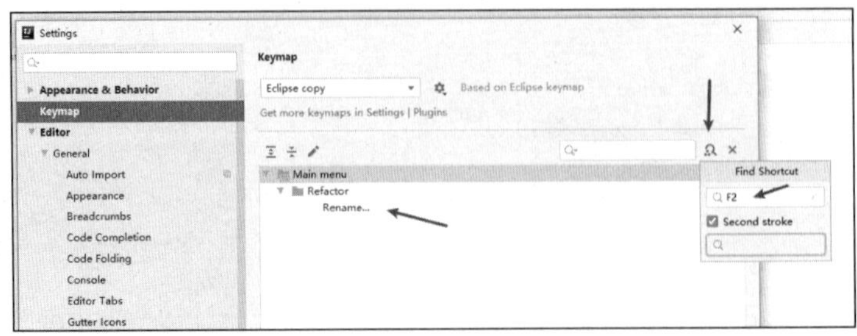

图 1-86

### 1.3.26 使用快捷键自动生成返回值或声明变量

先把光标定位在“)”和“;”之间，然后按“Alt+Enter”快捷键弹出菜单并执行，如图 1-87 所示，生成返回值的效果如图 1-88 所示。

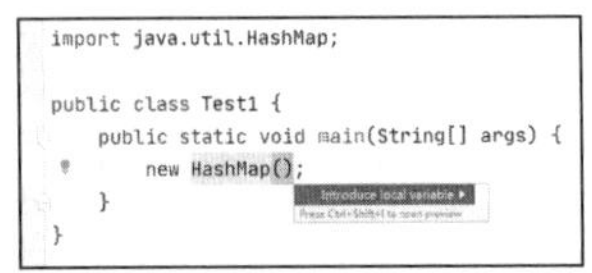

图 1-87

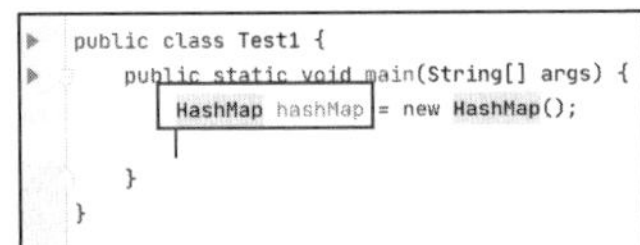

图 1-88

### 1.3.27 使用快捷键实现代码导航

在分析源代码时，代码导航属于高频使用的快捷键，其可以对调用的流程进行“前进”或“后退”的操作，快捷键分别是“Alt+←”和“Alt+→”。

### 1.3.28 使用“F4”键查看类继承结构

使用“F4”键可以查看类继承结构，如图 1-89 所示。

### 1.3.29 使用快捷键查看类大纲结构

在类的源代码中，使用“Ctrl+O”快捷键可以查看类大纲结构，如图 1-90 所示。

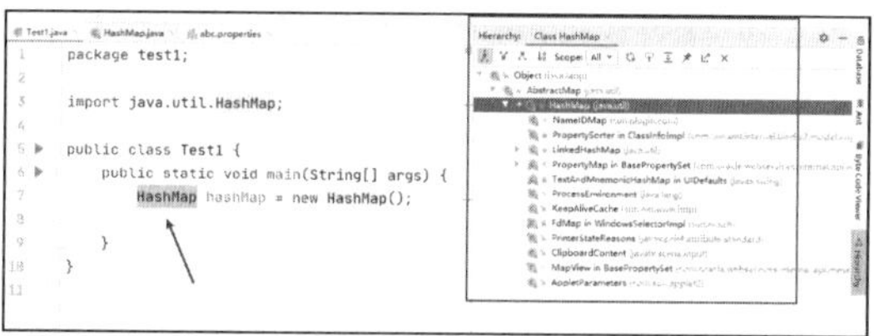

图 1-89

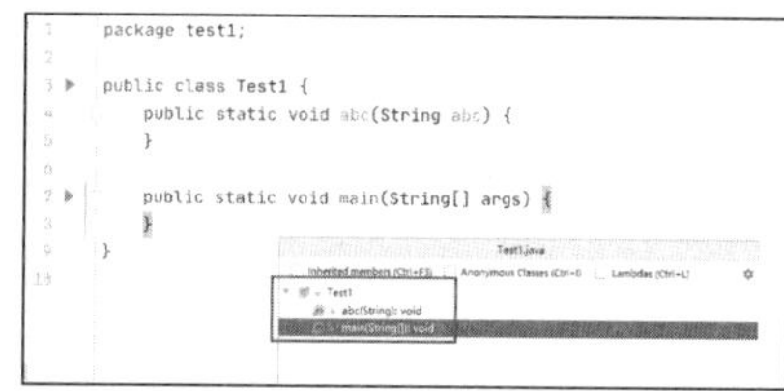

图 1-90

## 1.3.30　使用快捷键搜索与替换文本功能

在当前文件中，使用“Ctrl+F”快捷键可以搜索与替换文本，如图 1-91 和图 1-92 所示。

图 1-91

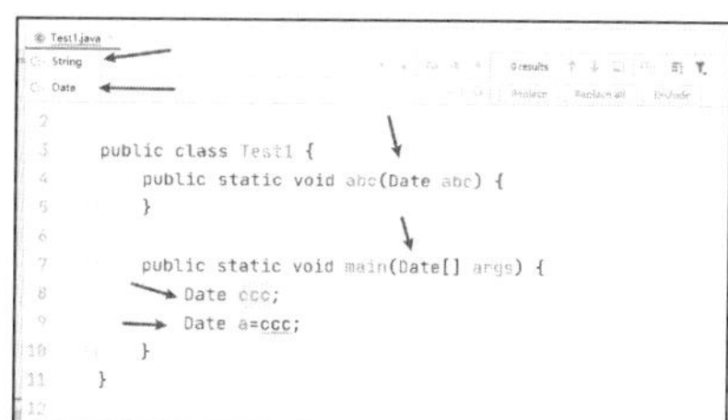

图 1-92

## 1.3.31　使用“Ctrl+H”快捷键实现查询更大范围的功能

“Ctrl+F”快捷键只可以在当前文件中进行查询，如果你想在项目、模块、文件夹范围中实现搜索与替换文件的功能，可以使用快捷键“Ctrl+H”，如图 1-93 所示。

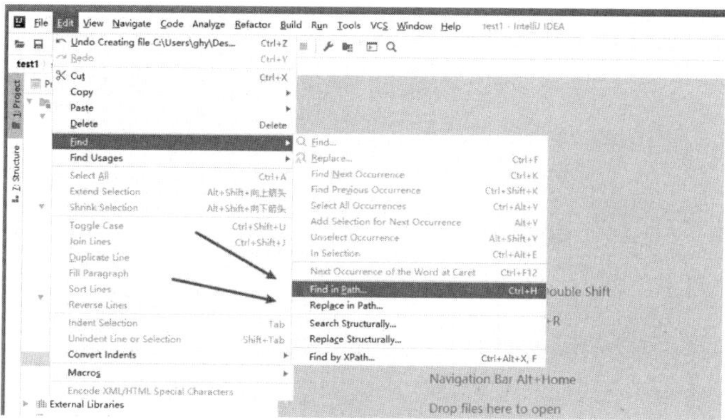

图 1-93

搜索的效果如图 1-94 所示。选项“Replace in Path”和“Find in Path”功能相似，都是在更大范围内执行替换操作，如图 1-95 所示。

图 1-94

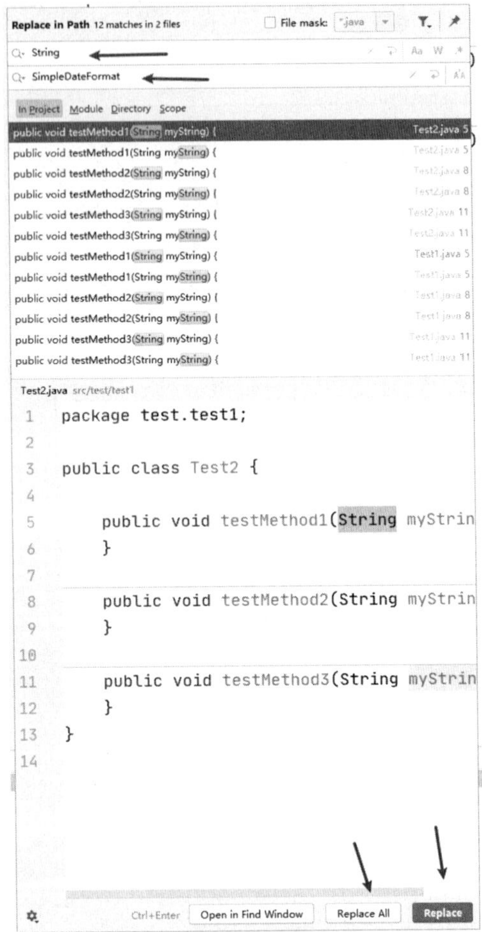

图 1-95

## 1.3.32 使用“Shift”键实现全局搜索

连续按两次“Shift”键会出现全局搜索对话框，如图 1-96 所示。

“Ctrl+F”快捷键适合在当前项目中查询，而此种方式是在全局范围内进行查询，当以文件名作为查询时比较方便，可以快速找到 JDK 中的 Java 类。

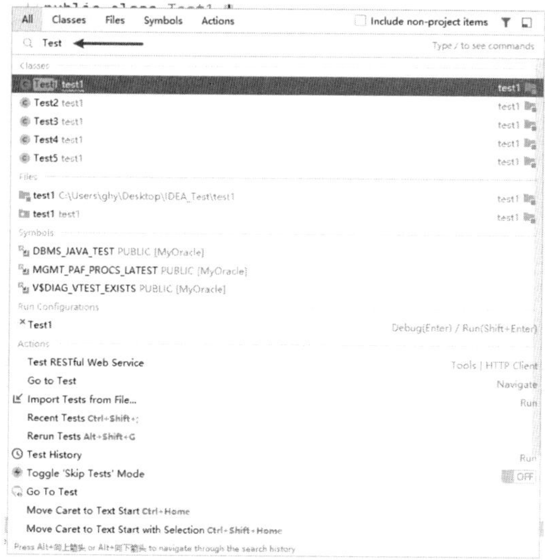


图 1-96

## 1.3.33　生成 UML 类图

使用“Ctrl+Alt+Shift+U”快捷键可以生成 UML 类图，类 ArrayList 的 UML 类图如图 1-97 所示。

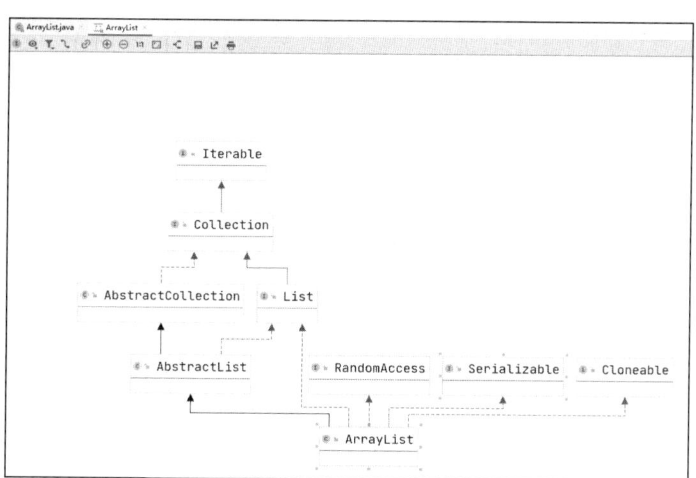


图 1-97

也可以单击“Show Diagram”选项生成 UML 类图，如图 1-98 所示。

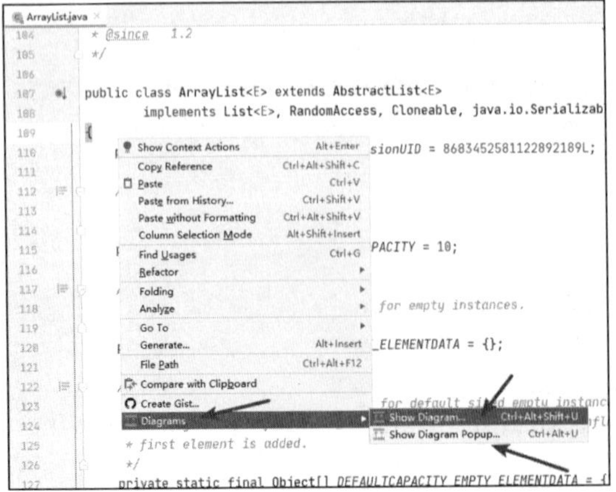


图 1-98

## 1.3.34 使用快捷键查看方法的重写或实现

使用“Ctrl+T”快捷键可以查看方法是在哪些类中被重写或实现的，如图 1-99 所示。

A1.java

```
package test;

public class A1 {
    public void testMethod() {
    }
}
```

Choose Implementation of A1.testMethod() (2 found)

A2 (test) test1

A3 (test) test1

A2.java

```
package test;

public class A2 extends A1{
    @Override
    public void testMethod() {
    }
}
```

A3.java

```
package test;

public class A3 extends A1 {
    @Override
    public void testMethod() {
    }
}
```

图 1-99

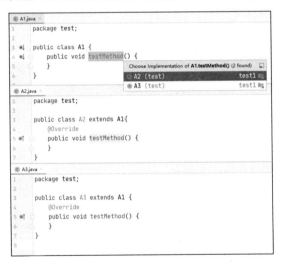

也可以使用“Ctrl+Shift+H”快捷键进行查看，其展示的信息更丰富，如图 1-100 所示。

## 1.3.35 使用快捷键查看文件

使用“Ctrl+E”快捷键可以查看最近编辑的文件，如图 1-101 所示。

A1.java

```
package test;

public class A1 {
    public void testMethod() {
    }
}
```

A2.java

```
package test;

public class A2 extends A1{
    @Override
    public void testMethod() {
    }
}
```

A3.java

```
package test;

public class A3 extends A1 {
    @Override
    public void testMethod() {
    }
}
```

Hierarchy: Method testMethod
Scope: All
A1 (test)
A3 (test)
A2 (test)

图 1-100

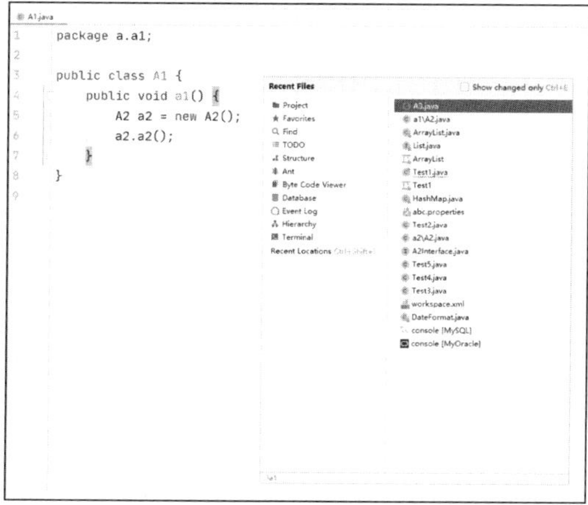


图 1-101

## 1.3.36　生成.jar 文件

创建名称为 test2 的 Java 项目，并创建两个 Java 类，代码如下：

```
package test;

public class Test1 {
   public static void main(String[] args) {
      System.out.println("test1 run !");
   }
}
```

```
package test;

public class Test2 {
    public static void main(String[] args) {
        System.out.println("test2 run !");
    }
}
```

首先单击“Project Structure”选项，如图 1-102 所示；然后配置“Artifacts”选项，如图 1-103 所示。

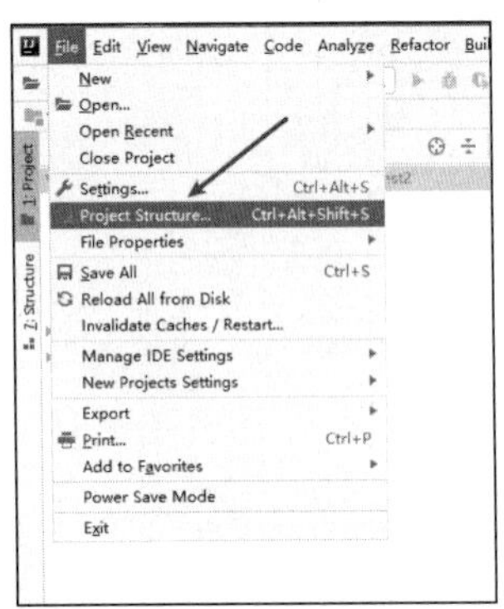


图 1-102

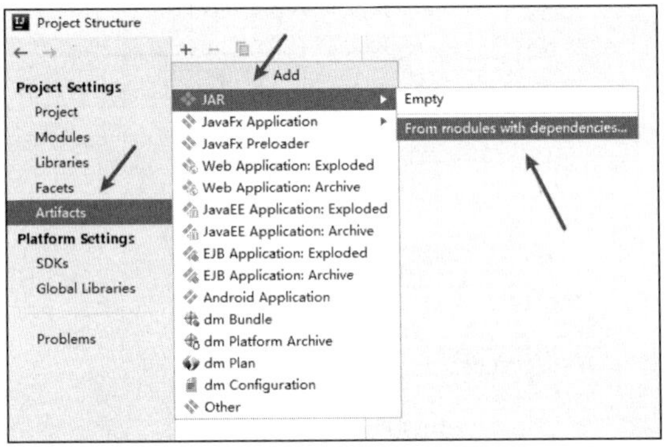

图 1-103

再配置“Main Class”，如图 1-104 所示。

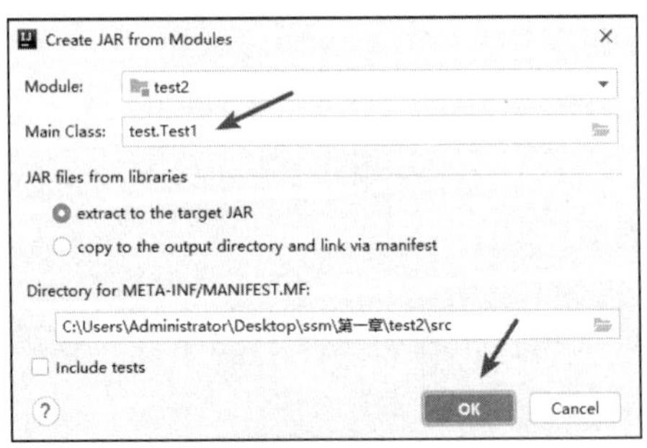


图 1-104

注意：文件“MANIFEST.MF”存放在 src 路径中。

单击“Build Artifacts...”选项，如图 1-105 所示，再单击“Build”选项，如图 1-106 所示，成功生成.jar 文件，如图 1-107 所示。

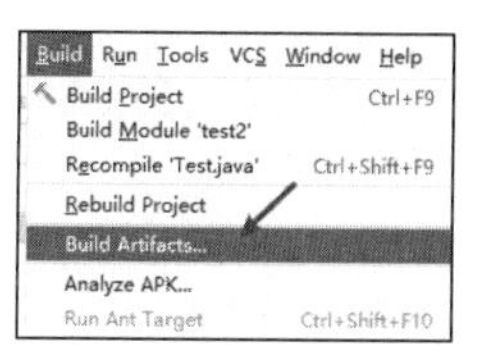


图 1-105

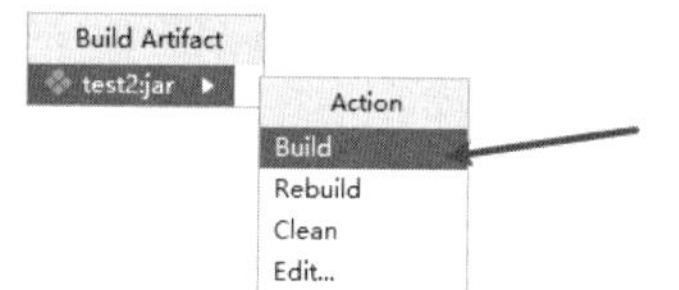


图 1-106

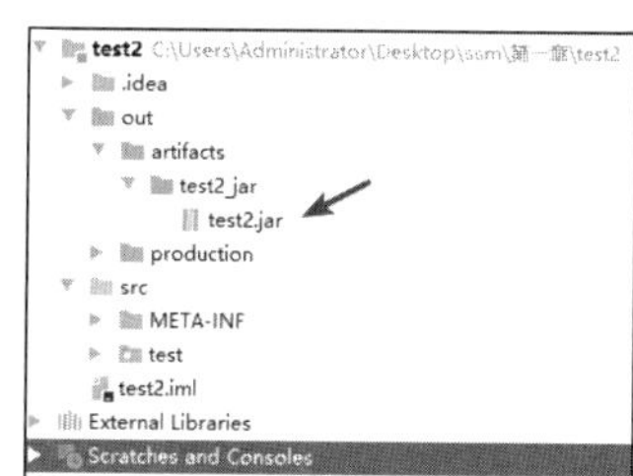


图 1-107

成功生成.jar 文件的 Test1 默认入口类，如图 1-108 所示。

```
C:\Users\Administrator\Desktop\ssm\第一章\test2\out\artifacts\test2_jar>java -jar test2.jar
test1 run !
```

图 1-108

如果想执行“Main Class”，则必须添加 JAR 参数；如果执行其他入口类，则不需要添加，效果如图 1-109 所示。

```
C:\Users\Administrator\Desktop\ssm\第一章\test2\out\artifacts\test2_jar>java -cp test2.jar test/Test2
test2 run !
```

图 1-109

## 1.3.37　设置显示 API 文档提示框的延迟时间

当光标放在方法名称上面时，会自动弹出 API 帮助文档，如图 1-110 所示。

默认是延迟 500 毫秒之后显示，你可以设置延迟显示的时间，如图 1-111 所示。

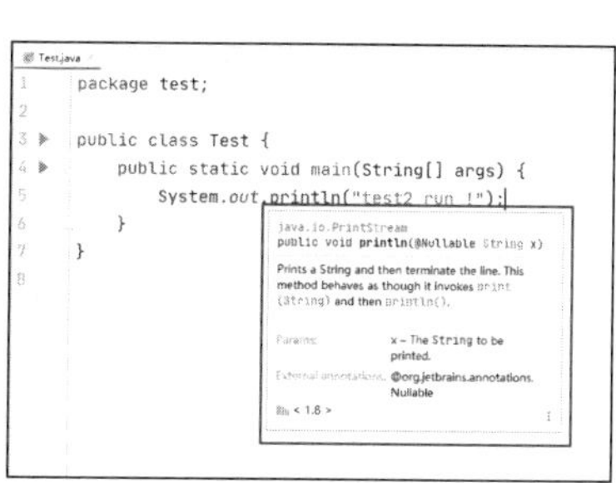


图 1-110

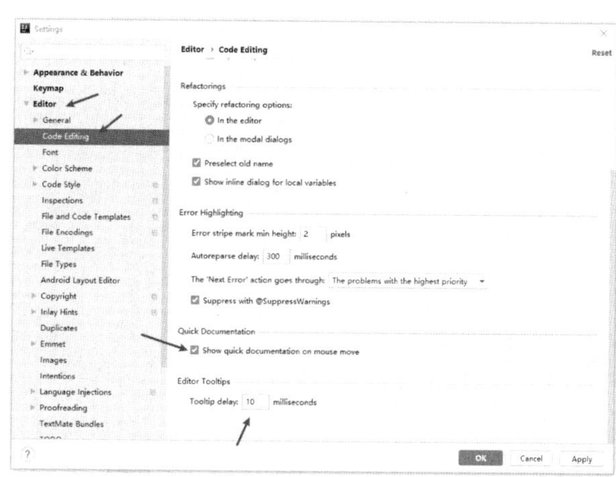


图 1-111

## 1.3.38 使用多行 Tab 显示

当打开多个文件时，可能会发生 Tab 隐藏的情况，如图 1-112 所示，这时，我们可以把所有的 Tab 标签多行显示，配置方法如图 1-113 所示。

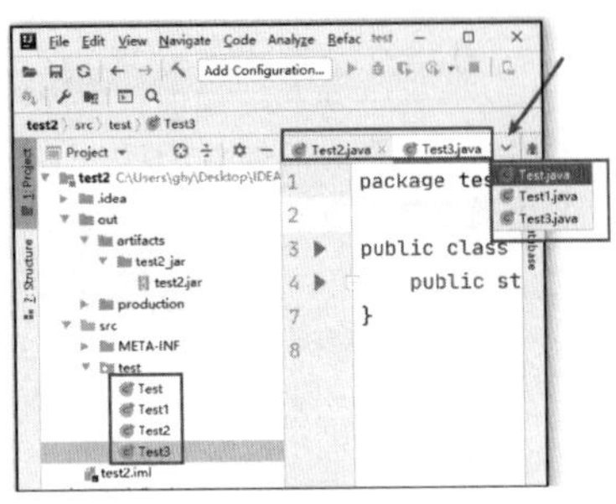


图 1-112

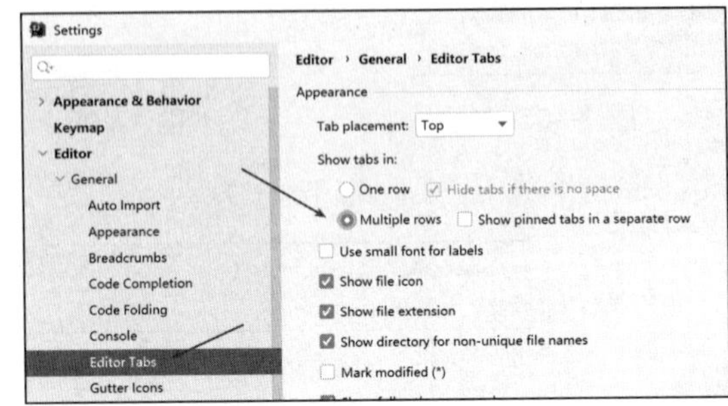


图 1-113

多行显示的效果如图 1-114 所示。

## 1.3.39 设置背景图片

设置背景图片的操作如图 1-115 所示。

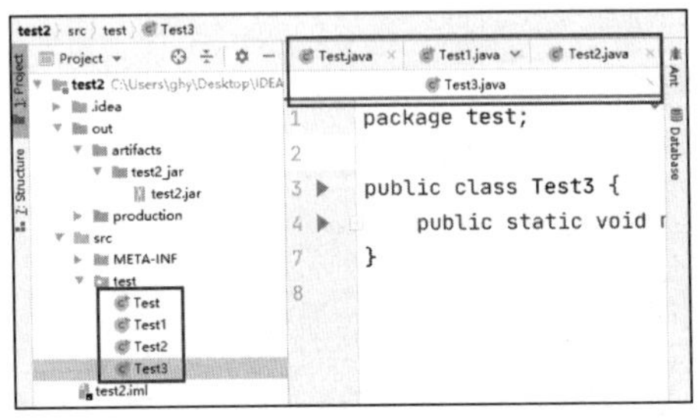


图 1-114

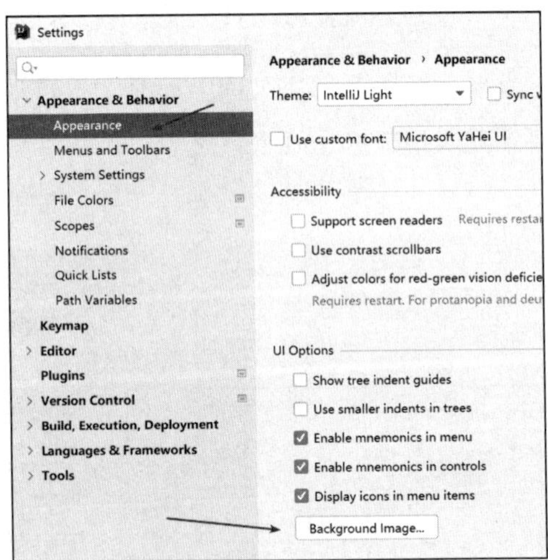


图 1-115

### 1.3.40　自动生成 serialVersionUID 属性

自动生成 serialVersionUID 属性的配置界面如图 1-116 所示。

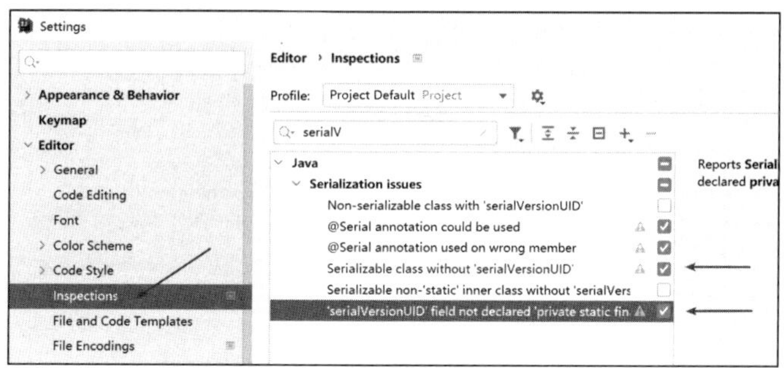

图 1-116

Java 类需要实现 Serializable 接口，先选中类名，然后按“Alt+Enter”快捷键，弹出如图 1-117 所示的菜单，成功生成 serialVersionUID 属性，如图 1-118 所示。

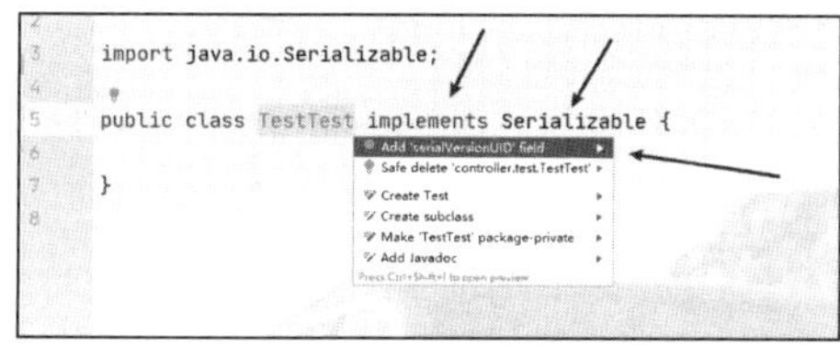

图 1-117

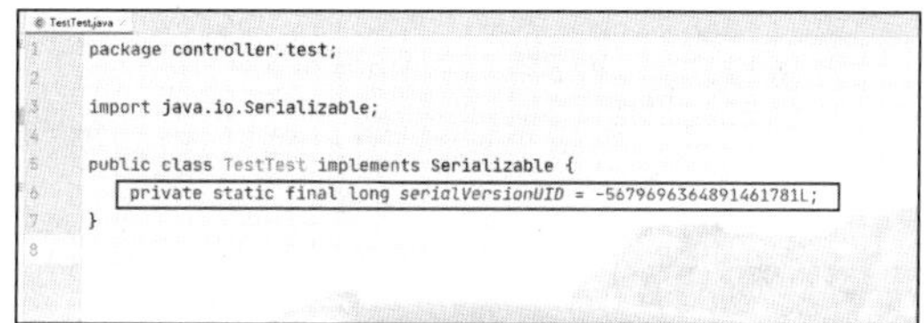

图 1-118

### 1.3.41　查看当前文件的位置

首先打开任意一个文件，如图 1-119 所示，然后单击准星图标定位文件在项目中的位置，如图 1-120 所示。

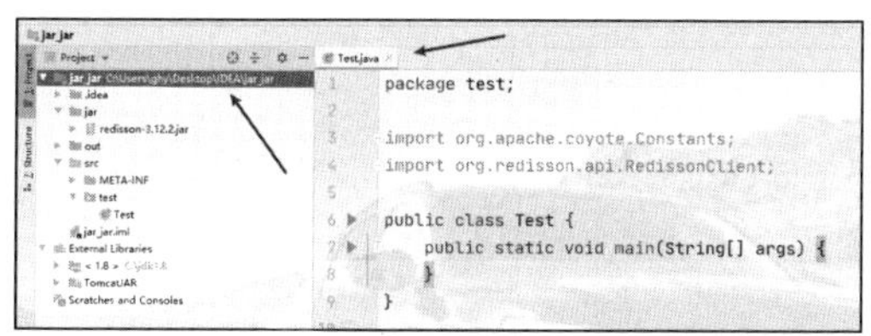

图 1-119

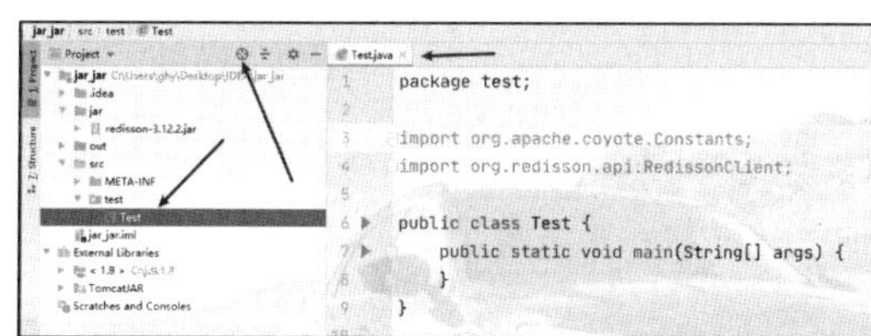

图 1-120

### 1.3.42 禁用 toString()方法查看对象内容

禁用 toString()方法查看对象内容的代码如下：

```
package test;

class Userinfo {
    private String username;
    private String password;

    public Userinfo() {
    }

    public Userinfo(String username, String password) {
        this.username = username;
        this.password = password;
    }

    public String getUsername() {
        return username;
    }

    public void setUsername(String username) {
        this.username = username;
    }

    public String getPassword() {
        return password;
    }

    public void setPassword(String password) {
        this.password = password;
    }

    @Override
    public String toString() {
        return "i am toString method show value username=" + username + "
password=" + password;
    }
}
```

```
public class Test {
    public static void main(String[] args) {
        Userinfo userinfo = new Userinfo();
        String testEnd = "end";
    }
}
```

调试效果如图 1-121 所示。

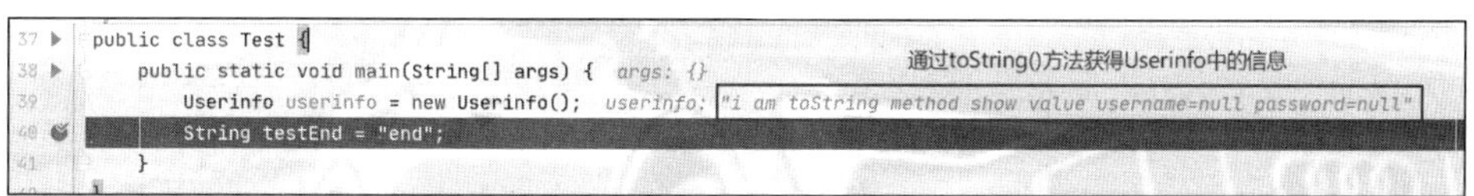


图 1-121

禁用 toString()方法查看数据的配置如图 1-122 所示，调试效果如图 1-123 所示。

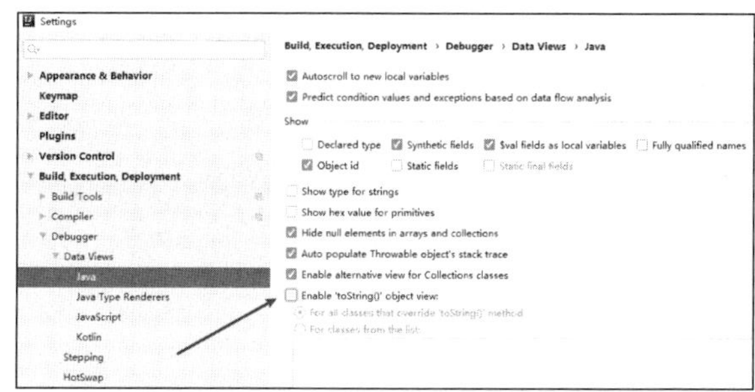

图 1-122

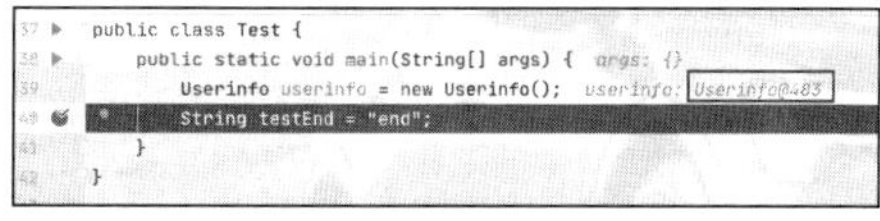

图 1-123

checkbox 选项可以解决在 Hibernate 或 MyBatis 中调试时延迟加载变成立即加载的问题。

## 1.3.43 设置注释不在最前面

按“Ctrl+/”快捷键可以对字符串“abc”所在行进行注释，如图 1-124 所示。

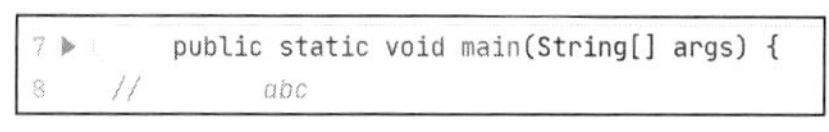

图 1-124

若不希望“//”在当前行的最前面，则需要做图 1-125 所示的设置。

此时，对字符串“abc”所在行形成注释，效果如图 1-126 所示。

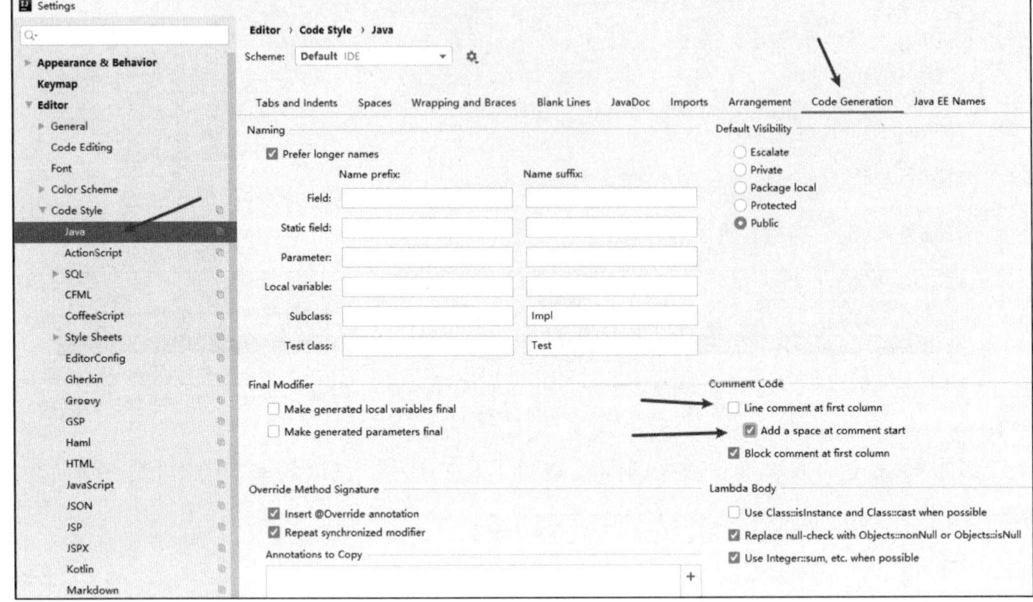


图 1-125

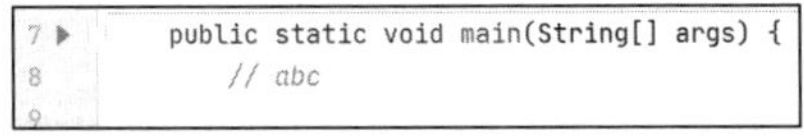


图 1-126

注意：此时，“//”不在当前行的最前面，并且“abc”前面有空格。

## 1.3.44 连接数据库的配置

你可以在 IntelliJ IDEA 中连接 Oracle 和 MySQL 数据库。

### 1. 连接Oracle数据库

首先单击“Database”标签，如图 1-127 所示，然后新建 Oracle 的连接，如图 1-128 所示。

设置连接 Oracle 数据库的参数，如果你没有 Oracle 的 JDBC 驱动，则可以在线下载，下载成功后单击“Test Connection”按钮进行测试，如图 1-129 所示。

如果连接成功，则会看到 Oracle 数据库中的表和序列，如图 1-130 所示。

打开执行 SQL 语句窗口，如图 1-131 所示。输入 SQL 语句并执行，效果如图 1-132 所示。

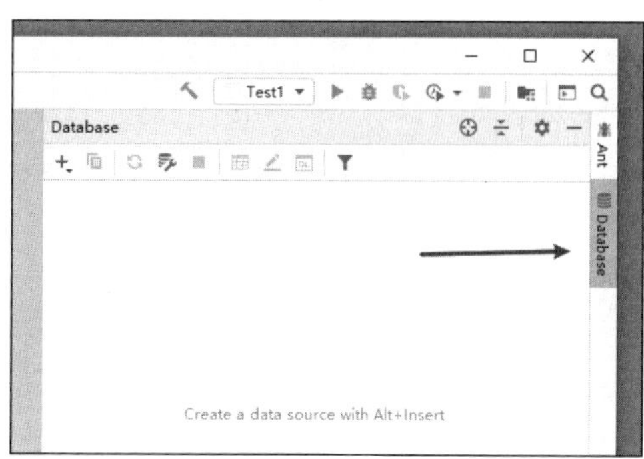


图 1-127

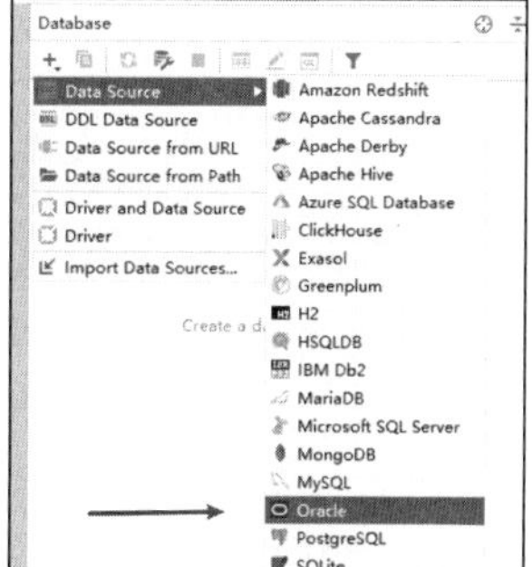


图 1-128

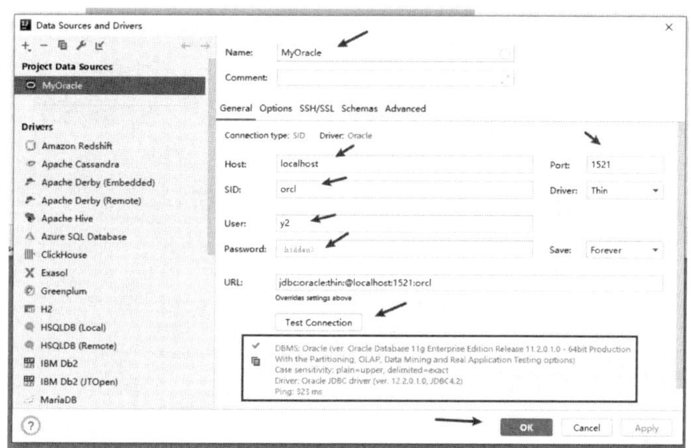


图 1-129

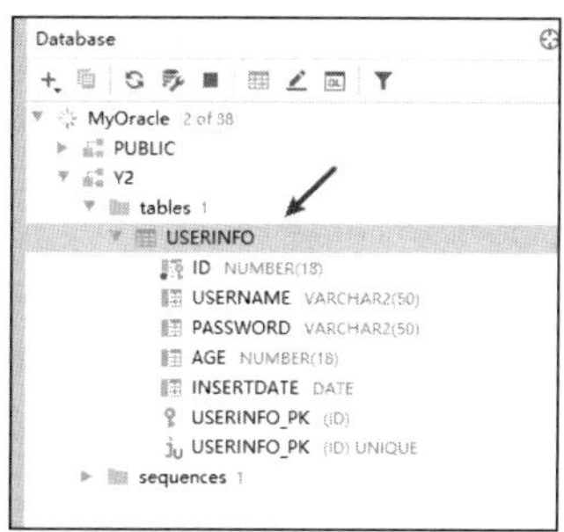


图 1-130

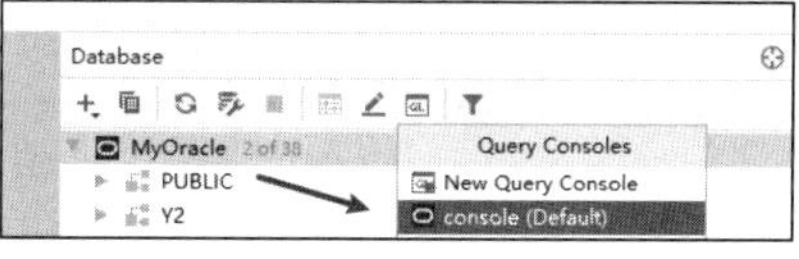


图 1-131

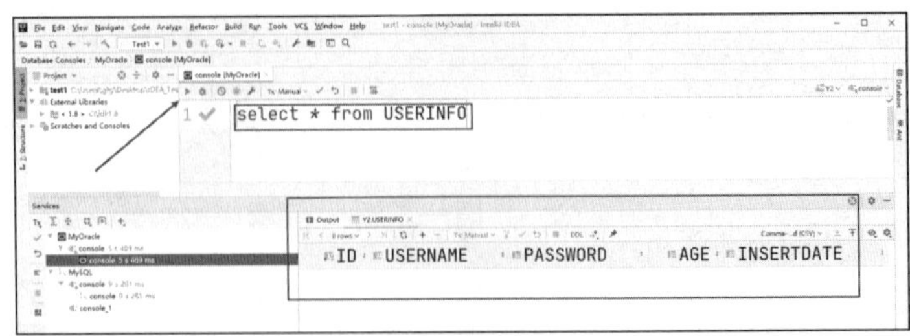


图 1-132

## 2. 连接MySQL数据库

当连接 MySQL 8.0 版本数据库时，需要在“my.ini”文件中配置时区。首先进入C:\ProgramData 文件夹，然后编辑“my.ini”文件，在现有[mysqld]节点下添加如下代码：

```
[mysqld]
default-time-zone='+08:00'
```

注意：C:\ProgramData 文件夹默认是隐藏的。文件“my.ini”存放路径为 C:\ProgramData\MySQL\MySQL Server 8.0。

重启 MySQL 服务，其中配置连接 MySQL 数据库的参数如图 1-133 所示。

连接成功后，数据库 y2 中的信息如图 1-134 所示。

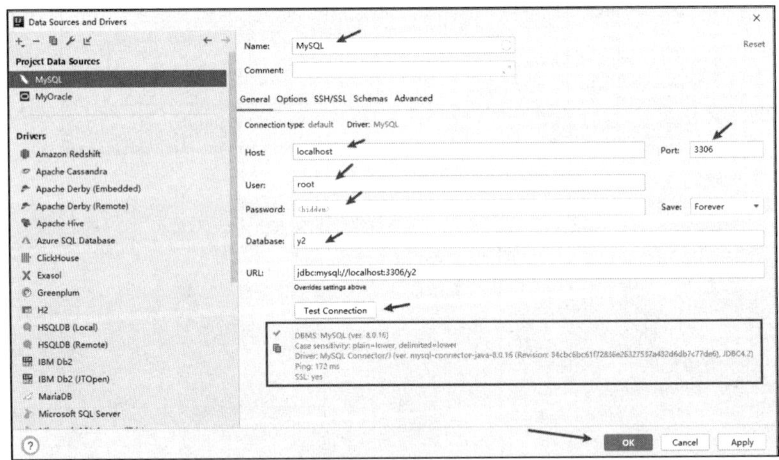


图 1-133

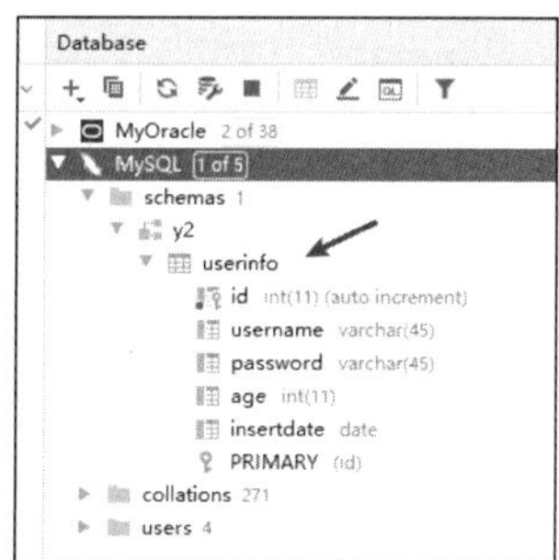


图 1-134

如果出现 serverTimezone 异常，则在 URL 中添加如下代码：

```
?serverTimezone=Asia/Shanghai
```

经过测试，MySQL 5.0 可以使用 MySQL 8.0 的 JDBC 驱动。

## 1.3.45 实现在控制台中进行查询

设置快捷键如图 1-135 所示，查询效果如图 1-136 所示。

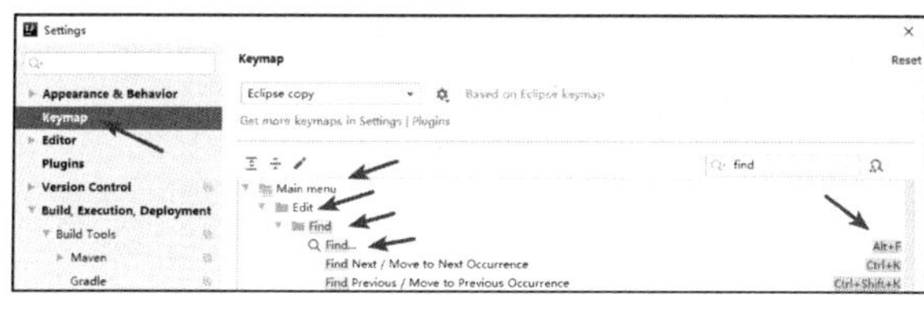

图 1-135

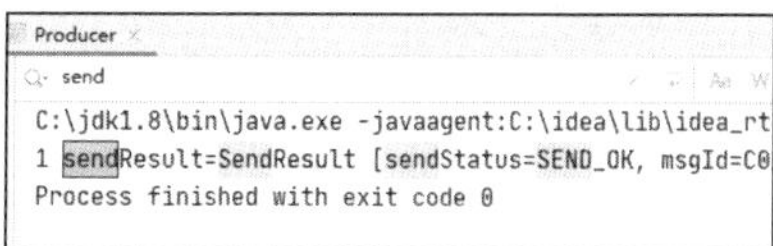


图 1-136

## 1.3.46 获取完整的包类路径

单击“Copy Path...”选项，如图 1-137 所示，弹出界面如图 1-138 所示。

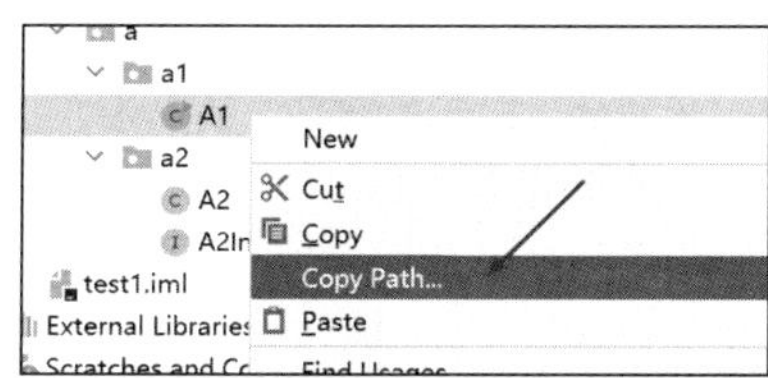


图 1-137

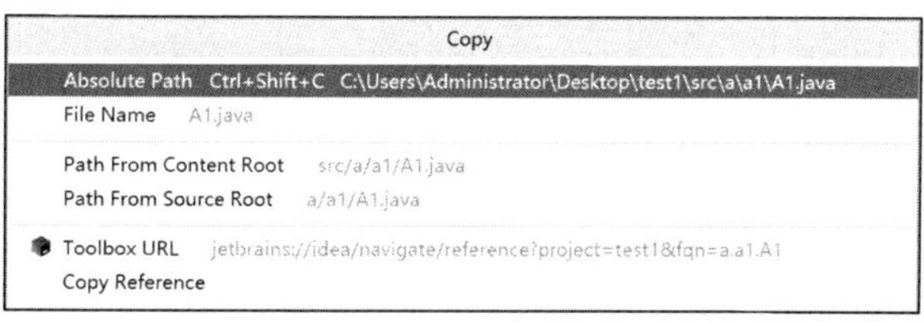


图 1-138

## 1.3.47　导入/导出 IntelliJ IDEA 的配置

IntelliJ IDEA 中的配置可以使用菜单进行导入（Import）/导出（Export），这在更换电脑或重做系统时经常被使用，可以避免每次安装新版本的 IntelliJ IDEA 都要重新配置。

对配置进行导入/导出的菜单，如图 1-139 所示。

导出的 Export Settings 配置是一个.zip 文件，如图 1-140 所示。

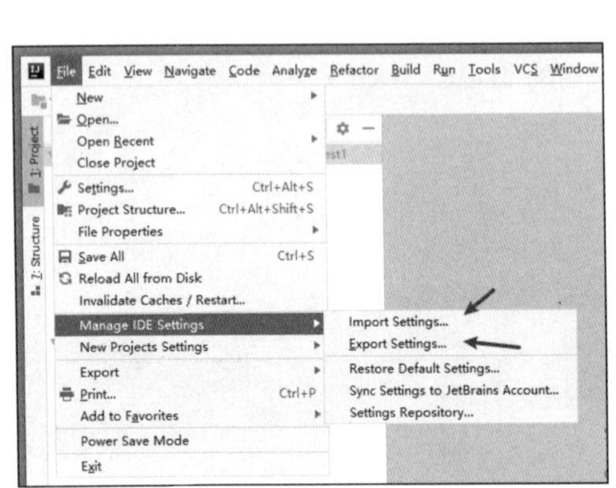


图 1-139

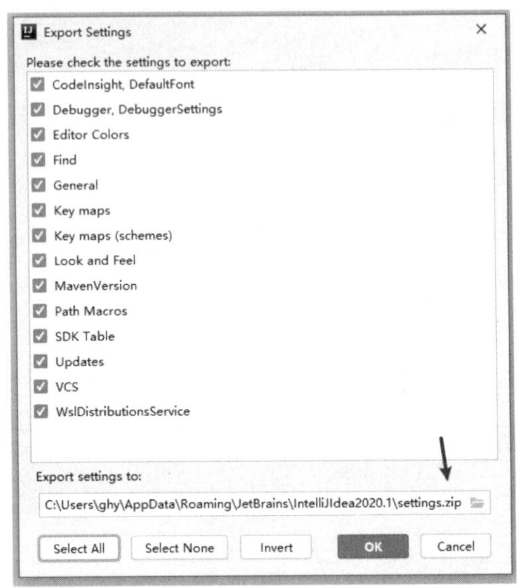


图 1-140

在导出成功后，再使用“Import Settings...”功能进行导入。

## 1.3.48　使用 Toolbox App 管理 IntelliJ IDEA

打开 Toolbox App 官网，单击“Toolbox App”，如图 1-141 所示。

选择 Windows(.exe)文件下载，如图 1-142 所示。

Toolbox App 可以在同一台电脑中维护不同的版本，如图 1-143 所示。

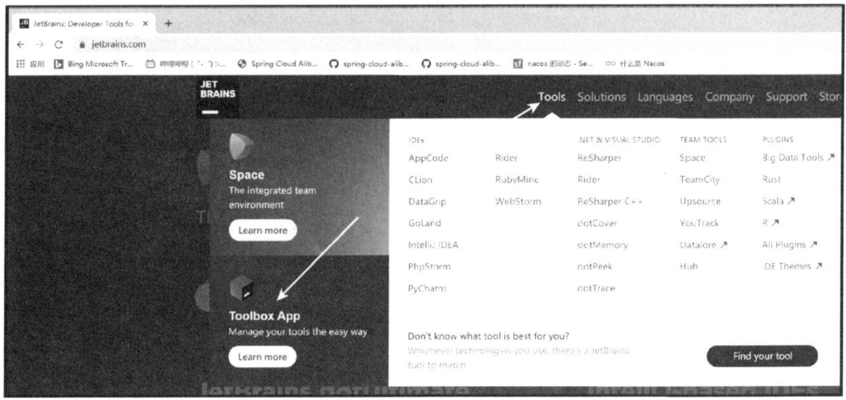


图 1-141

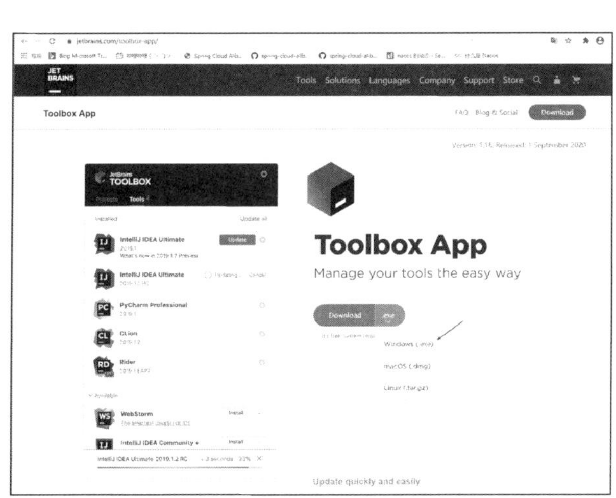


图 1-142

图 1-143

## 1.3.49 还原默认界面布局

单击“Restore Default Layout”选项可以还原默认界面，如图 1-144 所示。

## 1.3.50 使用内置控制台

使用内置控制台的效果如图 1-145 所示。

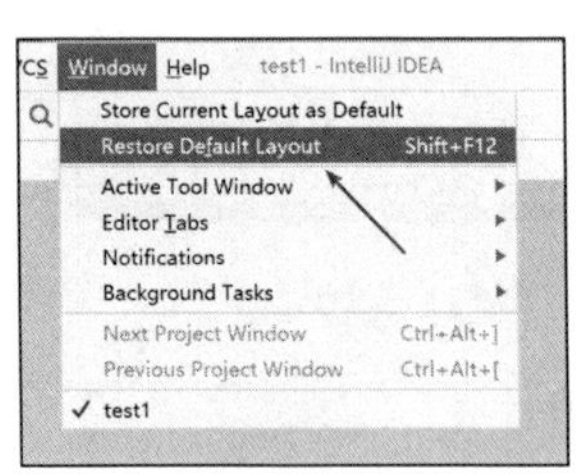


图 1-144

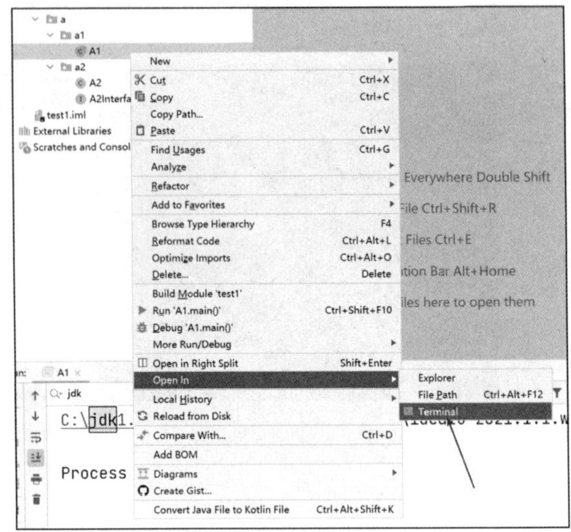


图 1-145

## 1.3.51 查看参数信息

查看参数信息设置的快捷键如图 1-146 所示。

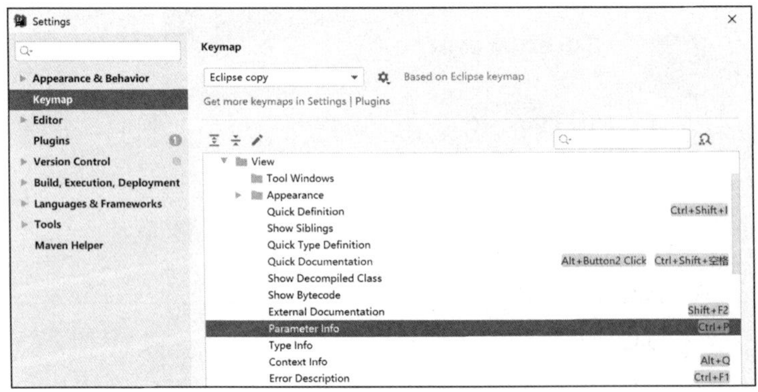


图 1-146

# 第 2 章 IntelliJ IDEA 核心技能

本章主要讲解 IntelliJ IDEA 的必备核心技能，属于重要知识点。

## 2.1 调试Java代码

本节介绍使用 IntelliJ IDEA 实现对 Java 代码的调试。

### 2.1.1 准备调试代码

测试 Test.java 类的代码如下：

```
package test;

public class Test {

    public static void testMethod() {
        System.out.println("1");
        System.out.println("2");
        System.out.println("3");
        System.out.println("4");
        for (int i = 0; i < 5; i++) {
            System.out.println("for " + (i + 1));
        }
        System.out.println("5");
        System.out.println("6");
        System.out.println("7");
        System.out.println("8");
    }
```

```
    public static void main(String[] args) {
        System.out.println("我是中国人");
        testMethod();
    }
}
```

## 2.1.2 设置断点

在 main( )方法的第一行设置断点，如图 2-1 所示。

## 2.1.3 调试

单击“Debug'Test.main()'”选项调试程序，如图 2-2 所示。这时，程序会在断点处停止，如图 2-3 所示。

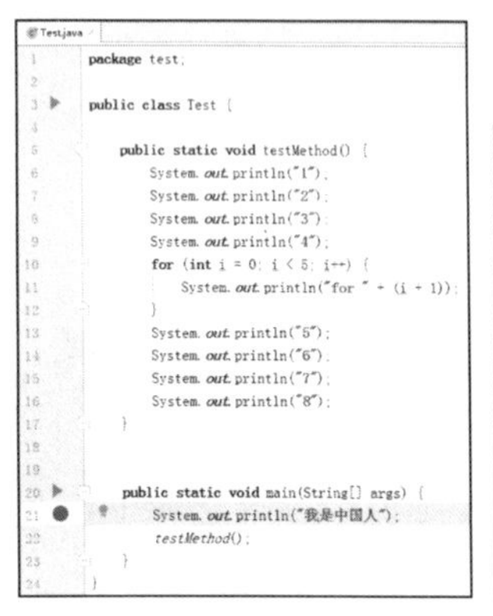


图 2-1

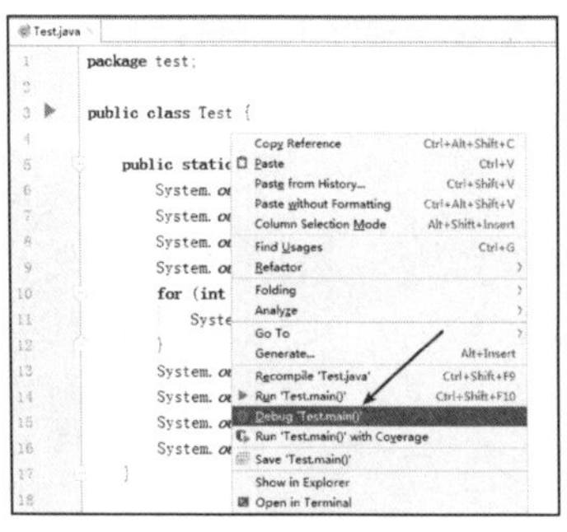


图 2-2

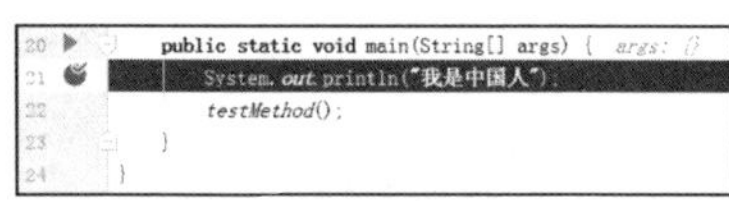


图 2-3

Alt+Shift+F7
force step into
强制单步进入

F6
step over
单步跳过

Ctrl+R
run to cursor
运行到光标处

单步进入
step into
F5

直接返回
step out
F7

图 2-4

## 2.1.4 跟踪按钮解释

进入调试模式后，会出现调试按钮如图 2-4 所示。

“step into”和“force step into”按钮的区别如下。

（1）“step into”按钮只能进入自己写的方法。

（2）“force step into”按钮能够进入所有的方法，包含 JDK 中的方法和第三方框架的源代码。

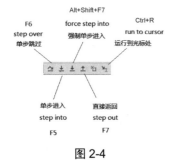

## 2.1.5 更改“force step into”按钮快捷键

在调试时，“force step into”按钮使用的频率很高，但其默认快捷键“Alt+ Shift+F7”

使用起来不方便，我们可以进行更改。

更改快捷键的配置路径为“Keymap”→“Main menu”→“Run”→“Debugging Actions”，配置界面如图 2-5 所示。

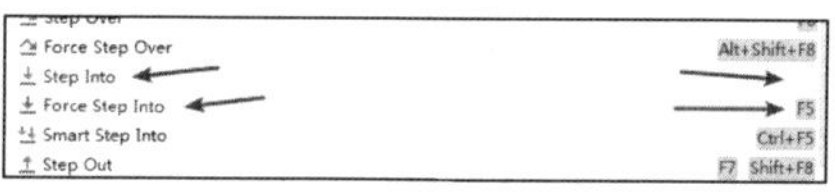


图 2-5

首先清除“Step Into”的快捷键，再清除“Force Step Into”的快捷键，最后对“Force Step Into”设置“F5”键，效果如图 2-6 所示。

经过更改之后，“F5”键是强制单步进入，“F6”键是单步跳过，“F7”键是返回，“F8”键是正常运行，方便进行调试。

通过对 Console 面板进行拖拽，可以实现控制台与调试的变量值一同显示的效果，如图 2-7 所示。

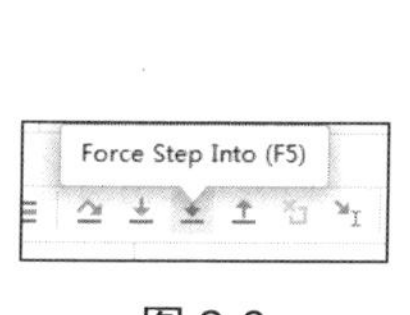


图 2-6

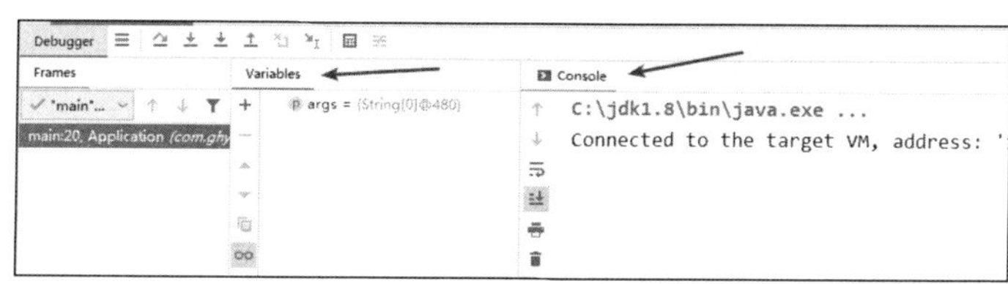


图 2-7

## 2.1.6 查看所有断点

单击图 2-8 中箭头所示的按钮，会弹出显示所有断点列表的窗口，如图 2-9 所示。

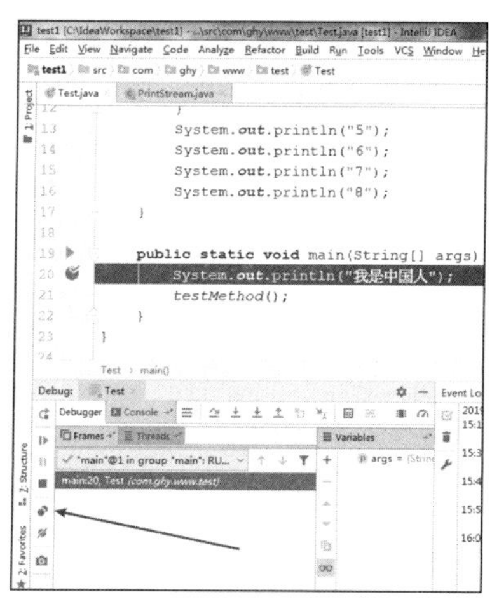


图 2-8

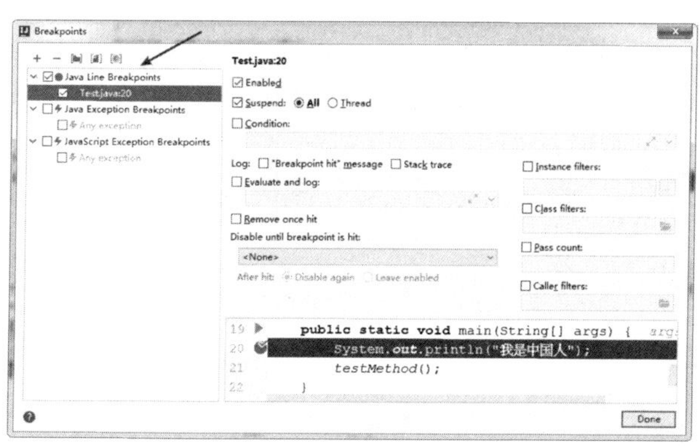


图 2-9

## 2.1.7 设置条件断点

创建测试用的代码如下：

```
package test;

public class Test {
    public static void main(String[] args) {
        for (int i = 0; i < 50; i++) {
            System.out.println("for " + (i + 1));
        }
    }
}
```

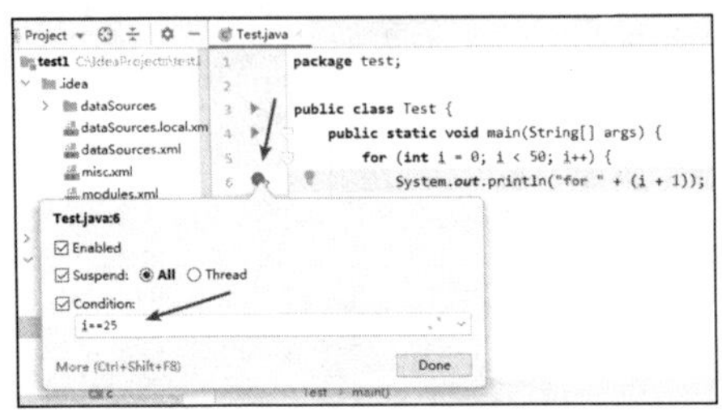


图 2-10

在第 6 行设置断点后，右击圆点，弹出如图 2-10 所示的界面，输入条件断点表达式。

这时，以 Debug 模式运行程序，当 i==25 时，会在断点处停下。

## 2.1.8 正确终止进程

当在调试的过程中找到了 Bug 出现的原因想立即终止进程时，在通常情况下是单击图 2-11 中箭头所指按钮，但这时断点后面的代码还是会被执行的，如图 2-12 所示。

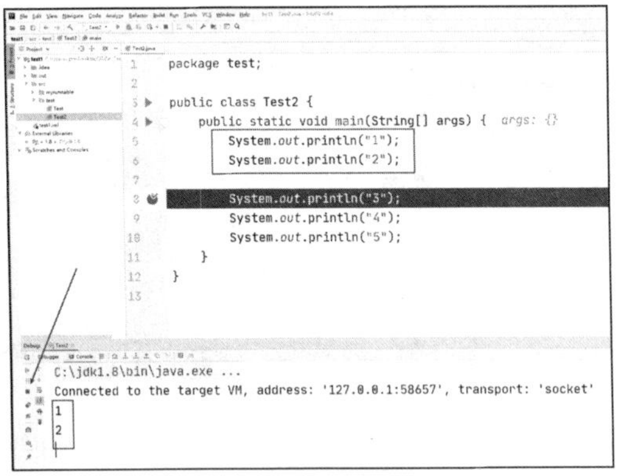

图 2-11

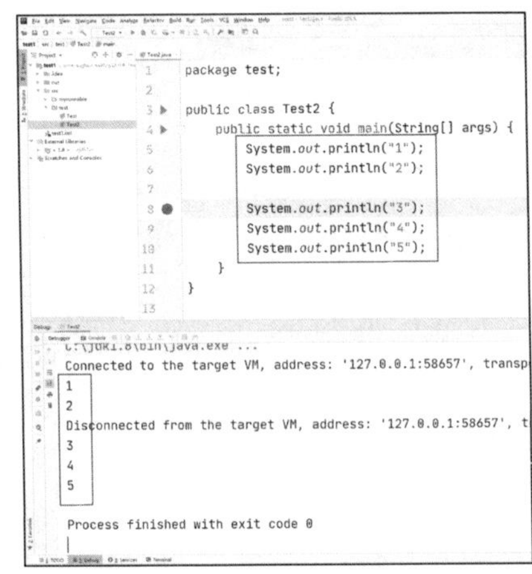


图 2-12

想要取消执行后面的 3 个步骤，则需要先单击“Force Return”按钮，如图 2-13 所示，再单击图 2-14 中所示的按钮。

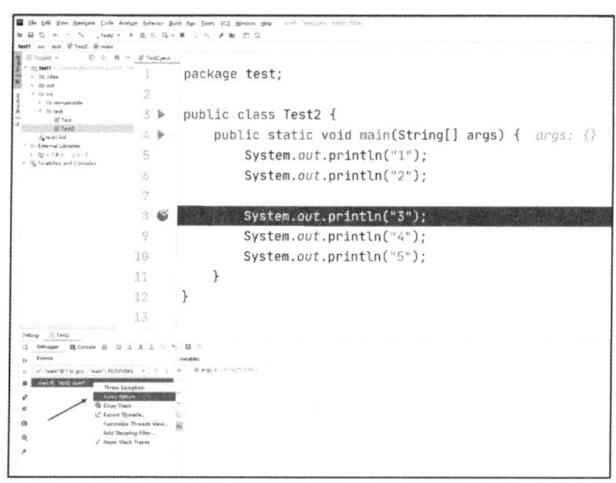


图 2-13

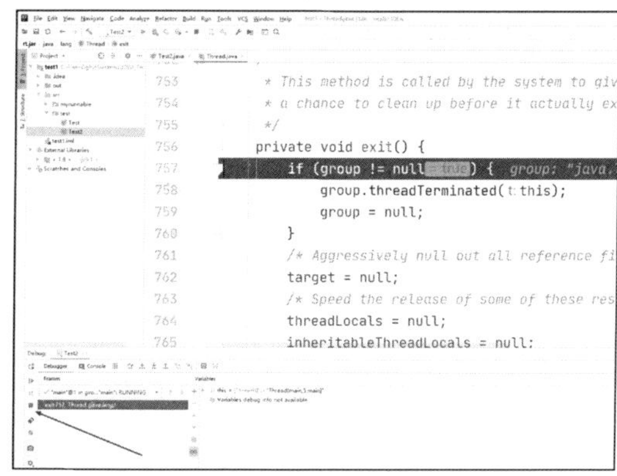


图 2-14

这样，后面的步骤没有被执行，如图 2-15 所示。

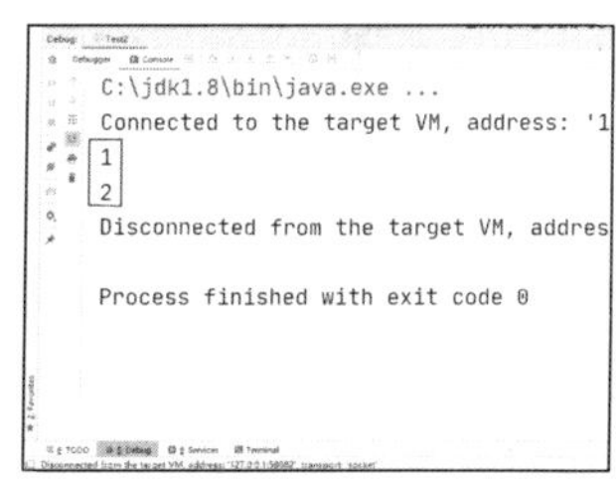


图 2-15

## 2.1.9　设置运行到光标处暂停

在调试时，可以设置将程序运行到光标处暂停，如图 2-16 所示。

在某代码行处右击，在弹出的菜单中选择“Run to Cursor”选项，如图 2-17 所示，运行结果如图 2-18 所示。

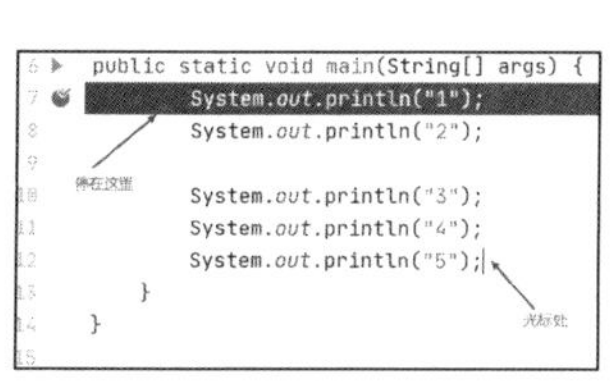


图 2-16

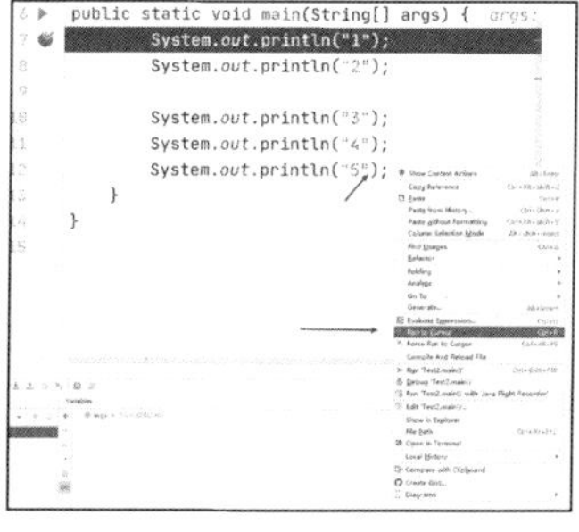


图 2-17

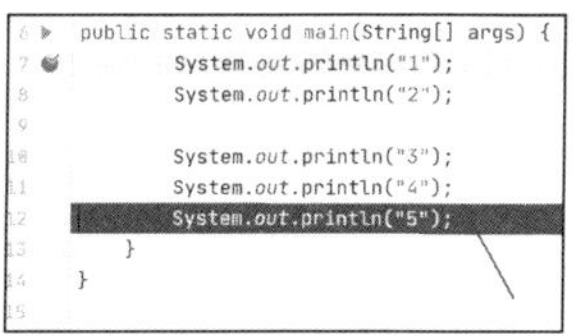


图 2-18

## 2.1.10　显示完整包路径

在默认情况下，调试代码时不显示完整的包路径，如图 2-19 所示。

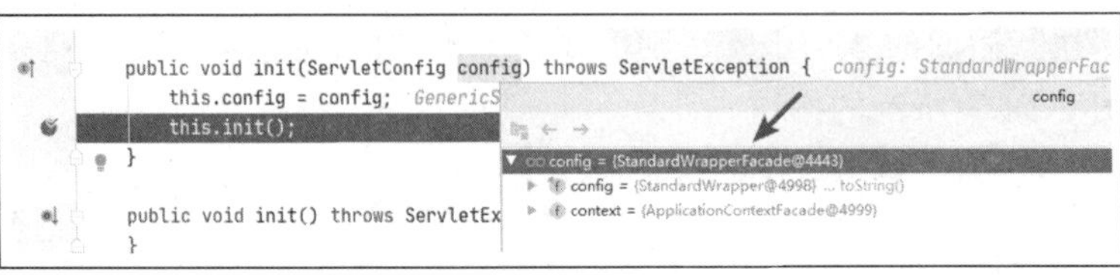


图 2-19

我们进行如图 2-20 所示的配置。

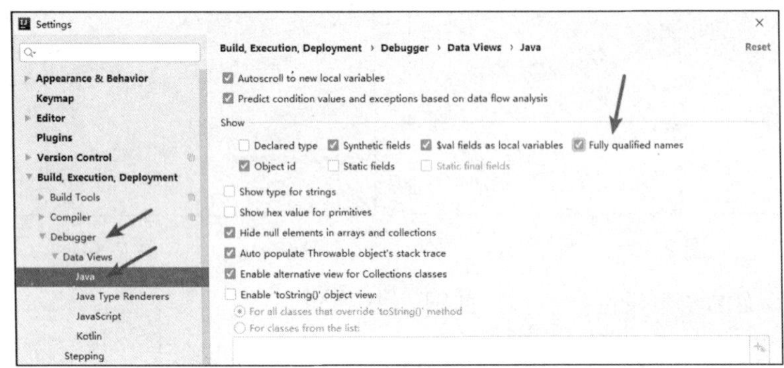


图 2-20

这时，即可显示完整的包路径，如图 2-21 所示。

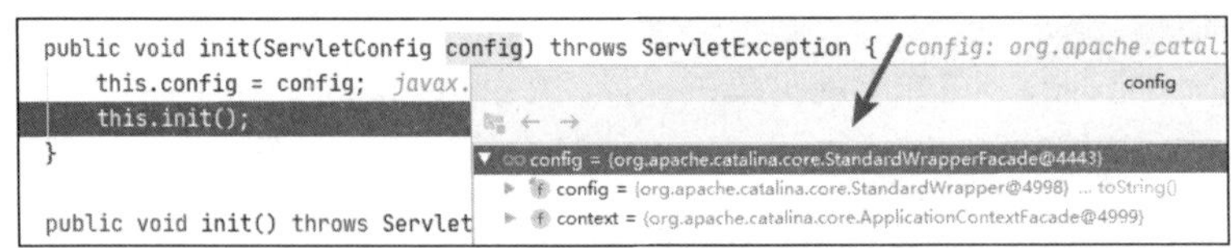


图 2-21

## 2.2 为Java项目引用jar包文件

本节将介绍如何在 IntelliJ IDEA 工具中引用 jar 包文件，有如下三种方式。

（1）使用 Global Libraries。

（2）使用 Libraries。

（3）将项目中的任意.jar 文件添加到 Libraries 中。

### 2.2.1 使用 Global Libraries 引用 jar 包文件

当学习一项新技术时，通常会创建 *n* 个项目作为案例来实现不同的功能，而这些项目中引用的 jar 包文件都是重复的。如果在每个项目中都存在 *n* 个 jar 包文件，则会非常占用空间，这时可以使用 IntelliJ IDEA 提供的 Global Libraries 全局库来解决这一问题。

创建名称为 addjar1 的测试项目，先单击“Project Structure...”选项，如图 2-22 所示，再单击“Global Libraries”选项，如图 2-23 所示。

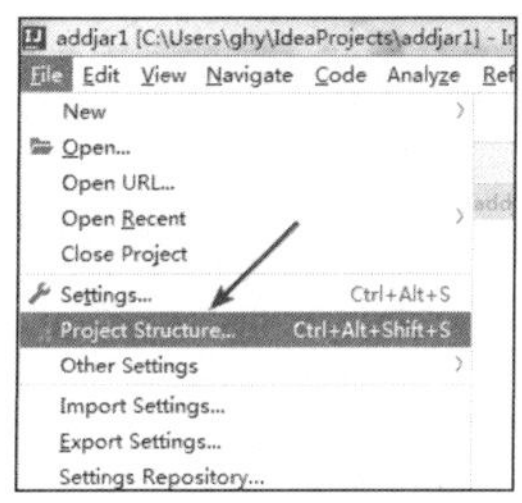


图 2-22

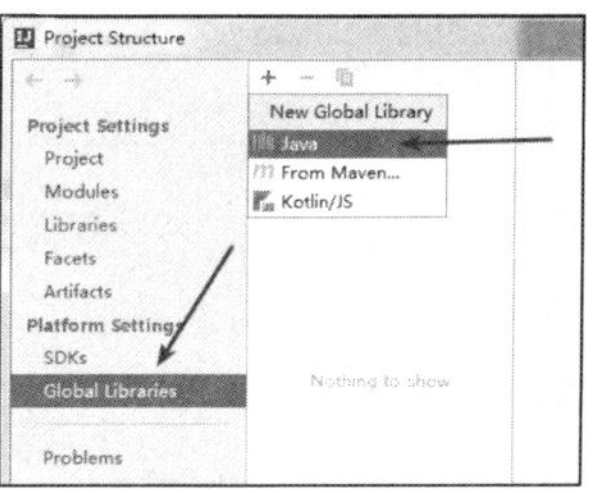


图 2-23

在弹出的界面中，选择 jar 包文件所在的文件夹，如图 2-24 所示，单击“OK”按钮。

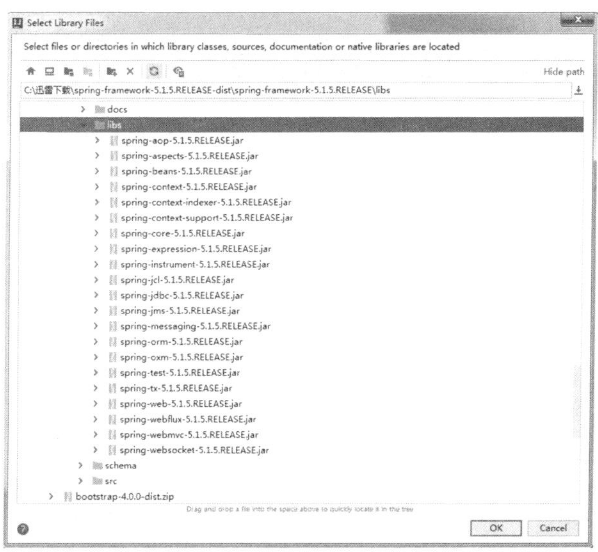


图 2-24

弹出的窗口如图 2-25 所示，询问是否把临时名字 libs 的 Library 库添加到 modules 中，为了演示更多的操作步骤，在此我们单击“Cancel”按钮，显示如图 2-26 所示界面。

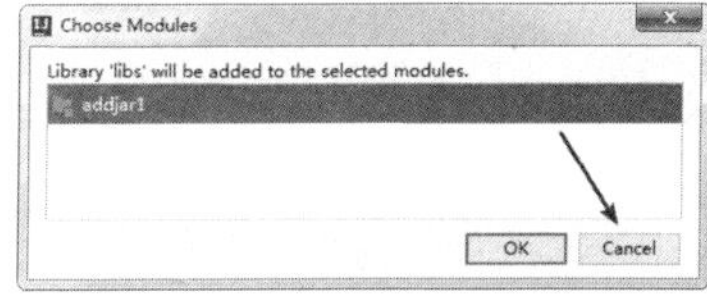


图 2-25

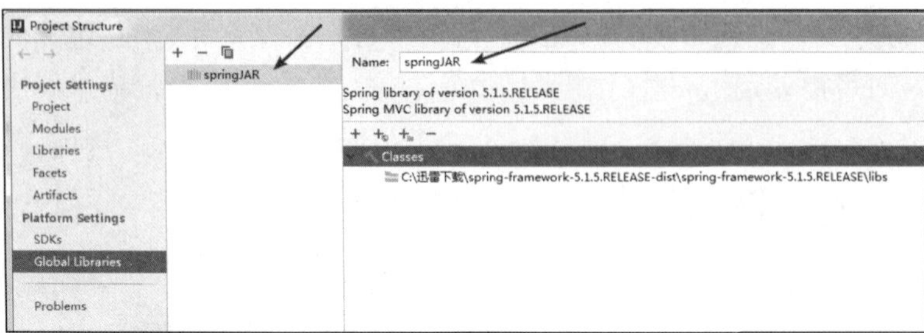


图 2-26

在更改 Library 库的名称为 springJAR 后，先以同样的操作添加 freemarker JAR 库，然后单击右下角的“Apply”按钮，如图 2-27 所示。

如果不在 Modules 中引用 Libraries 库，则项目中无法使用 jar 包文件中的类，相当于没有引用 jar 包文件，只是创建了 Libraries 库。故需要在 Modules 中引用 Libraries 库，操作如图 2-28 所示。

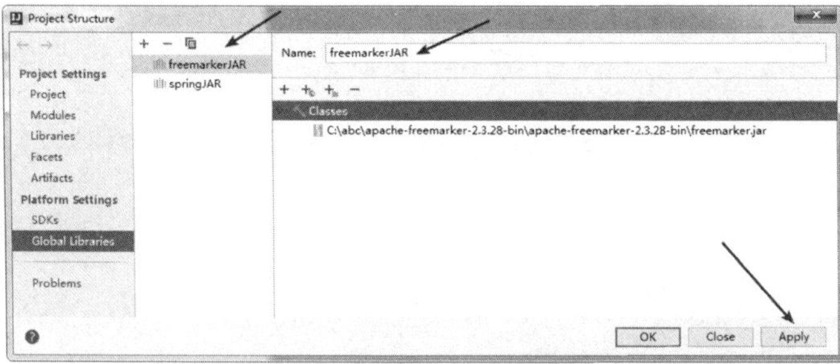


图 2-27

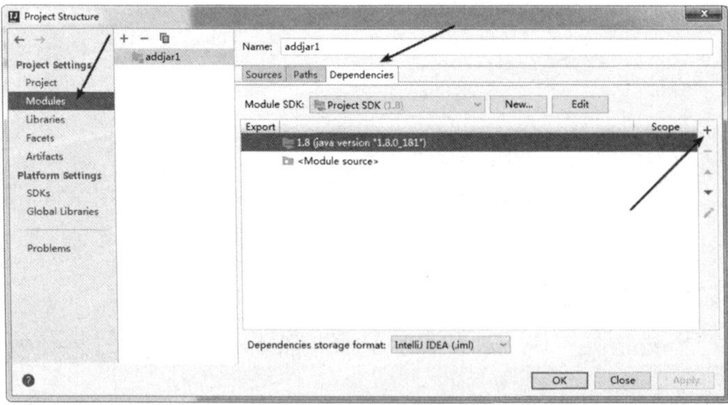


图 2-28

单击弹出框中的“Library...”菜单，如图 2-29 所示。

选择前面创建的两个 Global Libraries，单击“Add Selected”按钮，如图 2-30 所示。

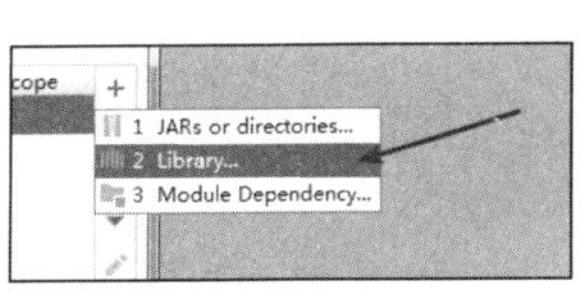


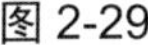

图 2-29

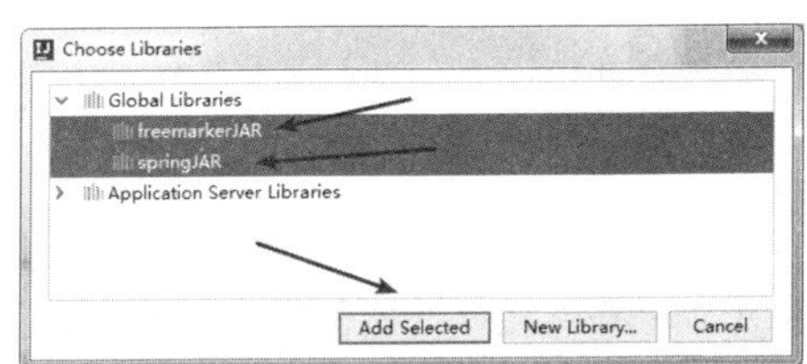


图 2-30

出现图 2-31 所示的界面。Export 列中的“Checkbox”选项的作用：例如 ModuleA 依赖了 jdbc.jar，ModuleB 依赖 ModuleA，如果不在 ModuleA 里 Dependencies 中 jdbc.jar 前面 Export 打上钩，ModuleB 无法用到 jdbc.jar 中的内容。在 Modules 中成功引用两个 Global Libraries，单击右下角的“OK”按钮关闭窗口。可见，项目中成功引用了 jar 包文件，如图 2-32 所示。

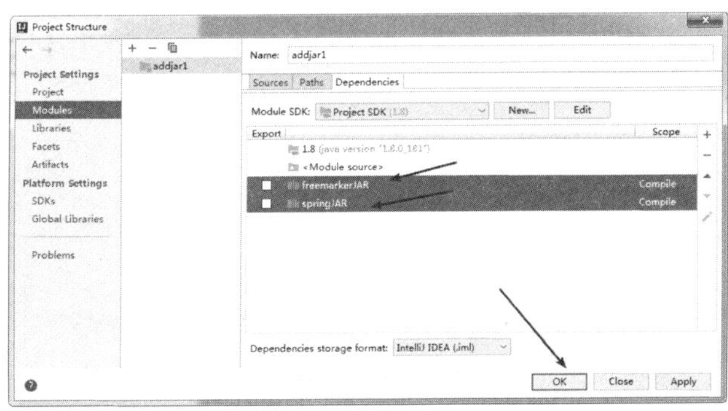


图 2-31

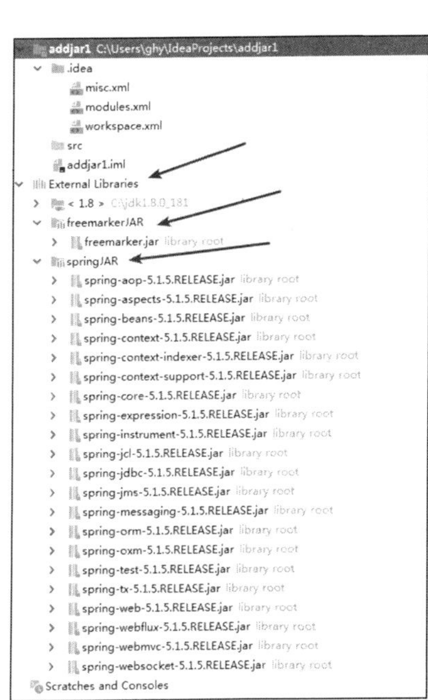


图 2-32

创建测试用的 Java 类，代码如下：

```
package test;
```

```
import org.springframework.context.ApplicationContext;
import org.springframework.context.support.ClassPathXmlApplicationContext;

public class Test {
    public static void main(String[] args) {
        ApplicationContext context = new ClassPathXmlApplicationContext();
        System.out.println(context);
    }
}
```

程序运行结果如下：

```
org.springframework.context.support.ClassPathXmlApplicationContext@533ddba
```

显示项目成功引用 jar 包文件，并且成功使用 jar 包文件中的类。

注意：创建的 Global Libraries 在所有的项目中都可以被引用。

## 2.2.2 使用 Libraries 引用 jar 包文件

我们知道，创建的 Global Libraries 在所有的项目中都可以被引用，但如果只有当前项目会使用某些 jar 包文件，则不需要为这些 jar 包文件单独创建 Global Libraries，只创建项目内的 Libraries 即可，其在不同项目中不能共享。

创建项目级别 Library 的目的是对当前项目中的 jar 包文件进行“分组”管理。

首先创建名称为 addjar2 的测试项目，然后进入“Project Structure”界面，单击“Java”菜单，如图 2-33 所示，选择 jar 包文件所在的路径，如图 2-34 所示，单击“OK”按钮。

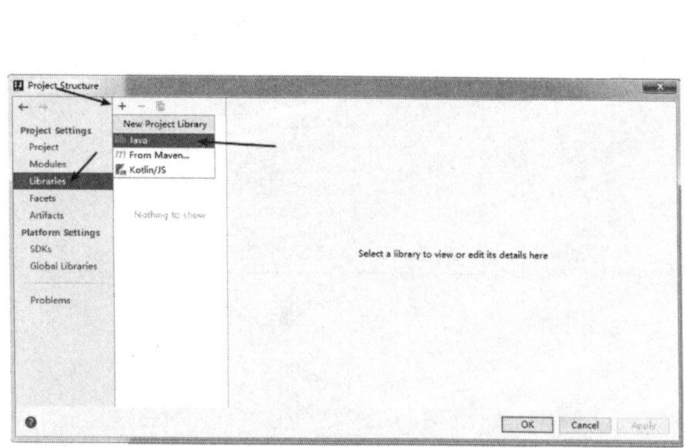


图 2-33

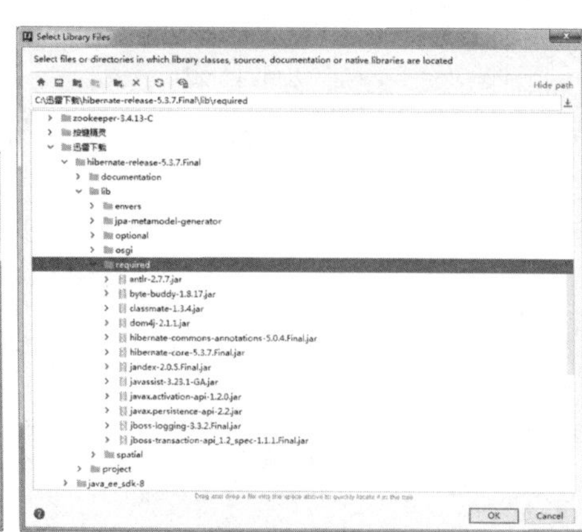


图 2-34

弹出对话框，询问是否将 Library 添加到 Modules 中，在这里依旧单击“Cancel”按钮，如图 2-35 所示。

更改 Library 库的名称为“hibernateJAR”，再单击“Apply”按钮，如图 2-36 所示。

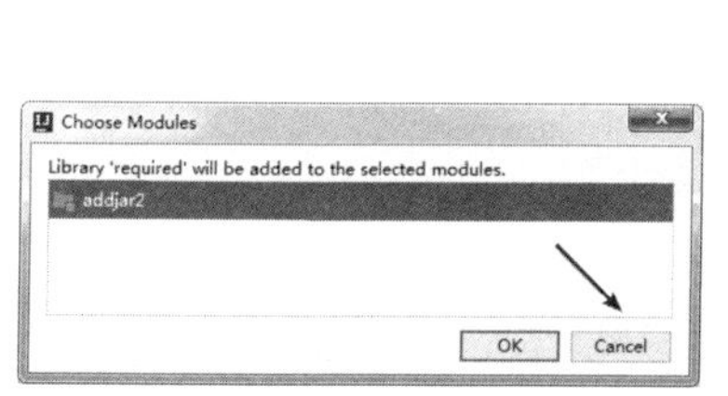


图 2-35

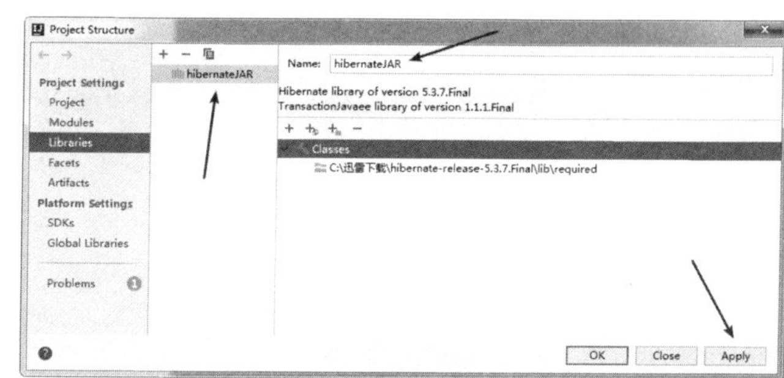


图 2-36

在 Modules 中引用 Library，如图 2-37 所示。选择 Project Libraries 中的 hibernateJAR 库，再单击“Add Selected”按钮，如图 2-38 所示。

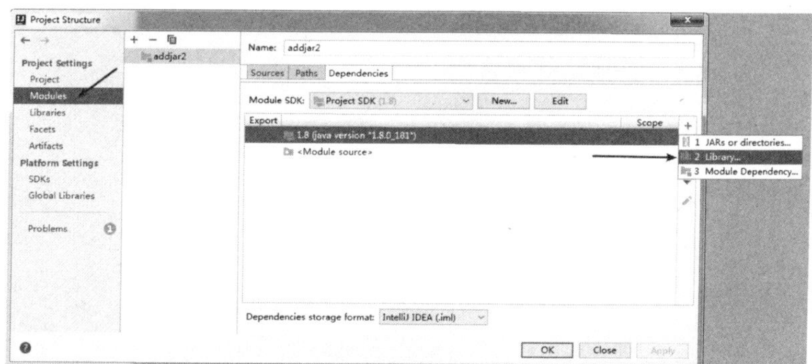


图 2-37

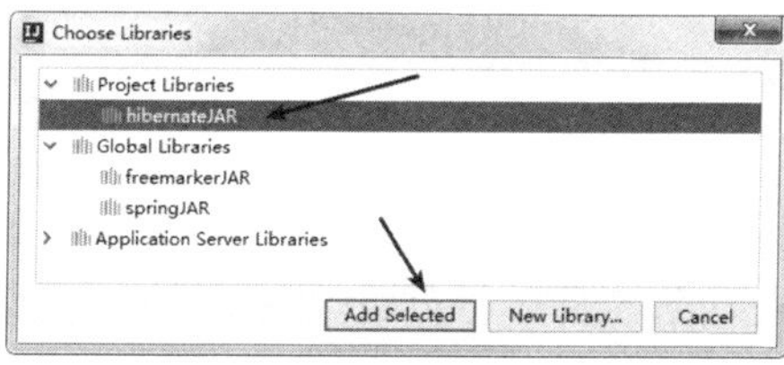


图 2-38

Modules 中成功引用 Library，如图 2-39 所示，项目中也成功引用 Library，如图 2-40 所示。

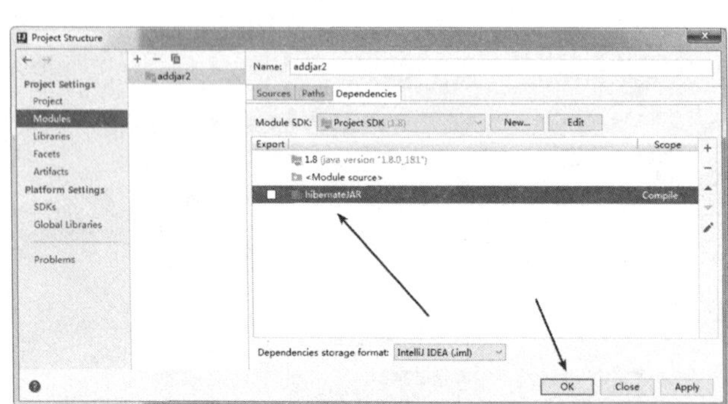

图 2-39

图 2-40

创建测试用的 Java 类，代码如下：

```
package test;

import org.hibernate.cfg.Configuration;

public class Test {
    public static void main(String[] args) {
        System.out.println(new Configuration());
    }
}
```

运行结果如下：

```
org.hibernate.cfg.Configuration@a38d7a3
```

注意：Project Libraries 只可以被当前的项目使用，不能共享，故使用 Project Libraries 可以对当前项目中的 jar 包文件进行分组管理。

## 2.2.3 使用 Add as Library 引用 jar 包文件

有时候，我们不需要对 jar 包文件进行分组，包括 Global Libraries 和 Project Libraries，就是想简单地把某一文件夹中的所有 jar 包文件都引入到项目中，进行快速开发，这时可以使用“Add as Library”菜单进行实现。

创建名称为“addjar3”的测试项目，并创建名称为 jar 的文件夹，如图 2-41 所示。

复制一个.jar 文件到 jar 文件夹，弹出对话框，单击“OK”按钮，如图 2-42 所示。

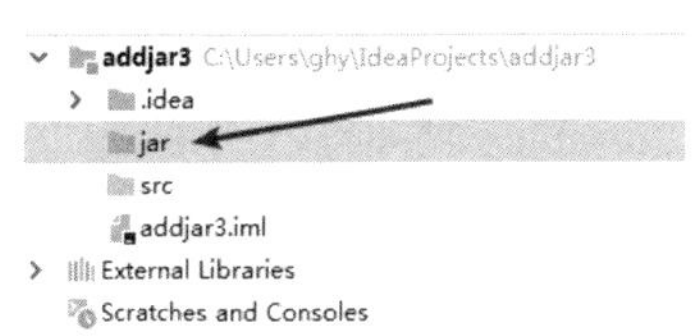

图 2-41

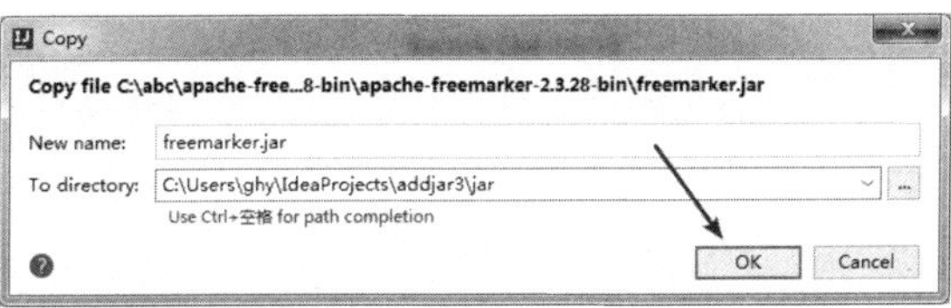

图 2-42

再复制多个.jar 文件到 jar 文件夹，如图 2-43 所示。对 jar 文件夹右击，弹出选择菜单，单击“Add as Library...”选项，如图 2-44 所示。

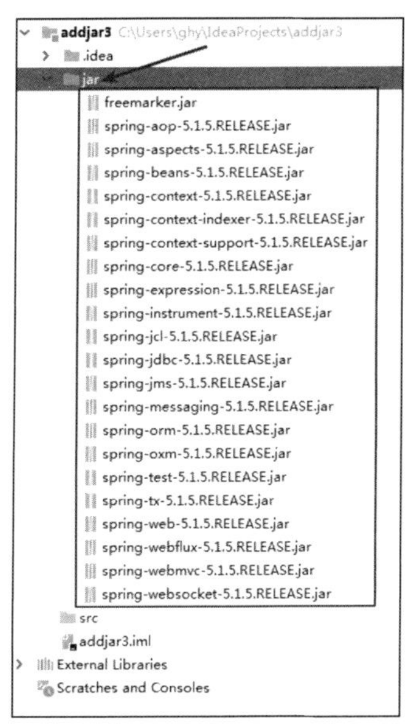

图 2-43

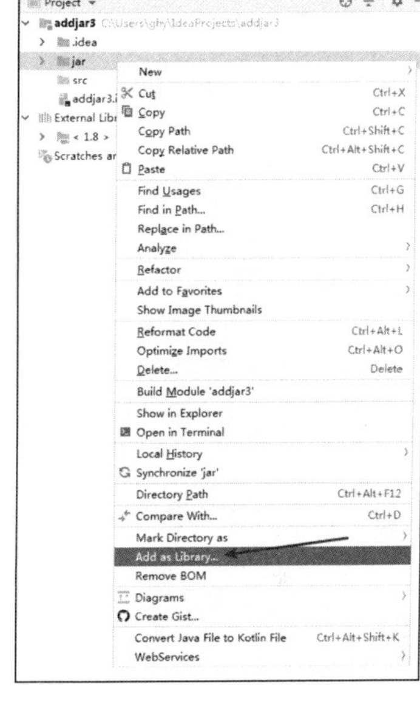

图 2-44

在弹出的界面中，设置 Library 库的名称和级别，如图 2-45 所示，单击“OK”按钮。

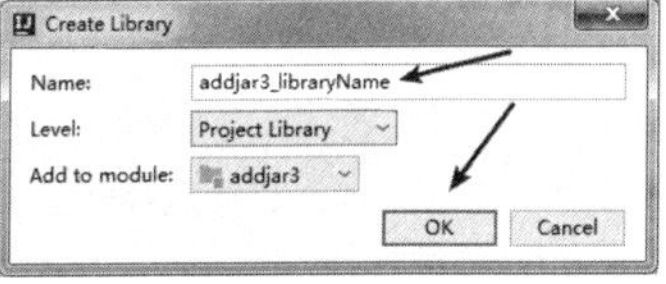

图 2-45

这时，项目中就可以使用 jar 包文件中的类了。

创建测试用的 Java 类，代码如下：

```
package test;

import org.springframework.context.support.ClassPathXmlApplicationContext;
```

```
public class Test {
    public static void main(String[] args) {
        System.out.println(new ClassPathXmlApplicationContext());
    }
}
```

运行结果如下：

```
org.springframework.context.support.ClassPathXmlApplicationContext@533ddba
```

## 2.3 对Java项目导出的.jar文件实现转换

首先创建 Java 项目并引用相关 jar 包文件，如图 2-46 所示，然后创建 Artifacts，如图 2-47 所示。

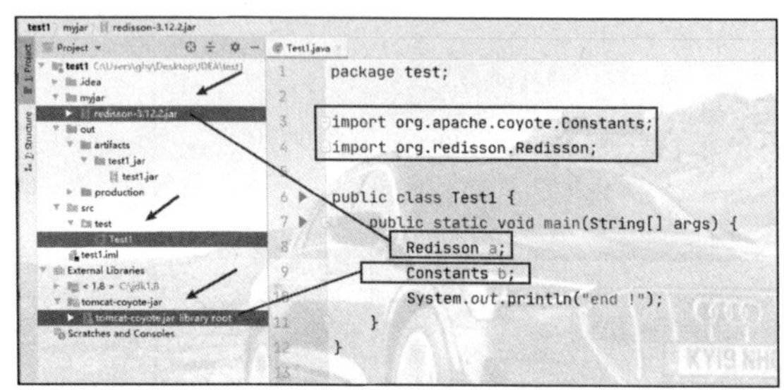

图 2-46

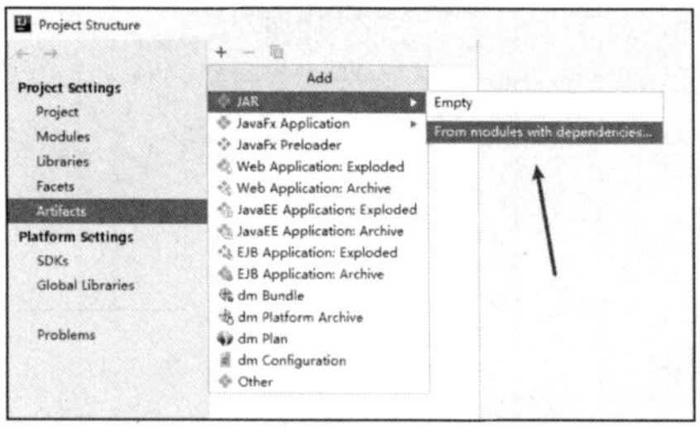

图 2-47

配置图 2-48 所示的界面，生成.jar 文件，如图 2-49 所示。

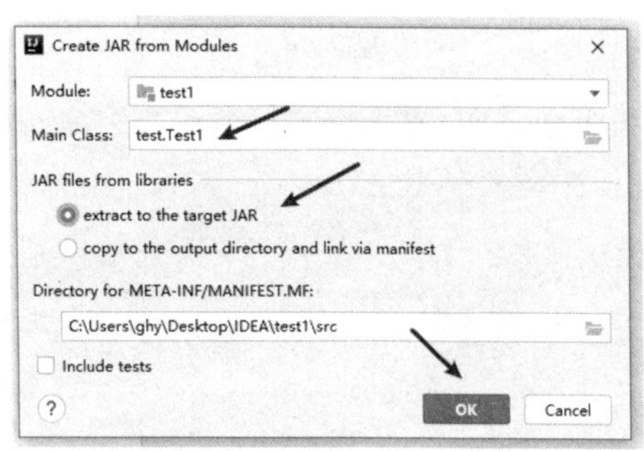


图 2-48

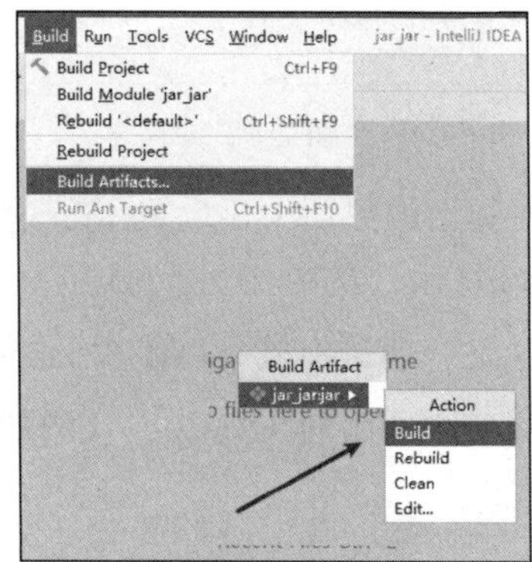


图 2-49

运行结果如图 2-50 所示。

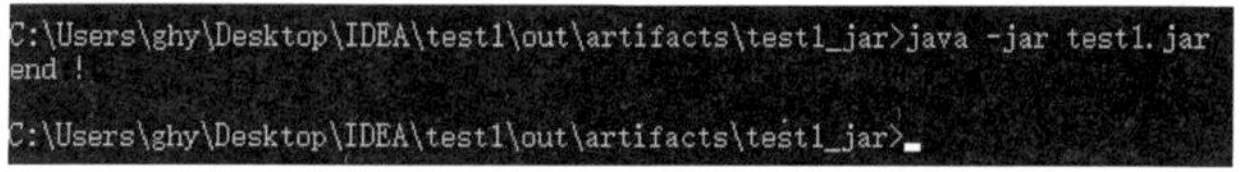


图 2-50

这时，解压 test1.jar 文件后会发现内部依赖的.jar 文件被转换成.class 文件，如图 2-51 所示。

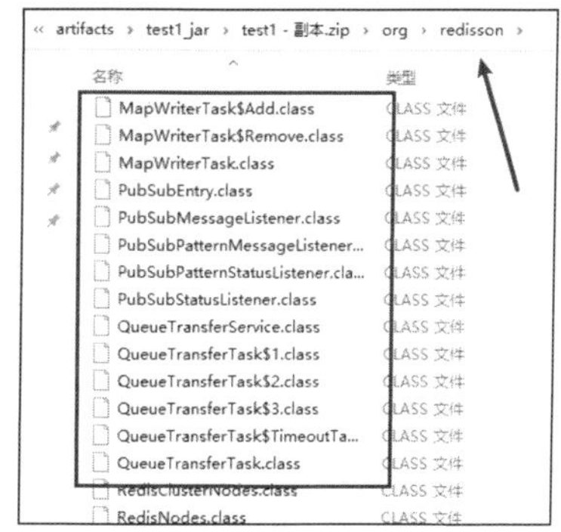


图 2-51

## 2.4　创建Web项目

本节介绍如何创建 Web 项目。

### 2.4.1　创建 Java 项目并配置 Web 环境

创建 Java 项目，如图 2-52 所示，单击“Next”按钮。

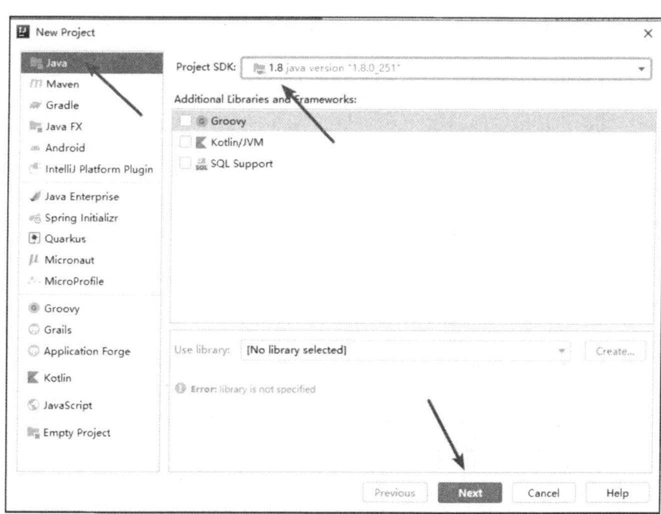


图 2-52

显示界面如图 2-53 所示，再单击“Next”按钮。

出现如图 2-54 所示的界面，单击“Finish”按钮完成 Java 项目的创建。

这时，单击如图 2-55 所示的菜单，显示界面如图 2-56 所示。

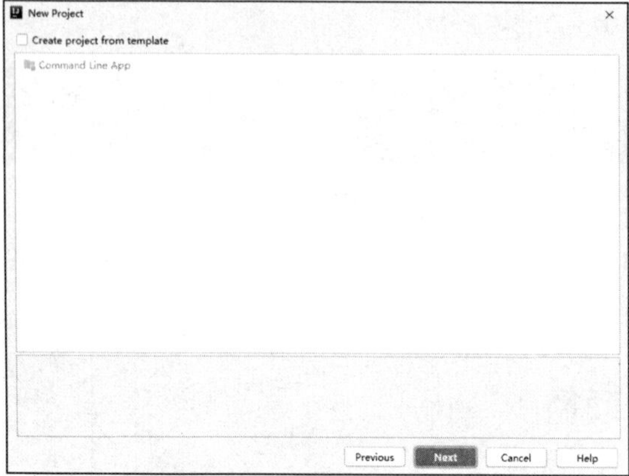


图 2-53

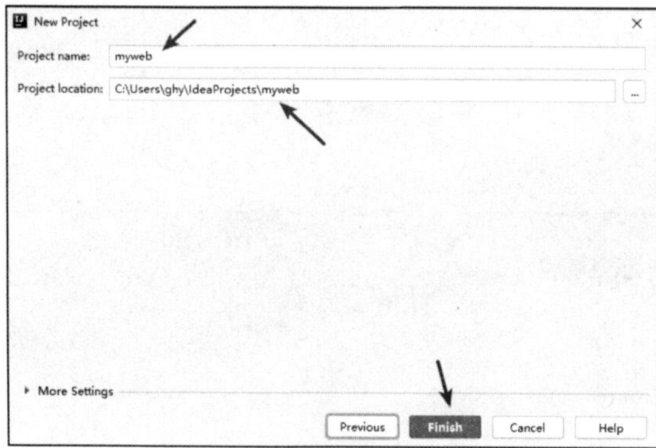


图 2-54

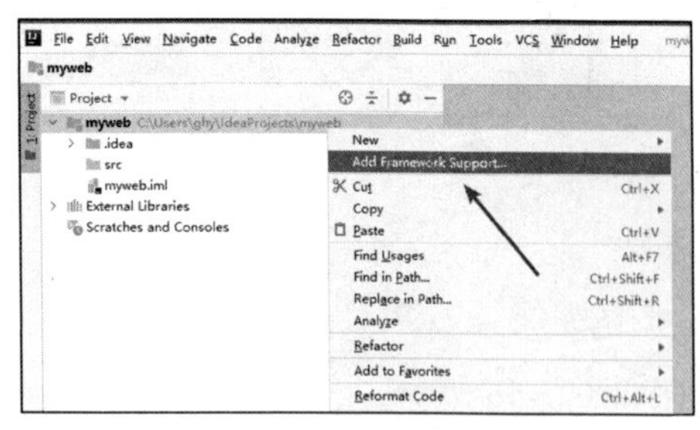


图 2-55

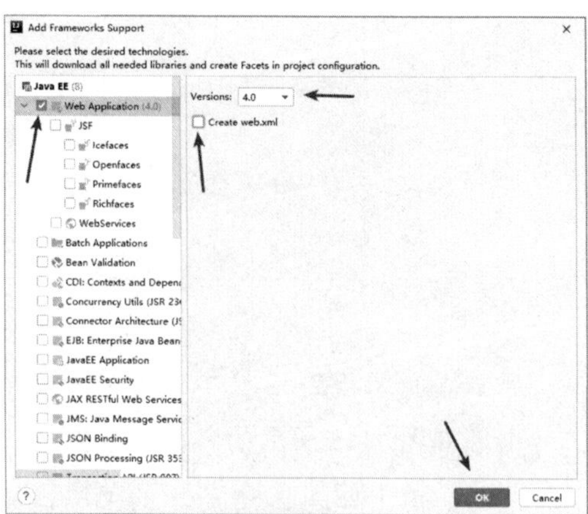


图 2-56

创建项目的结构如图 2-57 所示。

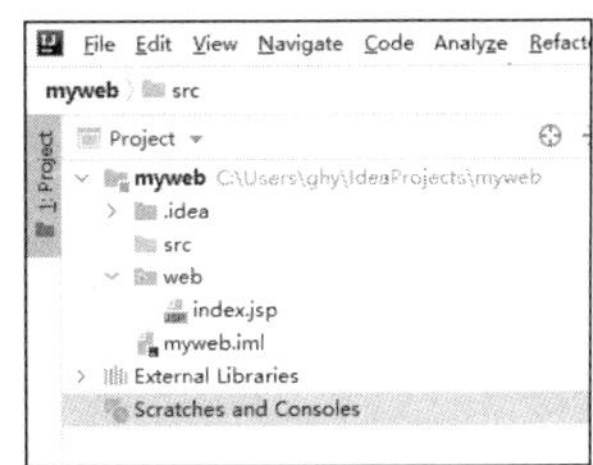


图 2-57

### 2.4.2 使用向导创建 Servlet

下面对 src 节点使用右击“Create New Servlet”菜单的方式创建新的 Servlet。首先在弹出窗口中输入“Servlet Name”、包名及类名，然后单击“OK”按钮完成创建。

### 2.4.3 添加与 Servlet 相关的.jar 文件依赖

由于创建的 Servlet 缺少 jar 包文件，因此代码颜色呈红色（图中的浅色内容），如图 2-58 所示。

```
Test1.java
package controller;

import java.io.IOException;

@javax.servlet.annotation.WebServlet(name = "Test1")
public class Test1 extends javax.servlet.http.HttpServlet {
    protected void doPost(javax.servlet.http.HttpServletRequest request, javax.servlet.http.HttpServletResponse response) throws javax.servlet.ServletException, IOException {

    }
    protected void doGet(javax.servlet.http.HttpServletRequest request, javax.servlet.http.HttpServletResponse response) throws javax.servlet.ServletException, IOException {

    }
}
```

图 2-58

因为与 Servlet 技术有关的 jar 包文件基本上都在 Tomcat 的 lib 文件夹中，所以创建一个全局 Library 只需要引用 Tomcat/lib 文件夹，然后在项目的 Modules 中配置即可。

*注意：如果添加 Tomcat/lib 全局库后，Servlet 代码依然出现红色文字，则需要在线下载 JavaEE6 相关的 JAR 文件。*

添加完 TomcatJAR 后代码不再出现红色，如图 2-59 所示。

```
Test1.java
package controller;

import java.io.IOException;

@javax.servlet.annotation.WebServlet(name = "Test1")
public class Test1 extends javax.servlet.http.HttpServlet {
    protected void doPost(javax.servlet.http.HttpServletRequest request, javax.servlet.http.HttpServletResponse response) throws javax.servlet.ServletException, IOException {

    }
    protected void doGet(javax.servlet.http.HttpServletRequest request, javax.servlet.http.HttpServletResponse response) throws javax.servlet.ServletException, IOException {

    }
}
```

图 2-59

### 2.4.4 完善 Servlet 代码

完善 Servlet 代码的操作如图 2-60 所示。

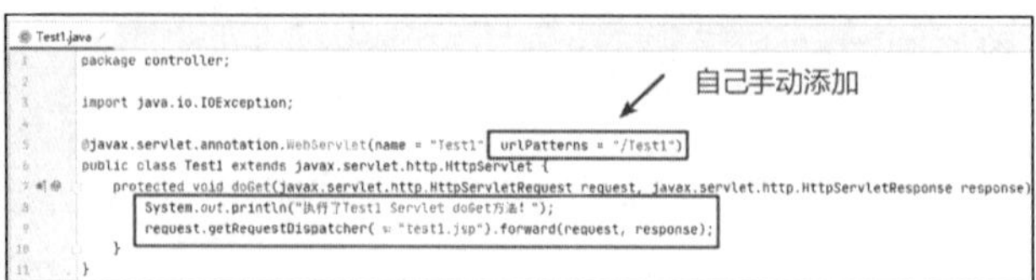


图 2-60

## 2.4.5 创建 test1.jsp 文件

在 Web 路径下创建视图 test1.jsp 文件，代码如下：

```
<%@ page contentType="text/html;charset=UTF-8" language="java" %>
<html>
<head>
    <title>Title</title>
</head>
<body>
显示 test1.jsp!
</body>
</html>
```

## 2.4.6 关联 Tomcat

首先，单击图 2-61 中箭头所指的按钮开始配置 Tomcat。

先选择“Tomcat Server”中的“Local”节点，再单击“Configure...”按钮配置 Tomcat 模板信息，如图 2-62 所示。

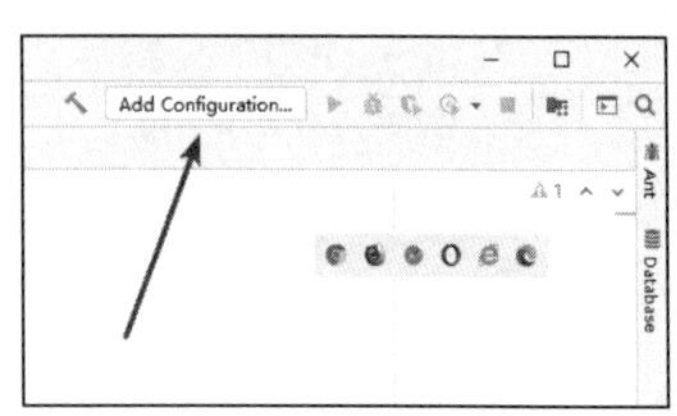


图 2-61

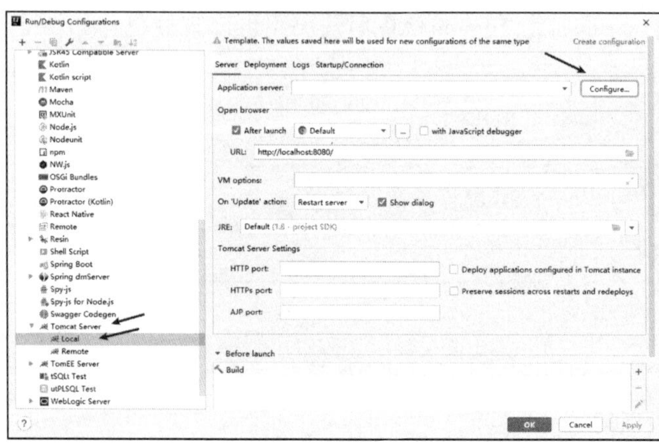


图 2-62

弹出如图 2-63 所示的界面，设置 Tomcat 路径，单击“OK”按钮完成配置，重新显示如图 2-64 所示的界面，可以在此界面中配置 Tomcat 的端口号，配置完成后单击右下角的“OK”按钮。

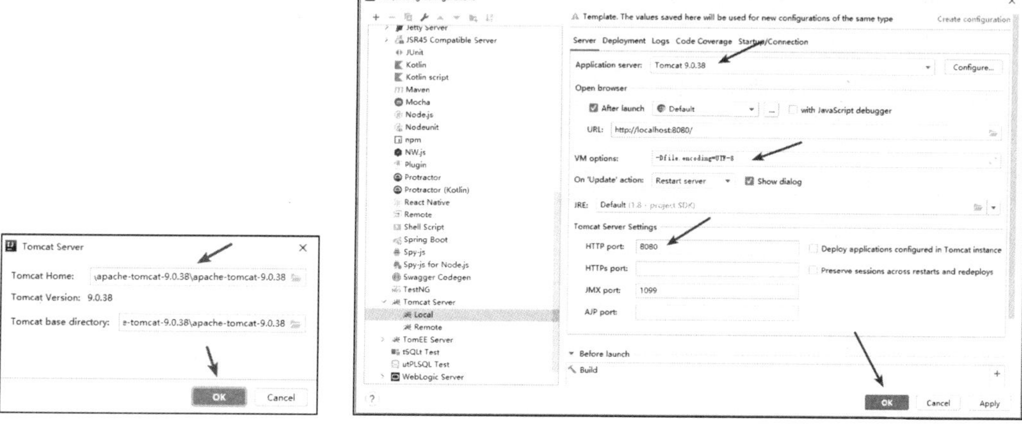


图 2-63　　　　图 2-64

在“VM options”中配置“-Dfile.encoding=UTF-8”的作用是防止 Servlet 使用如下语句：

```
System.out.println("执行了 Test1 Servlet! ");
```

输出中文为乱码。

单击左上角的加号按钮，在弹出的菜单中依次选择“Tomcat Server”→“Local”选项，如图 2-65 所示。

以 Tomcat Local 配置作为模板创建 Tomcat 配置，显示界面如图 2-66 所示。在界面的下方显示警告信息，可以通过单击“Fix”按钮解决。

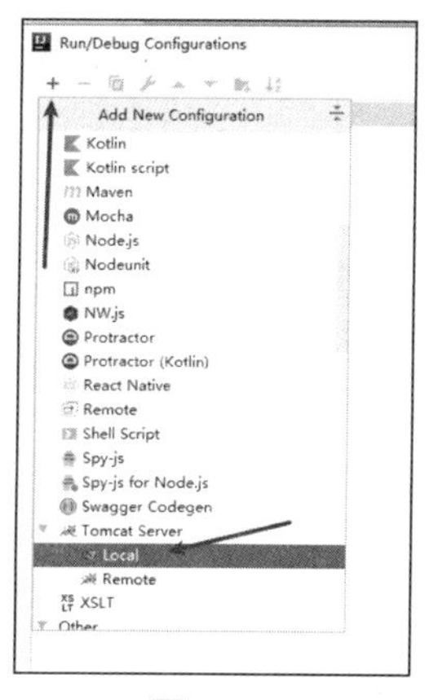


图 2-65

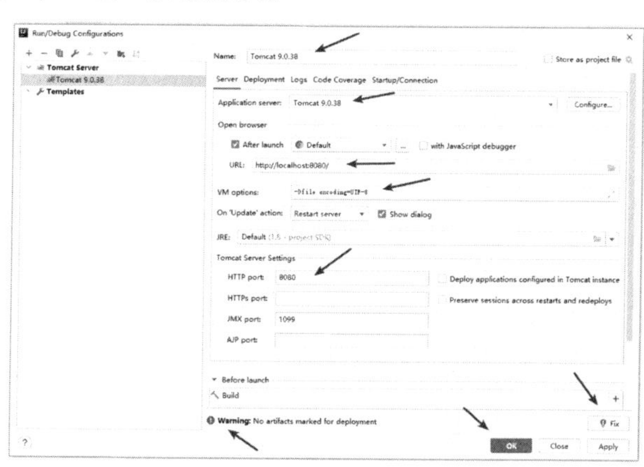


图 2-66

显示界面如图 2-67 所示，单击“OK”按钮成功配置 Tomcat。

注意：在 Servlet 中，通过执行代码 request.getServletContext().getRealPath("/")来获得 Web 项目部署的路径，而“Application context: /myweb”路径的作用是在浏览器中通过此 URL 访问 Web 项目，两者的作用不一样。

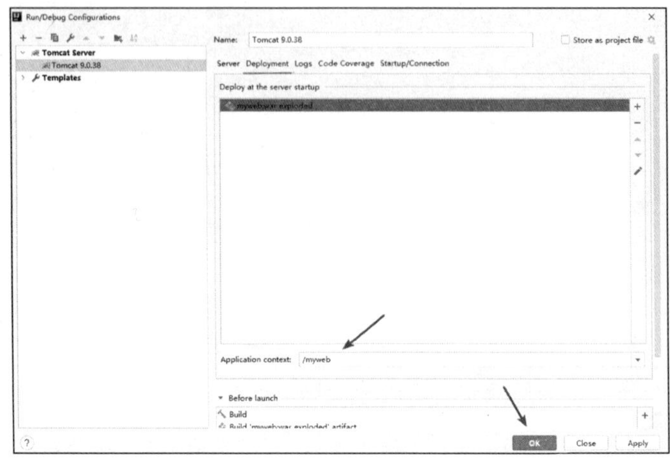


图 2-67

## 2.4.7 运行项目并解决 Tomcat 启动乱码

单击“运行”按钮，如图 2-68 所示，这时控制台输出乱码，如图 2-69 所示。

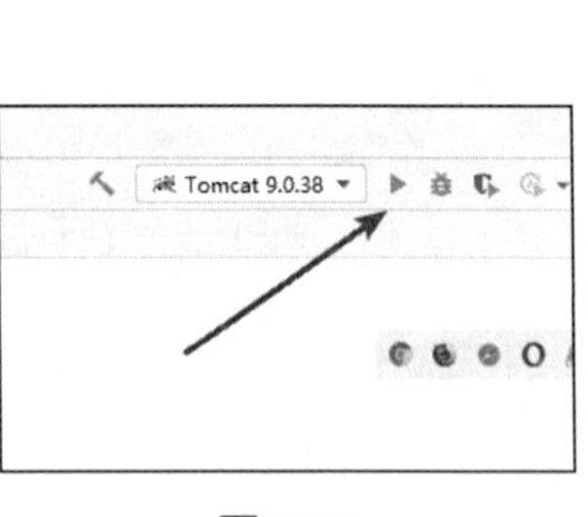


图 2-68

图 2-69

要想解决这一问题，需要更改配置文件：C:\Users\ghy\AppData\Roaming\JetBrains\IntelliJIdea2020.2\idea64.exe.vmoptions，即在最后添加配置：-Dfile.encoding=UTF-8。

这时，重新启动 Tomcat 后不再出现乱码，如图 2-70 所示。

如果打开网址：http://localhost:8080/myweb/Test1，则会在控制台和.jsp 文件中显示正确的中文，如图 2-71 所示。

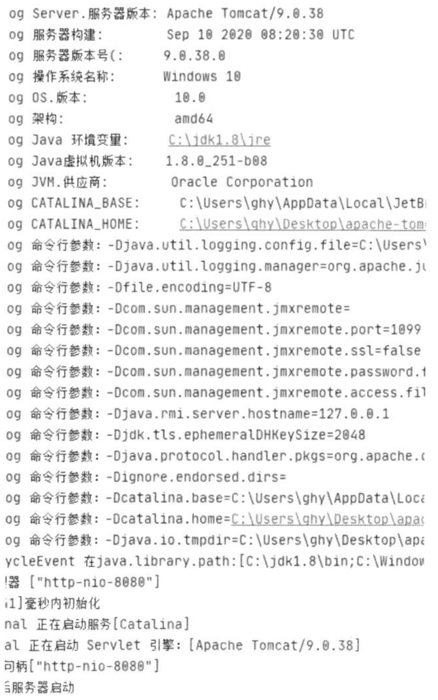
og Server.服务器版本: Apache Tomcat/9.0.38
og 服务器构建: Sep 10 2020 08:20:30 UTC
og 服务器版本号(: 9.0.38.0
og 操作系统名称: Windows 10
og OS.版本: 10.0
og 架构: amd64
og Java 环境变量: C:\jdk1.8\jre
og Java虚拟机版本: 1.8.0_251-b08
og JVM.供应商: Oracle Corporation
og CATALINA_BASE: C:\Users\ghy\AppData\Local\JetB
og CATALINA_HOME: C:\Users\ghy\Desktop\apache-tom
og 命令行参数: -Djava.util.logging.config.file=C:\Users
og 命令行参数: -Djava.util.logging.manager=org.apache.ju
og 命令行参数: -Dfile.encoding=UTF-8
og 命令行参数: -Dcom.sun.management.jmxremote=
og 命令行参数: -Dcom.sun.management.jmxremote.port=1099
og 命令行参数: -Dcom.sun.management.jmxremote.ssl=false
og 命令行参数: -Dcom.sun.management.jmxremote.password.f
og 命令行参数: -Dcom.sun.management.jmxremote.access.fil
og 命令行参数: -Djava.rmi.server.hostname=127.0.0.1
og 命令行参数: -Djdk.tls.ephemeralDHKeySize=2048
og 命令行参数: -Djava.protocol.handler.pkgs=org.apache.c
og 命令行参数: -Dignore.endorsed.dirs=
og 命令行参数: -Dcatalina.base=C:\Users\ghy\AppData\Loca
og 命令行参数: -Dcatalina.home=C:\Users\ghy\Desktop\apac
og 命令行参数: -Djava.io.tmpdir=C:\Users\ghy\Desktop\apa
ycleEvent 在java.library.path:[C:\jdk1.8\bin;C:\Window
器 ["http-nio-8080"]
1]毫秒内初始化
nal 正在启动服务[Catalina]
al 正在启动 Servlet 引擎: [Apache Tomcat/9.0.38]
句柄["http-nio-8080"]
服务器启动

图 2-70

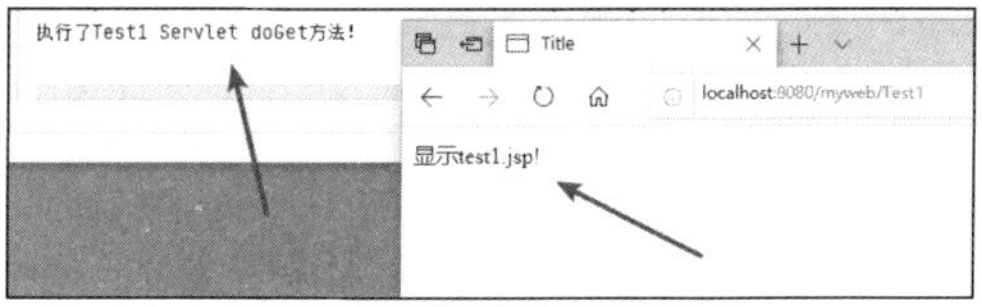
执行了Test1 Servlet doGet方法!
Title
localhost:8080/myweb/Test1
显示test1.jsp!

图 2-71

## 2.4.8　实现热部署

在默认情况下，要修改静态文件或 Java 类，需要重新启动 Tomcat 才可以加载最新版，但我们也可以通过配置实现热部署，如图 2-72 所示。

注意：只有以 debug 模式启动，热部署功能才能生效。

提示：在更改.java 文件时，建议还是采用重新启动 Tomcat 的方式来加载最新版本的.class 文件，不要过于依赖热部署。

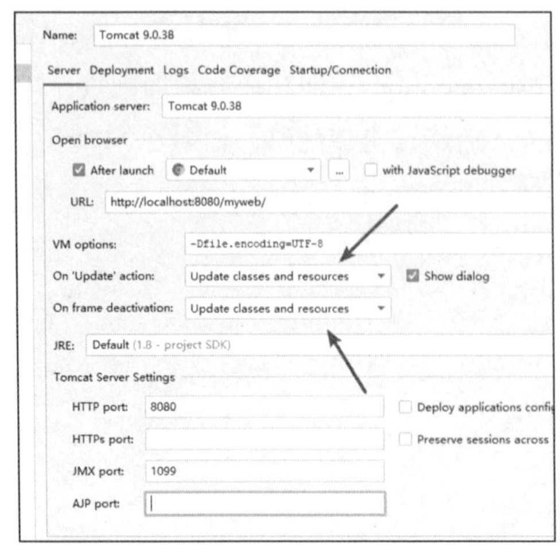


图 2-72

## 2.4.9 上传文件路径的小提示

假设“test1”是一个上传文件的测试项目，项目中包含 upload 文件夹用来存放上传的图片，upload 文件夹所在硬盘的路径为 C:\IDEAProjects\test1\web\upload。此路径就是 Web 项目保存在硬盘上的路径。

项目被部署到 Tomcat 并启动运行后，upload 文件夹所在路径为 C:\IDEAProjects\test1\out\artifacts\test1_war_exploded\upload。此路径就是 Web 项目运行的路径。

当实现文件上传时，图片文件需要在两个路径中分别进行存储：C:\IDEAProjects\test1\web\upload 和 C:\IDEAProjects\test1\out\artifacts\test1_war_exploded\upload。

尽量采用“双写”的方式进行文件上传。因为当重启 Tomcat 时，路径：C:\IDEAProjects\test1\out\ artifacts\test1_war_exploded\upload 中的图片会被自动清空，但路径：C:\IDEAProjects\test1\web\upload 中的文件会随着 Tomcat 重启被重新部署到路径：C:\IDEAProjects\test1\out\artifacts\test1_war_exploded\upload 中，所以上传的图片并未丢失。

## 2.4.10 禁止启动 Tomcat 完成后弹出网页

禁止启动 Tomcat 完成后弹出网页的配置如图 2-73 所示。

## 2.4.11 设置.html 文件自动缩进

.html 文件默认为不自动缩进，如图 2-74 所示。

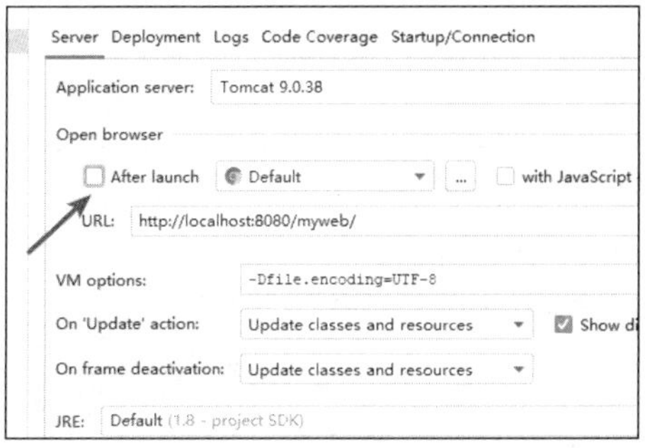


图 2-73

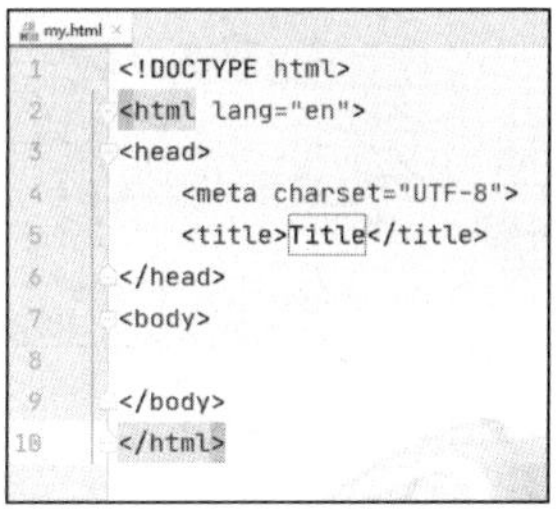

```
<!DOCTYPE html>
<html lang="en">
<head>
    <meta charset="UTF-8">
    <title>Title</title>
</head>
<body>

</body>
</html>
```

图 2-74

实现自动缩进的配置如图 2-75 所示，这时，格式化代码成功缩进，如图 2-76 所示。

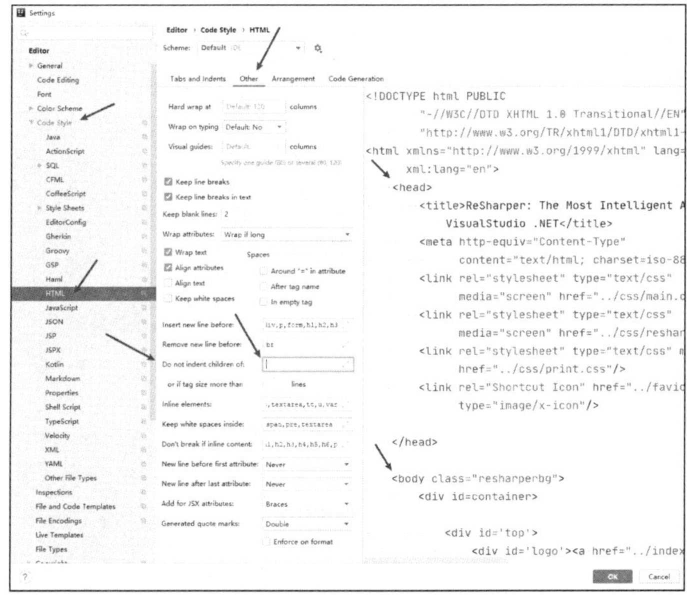


图 2-75

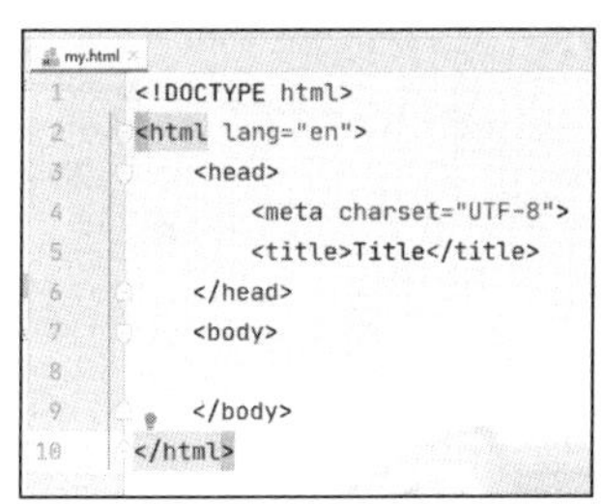


图 2-76

## 2.4.12　导出 war 包文件

首先创建 Web 项目，然后关联依赖的库。

创建 Web Application:Archive 文件，如图 2-77 所示。双击 Available Elements 中的.jar 文件，移到左侧的窗口中，以实现.war 文件包含指定的.jar 文件，如图 2-78 所示。

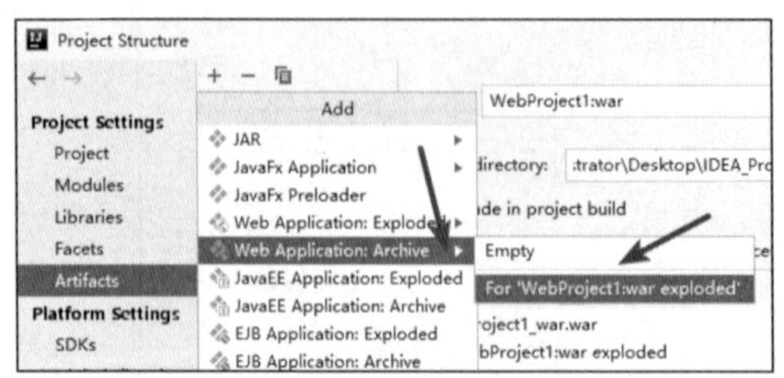


图 2-77

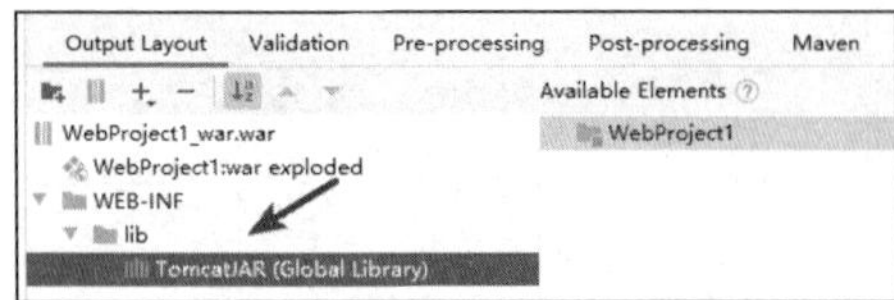


图 2-78

单击“Build”菜单，如图 2-79 所示，成功导出.war 文件的效果如图 2-80 所示。

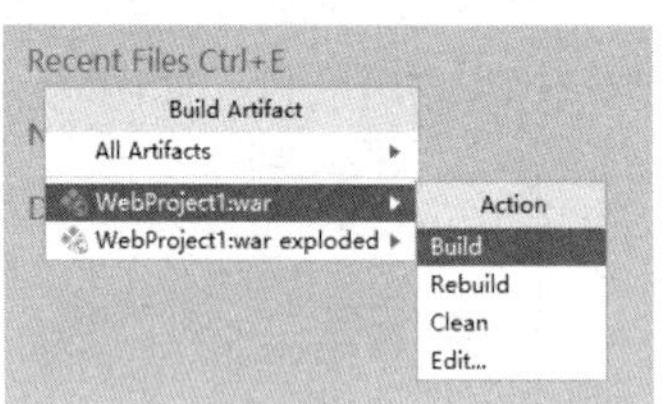


图 2-79

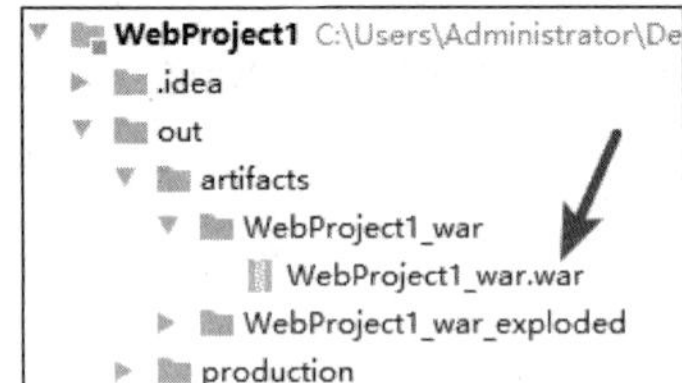


图 2-80

这时，可看到.war 文件中包含第三方依赖的.jar 文件，如图 2-81 所示。

图 2-81

## 2.5　创建Maven Web项目

本节介绍使用 IntelliJ IDEA 创建 Maven Web 项目。

## 2.5.1　搭建 Maven 环境

### 1. Maven介绍

Maven 是优秀的构建工具，支持自动化构建过程，包括从清理、编译、测试到生成报告，再到打包和部署的整个过程。Maven 可跨平台使用，无论是 Windows，Linux 还是 macOS 操作系统，使用的命令都相同。Maven 除了可以帮助我们管理构建过程，还提供依赖管理工具，包括中央仓库、自动下载 source 和 doc 等功能组件。Maven 仓库就是放置所有.jar 文件的服务器，Maven 项目可以从 Maven 仓库中获取自己所需要的依赖 jar 包文件。

Maven 项目的核心配置文件是 pom.xml，其中 pom 全称是 Project Object Model（项目对象模型）。

### 2. 如何搭建Maven环境

（1）下载 Maven。进入 Maven 官网下载，下载成功后解压到硬盘。

（2）配置 Maven 本地仓库路径。本地仓库是指存在于当前电脑中的仓库，如果你指定电脑中的某一个文件夹来存放.jar 文件，则该文件夹就可以被理解成本地仓库。

在开发项目时，需要加入依赖的.jar 文件。Maven 会先去本地仓库文件夹里查找，如果找不到.jar 文件，则会到远程仓库中进行查找。

远程仓库是指在其他服务器上的仓库，包括全球中央仓库、公司内部的私服、其他公司提供的公共库。

在默认情况下，打开 C:\apache-maven-3.6.1\conf\settings.xml 文件可以看到 Maven 本地仓库路径，如图 2-82 所示。

```
<!-- localRepository
  | The path to the local repository maven will use to store artifacts.
  |
  | Default: ${user.home}/.m2/repository
<localRepository>/path/to/local/repo</localRepository>
-->
```

图 2-82

在 settings.xml 配置文件中已经说明，默认路径在“${user.home}/.m2/repository”中，变量${user.home}可以使用代码进行获取，创建测试用的项目 test1，创建运行类代码如下：

```
package test;

public class Test {
```

```
    public static void main(String[] args) {
        String path = System.getProperty("user.home");
        System.out.println(path);
    }
}
```

控制台输出结果为 C:\Users\gaohongyan。

说明：在默认的情况下，本地仓库路径为“C:\Users\gaohongyan\.m2\repository”，但此路径可以更改。更改 settings.xml 配置文件的代码如下：

```
<localRepository>C:\mvnrepository</localRepository>
```

此时，本地仓库路径被设置在 C:\mvnrepository 中。

（3）配置 Maven 镜像。当本地仓库没有所需要的.jar 文件时，你需要访问国外的仓库下载，但由于网络环境等原因，快速下载.jar 文件成了奢望，因此可以使用国内阿里云提供的 Maven 仓库的镜像，可在 settings.xml 配置文件中添加如下代码：

```
<mirrors>
    <mirror>
        <id>aliyun</id>
        <name>aliyun Maven</name>
        <mirrorOf>central</mirrorOf>
        <url>http://maven.aliyun.com/nexus/content/groups/public/</url>
    </mirror>
</mirrors>
```

配置代码：

```
<id>aliyun</id>
```

其代表镜像库在当前 settings.xml 配置文件中的 id 值，是镜像库的唯一标识，它可以是任意的，但要有意义。

配置代码：

```
<name>aliyun Maven</name>
```

其代表镜像库的名称或者<id>的解释，该值可以是任意的，但要有意义。

配置代码：

```
<mirrorOf>central</mirrorOf>
```

其代表该配置为中央仓库的镜像，任何对于中央仓库的请求都会转至该镜像。

配置代码：

```
<url>http://maven.aliyun.com/nexus/content/groups/public/</url>
```

其代表镜像库的 URL 地址。

至此，更改 Maven 的配置结束。

## 2.5.2 创建 Maven Web 项目

首先创建 Maven 项目，并选择 maven-archetype-webapp 类型的项目，如图 2-83 所示，然后单击“Next”按钮。

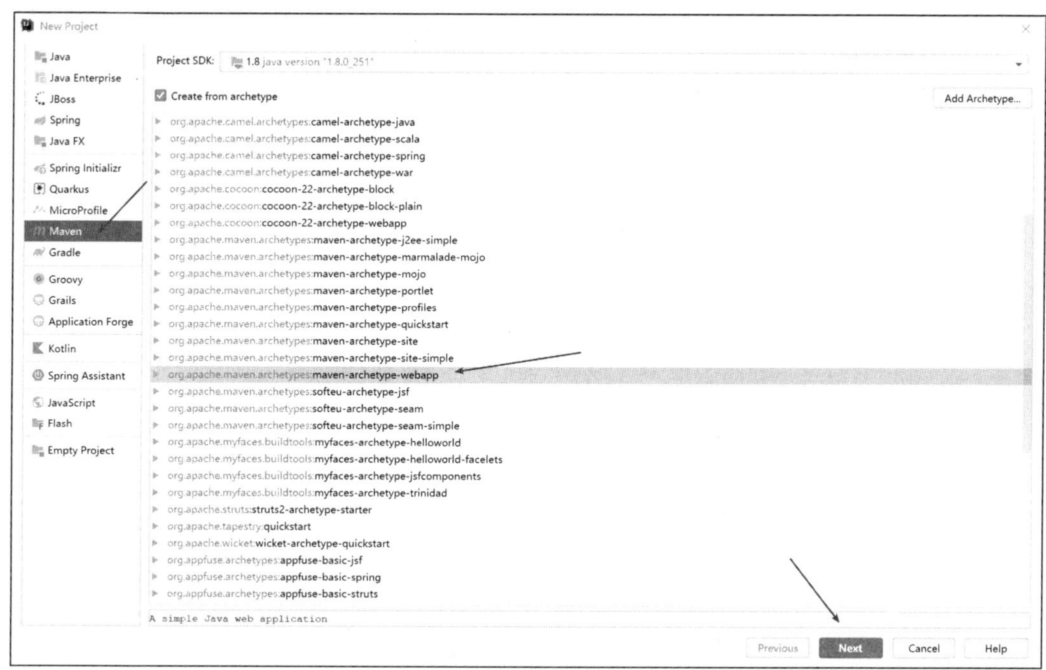

图 2-83

设置 Maven 项目属性，如图 2-84 所示，单击“Next”按钮。

配置 Maven 安装路径，如图 2-85 所示。

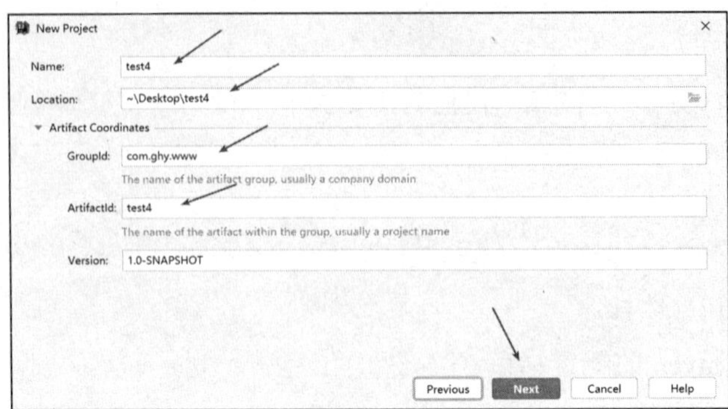


图 2-84

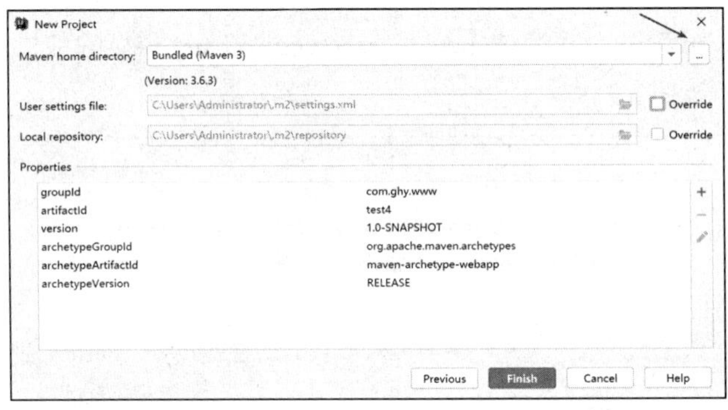


图 2-85

设置关联的 settings.xml 配置文件，Override 复选项需要打钩，以进行配置覆盖，如图 2-86 所示，单击“Finish”按钮完成 Maven Web 项目的创建。IntelliJ IDEA 开始下载相关的资源，但需要一些时间。

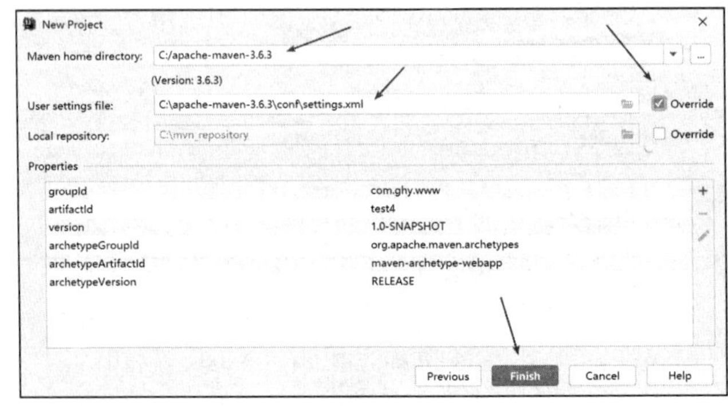


图 2-86

## 2.5.3 改变 JDK 版本

根据当前项目使用 JDK 的版本情况，需要更改 JDK 的版本。比如，如果需要将项目中的 JDK1.7 版本改成 JDK1.8 版本，则需要在 pom.xml 文件中配置如下代码：

```
<properties>
    <project.build.sourceEncoding>UTF-8</project.build.sourceEncoding>
    <maven.compiler.source>1.8</maven.compiler.source>
    <maven.compiler.target>1.8</maven.compiler.target>
</properties>
```

如果使用 JDK14 版本，则可以使用如下配置：

```
<properties>
    <project.build.sourceEncoding>UTF-8</project.build.sourceEncoding>
    <maven.compiler.source>14</maven.compiler.source>
    <maven.compiler.target>14</maven.compiler.target>
</properties>
```

## 2.5.4 创建 java 和 resource 文件夹

创建 java 和 resource 文件夹，如图 2-87 所示。

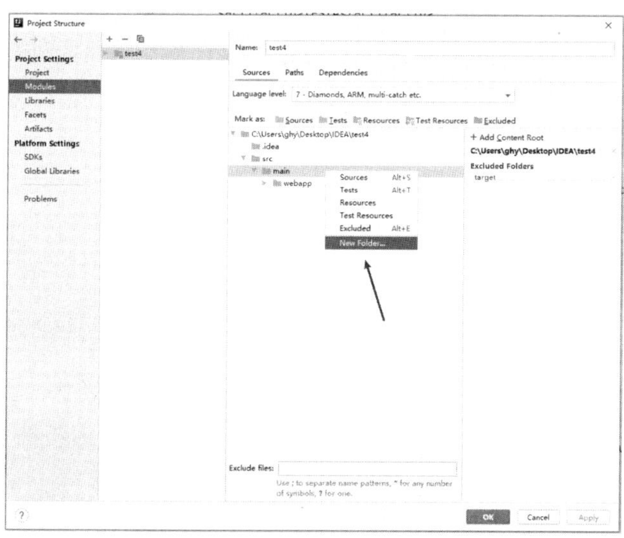


图 2-87

创建成功，如图 2-88 所示。

由于 java 文件夹并不是一个普通的文件夹，要作为 Sources 源代码的存放处，所以对 java 文件夹设置 Sources，如图 2-89 所示。

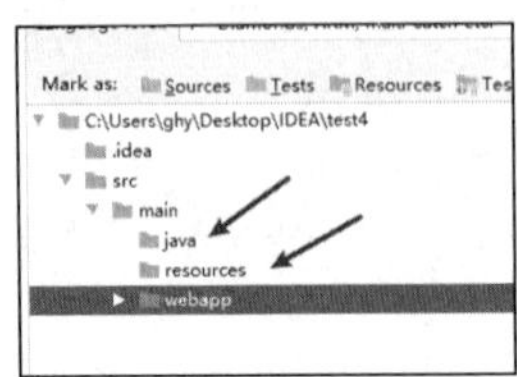


图 2-88

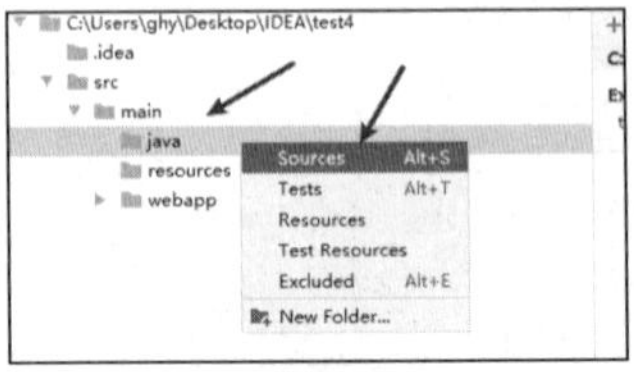

图 2-89

resource 文件夹作为 Resources 资源文件夹，里面要存放视图或相关的配置文件，其设置方法如图 2-90 所示。设置完成后，文件夹图标发生变化，如图 2-91 所示。

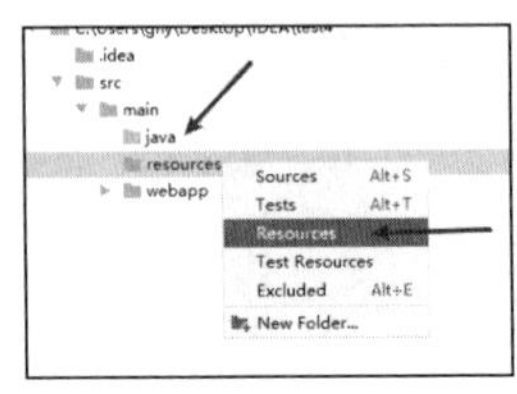


图 2-90

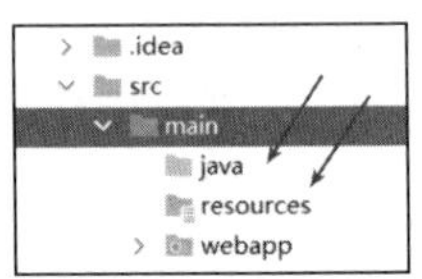


图 2-91

## 2.5.5 其他方式创建 java 和 resources 文件夹

创建 Sources 和 Resources 类型的文件夹也可以直接在项目中进行配置，把 java 和 resources 文件夹删除，重置实验环境，如图 2-92 所示。

实现的方式是单独创建文件夹，如图 2-93 所示。

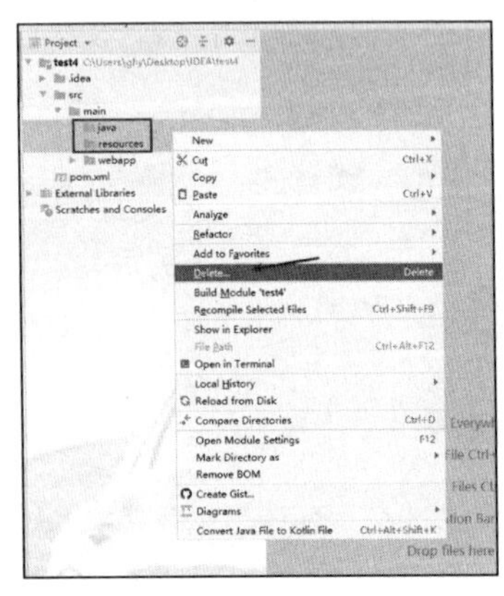


图 2-92

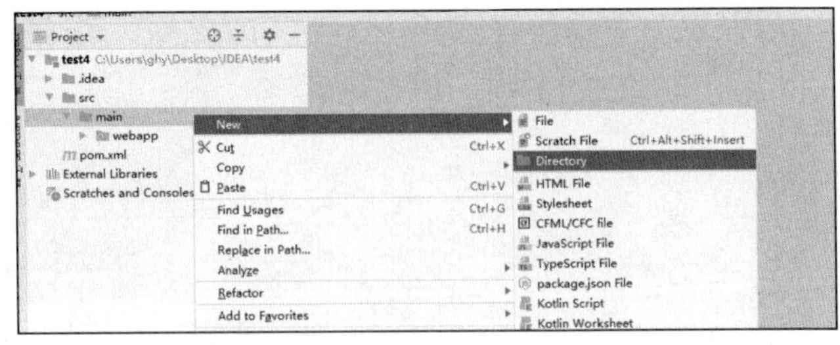


图 2-93

创建文件夹 a 和 b，对 a 文件夹设置 Sources 类型，设置方法如图 2-94 所示。

对 b 文件夹设置 Resources 类型，设置方法如图 2-95 所示。

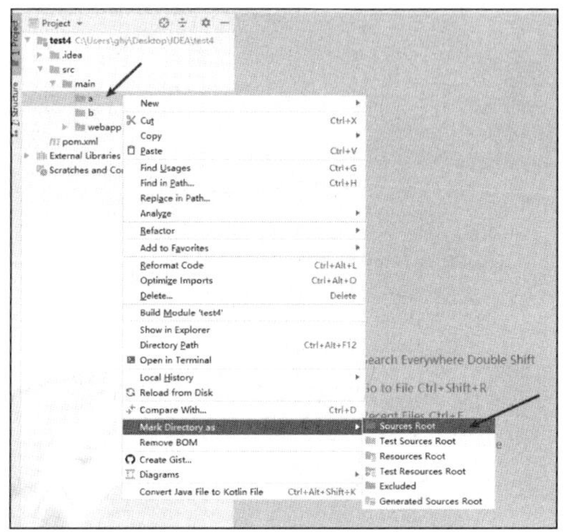


图 2-94

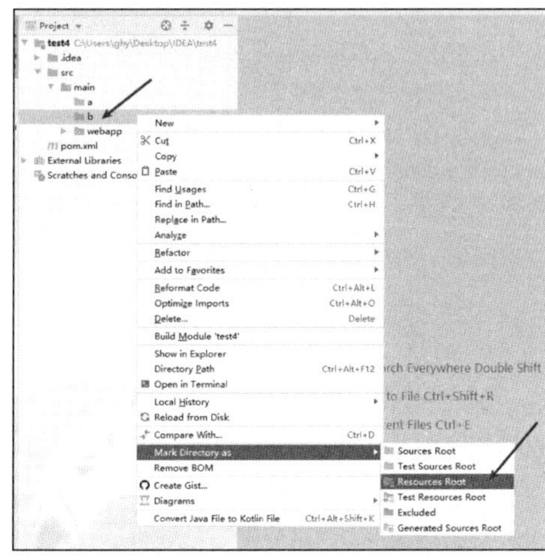


图 2-95

## 2.5.6　创建 Servlet 类

当创建 Servlet 时，如果没有出现“Create New Servlet”菜单，则可以右击，在弹出的“Reload project”菜单中重新导入，如图 2-96 所示。

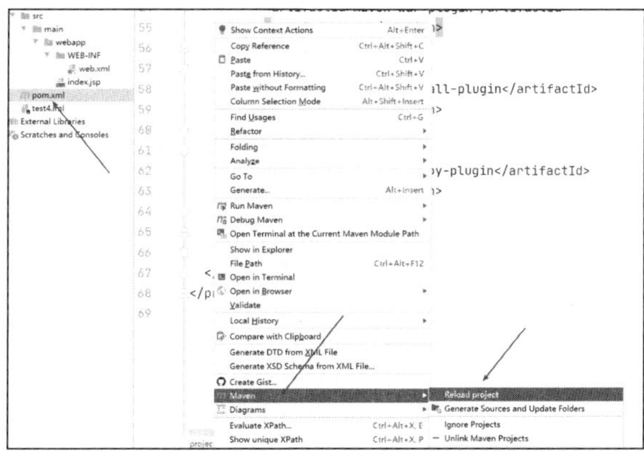


图 2-96

创建 Servlet 的代码如下：

```
package controller;
```

```
import java.io.IOException;

@javax.servlet.annotation.WebServlet(name = "Test1", urlPatterns = "/Test1")
public class Test1 extends javax.servlet.http.HttpServlet {
   protected void doGet(javax.servlet.http.HttpServletRequest request,
javax.servlet.http.HttpServletResponse response) throws
javax.servlet.ServletException, IOException {
       System.out.println("Test1 run 运行了！");
       request.setAttribute("myString", "中国人");
       request.getRequestDispatcher("Test1.jsp").forward(request,
response);
   }
}
```

如果找不到 Servlet 相关的类，则可以在 pom.xml 文件中添加依赖配置：

```
<dependency>
   <groupId>javax.servlet</groupId>
   <artifactId>javax.servlet-api</artifactId>
   <version>4.0.1</version>
</dependency>
<dependency>
   <groupId>org.glassfish.web</groupId>
   <artifactId>jstl-impl</artifactId>
   <version>1.2</version>
</dependency>
<dependency>
  <groupId>javax</groupId>
  <artifactId>javaee-api</artifactId>
  <version>8.0.1</version>
</dependency>
```

### 2.5.7 创建 JSP 文件

创建 Test1.jsp 文件的代码如下：

```
<%@ page contentType="text/html;charset=UTF-8" language="java" %>
<%@ taglib prefix="c" uri="http://java.sun.com/jsp/jstl/core" %>
<html>
   <head>
      <title>Title</title>
```

```
    </head>
    <body>
        信息：<c:out value="${myString}"/>
    </body>
</html>
```

## 2.5.8　启动项目并解决 Servlet 打印乱码的问题

如果启动 Tomcat 并运行 Servlet 后出现图 2-97 所示的乱码，则需要对 Tomcat 设置 VM 属性值：-Dfile.encoding=UTF-8，如图 2-98 所示。

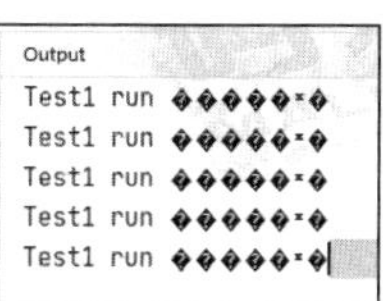


图 2-97

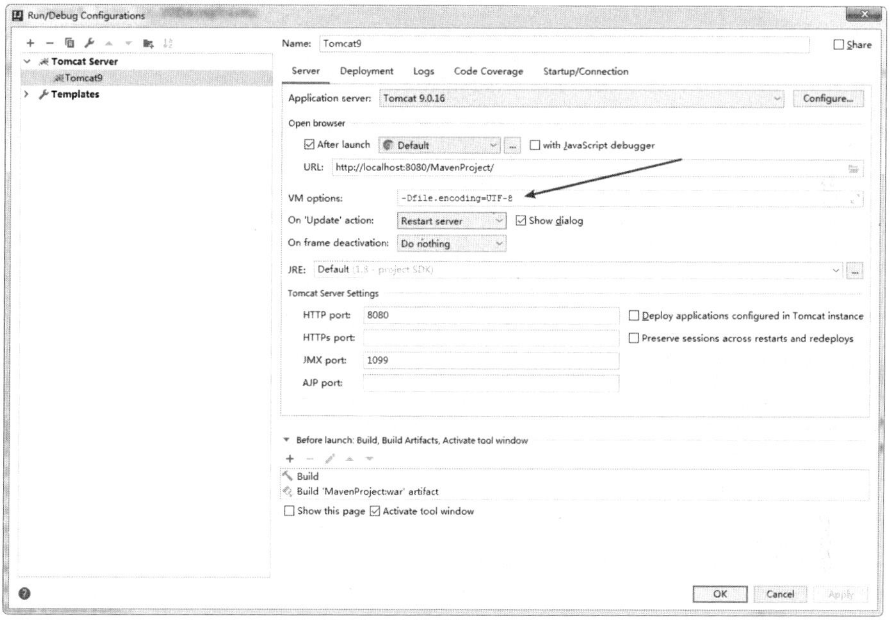

图 2-98

这时，重启 IntelliJ IDEA 再执行 Servlet，如图 2-99 所示。

## 2.5.9　解决 EL 表达式无效的问题

由于 JSP 输出的内容并没有识别 EL 表达式，如图 2-100 所示，因此需要更改 web.xml 中的配置，代码如下：

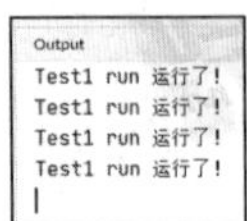

图 2-99

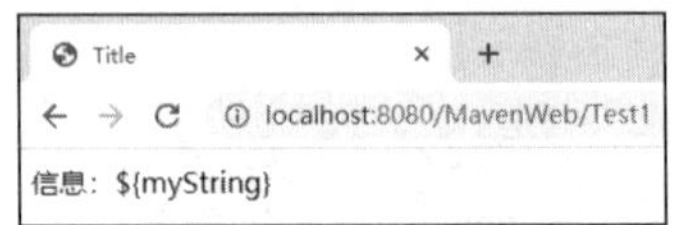


图 2-100

```
<?xml version="1.0" encoding="UTF-8"?>
<web-app version="2.5"
       xmlns="http://java.sun.com/xml/ns/javaee"
       xmlns:xsi="http://www.w3.org/2001/XMLSchema-instance"
       xsi:schemaLocation="http://java.sun.com/xml/ns/javaee
  http://java.sun.com/xml/ns/javaee/web-app_2_5.xsd">
   <display-name>Archetype Created Web Application</display-name>
</web-app>
```

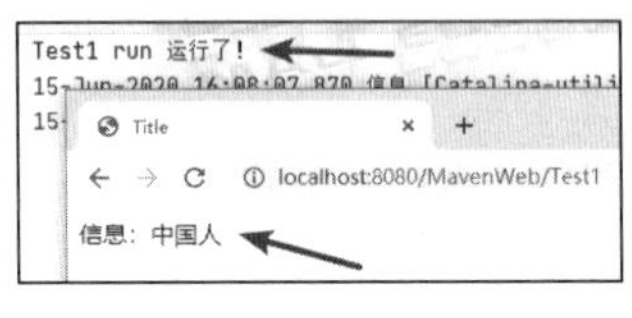


图 2-101

这里，使用 2.5 版本的 Web，默认为 2.3 版本。

重启 Tomcat，在 JSP 文件中正确使用 EL 表达式输出字符串，如图 2-101 所示。

### 2.5.10 导出 war 包文件

为了实现导出的 war 包文件中包含依赖的.jar 文件，可以更改 pom.xml 配置文件，添加如下代码：

```
<dependencies>
<dependency>
   <groupId>org.springframework</groupId>
   <artifactId>spring-context</artifactId>
   <version>5.2.2.RELEASE</version>
</dependency>
</dependencies>
```

项目中有依赖 Spring 框架的.jar 文件，如图 2-102 所示。

Maven: org.springframework:spring-aop:5.2.2.RELEASE
Maven: org.springframework:spring-beans:5.2.2.RELEASE
Maven: org.springframework:spring-context:5.2.2.RELEASE
Maven: org.springframework:spring-core:5.2.2.RELEASE
Maven: org.springframework:spring-expression:5.2.2.RELEASE
Maven: org.springframework:spring-jcl:5.2.2.RELEASE

图 2-102

选择“package”菜单开始生成 war 包文件，如图 2-103 所示。

成功生成 war 包文件，如图 2-104 所示。

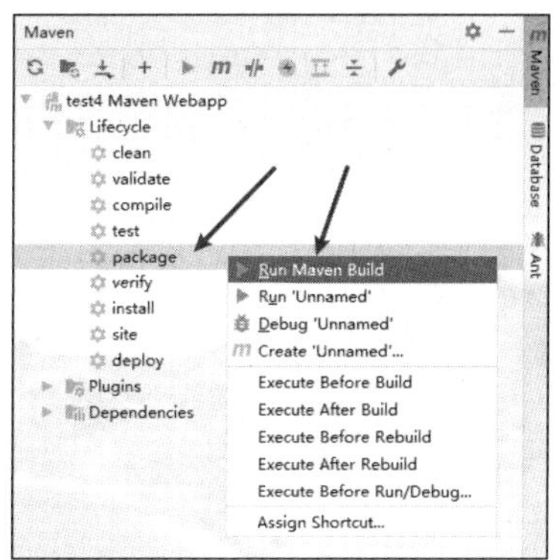


图 2-103

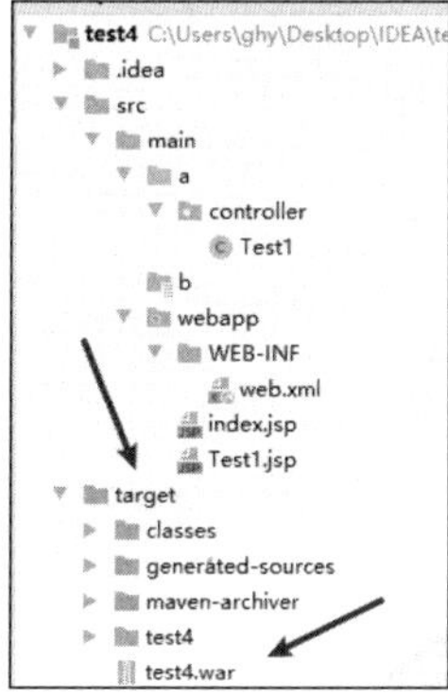


图 2-104

生成的 war 包文件中有 Spring 框架的.jar 文件，如图 2-105 所示。

图 2-105

# 2.6　创建Maven Java项目并导出.jar文件

本节介绍创建 Maven Java 项目并导出包含依赖 jar 包文件的可执行.jar 文件的方法。

## 2.6.1　创建 Maven Java 项目

创建 Maven Java 项目，如图 2-106 所示，单击“Next”按钮。

设置 Maven 项目属性，如图 2-107 所示，单击“Next”按钮。

设置 Maven 位置，如图 2-108 所示，单击“Finish”按钮完成项目的创建。

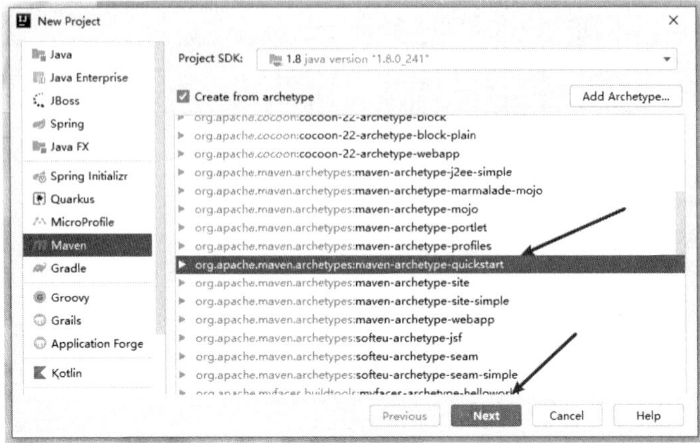


图 2-106

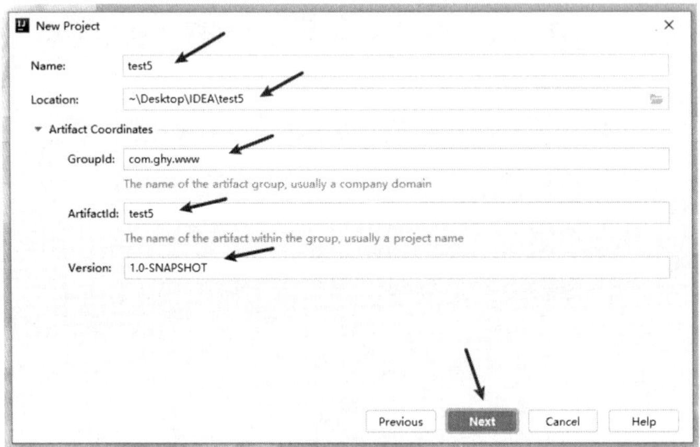


图 2-107

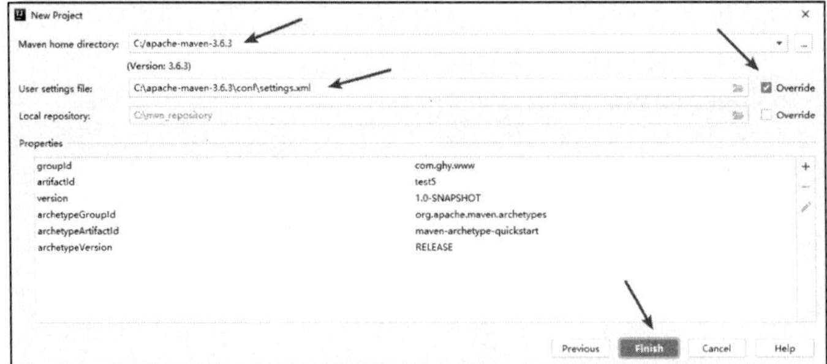


图 2-108

### 2.6.2 编辑 pom.xml 文件

在 pom.xml 文件中，添加如下代码：

```
<dependencies>
    <dependency>
        <groupId>org.springframework</groupId>
        <artifactId>spring-context</artifactId>
        <version>5.2.2.RELEASE</version>
    </dependency>
</dependencies>
```

### 2.6.3 创建运行类

创建 Test1 类的代码如下：

```
package com.ghy.www;

import org.springframework.context.support.ClassPathXmlApplicationContext;

public class Test1 {
    public static void main(String[] args) {
        System.out.println("test1 " + new ClassPathXmlApplicationContext());
    }
}
```

创建 Test2 类的代码如下：

```
package com.ghy.www;

import org.springframework.context.support.ClassPathXmlApplicationContext;

public class Test2 {
    public static void main(String[] args) {
        System.out.println("test2 " + new ClassPathXmlApplicationContext());
    }
}
```

### 2.6.4 执行运行类

程序运行结果如下：

test1 org.springframework.context.support.ClassPathXmlApplicationContext@7e6cbb7a 和 test2 org.springframework.context.support.ClassPathXmlApplicationContext@7e6cbb7a

### 2.6.5 打包并在 CMD 中运行类

创建新的 Artifacts 配置，如图 2-109 所示，配置细节如图 2-110 所示。

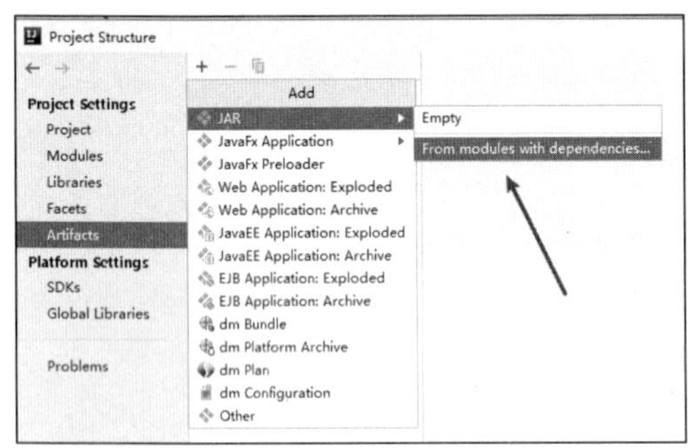


图 2-109

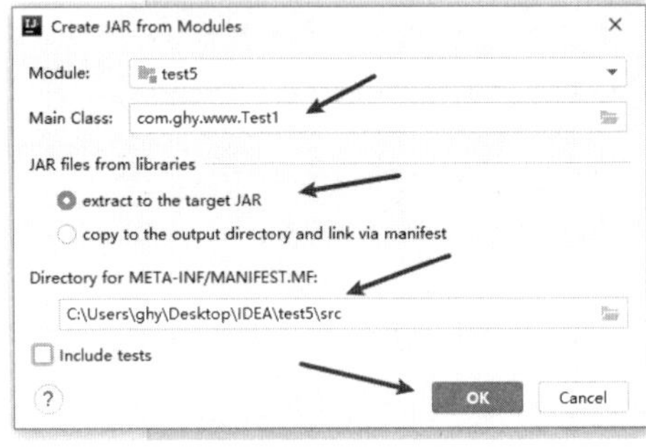


图 2-110

进行 Build 操作后，成功生成.jar 文件，如图 2-111 所示。

这时，文件 test5.jar 中包含依赖的.jar 文件，如图 2-112 所示。

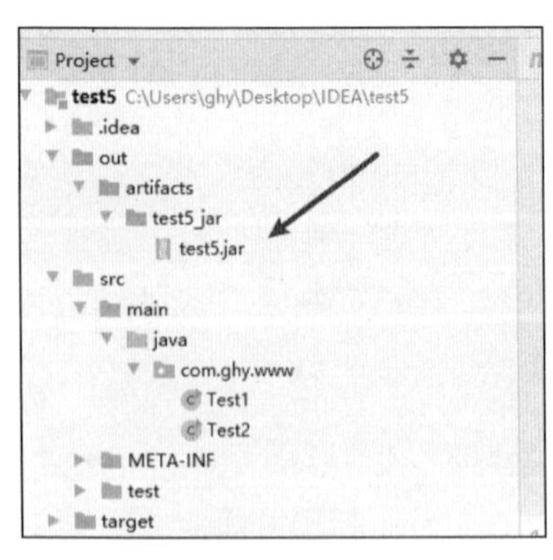


图 2-111

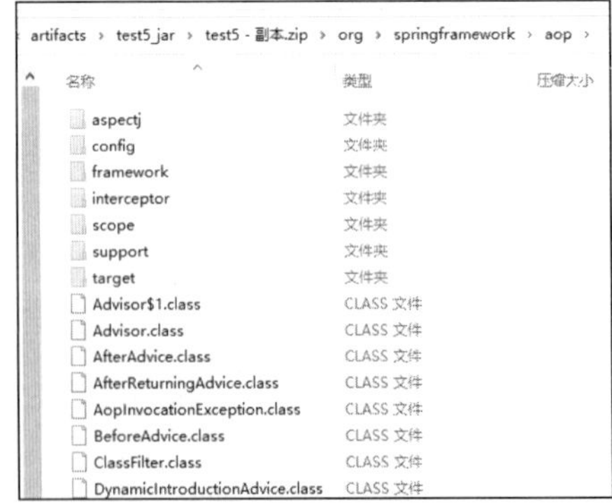


图 2-112

在 CMD 中执行 Test1.java 和 Test2.java 两个 Java 类，如图 2-113 所示。

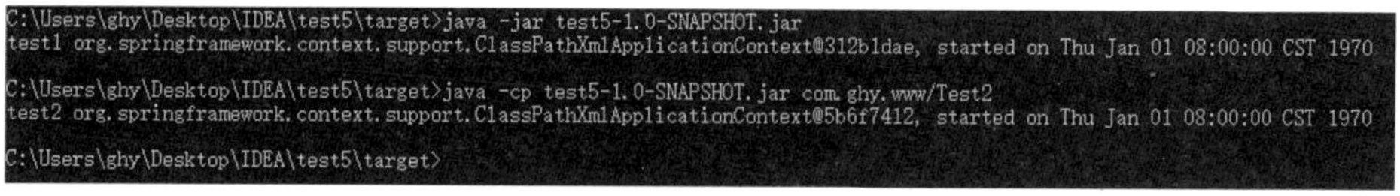

```
C:\Users\ghy\Desktop\IDEA\test5\target>java -jar test5-1.0-SNAPSHOT.jar
test1 org.springframework.context.support.ClassPathXmlApplicationContext@312b1dae, started on Thu Jan 01 08:00:00 CST 1970

C:\Users\ghy\Desktop\IDEA\test5\target>java -cp test5-1.0-SNAPSHOT.jar com.ghy.www/Test2
test2 org.springframework.context.support.ClassPathXmlApplicationContext@5b6f7412, started on Thu Jan 01 08:00:00 CST 1970

C:\Users\ghy\Desktop\IDEA\test5\target>
```

图 2-113

# 2.7　配置全局Maven

当创建 Maven 项目时，由于每次都要重新配置 Maven 的选项，比较烦琐，因此我们可以预定义一个 Maven 配置。

单击图 2-114 所示的菜单。

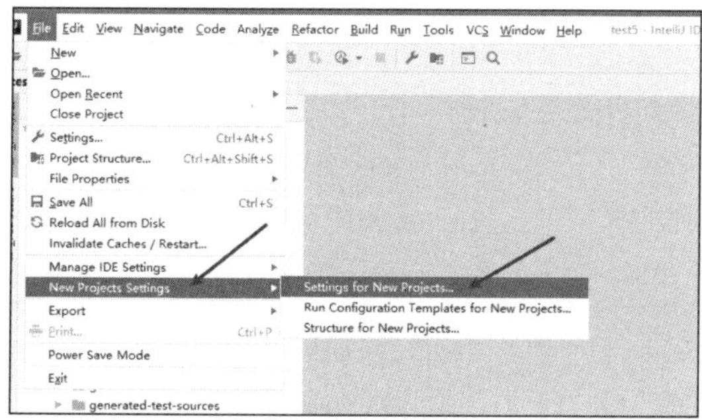


图 2-114

配置 Maven 路径，如图 2-115 所示。

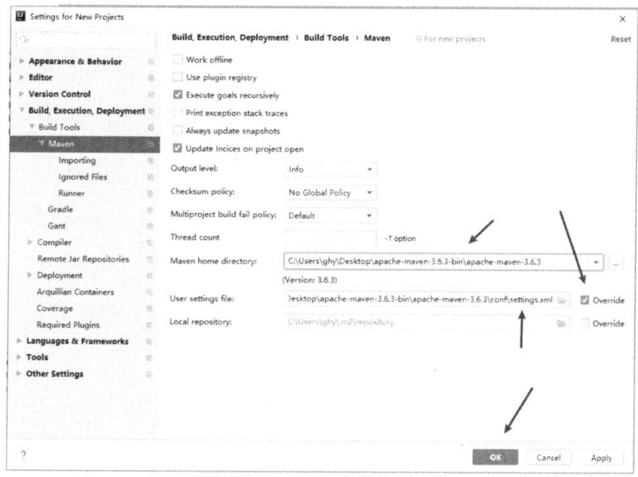


图 2-115

这时，再创建新的项目，Maven 的路径就是自定义的配置，如图 2-116 所示。

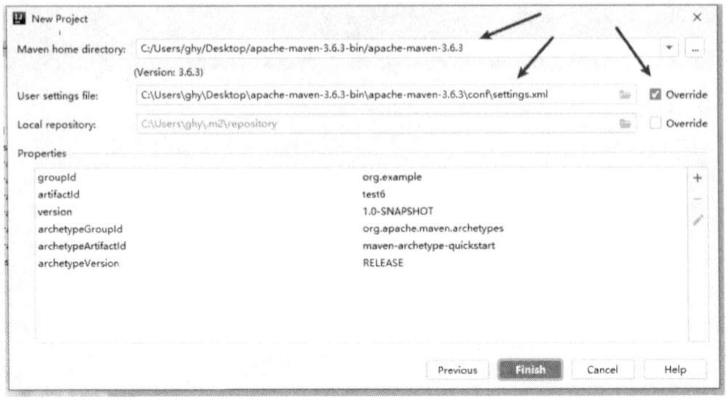

图 2-116

## 2.8　多Modules模块Web环境的搭建——非Maven环境

在 Eclipse 中有工作空间（WorkSpace）和项目（Project）的概念，而在 IntelliJ IDEA 中有项目（Project）和模块（Module）的概念，它们之间的对应关系如下。

（1）Eclipse 中的工作空间相当于 IntelliJ IDEA 中的项目。

（2）Eclipse 中的项目相当于 IntelliJ IDEA 中的模块。

在 IntelliJ IDEA 中，一个大型的项目由多个模块组成，如图 2-117 所示。

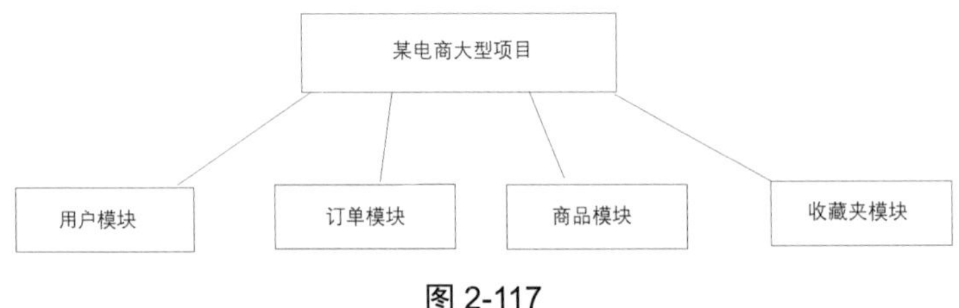


图 2-117

在开发项目时，从分工及后期维护等角度考虑，会把一个大型项目拆分成若干个子模块，每个人开发自己的子模块，在项目最后进行整合。

项目一般是由一个总项目或称父项目和若干个子模块组成的，本节就来搭建这样的开发环境。

### 2.8.1　创建父项目

创建一个 Empty Project 空项目，如图 2-118 所示，单击“Next”按钮继续配置，出

现图 2-119 所示的界面，单击“Finish”按钮完成项目的创建。

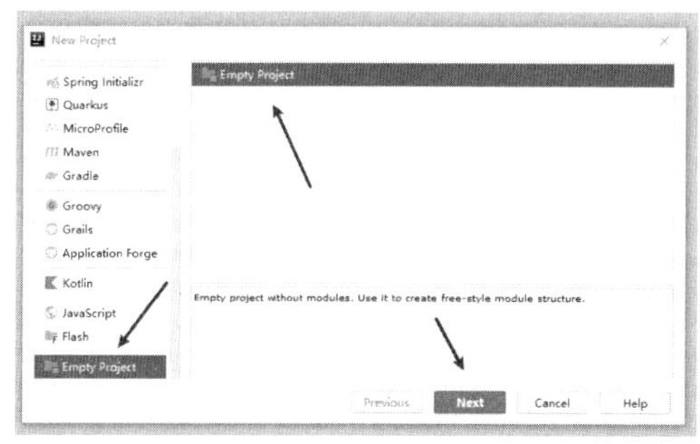

图 2-118

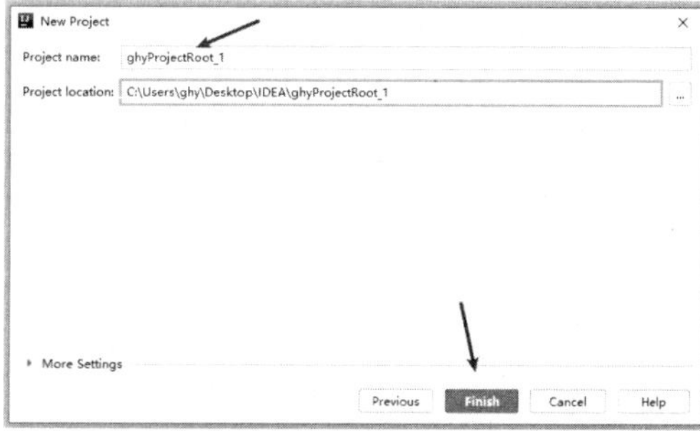

图 2-119

## 2.8.2　创建 DAO 子模块

新建子模块，如图 2-120 所示。

创建没有 Web 环境的 Java 模块，如图 2-121 所示，单击“Next”按钮，设置子模块名称，如图 2-122 所示。

成功添加一个子模块后的项目结构如图 2-123 所示。

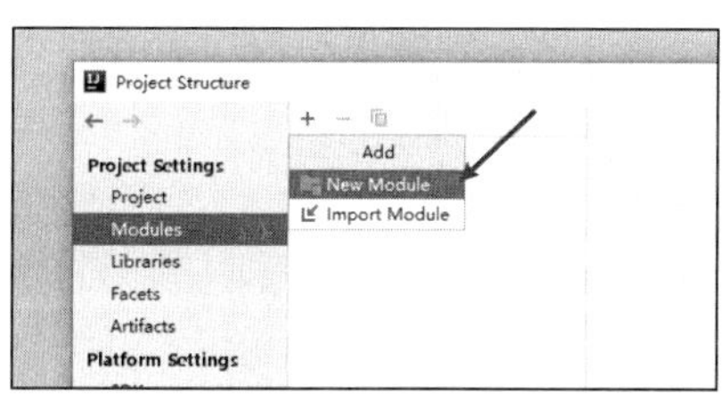

图 2-120

图 2-121

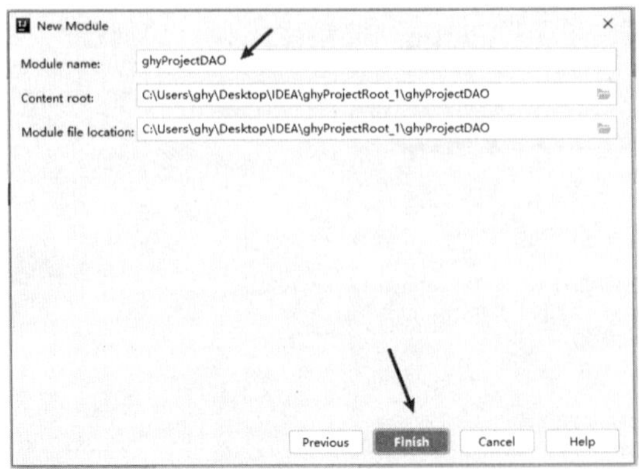


图 2-122

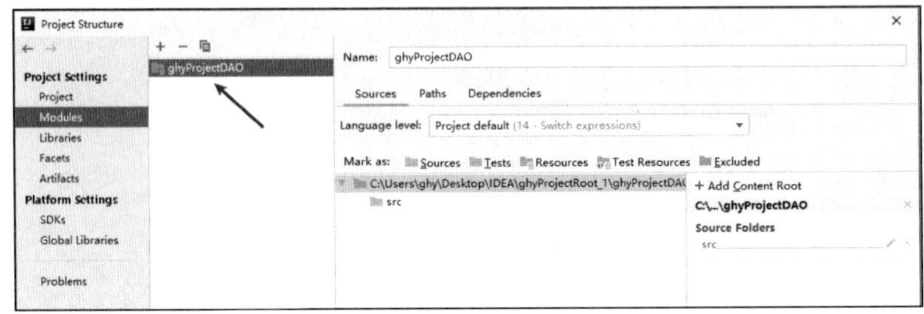


图 2-123

## 2.8.3 创建 Service 子模块

以同样的方式添加新的 Service 子模块，项目结构如图 2-124 所示。

## 2.8.4 创建 Web 子模块

创建 Java 类型的模块，如图 2-125 所示。当前项目中有 3 个子模块，如图 2-126 所示。最终创建的项目结构如图 2-127 所示。

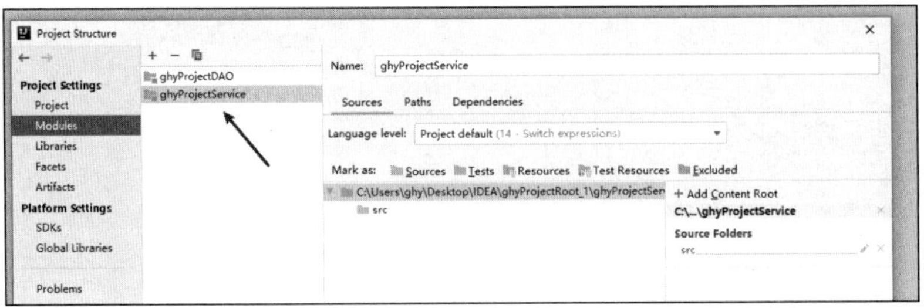


图 2-124

图 2-125

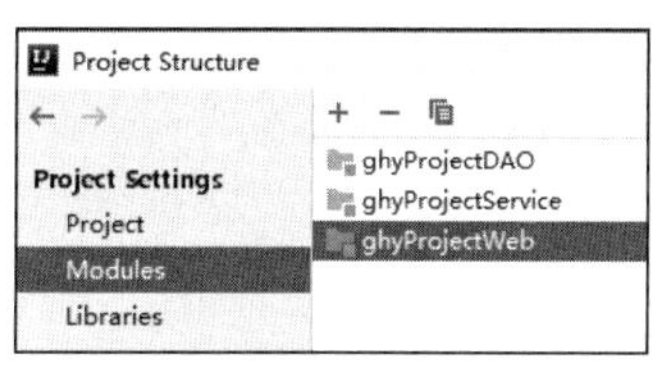


图 2-126

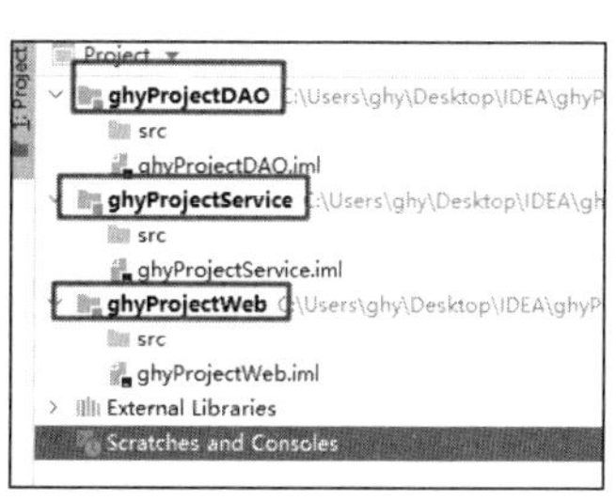


图 2-127

对 Web 模块右击并选择“Add Framework Suppport...”菜单，添加 Web 环境支持，如图 2-128 所示。

## 2.8.5　在 DAO 模块中创建实体类及 DAO 类

在 DAO 模块的 src 中创建实体类和 DAO 类，如果需要 resources 路径，则需要自行创建，如图 2-129 所示。

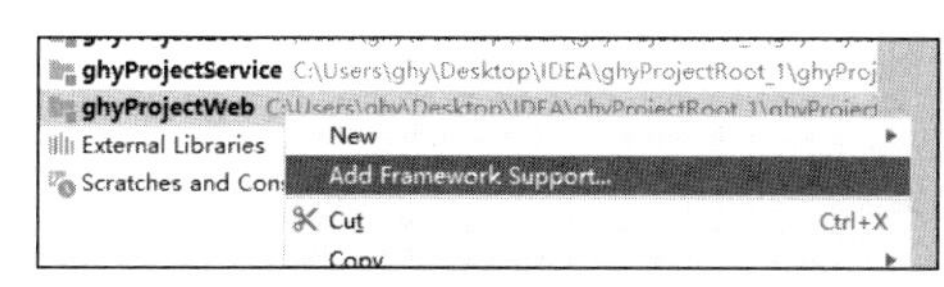


图 2-128

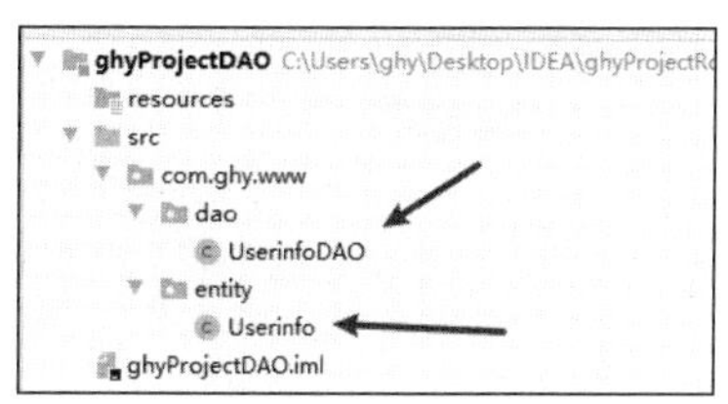


图 2-129

创建实体类的代码如下：

```
package com.ghy.www.entity;

public class Userinfo {
    private int id;
    private String username;
    private String password;

    public Userinfo() {
    }

    public Userinfo(int id, String username, String password) {
        this.id = id;
        this.username = username;
        this.password = password;
    }

    //此处省略set()和get()方法
}
```

创建 DAO 类的代码如下：

```
package com.ghy.www.dao;

import com.ghy.www.entity.Userinfo;

public class UserinfoDAO {
    public Userinfo getUserinfoById(int userId) {
        Userinfo userinfo = new Userinfo();
        userinfo.setId(123);
        userinfo.setUsername("username");
        userinfo.setPassword("password");
        return userinfo;
    }
}
```

### 2.8.6 在 Service 模块中引用 DAO 模块

在默认的情况下，不同模块之间的资源不可以共享，比如 Java 类，故当在 Service 模块中使用 DAO 模块中的类时就会出现异常，如图 2-130 所示。

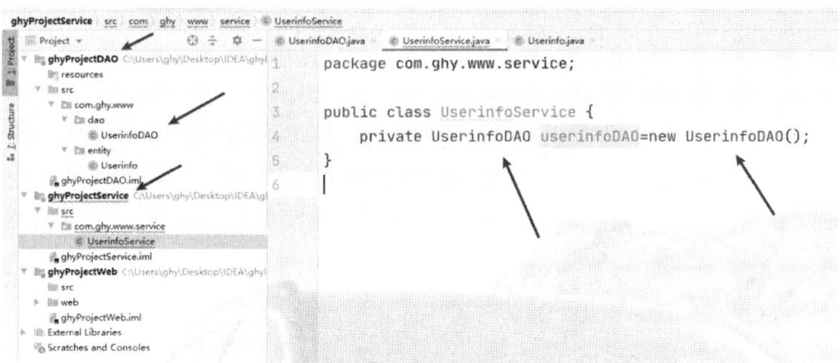

图 2-130

这时，需要使用“Alt+Enter”快捷键实现模块之间的依赖配置，如图 2-131 所示。

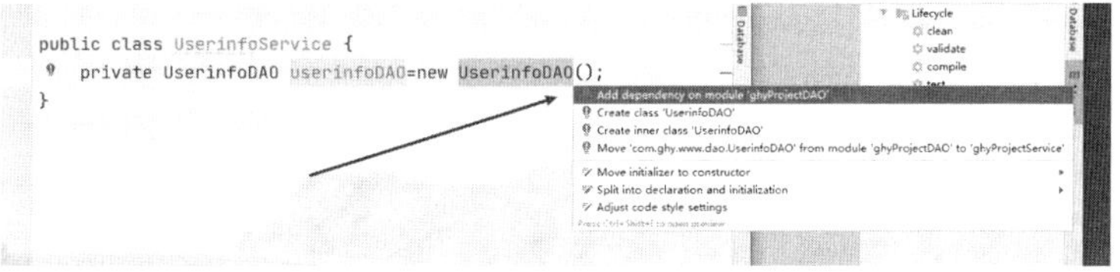

图 2-131

添加完模块间的依赖配置后，代码是正确的，如图 2-132 所示，说明 Service 模块成功引用 DAO 模块。完整的 UserinfoService.java 代码如下：

```
package com.ghy.www.service;

import com.ghy.www.dao.UserinfoDAO;
import com.ghy.www.entity.Userinfo;

public class UserinfoService {
    private UserinfoDAO userinfoDAO = new UserinfoDAO();

    public Userinfo getUserinfoById(int userId) {
        return userinfoDAO.getUserinfoById(123);
    }
}
```

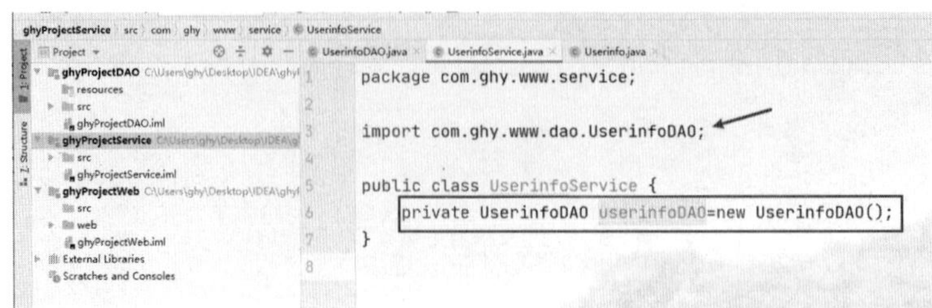

图 2-132

## 2.8.7 创建 Global Libraries 全局库

创建 Global Libraries 全局库的目的是使当前项目可以引用第三方依赖的.jar 文件，如图 2-133 所示。

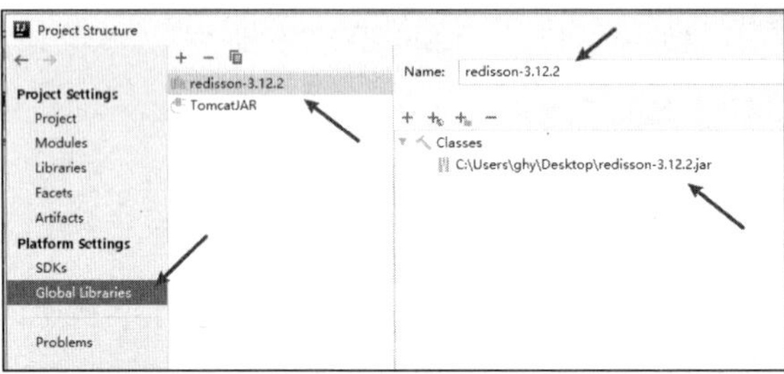

图 2-133

设置 Web 模块引用全局库，并访问第三方依赖.jar 文件中的类，如图 2-134 所示。

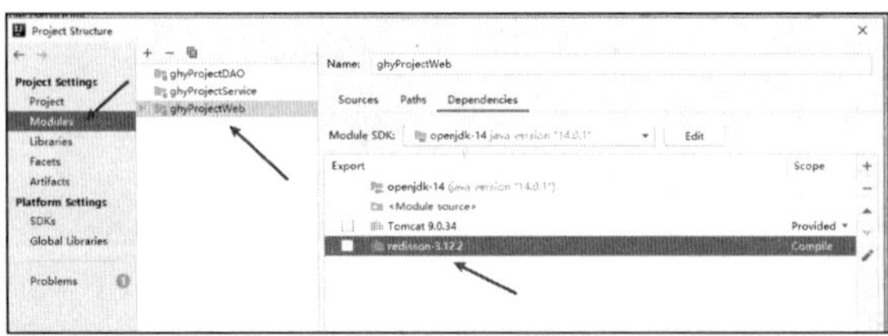

图 2-134

## 2.8.8 在 Web 模块中创建 Servlet

创建 Servlet 的代码如下：

```
package com.ghy.www.controller;

import com.ghy.www.entity.Userinfo;
```

```
import com.ghy.www.service.UserinfoService;
import org.redisson.api.BatchOptions;

import javax.servlet.ServletException;
import javax.servlet.annotation.WebServlet;
import javax.servlet.http.HttpServlet;
import javax.servlet.http.HttpServletRequest;
import javax.servlet.http.HttpServletResponse;
import java.io.IOException;

@WebServlet(name = "Test1", urlPatterns = "/Test1")
public class Test1 extends HttpServlet {
   protected void doGet(HttpServletRequest request, HttpServletResponse
response) throws ServletException, IOException {
       System.out.println("test1 run !");
       System.out.println(BatchOptions.defaults());
       Userinfo userinfo = new UserinfoService().getUserinfoById(123);
       System.out.println(userinfo);
       request.setAttribute("u", userinfo);
       request.getRequestDispatcher("Test1.jsp").forward(request,
response);
   }
}
```

## 2.8.9　创建 JSP 视图

为了在 JSP 文件中使用 JSTL 标签，我们要创建全局库并配置 Web 模块引用 JSTL 库，如图 2-135 所示。

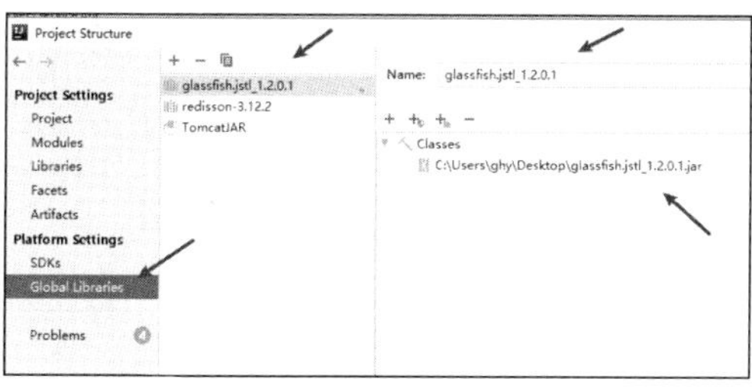


图 2-135

JSP 视图文件 Test1.jsp 的代码如下：

```
<%@ page contentType="text/html;charset=UTF-8" language="java" %>
```

```
<%@ taglib uri="http://java.sun.com/jsp/jstl/core" prefix="c" %>
<html>
    <head>
        <title>Title</title>
    </head>
    <body>
        ID：<c:out value="${u.id}"/><br/>
        账号：<c:out value="${u.username}"/><br/>
        密码：<c:out value="${u.password}"/><br/>
    </body>
</html>
```

如果 Maven 本地仓库没有 JSTL 相关的.jar 文件，则配置代码“http://java.sun.com/jsp/jstl/core”呈红色，这时可以先创建新的 Maven 项目并使用如下依赖下载 JSTL 相关的.jar 文件：

```
<dependency>
    <groupId>org.glassfish.web</groupId>
    <artifactId>jstl-impl</artifactId>
    <version>1.2</version>
</dependency>
```

然后创建 JSTL 全局库，最后在 Web 模块中引入 JSTL 全局库即可。

### 2.8.10 启动 Tomcat 并解决类找不到的异常

启动 Tomcat 并打开网址：http://localhost:8080/ghyProjectWeb_war_exploded/Test1，运行效果如图 2-136 所示。

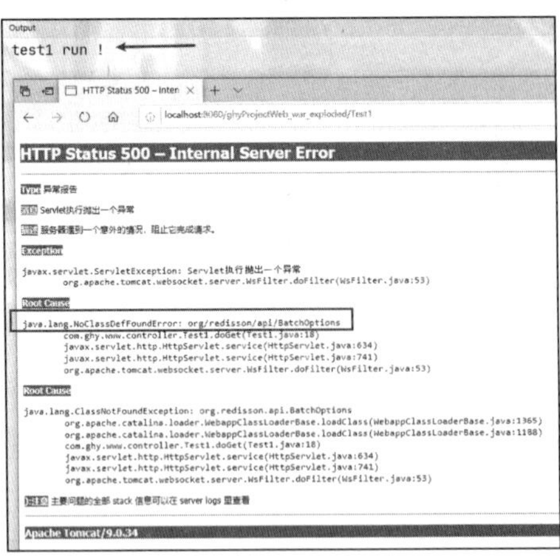


图 2-136

这说明 Servlet 能被正常访问，但不能找到第三方依赖.jar 文件中的类，需要进行如图 2-137 所示的配置。

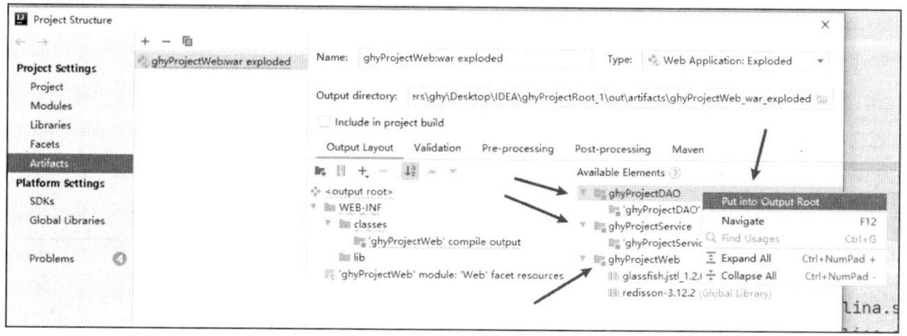

图 2-137

将 3 个模块分别执行“Put into Output Root”菜单，完成后的效果如图 2-138 所示。重新启动 Tomcat，再次打开上面的网址，运行效果如图 2-139 所示 。

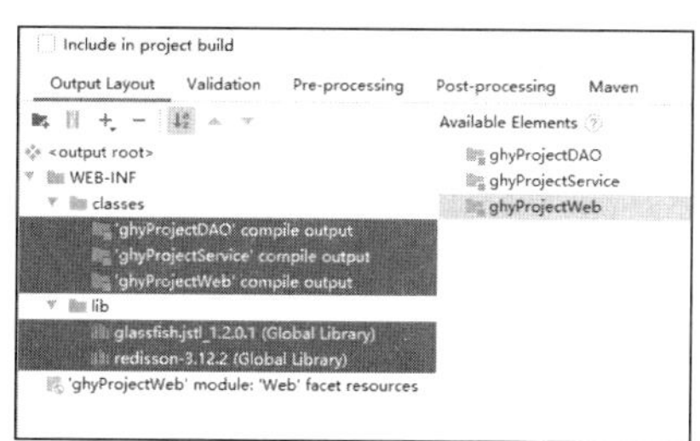

图 2-138

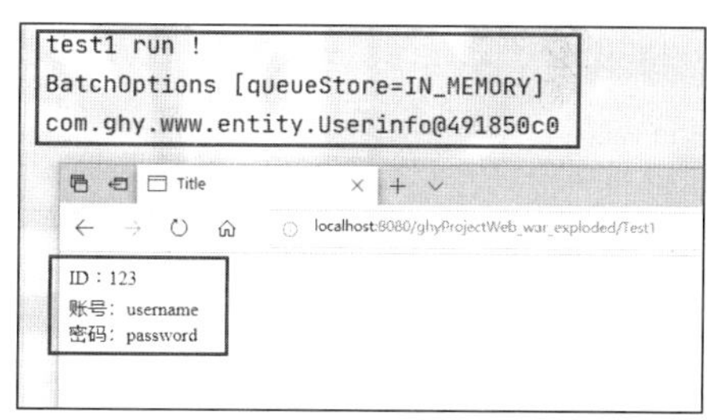


图 2-139

## 2.8.11 导出.war 文件

创建新的 Artifacts 配置，如图 2-140 所示。

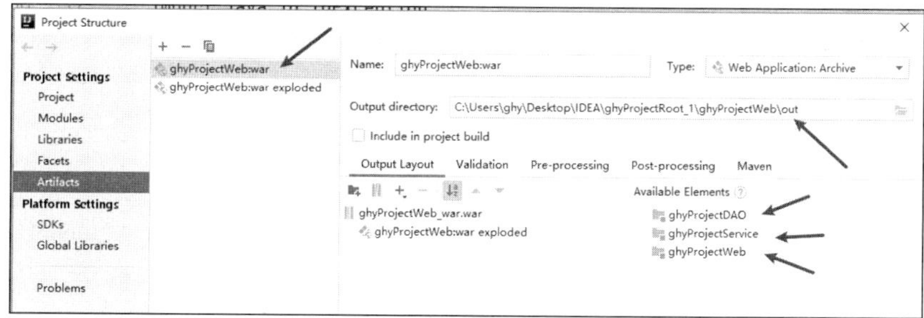

图 2-140

成功生成.war 文件，如图 2-141 所示。.war 文件中包含依赖的.jar 文件，如图 2-142 所示。

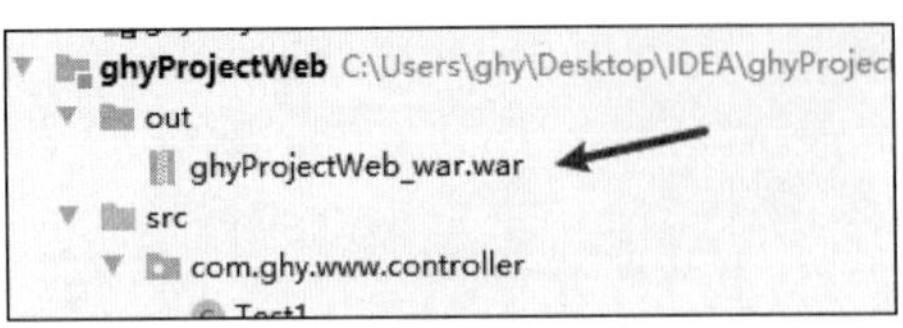


图 2-141

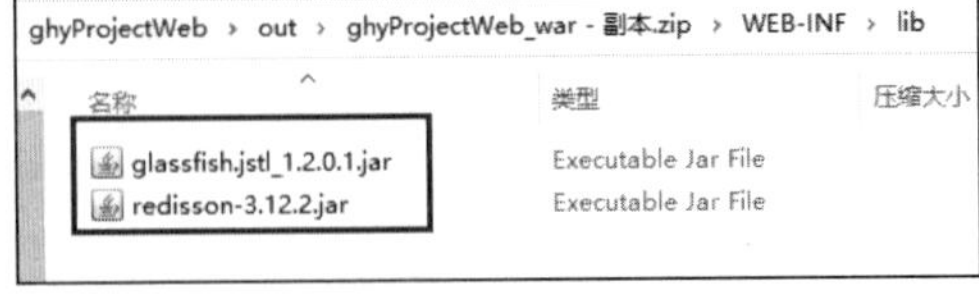


图 2-142

## 2.9 多Modules模块Web环境的搭建——Maven环境

本节搭建基于 Maven 的多模块开发环境 。

### 2.9.1 创建父项目

以 Empty Project 的方式创建父项目，如图 2-143 所示，单击“Next”按钮。

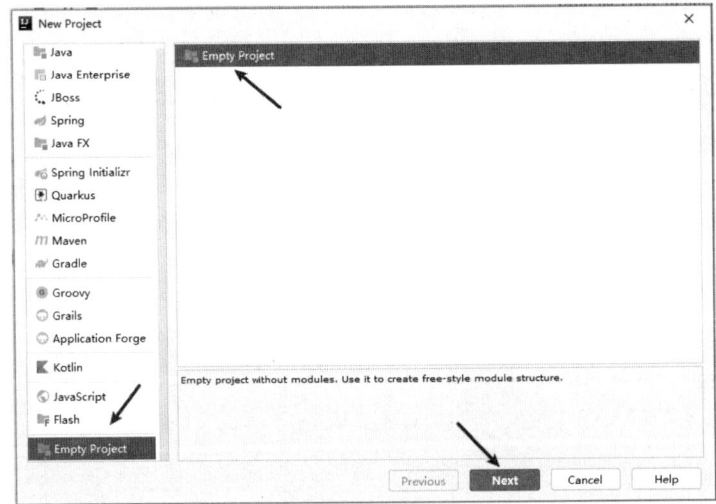


图 2-143

设置项目名称，如图 2-144 所示，单击“Finish”按钮完成创建父项目。

### 2.9.2 创建 DAO 子模块

IDEA 自动弹出界面如图 2-145 所示，新建一个子模块。配置子模块，如图 2-146

所示，单击“Next”按钮。

图 2-144

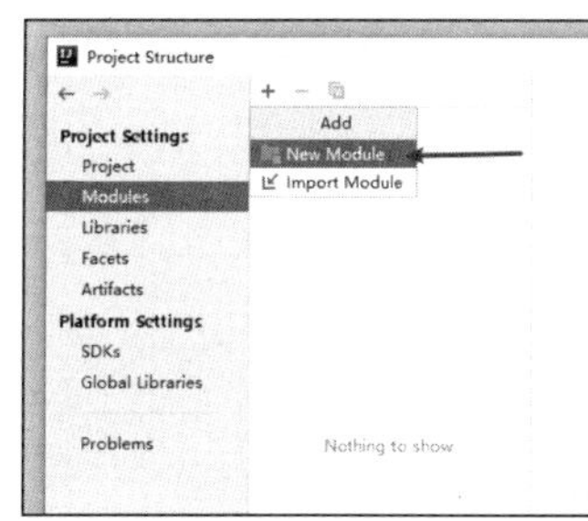


图 2-145

图 2-146

配置项目属性，如图 2-147 所示，单击“Next”按钮继续配置。

如图 2-148 所示，单击“Finish”按钮完成 DAO 子模块创建。

## 2.9.3　创建 Service 子模块和 Web 子模块

按照创建 DAO 子模块的步骤创建基于 Maven 的 Service 模块和 Web 模块。

这时的结构如图 2-149 所示，完成后的项目结构如图 2-150 所示。

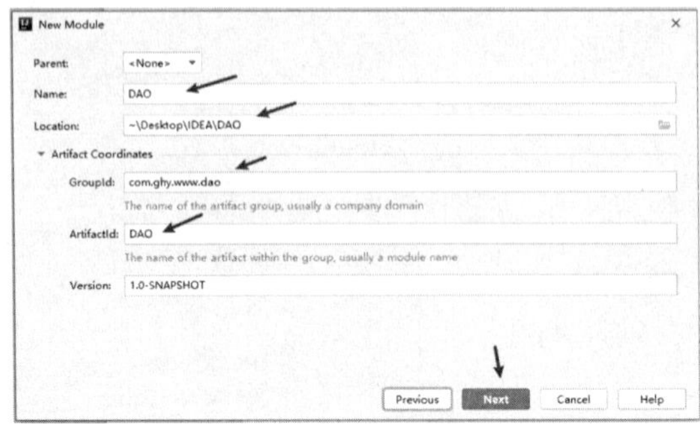


图 2-147

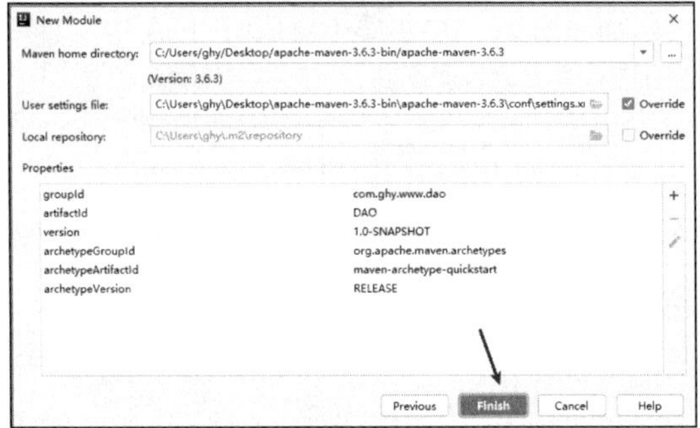


图 2-148

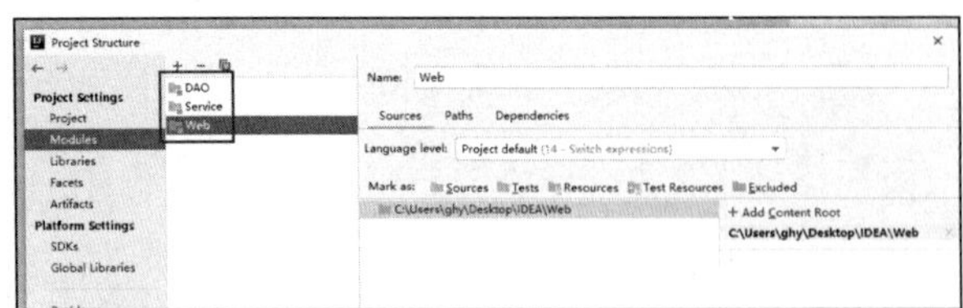


图 2-149

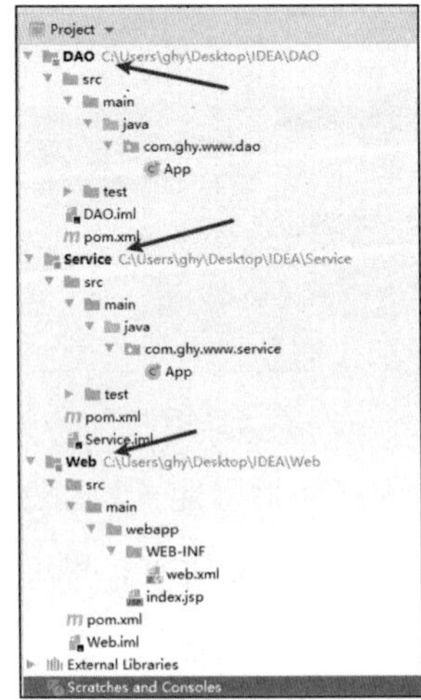


图 2-150

### 2.9.4　给 DAO 模块中的 pom.xml 文件添加依赖

添加依赖的代码如下：

```
<dependency>
    <groupId>org.springframework</groupId>
    <artifactId>spring-context</artifactId>
    <version>5.2.2.RELEASE</version>
</dependency>
```

### 2.9.5　在 DAO 模块中创建 DAO 类

在 DAO 模块中创建 DAO 类的代码如下：

```
package com.ghy.www.dao;

import org.springframework.context.support.ClassPathXmlApplicationContext;

public class UserinfoDAO {
    public String getUsernameByUserId(long userId) {
        return "返回的 username 值=" + new ClassPathXmlApplicationContext();
    }
}
```

### 2.9.6　在 Service 模块中创建业务类并引用 DAO 模块

在 Service 模块中创建业务类并引用 DAO 模块的代码如下：

```
package com.ghy.www.service;

import com.ghy.www.dao.UserinfoDAO;

public class UserinfoService {
    private UserinfoDAO userinfoDAO = new UserinfoDAO();

    public String getUsernameByUserIdService() {
        return userinfoDAO.getUsernameByUserId(123);
    }
}
```

## 2.9.7 在 Web 模块中配置 Sources Root 文件夹

在 Web 模块中配置 Sources Root 文件夹，如图 2-151 所示。

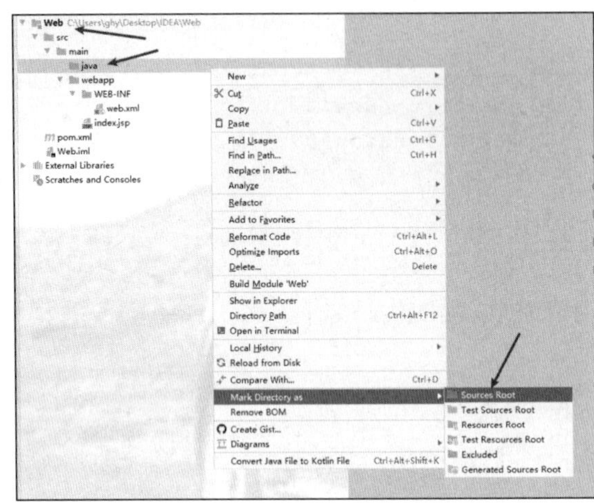

图 2-151

## 2.9.8 给 Web 模块中的 pom.xml 文件添加依赖

在 Web 模块中，给 pom.xml 配置文件添加 Servlet 依赖的代码如下：

```
<dependency>
   <groupId>javax.servlet</groupId>
   <artifactId>javax.servlet-api</artifactId>
   <version>4.0.1</version>
</dependency>
<dependency>
   <groupId>org.glassfish.web</groupId>
   <artifactId>jstl-impl</artifactId>
   <version>1.2</version>
</dependency>
```

## 2.9.9 创建 Servlet

创建 Servlet 的代码如下：

```
package com.ghy.www.controller;
```

```
import com.ghy.www.service.UserinfoService;

import javax.servlet.ServletException;
import javax.servlet.annotation.WebServlet;
import javax.servlet.http.HttpServlet;
import javax.servlet.http.HttpServletRequest;
import javax.servlet.http.HttpServletResponse;
import java.io.IOException;

@WebServlet(name = "Test1", urlPatterns = "/Test1")
public class Test1 extends HttpServlet {
   protected void doGet(HttpServletRequest request, HttpServletResponse
response) throws ServletException, IOException {
       UserinfoService userinfoService = new UserinfoService();
       System.out.println("Test  doGet() 方法执行了！ = " +
userinfoService.getUsernameByUserIdService());
   }
}
```

## 2.9.10　运行项目

启动 Tomcat 后执行 Servlet，程序成功运行，如图 2-152 所示。

```
Output
Test doGet 方法执行了！ = 返回的username值=org.springframework.context.support.ClassPathXmlApplicationContext@444ffe00
```

图 2-152

## 2.9.11　导出.war 文件

执行 Build 生成.war 文件，如图 2-153 所示。

.war 文件中包含第三方依赖的.jar 文件，如图 2-154 所示。

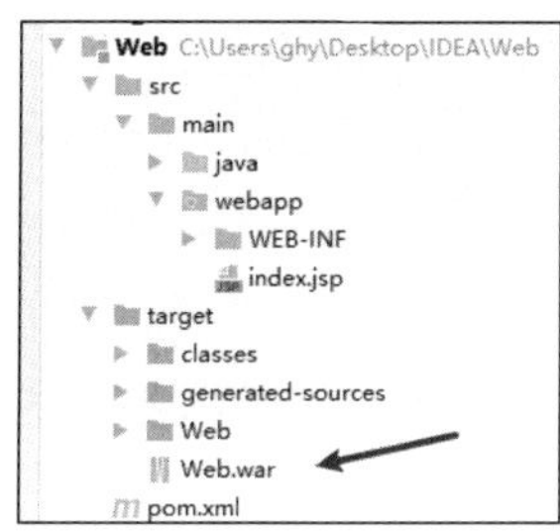


图 2-153

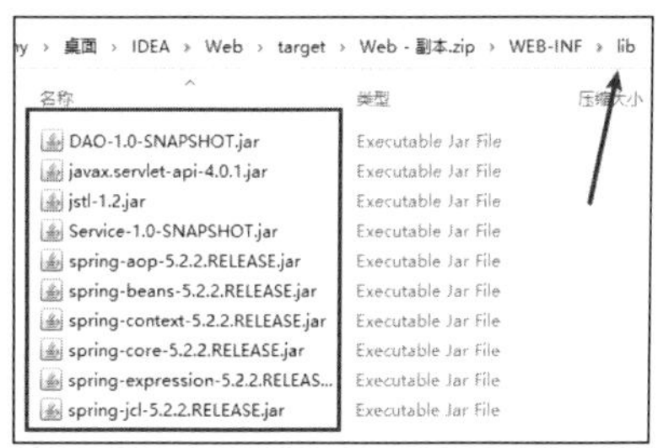


图 2-154

# 2.10 多Web Modules模块环境的搭建——非Maven环境

本节搭建多个 Web 模块的非 Maven 环境。

## 2.10.1 创建 Empty Project

创建 Empty Project，如图 2-155 所示；设置项目名称及存储路径，如图 2-156 所示。

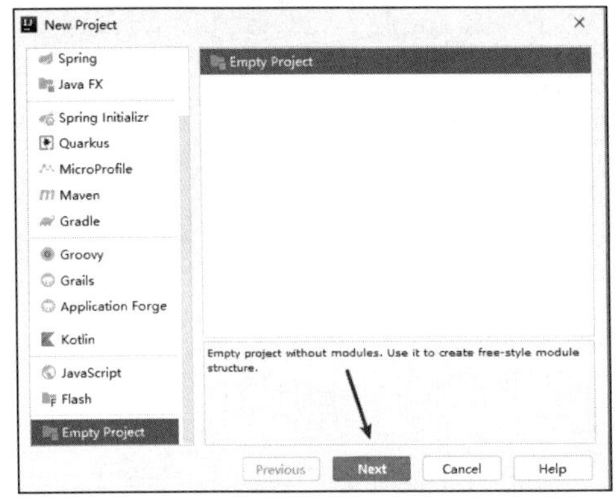


图 2-155

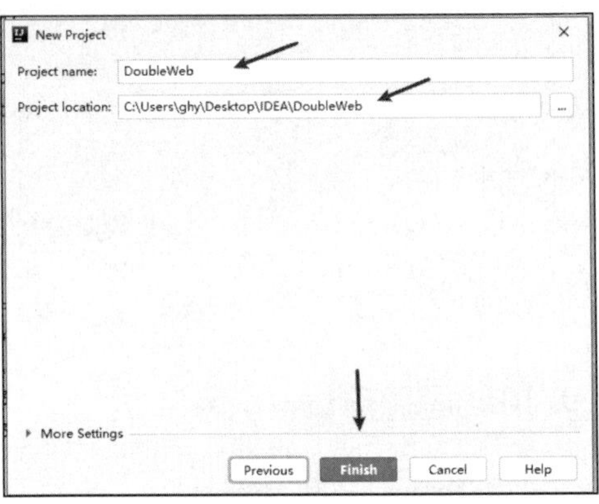


图 2-156

## 2.10.2 创建 Web 模块

创建新的模块，如图 2-157 所示。创建两个 Java 模块后利用“Add Framework Suppport”菜单配置 Web 环境，成功后的项目结构如图 2-158 所示。

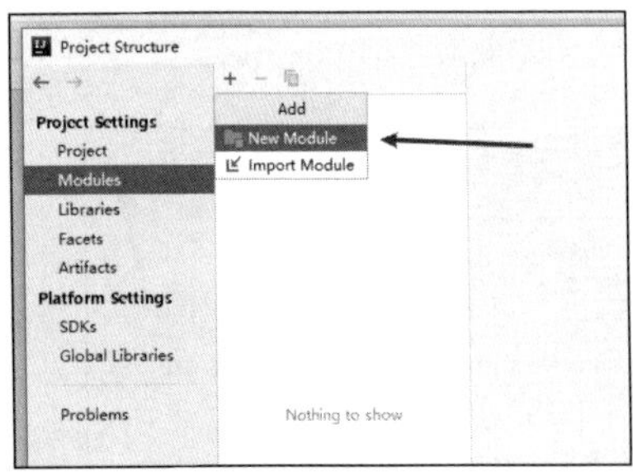


图 2-157

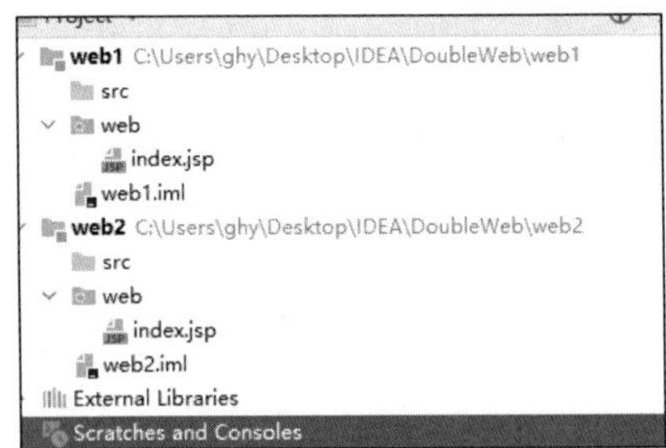


图 2-158

## 2.10.3　创建 Servlet

在 web1 模块中创建 Servlet，代码如图 2-159 所示。

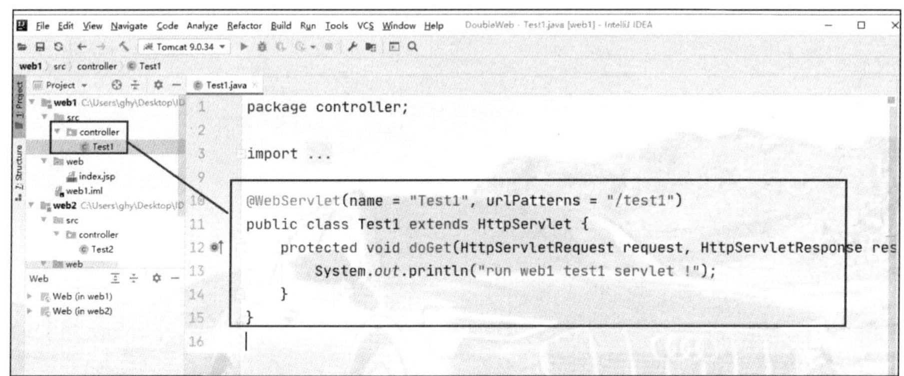

图 2-159

在 web2 模块中创建 Servlet，代码如图 2-160 所示。

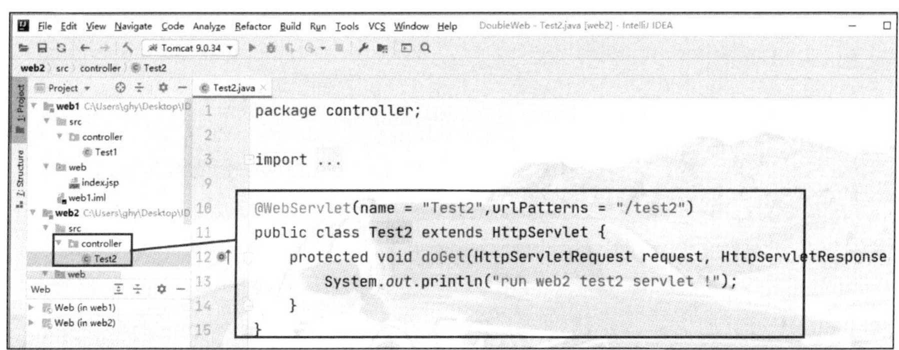

图 2-160

## 2.10.4　配置 Tomcat

如何实现不同的 Web 模块被不同的 Tomcat 进程执行呢？只需要对不同的模块设置不同的 URL 路径及 PORT 端口即可，其中 web2 模块对应的 Tomcat 配置，如图 2-161 所示。

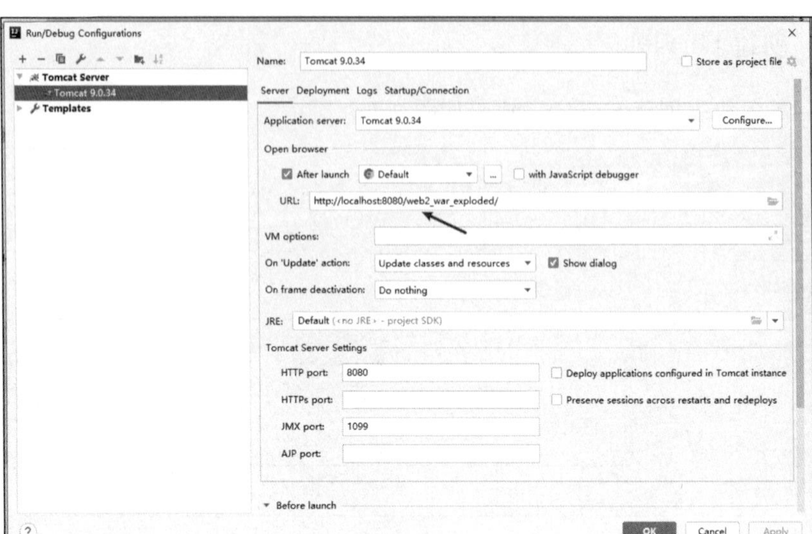


图 2-161

更改 web2 的 Application Context 路径，如图 2-162 所示。

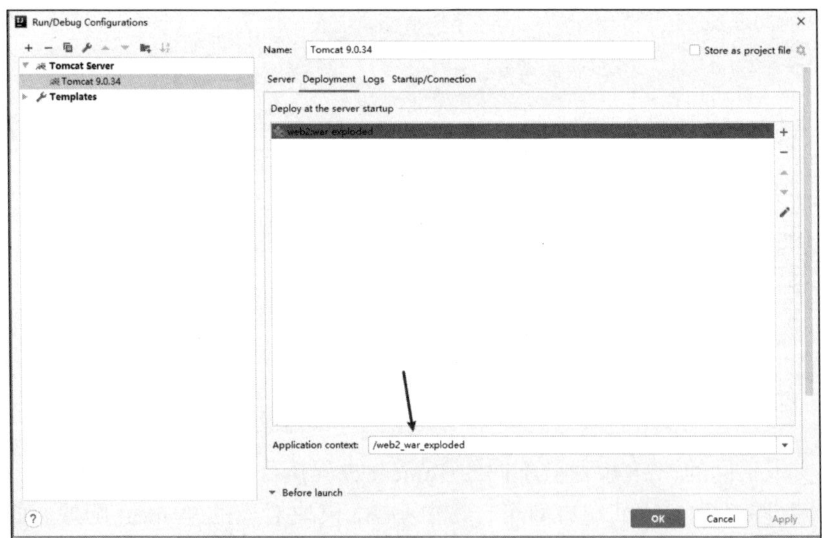


图 2-162

更改后的 Application Context 路径，如图 2-163 所示。

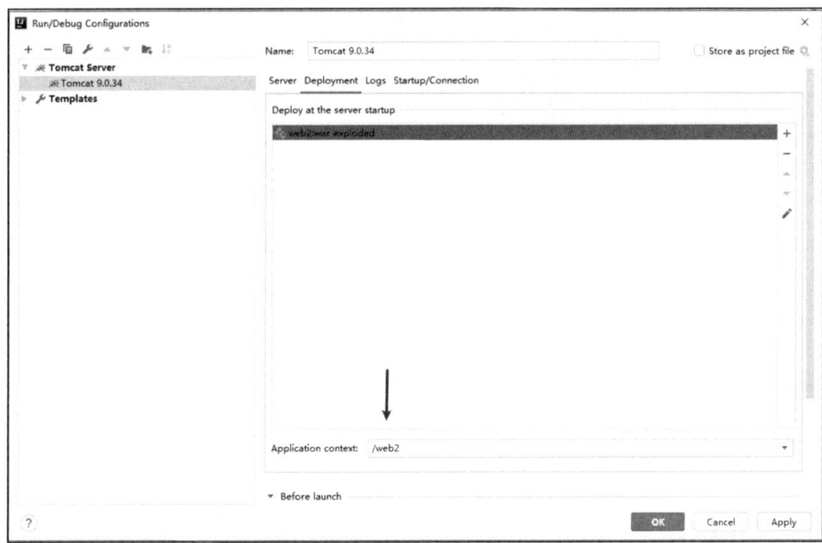


图 2-163

针对模块 web2 的 Tomcat 完整配置，如图 2-164 所示。

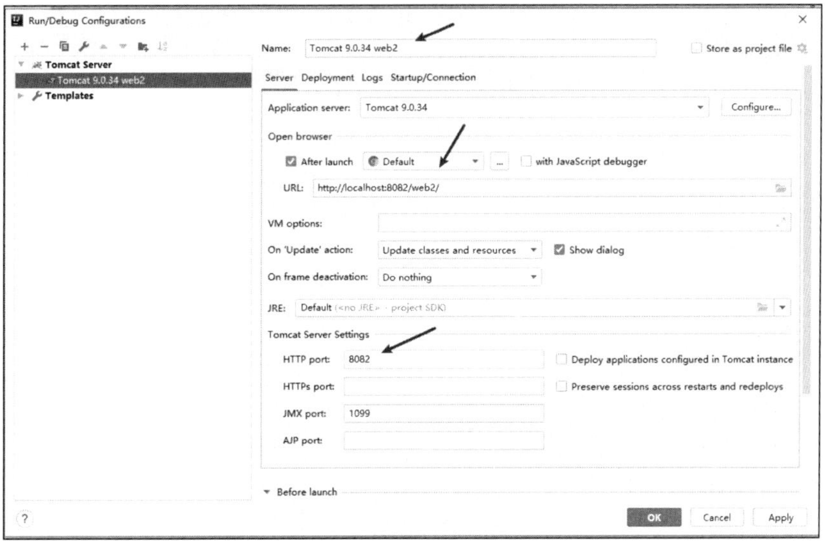

图 2-164

创建针对 web1 模块的 Tomcat 配置，如图 2-165 所示。选择“web1:war exploded”选项，如图 2-166 所示。

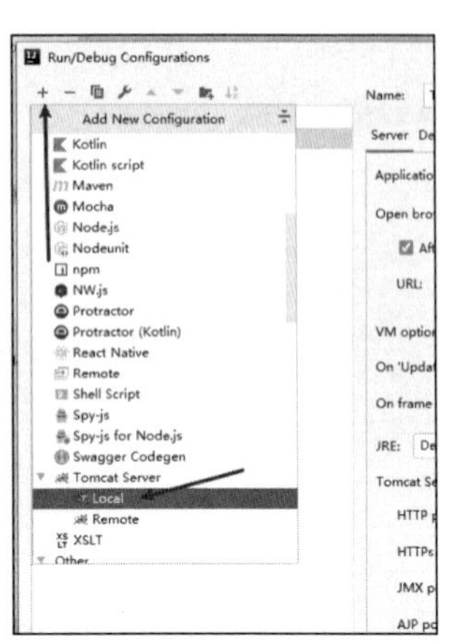


图 2-165

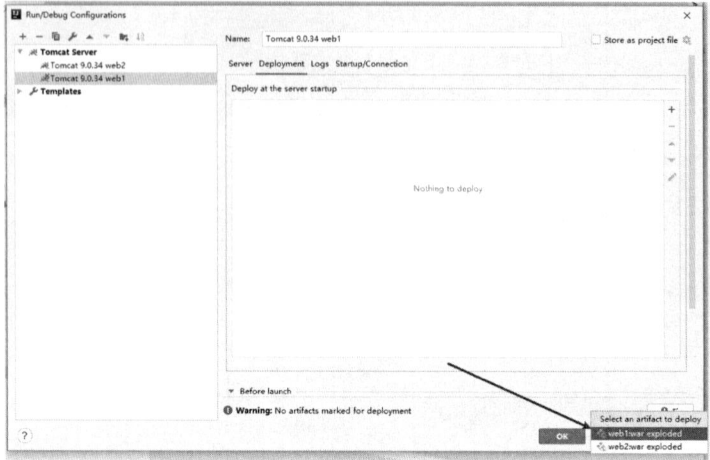


图 2-166

配置模块 web1 对应的 Application Context 路径，如图 2-167 所示。

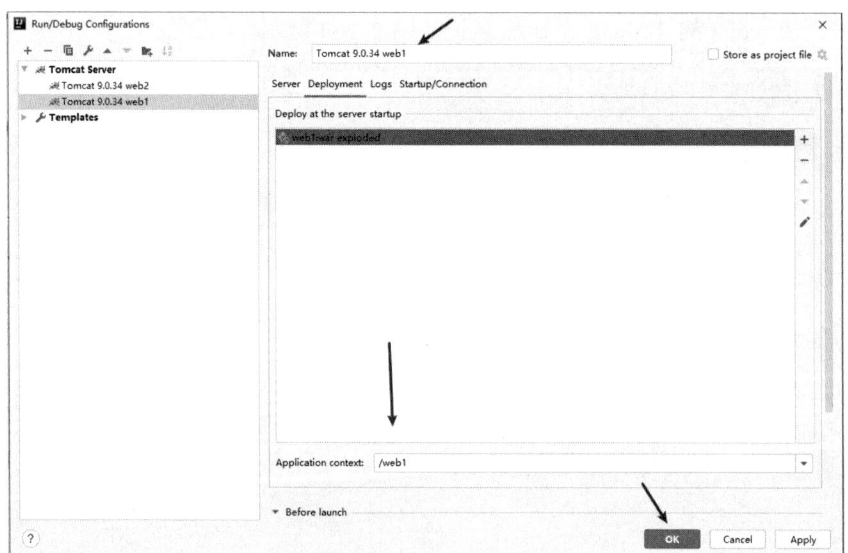


图 2-167

web1 模块对应的完整 Tomcat 配置，如图 2-168 所示。

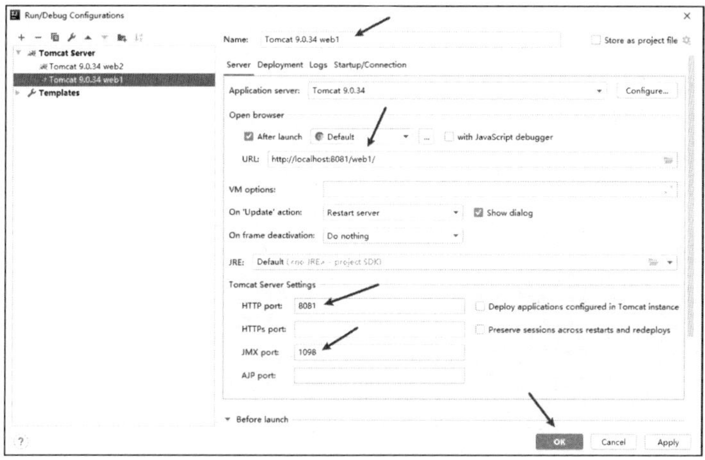

图 2-168

## 2.10.5 启动 Tomcat

在下拉选项中，分别启动 web1 和 web2 模块对应的 Tomcat 进程，如图 2-169 所示。

打开网址 http://localhost:8081/web1/test1，运行的结果如图 2-170 所示。

打开网址 http://localhost:8082/web2/test2，运行的结果如图 2-171 所示。

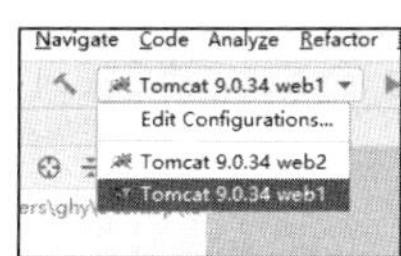

图 2-169

图 2-170

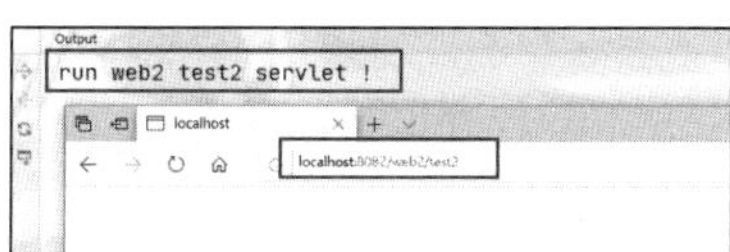

图 2-171

## 2.10.6 导出.war 文件

要想导出.war 文件，需要先创建 Artifacts 选项，如图 2-172 所示。

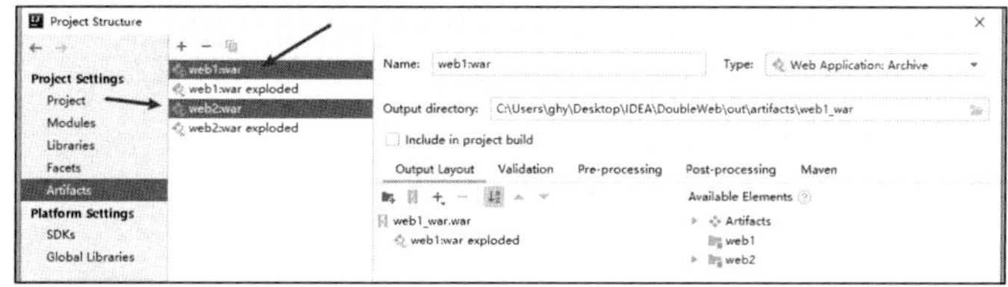

图 2-172

再执行 Build，就成功生成了两个.war 文件。

# 第 3 章 JDBC 核心技术

本章将开启使用 Java 操作数据库的旅程，目的是联合使用 Java 语言与数据库技术，这有助于我们理解 Mybatis 等 ORM 框架的原理。掌握 JDBC 技术较好的方法是多写代码，要熟练写出针对数据库的 CURD 操作，锻炼编写源代码的技能，提高常见的排错能力。

## 3.1 什么是JDBC

我们先来看看百度百科是如何解释 JDBC 的，如图 3-1 所示。

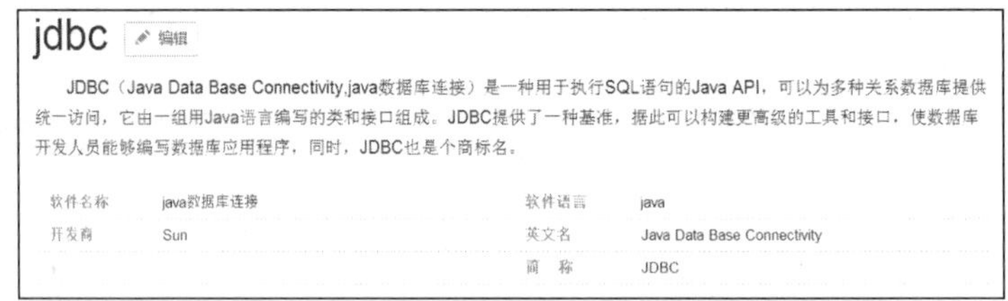

jdbc 编辑

JDBC（Java Data Base Connectivity,java数据库连接）是一种用于执行SQL语句的Java API，可以为多种关系数据库提供统一访问，它由一组用Java语言编写的类和接口组成。JDBC提供了一种基准，据此可以构建更高级的工具和接口，使数据库开发人员能够编写数据库应用程序，同时，JDBC也是个商标名。

| | | | |
|---|---|---|---|
| 软件名称 | java数据库连接 | 软件语言 | java |
| 开发商 | Sun | 英文名 | Java Data Base Connectivity |
| | | 简　称 | JDBC |

图 3-1

JDBC（Java DataBase Connectivity，Java 数据库连接），但不要被字面意思所迷惑，它的主要作用并不只是连接数据库，更加强大的功能在于，只要数据库支持 SQL 语句，就可以使用 JDBC 技术在 Java 语言中执行 SQL 语句，从而达到与数据库进行通信的目的。JDBC 还可以获取数据库、数据表及其他数据库对象的基本信息，但其最常使用的功能还是对数据库进行增删改查操作。

## 3.2 为什么要使用JDBC

先来看一幅图，如图 3-2 所示。

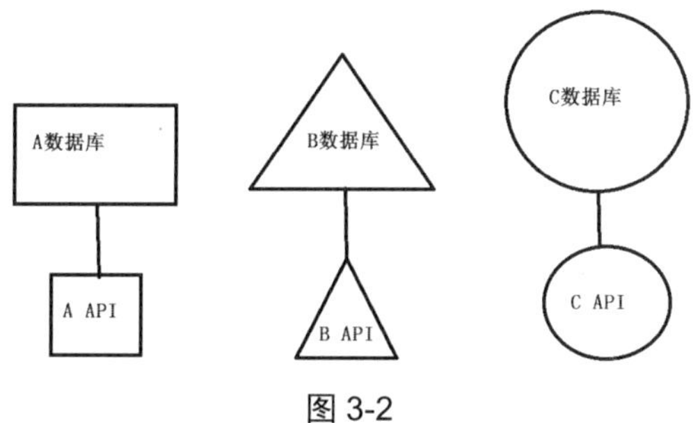


图 3-2

在图 3-2 中，可以发现有三种主流的数据库产品，分别是 A、B 和 C。我们现在想要使用 Java 语言去操作这些数据库，但出现了一个重要的问题：Java 是 Sun 公司设计的程序开发语言，而 Oracle 是甲骨文公司设计的数据库产品，在通常情况下，使用 A 厂商的程序语言去操作 B 厂商的数据库是不可能的事情。想要解决这一难题，可以参考以下两种途径。

（1）B 厂商提供数据库的源代码给 A 厂商，A 厂商通过数据库的源代码来分析如何与 B 厂商的数据库进行通信。但开放数据库的源代码除了 MySQL，其他的厂商由于商业利益等原因是不可能做到的，所以让 B 厂商开放源代码难以实现。

（2）B 厂商不需要公开数据库的源代码，而是自行设计 Java 类库并结合自家数据库产品的通信协议实现利用 Java 操作数据库。由于数据库系统的通信协议是相当私密的技术内容，因此这三个数据库厂商分别根据自己的数据库产品提供了操作数据库的私有 Java API 类库。这些私有 API 都是数据库厂商提供的，程序员使用它们就可以操作这三种数据库了。

但是，问题出现了：程序员必须要学习三种数据库的 Java API 才能对这三种数据库进行操作，这就给程序员带来极大的学习成本，也给项目在数据库移植时带来很大的不便。

Sun 公司发现了这个问题，并发布了一个标准来解决这个问题，它就是 JDBC。程序员在使用 JDBC 后，上面遇到的问题得到了解决，其原理如图 3-3 所示。

由于 Sun 的 JDBC 规范被各个数据库厂商所实现（注意：是数据库厂商实现 JDBC 规范，而不是 Sun 公司），因此程序员只需要学习一种 Java API，即 JDBC，就可以操作三种不同的数据库，这统一了操作数据库的写法，是一件让程序员十分受益的事情。

JDBC 中的 API 很多，使用这些 API 可以操作不同的数据库，但 JDBC 中的 API 不绑定指定数据库，比如当调用 JDBC 的 save()方法时，可以将数据保存到 A 数据库或者 B 数据库中，如图 3-4 所示。

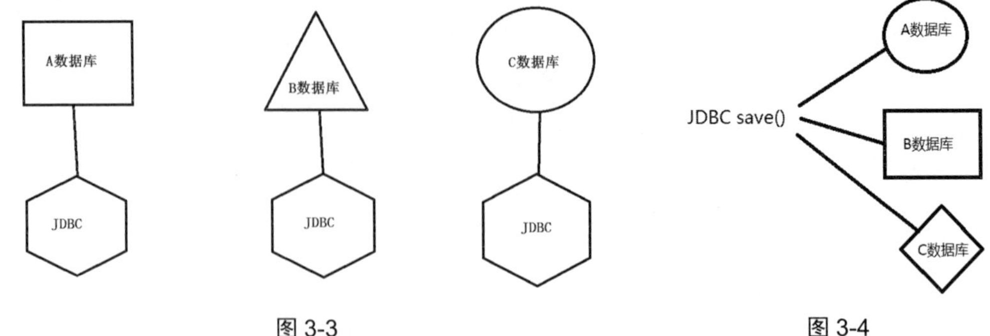


图 3-3　　　　图 3-4

有了 JDBC 技术，向各种关系数据库发送 SQL 语句就是一件很容易的事情了，程序员只需要使用 JDBC 的 API 写一个程序就够了。因为基本的 SQL 语句都被标准化了，所以使用 JDBC 可以向不同厂商的数据库发送 SQL 调用。

总结一下什么是 JDBC：

（1）JDBC 是 Sun 公司为统一操作数据库而发布的一个标准，并为这个标准定义了一套使用 Java 操作数据库的 API。

（2）使用 JDBC 可以执行 SQL 语句。

（3）JDBC 由一组使用 Java 语言定义的接口和类组成，程序员需要学习这些接口和类。

（4）JDBC 是 Java 语言与数据库之间通信的桥梁。

（5）JDBC 是众多 ORM 框架的基础，比如 Hibernate，MyBatis 等框架，内部的原理都是基于 JDBC 的。

（6）JDBC 属于 JavaEE 规范中的一种。

（7）开发 JDBC 程序需要使用 java.sql 和 javax.sql 包中的 API。

简单地说，JDBC 可做三件事：

（1）与数据库建立连接。

（2）发送操作数据库的 SQL 语句。

（3）获得返回的数据。

## 3.3 什么是JDBC驱动

在对电脑重新安装操作系统时，需要安装显卡驱动、声卡驱动、网卡驱动等，若不安装这些驱动则硬件设备不能很好地运行，甚至会出现不运行的情况，而 JDBC 驱动也有类似的功能。如果在使用 Java 操作数据库时没有安装 JDBC 驱动，则无法运行，那么，这个 JDBC 驱动到底是什么呢？简单来讲，它就是一个以.jar 为扩展名的包，如图 3-5 所示。这个文件是由数据库厂商提供的，它们最清楚如何用 Java 连接到本公司的数据库产品，重点是.class 中的部分文件实现了 JDBC 规范中的接口，从而我们可以使用类似于下面的 JDBC 代码对数据库进行操作：

ojdbc6.jar

图 3-5

```
JDBC 的标准接口对象=new jar 包文件中的实现类();
```

由于 jar 包文件中的类实现了 JDBC 的标准接口，属于实现关系，也是多态关系，所以可以直接赋值。

程序员与 JDBC 接口、JDBC 驱动和数据库之间的关系如图 3-6 所示。

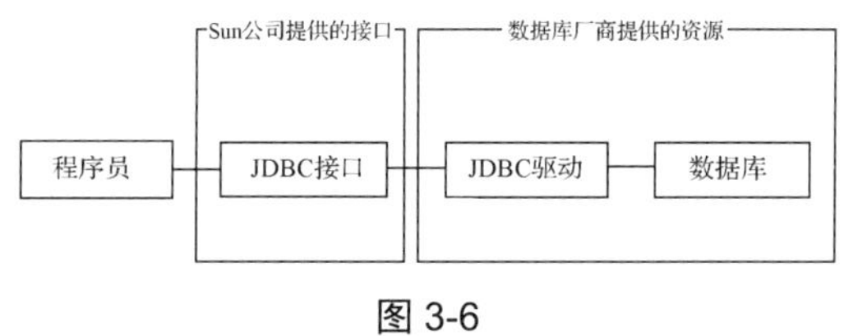


图 3-6

程序员“直接”使用 JDBC 接口，而“间接”地使用了 JDBC 接口的实现类，也就是 JDBC 驱动，这些实现类可以驱动操作数据库。

## 3.4　JDBC核心接口介绍

JDBC 规范提供了若干接口，而掌握这些接口的使用方法是掌握 JDBC 技术的关键，下面介绍常用的 JDBC 接口。

### 3.4.1　Driver

Driver 的声明信息如图 3-7 所示，API 声明如图 3-8 所示。

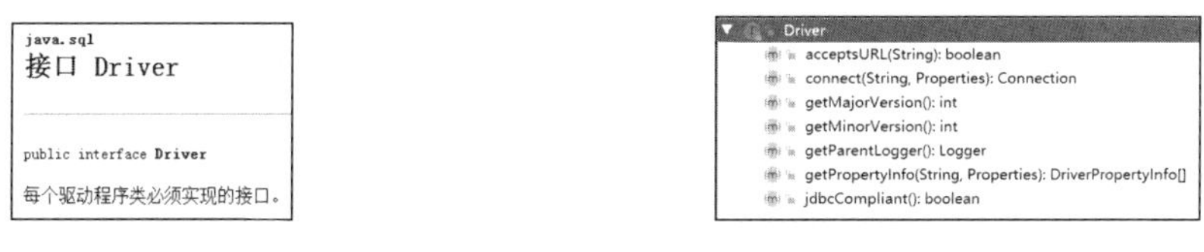
java.sql
接口 Driver

public interface Driver

每个驱动程序类必须实现的接口。



图 3-7　　　　图 3-8

Driver 接口是驱动类必须要实现的接口，在驱动 jar 包文件中有很多.class 文件，利用“Ctrl+T”快捷键可以找到所有实现或继承此 Driver 接口的子接口或实现类，如图 3-9 所示。

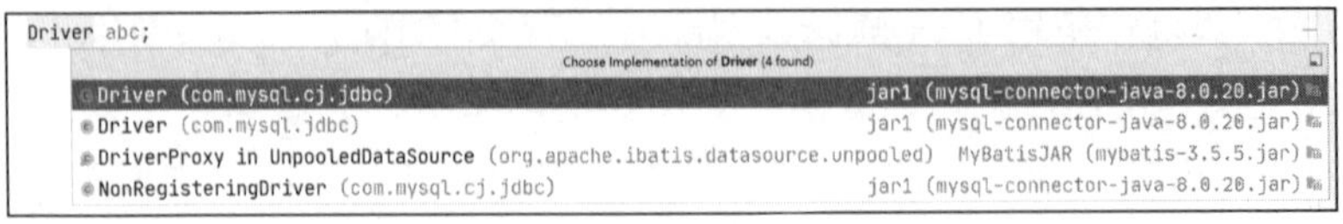
Driver abc;
Choose Implementation of Driver (4 found)
Driver (com.mysql.cj.jdbc) jar1 (mysql-connector-java-8.0.20.jar)
Driver (com.mysql.jdbc) jar1 (mysql-connector-java-8.0.20.jar)
DriverProxy in UnpooledDataSource (org.apache.ibatis.datasource.unpooled) MyBatisJAR (mybatis-3.5.5.jar)
NonRegisteringDriver (com.mysql.cj.jdbc) jar1 (mysql-connector-java-8.0.20.jar)

图 3-9

成功找到 Driver 接口的实现类 com.mysql.cj.jdbc.Driver，该类是由 MySQL 数据库驱动 jar 包文件提供的，源代码如图 3-10 所示。

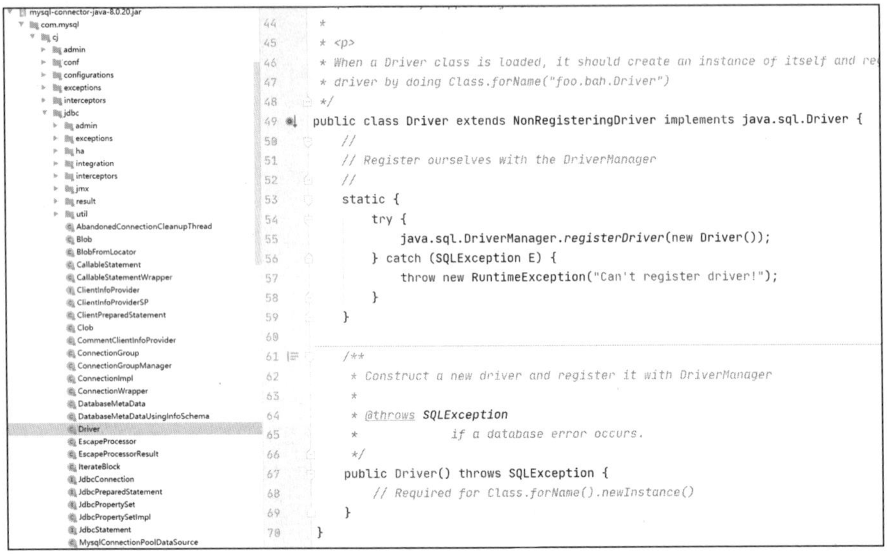
mysql-connector-java-8.0.20.jar
com.mysql
cj
admin
conf
configurations
exceptions
interceptors
jdbc
admin
exceptions
ha
integration
interceptors
jmx
result
util
AbandonedConnectionCleanupThread
Blob
BlobFromLocator
CallableStatement
CallableStatementWrapper
ClientInfoProvider
ClientInfoProviderSP
ClientPreparedStatement
Clob
CommentClientInfoProvider
ConnectionGroup
ConnectionGroupManager
ConnectionImpl
ConnectionWrapper
DatabaseMetaData
DatabaseMetaDataUsingInfoSchema
Driver
EscapeProcessor
EscapeProcessorResult
IterateBlock
JdbcConnection
JdbcPreparedStatement
JdbcPropertySet
JdbcPropertySetImpl
JdbcStatement
MysqlConnectionPoolDataSource
*
* <p>
* When a Driver class is loaded, it should create an instance of itself and re
* driver by doing Class.forName("foo.bah.Driver")
*/
public class Driver extends NonRegisteringDriver implements java.sql.Driver {
//
// Register ourselves with the DriverManager
//
static {
try {
java.sql.DriverManager.registerDriver(new Driver());
} catch (SQLException E) {
throw new RuntimeException("Can't register driver!");
}
}
/**
* Construct a new driver and register it with DriverManager
*
* @throws SQLException
* if a database error occurs.
*/
public Driver() throws SQLException {
// Required for Class.forName().newInstance()
}
}

图 3-10

源代码中与注册驱动相关的代码如图 3-11 所示。

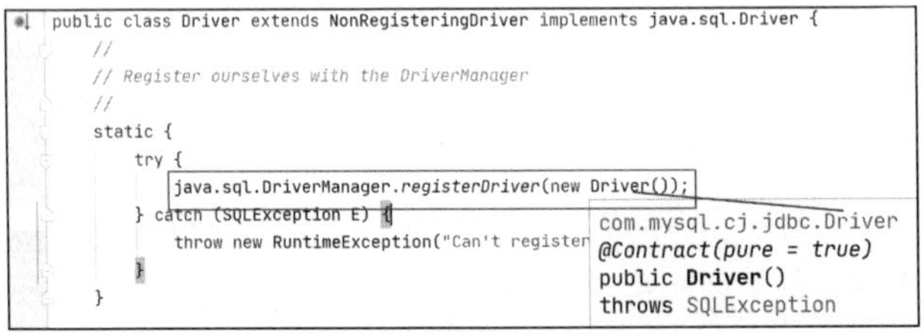
public class Driver extends NonRegisteringDriver implements java.sql.Driver {
//
// Register ourselves with the DriverManager
//
static {
try {
java.sql.DriverManager.registerDriver(new Driver());
} catch (SQLException E) {
throw new RuntimeException("Can't register
}
}
com.mysql.cj.jdbc.Driver
@Contract(pure = true)
public Driver()
throws SQLException

图 3-11

该类中 static 静态代码块具有将自身注册到系统中的功能。

需要说明的是，因为 Oracle 的 JDBC 驱动并不是开源的，所以在这里只能参考针对 MySQL 数据库的 JDBC 驱动来研究内部的代码实现，它们的原理都是相通的。

所谓“注册驱动”就是在调用 java.sql.DriverManager.java 类中的 registerDriver()方法时，将 Driver 驱动类放入 ArrayList 集合，在后面的操作中可以从此集合中获取驱动信息，如图 3-12 所示。

```
public class DriverManager {

    // List of registered JDBC drivers
    private final static CopyOnWriteArrayList<DriverInfo> registeredDrivers = new CopyOnWriteArrayList<>();

    @   public static synchronized void registerDriver(java.sql.Driver driver,
            DriverAction da)
        throws SQLException {

        /* Register the driver if it has not already been added to our list */
        if(driver != null) {
            registeredDrivers.addIfAbsent(new DriverInfo(driver, da));
        } else {
            // This is for compatibility with the original DriverManager
            throw new NullPointerException();
        }

        println("registerDriver: " + driver);

    }
```

图 3-12

如果第三方的厂商想要将驱动注册到 JVM 中，则其驱动类就要实现 Driver 接口，这就是 Driver 接口的作用。

## 3.4.2　Connection

Connection 接口的主要作用正如它的名称一样，是连接数据库。使用此接口去连接数据库，只有 Java 先连接到数据库后，才可以操作数据库，其相当于 Java 程序与数据库之间的桥梁。

Connection 的声明信息如图 3-13 所示，API 声明如图 3-14 所示。

java.sql
**接口 Connection**

**所有超级接口：**
Wrapper

public interface **Connection**
extends Wrapper

与特定数据库的连接（会话）。在连接上下文中执行 SQL 语句并返回结果。

Connection 对象的数据库能够提供描述其表、所支持的 SQL 语法、存储过程、此连接功能等等的信息。此信息是使用 getMetaData 方法获得的。

图 3-13

图 3-14

## 3.4.3 Statement

Statement 接口负责封装和发送 SQL 语句，其声明信息如图 3-15 所示。

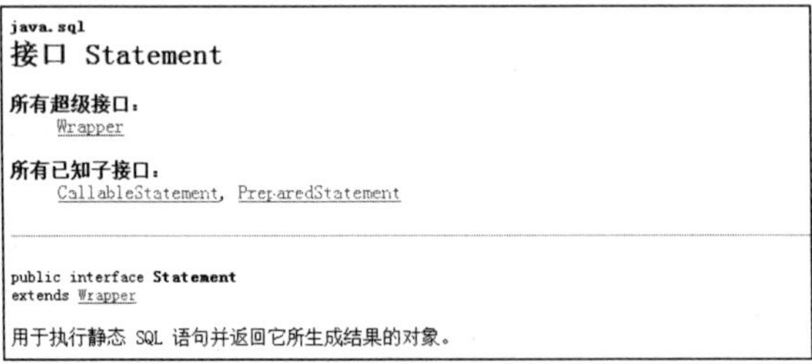

java.sql

**接口 Statement**

**所有超级接口：**

Wrapper

**所有已知子接口：**

CallableStatement, PreparedStatement

```
public interface Statement
extends Wrapper
```

用于执行静态 SQL 语句并返回它所生成结果的对象。

图 3-15

Statement 接口的 API 声明如图 3-16 所示。

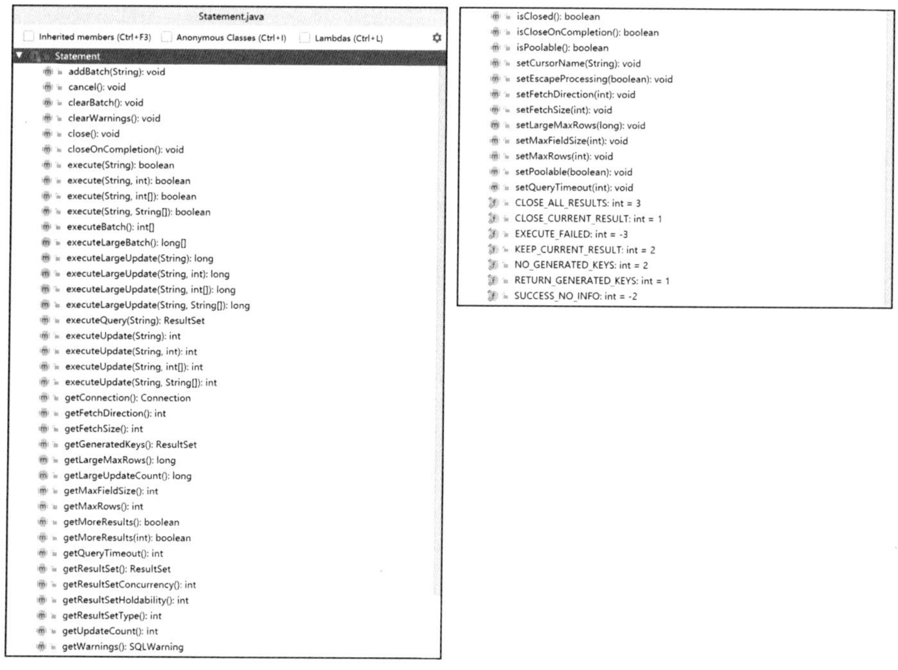


图 3-16

## 3.4.4 ResultSet

当使用 Statement 接口发送查询的 SQL 语句后，我们需要获得查询的结果返回值，而 ResultSet 接口的功能就是封装返回值的结果，它的数据组织形式类似于二维表格，在二维表格中存储查询到的数据。

ResultSet 接口的声明信息如图 3-17 所示，API 声明如图 3-18 ~ 图 3-20 所示。

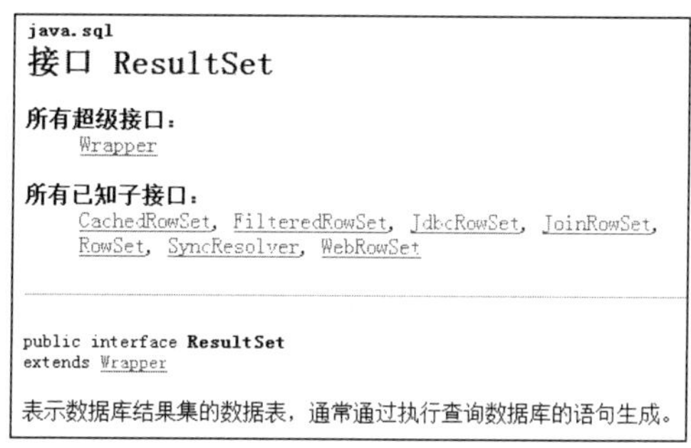

java.sql

接口 ResultSet

**所有超级接口：**

Wrapper

**所有已知子接口：**

CachedRowSet, FilteredRowSet, JdbcRowSet, JoinRowSet, RowSet, SyncResolver, WebRowSet

```
public interface ResultSet
extends Wrapper
```

表示数据库结果集的数据表，通常通过执行查询数据库的语句生成。

图 3-17

ResultSet.java

Inherited members (Ctrl+F3) Anonymous Classes (C

- ResultSet
- absolute(int): boolean
- afterLast(): void
- beforeFirst(): void
- cancelRowUpdates(): void
- clearWarnings(): void
- close(): void
- deleteRow(): void
- findColumn(String): int
- first(): boolean
- getArray(int): Array
- getArray(String): Array
- getAsciiStream(int): InputStream
- getAsciiStream(String): InputStream
- getBigDecimal(int): BigDecimal
- ~~getBigDecimal(int, int): BigDecimal~~
- getBigDecimal(String): BigDecimal
- ~~getBigDecimal(String, int): BigDecimal~~
- getBinaryStream(int): InputStream
- getBinaryStream(String): InputStream
- getBlob(int): Blob
- getBlob(String): Blob
- getBoolean(int): boolean
- getBoolean(String): boolean
- getByte(int): byte
- getByte(String): byte
- getBytes(int): byte[]
- getBytes(String): byte[]
- getCharacterStream(int): Reader
- getCharacterStream(String): Reader
- getClob(int): Clob
- getClob(String): Clob
- getConcurrency(): int
- getCursorName(): String
- getDate(int): Date
- getDate(int, Calendar): Date
- getDate(String): Date
- getDate(String, Calendar): Date
- getDouble(int): double
- getDouble(String): double
- getFetchDirection(): int
- getFetchSize(): int
- getFloat(int): float
- getFloat(String): float

ResultSet.java

Inherited members (Ctrl+F3) Anonymous Classes (Ctrl+I)

- getHoldability(): int
- getInt(int): int
- getInt(String): int
- getLong(int): long
- getLong(String): long
- getMetaData(): ResultSetMetaData
- getNCharacterStream(int): Reader
- getNCharacterStream(String): Reader
- getNClob(int): NClob
- getNClob(String): NClob
- getNString(int): String
- getNString(String): String
- getObject(int): Object
- getObject(int, Class<T>): T
- getObject(int, Map<String, Class<?>>): Object
- getObject(String): Object
- getObject(String, Class<T>): T
- getObject(String, Map<String, Class<?>>): Object
- getRef(int): Ref
- getRef(String): Ref
- getRow(): int
- getRowId(int): RowId
- getRowId(String): RowId
- getShort(int): short
- getShort(String): short
- getSQLXML(int): SQLXML
- getSQLXML(String): SQLXML
- getStatement(): Statement
- getString(int): String
- getString(String): String
- getTime(int): Time
- getTime(int, Calendar): Time
- getTime(String): Time
- getTime(String, Calendar): Time
- getTimestamp(int): Timestamp
- getTimestamp(int, Calendar): Timestamp
- getTimestamp(String): Timestamp
- getTimestamp(String, Calendar): Timestamp
- getType(): int
- ~~getUnicodeStream(int): InputStream~~
- ~~getUnicodeStream(String): InputStream~~
- getURL(int): URL
- getURL(String): URL
- getWarnings(): SQLWarning
- insertRow(): void

图 3-18

ResultSet.java

Inherited members (Ctrl+F3) Anonymous Classes (Ctrl+I)

- isAfterLast(): boolean
- isBeforeFirst(): boolean
- isClosed(): boolean
- isFirst(): boolean
- isLast(): boolean
- last(): boolean
- moveToCurrentRow(): void
- moveToInsertRow(): void
- next(): boolean
- previous(): boolean
- refreshRow(): void
- relative(int): boolean
- rowDeleted(): boolean
- rowInserted(): boolean
- rowUpdated(): boolean
- setFetchDirection(int): void
- setFetchSize(int): void
- updateArray(int, Array): void
- updateArray(String, Array): void
- updateAsciiStream(int, InputStream): void
- updateAsciiStream(int, InputStream, int): void
- updateAsciiStream(int, InputStream, long): void
- updateAsciiStream(String, InputStream): void
- updateAsciiStream(String, InputStream, int): void
- updateAsciiStream(String, InputStream, long): void
- updateBigDecimal(int, BigDecimal): void
- updateBigDecimal(String, BigDecimal): void
- updateBinaryStream(int, InputStream): void
- updateBinaryStream(int, InputStream, int): void
- updateBinaryStream(int, InputStream, long): void
- updateBinaryStream(String, InputStream): void
- updateBinaryStream(String, InputStream, int): void
- updateBinaryStream(String, InputStream, long): void
- updateBlob(int, InputStream): void
- updateBlob(int, InputStream, long): void
- updateBlob(int, Blob): void
- updateBlob(String, InputStream): void
- updateBlob(String, InputStream, long): void
- updateBlob(String, Blob): void
- updateBoolean(int, boolean): void
- updateBoolean(String, boolean): void
- updateByte(int, byte): void
- updateByte(String, byte): void
- updateBytes(int, byte[]): void
- updateBytes(String, byte[]): void

ResultSet.java

Inherited members (Ctrl+F3) Anonymous Classes (Ctrl+I)

- updateCharacterStream(int, Reader): void
- updateCharacterStream(int, Reader, int): void
- updateCharacterStream(int, Reader, long): void
- updateCharacterStream(String, Reader): void
- updateCharacterStream(String, Reader, int): void
- updateCharacterStream(String, Reader, long): void
- updateClob(int, Clob): void
- updateClob(int, Reader): void
- updateClob(int, Reader, long): void
- updateClob(String, Clob): void
- updateClob(String, Reader): void
- updateClob(String, Reader, long): void
- updateDate(int, Date): void
- updateDate(String, Date): void
- updateDouble(int, double): void
- updateDouble(String, double): void
- updateFloat(int, float): void
- updateFloat(String, float): void
- updateInt(int, int): void
- updateInt(String, int): void
- updateLong(int, long): void
- updateLong(String, long): void
- updateNCharacterStream(int, Reader): void
- updateNCharacterStream(int, Reader, long): void
- updateNCharacterStream(String, Reader): void
- updateNCharacterStream(String, Reader, long): void
- updateNClob(int, NClob): void
- updateNClob(int, Reader): void
- updateNClob(int, Reader, long): void
- updateNClob(String, NClob): void
- updateNClob(String, Reader): void
- updateNClob(String, Reader, long): void
- updateNString(int, String): void
- updateNString(String, String): void
- updateNull(int): void
- updateNull(String): void
- updateObject(int, Object): void
- updateObject(int, Object, int): void
- updateObject(int, Object, SQLType): void
- updateObject(int, Object, SQLType, int): void
- updateObject(String, Object): void
- updateObject(String, Object, int): void
- updateObject(String, Object, SQLType): void
- updateObject(String, Object, SQLType, int): void
- updateRef(int, Ref): void

图 3-19

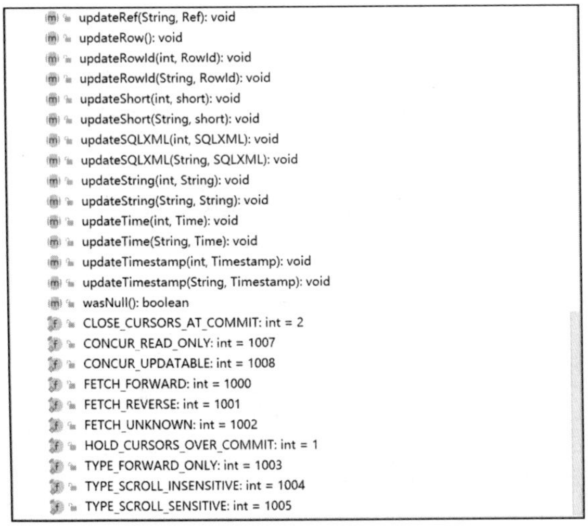

- updateRef(String, Ref): void
- updateRow(): void
- updateRowId(int, RowId): void
- updateRowId(String, RowId): void
- updateShort(int, short): void
- updateShort(String, short): void
- updateSQLXML(int, SQLXML): void
- updateSQLXML(String, SQLXML): void
- updateString(int, String): void
- updateString(String, String): void
- updateTime(int, Time): void
- updateTime(String, Time): void
- updateTimestamp(int, Timestamp): void
- updateTimestamp(String, Timestamp): void
- wasNull(): boolean
- CLOSE_CURSORS_AT_COMMIT: int = 2
- CONCUR_READ_ONLY: int = 1007
- CONCUR_UPDATABLE: int = 1008
- FETCH_FORWARD: int = 1000
- FETCH_REVERSE: int = 1001
- FETCH_UNKNOWN: int = 1002
- HOLD_CURSORS_OVER_COMMIT: int = 1
- TYPE_FORWARD_ONLY: int = 1003
- TYPE_SCROLL_INSENSITIVE: int = 1004
- TYPE_SCROLL_SENSITIVE: int = 1005

图 3-20

前面的 4 个接口都在 java.sql 包中，如图 3-21 所示，它们完整地组成了使用 JDBC 操作数据库的基本步骤，首先注册 Driver 驱动，然后建立连接，再使用 Statement 发送 SQL 语句，最后使用 ResultSet 接口获得查询的结果。这个步骤已经成为写作代码的模式，在 Java 语言中使用 JDBC 操作数据库一般都这么做，是掌握 JDBC 的重中之重。

```
import java.sql.Connection;
import java.sql.Driver;
import java.sql.ResultSet;
import java.sql.Statement;
```

图 3-21

## 3.4.5 PreparedStatement

Statement 接口提供了执行 SQL 语句的功能，但在执行效率及保证执行的 SQL 语句运行的安全性上，却不能满足需求，所以出现了子接口 PreparedStatement，其声明信息如图 3-22 所示。

java.sql
接口 PreparedStatement

所有超级接口：
Statement, Wrapper

所有已知子接口：
CallableStatement

public interface PreparedStatement
extends Statement

表示预编译的 SQL 语句的对象。

SQL 语句被预编译并存储在 PreparedStatement 对象中。然后可以使用此对象多次高效地执行该语句。

图 3-22

预编译是在执行 SQL 语句之前就对 SQL 语句进行分析、校验和优化，只执行一次预编译，后面可以多次复用预编译的对象，提升了程序运行的效率。

PreparedStatement 接口的 API 声明如图 3-23 所示。

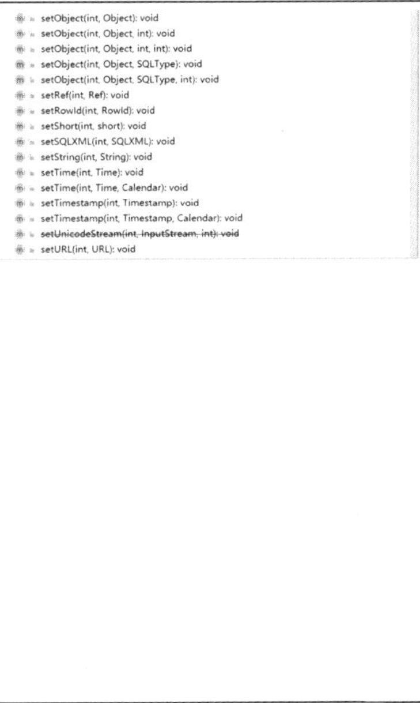


图 3-23

## 3.5 创建Driver对象

使用 JDBC 操作数据库的第一步就是注册驱动，而注册驱动之前需要先创建 Driver 对象，创建 Driver 对象的示例代码如图 3-24 所示。

在实例化 new com.mysql.cj.jdbc.Driver()类时，其实在源代码中就已经实现了注册驱动的过程，源代码如图 3-25 所示。

```
package test;

import java.sql.SQLException;

public class Test1 {
    public static void main(String[] args) throws SQLException {
        com.mysql.cj.jdbc.Driver driver = new com.mysql.cj.jdbc.Driver();
        System.out.println(driver);
    }
}
Test1
C:\jdk1.8\bin\java.exe ...
com.mysql.cj.jdbc.Driver@49097b5d
```

图 3-24

```
package com.mysql.cj.jdbc;

import java.sql.SQLException;

/** The Java SQL framework allows for multiple database drivers. Each driver s
public class Driver extends NonRegisteringDriver implements java.sql.Driver {
    //
    // Register ourselves with the DriverManager
    //
    static {
        try {
            java.sql.DriverManager.registerDriver(new Driver());
        } catch (SQLException E) {
            throw new RuntimeException("Can't register driver!");
        }
    }
```

图 3-25

## 3.6 创建Connection对象

想要获得 Connection 对象，需要使用 Driver 类的 connect(String url, java.util.Properties info)方法。在使用 connect()方法时，需要先在 Properties 对象中存储 user 和 password 信息，然后再传递给 connect()方法，文档中的说明如图 3-26 所示。

获得 Connection 对象的完整代码，如图 3-27 所示。

Connection **connect**(String url,
Properties info)
throws SQLException

试图创建一个到给定 URL 的数据库连接。如果驱动程序识别出它是连接到给定 URL 的错误类型的驱动程序，那么它应该返回 "null"。这会很常见，因为在要求 JDBC 驱动程序管理器连接到给定 URL 时，它会将该 URL 依次传递给每个已加载的驱动程序。

如果要连接到给定 URL 的驱动程序是正确的驱动程序，但却无法连接到数据库，则该驱动程序应该抛出 SQLException。

可以使用 java.util.Properties 参数将任意字符串标记/值对作为连接参数传递。通常 Properties 对象中至少应该包括 "user" 和 "password" 属性。

**参数：**
url - 要连接到的数据库的 URL
info - 做为连接参数的任意字符串标记/值对的列表。通常至少应该包括 "user" 和 "password" 属性。
**返回：**
表示到 URL 的连接的 Connection 对象
**抛出：**
SQLException - 如果发生数据库访问错误

图 3-26

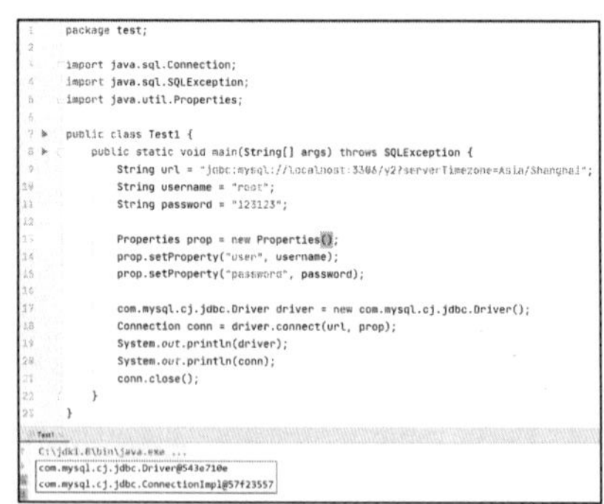

```
package test;

import java.sql.Connection;
import java.sql.SQLException;
import java.util.Properties;

public class Test1 {
    public static void main(String[] args) throws SQLException {
        String url = "jdbc:mysql://localhost:3306/y2?serverTimezone=Asia/Shanghai";
        String username = "root";
        String password = "123123";

        Properties prop = new Properties();
        prop.setProperty("user", username);
        prop.setProperty("password", password);

        com.mysql.cj.jdbc.Driver driver = new com.mysql.cj.jdbc.Driver();
        Connection conn = driver.connect(url, prop);
        System.out.println(driver);
        System.out.println(conn);
        conn.close();
    }
}
Test1
C:\jdk1.8\bin\java.exe ...
com.mysql.cj.jdbc.Driver@543e710e
com.mysql.cj.jdbc.ConnectionImpl@57f23557
```

图 3-27

需要注意的是，使用完毕的 Connection 对象一定要调用 close()方法关闭连接，不然就像占着窗口不买票一样，会导致其他人无法获得 Connection 对象，因为可用的连接数

是有限的。

在源代码的第 9 行有一个 url 变量值：

```
jdbc:mysql://localhost:3306/y2?serverTimezone=Asia/Shanghai
```

该值指向 MySQL 数据库的地址，字符串“jdbc:mysql://localhost:3306/y2”包含的信息有安装数据库服务器的 IP、访问数据库的端口，以及数据库的名称。

那么，这个字符串是如何确定的呢？每一种数据库的 JDBC 对应的 URL 值都不一样，其中 MySQL 的 URL 值在 com.mysql.cj.jdbc.NonRegisteringDriver.java 类中的 public java.sql.Connection connect (String url, Properties info)方法上方的说明中，如图 3-28 所示。

```
NonRegisteringDriver.java
     * <p>
     * The java.util.Properties argument can be used to pass arbitrary string tag/value pai
     * properties sent in the URL.
     * </p>
     *
     * <p>
     * MySQL protocol takes the form: jdbc:mysql://host:port/database
     * </p>
     *
     * @param url
     *            the URL of the database to connect to
     * @param info
     *            a list of arbitrary tag/value pairs as connection arguments
     *
     * @return a connection to the URL or null if it isn't us
     *
     * @exception SQLException
     *                if a database access error occurs or the url is {@code null}
     */
    @Override
    public java.sql.Connection connect(String url, Properties info) throws SQLException {
```

图 3-28

其实，这个 URL 值不需要死记硬背，因为第三方的数据库浏览工具已经提供了 URL 值，比如 Dbeaver，IntelliJ IDEA。

## 3.7　创建Statement对象

先使用 Connection 创建 Statement 对象，然后使用 Statement 对象发送 SQL 语句，进行数据库的操作，示例代码如下：

```
public class Test3 {
    public static void main(String[] args) throws SQLException {
        String url = "jdbc:oracle:thin:@localhost:1521:orcl";
        Properties prop = new Properties();
        prop.setProperty("user", "y2");
        prop.setProperty("password", "123");
```

```
        Driver driver = new OracleDriver();
        Connection connection = null;
        connection = driver.connect(url, prop);
        Statement statement = connection.createStatement();
       statement.executeUpdate("insert into userinfo(id,username,password)
values(idauto.nextval,'JDBC账号','JDBC密码')");
        statement.close();
        connection.close();
    }
}
```

程序运行后，成功插入数据库，如图 3-29 所示。

| ID | USERNAME | PASSWORD | AGE | ADDRESS | INSERTDATE |
| --- | --- | --- | --- | --- | --- |
| 242 | JDBC账号 | JDBC密码 | | | |

图 3-29

## 3.8 创建ResultSet对象

在查询数据时一般要通过 ResultSet 遍历查询到的结果集，示例代码如下：

```
public class Test4 {
    public static void main(String[] args) throws SQLException {
        String url = "jdbc:oracle:thin:@localhost:1521:orcl";
        Properties prop = new Properties();
        prop.setProperty("user", "y2");
        prop.setProperty("password", "123");

        Driver driver = new OracleDriver();
        Connection connection = null;
        connection = driver.connect(url, prop);
        Statement statement = connection.createStatement();
        ResultSet rs = statement.executeQuery("select * from userinfo");
        while (rs.next()) {
            long id = rs.getLong("id");
            String username = rs.getString("username");
            String password = rs.getString("password");
            System.out.println(id + " " + username + " " + password);
        }
        rs.close();
```

```
        statement.close();
        connection.close();
    }
}
```

ResultSet（结果集）封装了一个从数据表中查询数据记录的二维表，在表中存储从数据库中查询到的结果，再使用 while()语句进行遍历，过程如图 3-30 所示。

当取得 ResultSet 对象时，指针在第一行的上面，执行 next()方法后指针移向第一行数据记录，然后 while()语句结合 next()方法一直遍历到最后一条，在最后一行执行 next()方法后，由于后面没有记录，因此就会退出 while()语句，执行 while()语句后面的程序代码。

程序运行后的结果如图 3-31 所示。

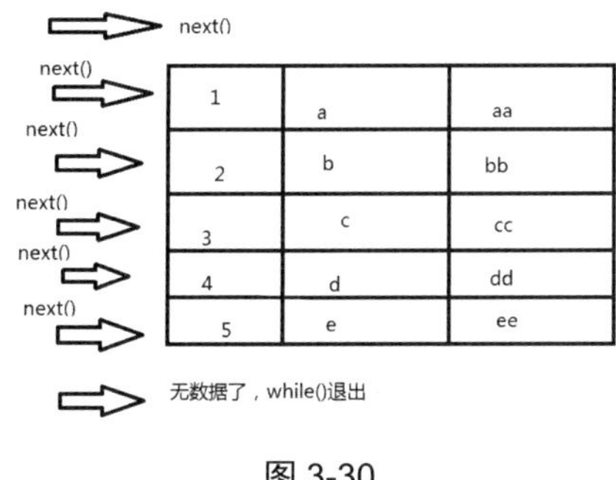


图 3-30

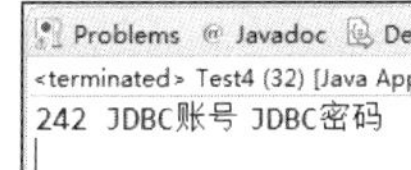


图 3-31

## 3.9　使用Statement接口有损软件的安全性

我们可以使用 Statement 接口对数据库进行查询操作，查询操作也是登录功能的实现原理，所以本节就实现这个功能。

数据表 userinfo 中的内容如图 3-32 所示。

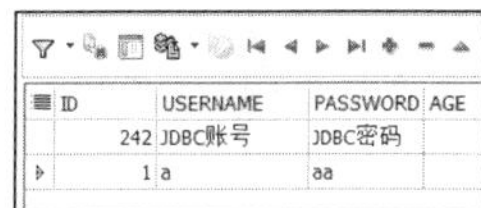

| ID | USERNAME | PASSWORD | AGE |
|---|---|---|---|
| 242 | JDBC账号 | JDBC密码 | |
| 1 | a | aa | |

图 3-32

登录功能的测试代码如下：

```
public class Test5 {
    public static void main(String[] args) throws SQLException {
        boolean isLoginSucces = false;
        String username = "a";
        String password = "aa";

        String url = "jdbc:oracle:thin:@localhost:1521:orcl";
        Properties prop = new Properties();
```

```
        prop.setProperty("user", "y2");
        prop.setProperty("password", "123");

        Driver driver = new OracleDriver();
        Connection connection = null;
        connection = driver.connect(url, prop);
        Statement statement = connection.createStatement();
        ResultSet rs = statement.executeQuery(
                "select * from userinfo where username='" + username + "' and
password='" + password + "'");
        while (rs.next()) {
            isLoginSucces = true;
        }
        rs.close();
        statement.close();
        connection.close();
        System.out.println("登录结果：" + isLoginSucces);
    }
}
```

程序运行后的结果如图 3-33 所示。

改变如下代码内容：

```
    String username = "a";
    String password = "xyzzzzzzzzzzzzzzzzzzzzzz";
```

即输入一个错误的密码，再次运行程序，如图 3-34 所示。

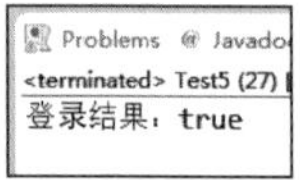

图 3-33

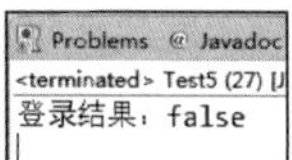


图 3-34

虽然登录失败了，但是上面的代码可以在不知道账号和密码的前提下实现成功登录，更改代码如下：

```
public class Test5 {
    public static void main(String[] args) throws SQLException {
        boolean isLoginSucces = false;

        // 最终的 SQL 语句格式为 select * from userinfo where username='abc' and
password='xyz' or '1'='1'
```

```
        String username = '任意内容';
        String password = '任意内容' or '1'='1';

        String url = "jdbc:oracle:thin:@localhost:1521:orcl";
        Properties prop = new Properties();
        prop.setProperty("user", "y2");
        prop.setProperty("password", "123");

        Driver driver = new OracleDriver();
        Connection connection = null;
        connection = driver.connect(url, prop);
        Statement statement = connection.createStatement();
        String sql = "select * from userinfo where username='" + username +
"' and password='" + password + "'";
        System.out.println("sql=" + sql);
        ResultSet rs = statement.executeQuery(sql);
        while (rs.next()) {
            isLoginSucces = true;
        }
        rs.close();
        statement.close();
        connection.close();
        System.out.println("登录结果：" + isLoginSucces);
    }
}
```

再次运行程序，在没有正确账号和密码的前提下，居然成功登录了。这种行为称为 SQL 注入，如图 3-35 所示。

```
sql=select * from userinfo where username='任意内容' and password='任意内容' or '1'='1'
登录结果：true
```

图 3-35

上面的程序存在安全漏洞，因为 Statement 接口使用的 SQL 语句是拼接的。那么，如何避免出现这种情况呢？答案是使用 PreparedStatement 接口。

## 3.10　创建PreparedStatement接口

PreparedStatement 接口是 Statement 接口的子接口，如图 3-36 所示。

```
public interface PreparedStatement extends Statement {
```

图 3-36

解决安全隐患的示例代码如下：

```
public class Test6 {
    public static void main(String[] args) throws SQLException {
        boolean isLoginSucces = false;

        String username = '任意内容';
        String password = '任意内容' or '1'='1';

        String url = "jdbc:oracle:thin:@localhost:1521:orcl";
        Properties prop = new Properties();
        prop.setProperty("user", "y2");
        prop.setProperty("password", "123");

        Driver driver = new OracleDriver();
        Connection connection = null;
        connection = driver.connect(url, prop);
        String sql = "select * from userinfo where username=? and password=?";
        PreparedStatement ps = connection.prepareStatement(sql);
        ps.setString(1, username);//向第一个问号传值
        ps.setString(2, password);//向第二个问号传值
        System.out.println("sql=" + sql);
        ResultSet rs = ps.executeQuery();
        while (rs.next()) {
            isLoginSucces = true;
        }
        rs.close();
        ps.close();
        connection.close();
        System.out.println("登录结果：" + isLoginSucces);
    }
}
```

需要注意的是，使用 PreparedStatement 接口后，方法 executeQuery()就不需要传入 SQL 语句了，这时，SQL 语句是通过 connection.prepareStatement(sql)方法传入的。

下列代码中的问号具有占位的含义，类似于形式参数，是等待被传入的数据。

```
String sql = "select * from userinfo where username=? and password=?";
```

下列代码中的第一个参数 1 代表向第一个问号传入 username 值，其他的传入参数依次类推即可。

```
ps.setString(1, username);
```

程序运行后出现登录失败的结果，如图 3-37 所示。

通过使用 PreparedStatement 接口，程序排除了安全隐患，这时必须输入正确的用户名和密码才可以登录，更改代码后运行结果如图 3-38 所示。

```
sql=select * from userinfo where username=? and password=?
登录结果：false
```

图 3-37

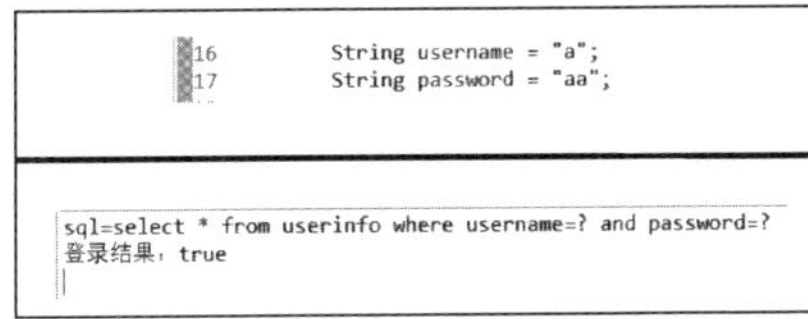


图 3-38

在使用 PreparedStatement 接口时，传入参数值使用的是 setXXX()方法（表示 set 的一系列方法），在 Java 中与 MySQL 数据类型的对应关系可以在“MySQL Connector/J 5.1 Developer Guide”文档中找到答案，如图 3-39 所示。

| MySQL Type Name | Return value of GetColumnClassName | Returned as Java Class |
|---|---|---|
| BIT(1) (new in MySQL-5.0) | BIT | java.lang.Boolean |
| BIT( > 1) (new in MySQL-5.0) | BIT | byte[] |
| TINYINT | TINYINT | java.lang.Boolean if the configuration property tinyInt1isBit is set to true (the default) and the storage size is 1, or java.lang.Integer if not. |
| BOOL, BOOLEAN | TINYINT | See TINYINT, above as these are aliases for TINYINT(1), currently. |
| SMALLINT[(M)] [UNSIGNED] | SMALLINT [UNSIGNED] | java.lang.Integer (regardless of whether it is UNSIGNED or not) |
| MEDIUMINT[(M)] [UNSIGNED] | MEDIUMINT [UNSIGNED] | java.lang.Integer (regardless of whether it is UNSIGNED or not) |
| INT,INTEGER[(M)] [UNSIGNED] | INTEGER [UNSIGNED] | java.lang.Integer, if UNSIGNED java.lang.Long |
| BIGINT[(M)] [UNSIGNED] | BIGINT [UNSIGNED] | java.lang.Long, if UNSIGNED java.math.BigInteger |
| FLOAT[(M,D)] | FLOAT | java.lang.Float |
| DOUBLE[(M,B)] | DOUBLE | java.lang.Double |
| DECIMAL[(M[,D])] | DECIMAL | java.math.BigDecimal |
| DATE | DATE | java.sql.Date |
| DATETIME | DATETIME | java.sql.Timestamp |
| TIMESTAMP[(M)] | TIMESTAMP | java.sql.Timestamp |
| TIME | TIME | java.sql.Time |
| YEAR[(2\|4)] | YEAR | If yearIsDateType configuration property is set to false, then the returned object type is java.sql.Short. If set to true (the default), then the returned object is of type java.sql.Date with the date set to January 1st, at midnight. |
| CHAR(M) | CHAR | java.lang.String (unless the character set for the column is BINARY, then byte[] is returned. |
| VARCHAR(M) [BINARY] | VARCHAR | java.lang.String (unless the character set for the column is BINARY, then byte[] is returned. |
| BINARY(M) | BINARY | byte[] |
| VARBINARY(M) | VARBINARY | byte[] |
| TINYBLOB | TINYBLOB | byte[] |
| TINYTEXT | VARCHAR | java.lang.String |
| BLOB | BLOB | byte[] |
| TEXT | VARCHAR | java.lang.String |
| MEDIUMBLOB | MEDIUMBLOB | byte[] |
| MEDIUMTEXT | VARCHAR | java.lang.String |
| LONGBLOB | LONGBLOB | byte[] |
| LONGTEXT | VARCHAR | java.lang.String |
| ENUM('value1','value2',...) | CHAR | java.lang.String |
| SET('value1','value2',...) | CHAR | java.lang.String |

图 3-39

在调用 PreparedStatement 接口中的 setXXX(int,value)方法时，int 参数值的含义是向第几个问号传递参数，而在调用 ResultSet 接口中的 getXXX(int)方法时，int 参数值的含义是获取查询的虚拟二维表的第几列，不是数据库中物理数据表的第几列。

与 Statement 接口相比，PreparedStatement 接口有三点优势。

（1）防止 SQL 注入，提高软件系统的安全性。

（2）方便开发，使用问号占位的方式传入参数值，替代了使用拼接字符串的方式。

（3）使用预编译机制提高程序运行效率。

## 3.11 PreparedStatement的预编译特性

除了可以提高软件系统的安全性，PreparedStatement 接口还提供了预编译 SQL 语句的特性。当使用 Statement 对象执行 SQL 语句时，需要先将 SQL 语句发送给 DBMS，由 DBMS 进行编译，然后再执行，即每次执行 SQL 语句都要经过“编译”和“运行”两个过程，运行效率较低。而当对 PreparedStatement 接口传入 SQL 语句时，立即将 SQL 语句发送给 DBMS 进行编译，然后将参数传给已经编译后的 SQL 语句进行执行。当重复执行相同的 SQL 语句时，比如批量插入添加功能，不需要每次都对 SQL 语句进行编译，只需要传入不同的参数值即可，大大提高了程序运行效率。

当执行如下代码：

```
PreparedStatement ps = conn.prepareStatement("select * from userinfo where
username=? and password=?");
```

时，在 MySQL 驱动内部执行了服务器端 SQL 预编译操作，另外想要实现服务器端 SQL 预编译，必须添加两个参数：

```
useServerPrepStmts=true&cachePrepStmts=true
```

完整的 url 字符串如下：

```
String url =
"jdbc:mysql://localhost:3306/y2?useServerPrepStmts=true&cachePrepStmts=true";
```

注意：当执行代码“PreparedStatement ps = conn.prepareStatement(sql);”时，驱动会将 SQL 语句通过 Socket 技术传给 MySQL 数据库引擎，其会对 SQL 语句进行校验、编译和优化。如果 SQL 语句是错误的，则在调试 MySQL 驱动时可以发现服务器返回错误的

结果，但因为在驱动内部会将异常“吃掉”，所以在执行代码“conn. prepareStatement(sql)”时并没有出现异常，直到执行代码“ps.executeQuery()”时异常才出现。

你可以使用如下测试代码对 MySQL 驱动源代码中的预编译处理进行调试验证：

```
public static void main(String[] args) throws SQLException {
    String                                url                                =
"jdbc:mysql://localhost:3306/y2?useServerPrepStmts=true&cachePrepStmts=t
rue&serverTimezone=Asia/Shanghai";
    Properties prop = new Properties();
    prop.setProperty("user", "root");
    prop.setProperty("password", "123123");

    com.mysql.cj.jdbc.Driver driver = new com.mysql.cj.jdbc.Driver();
    Connection conn = driver.connect(url, prop);
    //当执行 prepareStatement()方法时，MySQL 驱动在内部就把 SQL 语句发送给 3306 端口
的 MySQL 服务，进行预编译处理了。
    PreparedStatement ps = conn.prepareStatement("select * from userinfo
where username like ?");
    ps.setString(1, "%中国%");
    ps.executeQuery();
    ps.close();
    conn.close();
}
```

## 3.12 PreparedStatement执行效率高

本实验在 Oracle 数据库环境中进行测试，如果使用 MySQL 数据库作为测试环境，则要在 URL 中令参数 useServerPrepStmts=true，开启服务器编译功能，还可以令参数 cachePrepStmts=true，开启 ps 命令缓存。

本实验尽量要在机械硬盘环境下测试，创建测试用的代码如下：

```
package test;

import java.sql.Connection;
import java.sql.PreparedStatement;
import java.sql.SQLException;
import java.util.Properties;
```

```
public class Test11 {
    public static void main(String[] args) throws SQLException {
        String url = "jdbc:mysql://localhost:3306/y2?useServerPrepStmts=
true&cachePrepStmts=true";
        Properties prop = new Properties();
        prop.setProperty("user", "root");
        prop.setProperty("password", "123123");

        com.mysql.cj.jdbc.Driver driver = new com.mysql.cj.jdbc.Driver();
        Connection conn = driver.connect(url, prop);
        PreparedStatement ps = null;
        long beginTime = System.currentTimeMillis();
        ps = conn.prepareStatement("select * from userinfo where id=?");
        for (int i = 0; i < 250000; i++) {
            ps.setObject(1, i + 1);
            ps.executeQuery();
        }
        long endTime = System.currentTimeMillis();
        ps.close();
        conn.close();
        System.out.println(endTime - beginTime);
    }
}
```

创建测试用的代码如下：

```
package test;

import java.sql.Connection;
import java.sql.SQLException;
import java.sql.Statement;
import java.util.Properties;
public class Test13 {
    public static void main(String[] args) throws SQLException {
        String url = "jdbc:mysql://localhost:3306/y2?cachePrepStmts=true";
        Properties prop = new Properties();
        prop.setProperty("user", "root");
        prop.setProperty("password", "123123");

        com.mysql.cj.jdbc.Driver driver = new com.mysql.cj.jdbc.Driver();
        Connection conn = driver.connect(url, prop);
        Statement stmt = null;
        long beginTime = System.currentTimeMillis();
```

```
        stmt = conn.createStatement();
        for (int i = 0; i < 250000; i++) {
            stmt.executeQuery("select * from userinfo where id=" + (i + 1));
        }
        long endTime = System.currentTimeMillis();
        stmt.close();
        conn.close();
        System.out.println(endTime - beginTime);
    }
}
```

根据运行时间可知，PreparedStatement 接口比 Statement 接口执行的效率高。

## 3.13　不同方式注册驱动的比较

下面介绍不同方式注册驱动流程之间的区别和分析方法。

### 3.13.1　方法 1

前面都是使用如下方法获得驱动及连接的：

```
Driver driver = new OracleDriver();
Connection connection = driver.connect(url, prop);
```

其实还有另外两种比较常用的方法也可以实现同样的效果。

### 3.13.2　方法 2

示例代码如下：

```
Class.forName("oracle.jdbc.OracleDriver");
Connection  connection  =  DriverManager.getConnection("jdbc:oracle:thin:
@localhost:1521:orcl", "y2", "123");
```

程序运行结果如图 3-40 所示。

```
sql=select * from userinfo where username=? and password=?
登录结果：true
```

图 3-40

### 3.13.3 方法 3

示例代码如下：

```
//Class.forName("oracle.jdbc.OracleDriver");
Connection connection = DriverManager.getConnection("jdbc:oracle:thin:
@localhost:1521:orcl", "y2", "123");
```

三种方法都可以获得连接并操作数据库。

### 3.13.4 读懂源代码的基本知识

在分析源代码之前，需要复习一下 static{}静态代码块相关的基本知识。

当实例化一个类时，该类的 static{}静态代码块会自动执行。

创建类 A 的代码如下：

```
package test;

public class A {
    static {
        System.out.println("自动注册驱动");
    }
}
```

创建类 A 的子类 B 的代码如下：

```
package test;

public class B extends A {
}
```

运行类代码如下：

```
package test;

import java.sql.SQLException;

public class Test1 {
    public static void main(String[] args) throws SQLException {
        new B();
    }
```

```
}
```

运行程序后，控制台输出结果如图 3-41 所示。

而当使用 Class.forName()方法时，会触发 static{}静态代码块，运行类代码如下：

```
package test;

public class Test1 {
    public static void main(String[] args) throws ClassNotFoundException {
        Class.forName("test.B");
    }
}
```

运行程序后，控制台输出结果同样如图 3-41 所示。

因为在系统中注册驱动的过程就是在 static{}静态代码块中完成的，所以只有知道这个知识点，在调试源代码时才会清楚执行流程。

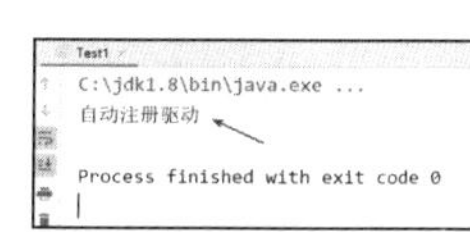

图 3-41

## 3.13.5　针对 MySQL 驱动源码分析

相比方法 3，方法 1 和方法 2 的源代码调试比较简单，故我们先来比较方法 1 和方法 2 的区别。

方法 1 的源代码如下：

```
java.sql.Driver driver = new com.mysql.jdbc.Driver();
```

当执行上面的程序代码时，会在 MySQL 驱动源代码中执行 com.mysql.jdbc.Driver 类的父类 com.mysql.cj.jdbc.Driver 中的 static{}静态代码块。

```
static {
    try {
        java.sql.DriverManager.registerDriver(new Driver());
    } catch (SQLException E) {
        throw new RuntimeException("Can't register driver!");
    }
}
```

可以发现，在源代码中驱动类 com.mysql.cj.jdbc.Driver 的 static{}静态代码块通过 DriverManager 类的 registerDriver()方法将驱动进行注册。registerDriver()方法具有重载

特性，registerDriver(java.sql.Driver driver)方法的源代码如下：

```
public static synchronized void registerDriver(java.sql.Driver driver)
    throws SQLException {
    registerDriver(driver, null);
}
```

继续执行两个参数的 registerDriver(java.sql.Driver driver, DriverAction da)方法，源代码如下：

```
public static synchronized void registerDriver(java.sql.Driver driver,
        DriverAction da)
    throws SQLException {
    /* Register the driver if it has not already been added to our list */
    if(driver != null) {
        registeredDrivers.addIfAbsent(new DriverInfo(driver, da));
    } else {
        // This is for compatibility with the original DriverManager
        throw new NullPointerException();
    }
    println("registerDriver: " + driver);
}
```

在源代码 MySQL 驱动类 com.mysql.cj.jdbc.Driver 向 DriverManager 类的 private final static CopyOnWriteArrayList<DriverInfo> registeredDrivers = new CopyOnWriteArrayList<>(); 对象中添加 Driver 驱动类，以完成驱动注册。

以上是方法 1 驱动注册源码执行流程分析，下面来看一下方法 2 驱动注册执行流程分析。

方法 2 的源代码如下：

```
Class.forName("com.mysql.cj.jdbc.Driver");
Connection connection = DriverManager.getConnection(url, username,
password);
```

当看到上面第一行代码时，你可能会立即执行 com.mysql.cj.jdbc.Driver()类 static{}静态代码块中的代码。

```
static {
    try {
        java.sql.DriverManager.registerDriver(new Driver());
    } catch (SQLException E) {
```

```
        throw new RuntimeException("Can't register driver!");
    }
}
```

后面的执行流程与方法 1 的一样。

两种方法代码的流程对比如图 3-42 所示。

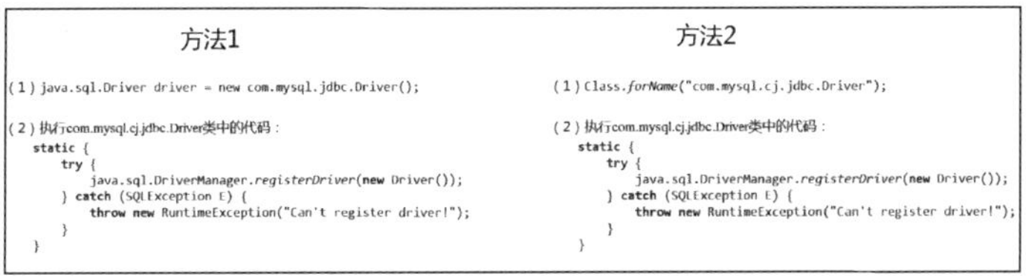

方法1

（1）java.sql.Driver driver = new com.mysql.jdbc.Driver();

（2）执行com.mysql.cj.jdbc.Driver类中的代码：

```
static {
    try {
        java.sql.DriverManager.registerDriver(new Driver());
    } catch (SQLException E) {
        throw new RuntimeException("Can't register driver!");
    }
}
```

方法2

（1）Class.forName("com.mysql.cj.jdbc.Driver");

（2）执行com.mysql.cj.jdbc.Driver类中的代码：

```
static {
    try {
        java.sql.DriverManager.registerDriver(new Driver());
    } catch (SQLException E) {
        throw new RuntimeException("Can't register driver!");
    }
}
```

图 3-42

通过流程对比可知，两者的执行流程差异很小，只是在第一步不一样，后面的流程都是相同的。

## 3.13.6　针对 Oracle 驱动源码分析

方法 1 的代码如下：

```
java.sql.Driver driver = new oracle.jdbc.OracleDriver();
Connection connection = driver.connect(url, prop);
```

当实例化 new oracle.jdbc.OracleDriver()类时，会执行 oracle.jdbc.OracleDriver 类的父类 oracle.jdbc.driver.OracleDriver 的静态代码块中的核心代码：

```
static
{
    try
    {
      if (defaultDriver == null)
      {
        defaultDriver = new oracle.jdbc.OracleDriver();
        DriverManager.registerDriver(defaultDriver);
    }
```

方法 2 的代码如下：

```
Class.forName("oracle.jdbc.OracleDriver");
Connection  connection  =  DriverManager.getConnection("jdbc:oracle:thin:
```

```
@localhost:1521:orcl", "y2", "123");
```

当执行 Class.forName("oracle.jdbc.OracleDriver")代码时，会执行 oracle.jdbc.OracleDriver 类的父类 oracle.jdbc.driver.OracleDriver 的静态代码块中的代码，流程和方法 1 的一样。

以上是针对 Oracle 注册驱动的流程分析，和针对 MySQL 注册驱动的流程是一样的。

### 3.13.7 方法 3 的原理

当使用 JDBC4 标准的驱动时，可以不执行如下代码也能实现驱动的注册。

```
Class.forName("oracle.jdbc.OracleDriver");
```

原因是在 java.sql.DriverManager 类的 static{}静态代码块中执行了如下代码：

```
static {
    loadInitialDrivers();
    println("JDBC DriverManager initialized");
}
```

loadInitialDrivers()方法中的部分核心源代码如下：

```
private static void loadInitialDrivers() {
    AccessController.doPrivileged(new PrivilegedAction<Void>() {
        public Void run() {
            ServiceLoader<Driver> loadedDrivers = ServiceLoader.load(Driver.
class);
            Iterator<Driver> driversIterator = loadedDrivers.iterator();
            try{
                while(driversIterator.hasNext()) {
                    driversIterator.next();
                }
            } catch(Throwable t) {
            }
            return null;
        }
    });
```

执行代码：

```
ServiceLoader<Driver> loadedDrivers = ServiceLoader.load(Driver.class);
```

会在 classpath 类路径的所有.jar 文件的 META-INF 文件夹中寻找 services 文件夹。如果存在 services 文件夹，则寻找其内部的文件，文件名称就是接口的全路径名称，比如 java.sql.Driver 文件的内容 oracle.jdbc.OracleDriver 就是接口的实现类全路径名。

Oracle 驱动中的 META-INF 文件夹结构，如图 3-43 所示。

MySQL 驱动中的 META-INF 文件夹结构，如图 3-44 所示。

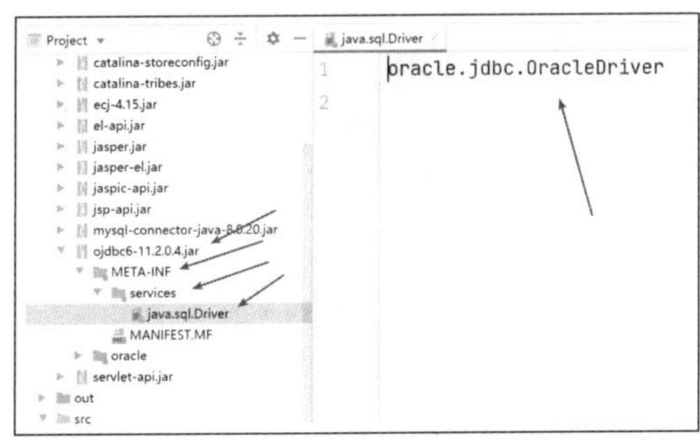


图 3-43

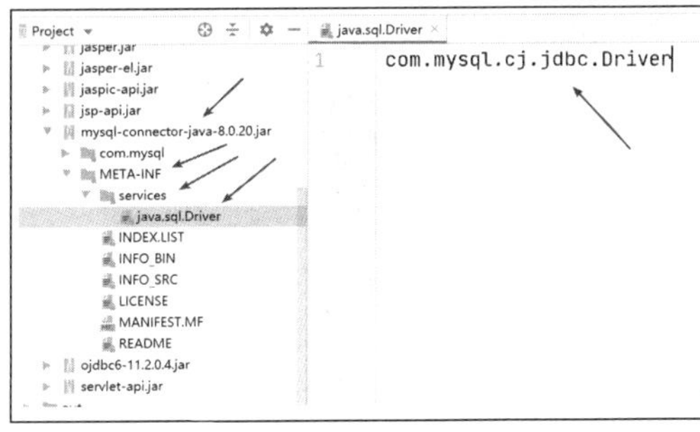


图 3-44

变量 ServiceLoader<Driver> loadedDrivers 存储的就是从 services 文件夹中获得的所有 java.sql.Driver 接口的实现类，也就是类路径中所有数据库的驱动类信息。变量 ServiceLoader<Driver> loadedDrivers 中存储的信息是借助 ServiceLoader 类从.jar 文件中获得的驱动类。

关键步骤“加载驱动”的过程就隐藏在 driversIterator.next()方法中，next()方法在 ServiceLoader.class 类的 public Iterator<S> iterator()方法的 public S next()方法中，源代码如下：

```
public S next() {
    if (knownProviders.hasNext())
        return knownProviders.next().getValue();
    return lookupIterator.next();
}
```

在执行 lookupIterator.next()方法后源代码如下：

```
public S next() {
    if (acc == null) {
        return nextService();
    } else {
        PrivilegedAction<S> action = new PrivilegedAction<S>() {
```

```
            public S run() { return nextService(); }
        };
        return AccessController.doPrivileged(action, acc);
    }
}
```

再执行 return nextService()方法，部分核心源代码如下：

```
private S nextService() {
    if (!hasNextService())
        throw new NoSuchElementException();
    String cn = nextName;
    nextName = null;
    Class<?> c = null;
    try {
        c = Class.forName(cn, false, loader);
    } catch (ClassNotFoundException x) {
        fail(service,
             "Provider " + cn + " not found");
    }
```

其中，“c = Class.forName(cn, false, loader);”中的变量 cn 就是驱动类的全路径名 oracle.jdbc.OracleDriver。当执行代码“c = Class.forName(cn, false, loader)”时，会先执行 oracle.jdbc.OracleDriver 类中的静态代码块，然后将当前的驱动类注册到 DriverManager 中。

至此，不需要执行 Class.forName(" ")方法也能实现获得 Connection 连接的原理介绍完毕。

### 3.13.8 三种方法的使用场景

这三种方法使用的优先级是什么呢？我们先来看看方法 1 的代码：

```
java.sql.Driver driver = new com.mysql.jdbc.Driver();
Connection connection = driver.connect(url, prop);
```

根据方法 1 的代码可知，由于其直接利用 new com.mysql.jdbc.Driver()方法实例化 MySQL 驱动中的类，因此在切换数据库时必须要更改目标数据库驱动类名，比如：

```
java.sql.Driver driver = new oracle.jdbc.OracleDriver();
```

这会造成.java 源代码的改动，不利于项目代码的后期维护。那么，是否可以实现在切换数据库时不改动.java 源代码呢？方法 2 可以解决这一问题。方法 2 的代码如下：

```
    Class.forName("com.mysql.cj.jdbc.Driver");
    Connection connection = DriverManager.getConnection(url, username,
password);
```

只需要将驱动类名"com.mysql.cj.jdbc.Driver"放入.properties 属性文件中，而在切换数据库时更改.properties 属性文件中的驱动类名即可。代码"Class.forName(driverNameVar)"中的 driverNameVar 参数是变量，读取.properties 属性文件中的内容并对 driverNameVar 变量赋值，也就实现了在切换数据库时.java 源代码不改动，便于代码的后期维护。

因此，方法 1 和方法 2 相比，优先使用方法 2，但方法 3 是最值得推荐使用的，因为其最为简洁，代码如下：

```
//Class.forName("com.mysql.cj.jdbc.Driver");
Connection connection = DriverManager.getConnection(url, username, password);
```

使用方法 3 的前提是 JDBC 驱动必须实现了 JDBC4 标准。

## 3.14　使用PreparedStatement实现记录的增删改

前面使用 PreparedStatement 接口的 executeQuery()方法来查询数据，而想要实现记录的增删改，则需要执行 executeUpdate()方法。

增加记录的示例代码如下：

```
package test4;

import java.sql.Connection;
import java.sql.DriverManager;
import java.sql.PreparedStatement;
import java.sql.SQLException;

public class Test1 {
    public static void main(String[] args) throws ClassNotFoundException,
SQLException, NoSuchFieldException,
            SecurityException,                    IllegalArgumentException,
IllegalAccessException {
        String usernameValue = "a";
        String passwordValue = "aa";

        String driverName = "oracle.jdbc.OracleDriver";
```

```
        String url = "jdbc:oracle:thin:@//localhost:1521/orcl";
        String usernameDB = "y2";
        String passwordDB = "123";

        String   sql   =   "insert   into   userinfo(id,username,password)
values(idauto.nextval,?,?)";

        Class.forName(driverName);
        Connection connection = DriverManager.getConnection(url, usernameDB,
passwordDB);
        PreparedStatement ps = connection.prepareStatement(sql);
        ps.setString(1, usernameValue);
        ps.setString(2, passwordValue);
        int resultRowCount = ps.executeUpdate();
        ps.close();
        connection.close();
        System.out.println("影响的行数：" + resultRowCount);
    }
}
```

修改记录的示例代码如下：

```
package test4;

import java.sql.Connection;
import java.sql.DriverManager;
import java.sql.PreparedStatement;
import java.sql.SQLException;

public class Test2 {
    public static void main(String[] args) throws ClassNotFoundException,
SQLException, NoSuchFieldException,
            SecurityException,                   IllegalArgumentException,
IllegalAccessException {
        String usernameValue = "a 新值";
        String passwordValue = "aa 新值";
        int idValue = 123;

        String driverName = "oracle.jdbc.OracleDriver";
        String url = "jdbc:oracle:thin:@//localhost:1521/orcl";
        String usernameDB = "y2";
        String passwordDB = "123";
```

```
        String sql = "update userinfo set username=?,password=? where id=?";

        Class.forName(driverName);
       Connection connection = DriverManager.getConnection(url, usernameDB,
passwordDB);
        PreparedStatement ps = connection.prepareStatement(sql);
        ps.setString(1, usernameValue);
        ps.setString(2, passwordValue);
        ps.setInt(3, idValue);
        int resultRowCount = ps.executeUpdate();
        ps.close();
        connection.close();
        System.out.println("影响的行数：" + resultRowCount);
    }
}
```

删除记录的示例代码如下：

```
package test4;

import java.sql.Connection;
import java.sql.DriverManager;
import java.sql.PreparedStatement;
import java.sql.SQLException;

public class Test3 {
    public static void main(String[] args) throws ClassNotFoundException,
SQLException, NoSuchFieldException,
            SecurityException, IllegalArgumentException, IllegalAccessException {
        int idValue = 123;

        String driverName = "oracle.jdbc.OracleDriver";
        String url = "jdbc:oracle:thin:@//localhost:1521/orcl";
        String usernameDB = "y2";
        String passwordDB = "123";

        String sql = "delete from userinfo where id=?";

        Class.forName(driverName);
       Connection connection = DriverManager.getConnection(url, usernameDB,
passwordDB);
        PreparedStatement ps = connection.prepareStatement(sql);
        ps.setInt(1, idValue);
```

```
        int resultRowCount = ps.executeUpdate();
        ps.close();
        connection.close();
        System.out.println("影响的行数: " + resultRowCount);
    }
}
```

本实验说明使用 PreparedStatement 接口的 executeQuery()方法只能做查询 select 的操作，而使用 executeUpdate()方法可以实现添加、删除和修改记录的操作。

## 3.15 建议释放资源放入finally块

好的释放资源的代码模式是将 close()方法放在 finally 块中，示例代码如下：

```
package testA;

import java.sql.Connection;
import java.sql.Driver;
import java.sql.DriverManager;
import java.sql.PreparedStatement;
import java.sql.ResultSet;
import java.sql.SQLException;

import oracle.jdbc.driver.OracleDriver;

public class Test16 {
    public static void main(String[] args) {
        Connection connection = null;
        PreparedStatement ps = null;
        ResultSet rs = null;
        try {
            Driver driver = new OracleDriver();
            connection = DriverManager.getConnection("jdbc:oracle:thin:
@localhost:1521:orcl", "y2", "123");

            String sql = "select * from userinfo order by id asc";
            ps = connection.prepareStatement(sql);
            rs = ps.executeQuery();
            while (rs.next()) {
                long id = rs.getLong("id");
                String username = rs.getString("username");
                String password = rs.getString("password");
```

```
                System.out.println(id + " " + username + " " + password);
            }
        } catch (SQLException e) {
            e.printStackTrace();
        } finally {
            try {
                if (rs != null) {
                    rs.close();
                }
                if (ps != null) {
                    ps.close();
                }
                if (connection != null) {
                    connection.close();
                }
                System.out.println("资源全部被释放");
            } catch (SQLException e) {
                e.printStackTrace();
            }
        }
    }
}
```

如果释放的资源不放在 finally 块中，当程序出现异常时，则不能正确释放资源，示例代码如下：

```
package testA;

import java.sql.Connection;
import java.sql.Driver;
import java.sql.DriverManager;
import java.sql.PreparedStatement;
import java.sql.ResultSet;
import java.sql.SQLException;

import oracle.jdbc.driver.OracleDriver;

public class Test17 {
    public static void main(String[] args) {
        Connection connection = null;
        PreparedStatement ps = null;
```

```
        ResultSet rs = null;
        try {
            Driver driver = new OracleDriver();
            DriverManager.registerDriver(driver);
            connection  =  DriverManager.getConnection("jdbc:oracle:thin:
@localhost:1521:orcl", "y2", "123");

            String sql = "select * from 没有这张表啊!!!!!!!!!!!! order by id
asc";
            ps = connection.prepareStatement(sql);
            rs = ps.executeQuery();
            while (rs.next()) {
                long id = rs.getLong("id");
                String username = rs.getString("username");
                String password = rs.getString("password");

                System.out.println(id + " " + username + " " + password);
            }
            rs.close();
            ps.close();
            connection.close();
            System.out.println("资源全部被释放");
        } catch (SQLException e) {
            e.printStackTrace();
        }
    }
}
```

程序运行结果如图 3-45 所示。

```
java.sql.SQLSyntaxErrorException: ORA-00933: SQL 命令未正确结束

	at oracle.jdbc.driver.T4CTTIoer.processError(T4CTTIoer.java:440)
	at oracle.jdbc.driver.T4CTTIoer.processError(T4CTTIoer.java:396)
	at oracle.jdbc.driver.T4C8Oall.processError(T4C8Oall.java:837)
	at oracle.jdbc.driver.T4CTTIfun.receive(T4CTTIfun.java:445)
	at oracle.jdbc.driver.T4CTTIfun.doRPC(T4CTTIfun.java:191)
	at oracle.jdbc.driver.T4C8Oall.doOALL(T4C8Oall.java:523)
	at oracle.jdbc.driver.T4CPreparedStatement.doOall8(T4CPreparedStatement.java:207)
	at oracle.jdbc.driver.T4CPreparedStatement.executeForDescribe(T4CPreparedStatement.java:863)
	at oracle.jdbc.driver.OracleStatement.executeMaybeDescribe(OracleStatement.java:1153)
	at oracle.jdbc.driver.OracleStatement.doExecuteWithTimeout(OracleStatement.java:1275)
	at oracle.jdbc.driver.OraclePreparedStatement.executeInternal(OraclePreparedStatement.java:3576)
	at oracle.jdbc.driver.OraclePreparedStatement.executeQuery(OraclePreparedStatement.java:3620)
	at oracle.jdbc.driver.OraclePreparedStatementWrapper.executeQuery(OraclePreparedStatementWrapper.java:1491)
	at testA.Test17.main(Test17.java:24)
```

图 3-45

控制台并未输出“资源全部被释放”的信息，说明实验代码由于出现异常并未执行后面的 close()方法来释放资源，Connection 对象没有被释放。如何证明 Connection 对象没有被释放呢？测试代码如下：

```
package com.ghy.www.test;

import java.sql.Connection;
import java.sql.DriverManager;
import java.sql.SQLException;

public class Test2 {
   public static void main(String[] args) throws InterruptedException,
SQLException {
      Connection[] connections = new Connection[50];
      try {
         String driverName = "oracle.jdbc.OracleDriver";
         String url = "jdbc:oracle:thin:@localhost:1521/orcl";
         String usernameDB = "y2";
         String passwordDB = "123123";

         Class.forName(driverName);
         for (int i = 0; i < 50; i++) {
            connections[i] = DriverManager.getConnection(url, usernameDB,
passwordDB);
            System.out.println("i=" + (i + 1) + "次");
            try {
               connections[i].prepareStatement("select * from 没有这个表
").executeUpdate();
            } catch (SQLException throwables) {
               System.out.println("error     message     :     "     +
throwables.getMessage());
            }
         }
      } catch (ClassNotFoundException e) {
         e.printStackTrace();
      } catch (SQLException e) {
         e.printStackTrace();
      } finally {
         for (int i = 0; i < connections.length; i++) {
            // connections[i].close();
         }
      }
      Thread.sleep(3000000);
      for (int i = 0; i < connections.length; i++) {
         System.out.println(connections[i]);
```

```
        }
    }
}
```

其中：

```
for (int i = 0; i < connections.length; i++) {
      System.out.println(connections[i]);
   }
```

这一代码的主要作用是设置每个 Connection 对象为强引用，否则当内存不够时，JVM 会将闲置无用的 Connection 对象进行 GC 垃圾回收释放，导致 Connection 连接数不准确，运行出来的结果是连接数不是 50，有可能是 12 或 39 等随机数字。

以上测试代码并没有执行 connection.close()方法，程序运行后，连接数一直呈自增加的状态，并且在 toad 工具中查看到非常多的 session 会话连接，说明没有执行 Connection 对象的 close()方法，导致连接没有被释放，如图 3-46 所示。

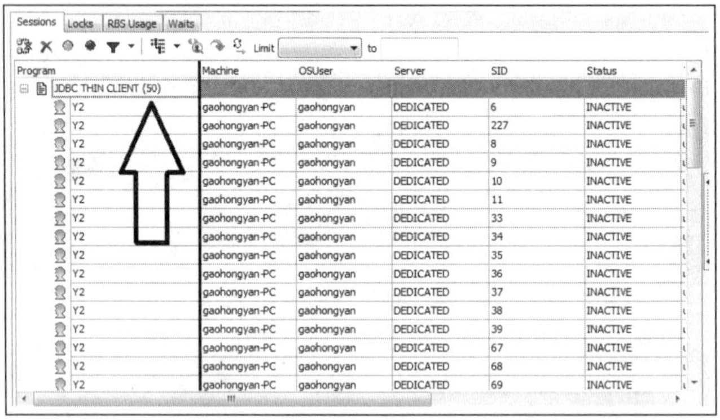

图 3-46

使用“Database”→“Monitor”→“Session Browser”菜单可以查看 session 会话列表。

重新启动 Oracle 服务。

如果在 finally 块中执行 conn.close()方法，连接被释放，则验证执行 close()方法释放连接的测试代码如下：

```
package com.ghy.www.test;

import java.sql.Connection;
```

```
import java.sql.DriverManager;
import java.sql.SQLException;

public class Test2 {
    public static void main(String[] args) throws InterruptedException,
SQLException {
        Connection[] connections = new Connection[50];
        try {
            String driverName = "oracle.jdbc.OracleDriver";
            String url = "jdbc:oracle:thin:@localhost:1521/orcl";
            String usernameDB = "y2";
            String passwordDB = "123123";

            Class.forName(driverName);
            for (int i = 0; i < 50; i++) {
                connections[i] = DriverManager.getConnection(url, usernameDB,
passwordDB);
                System.out.println("i=" + (i + 1) + "次");
                try {
                    connections[i].prepareStatement("select * from 没有这个表
").executeUpdate();
                } catch (SQLException throwables) {
                    System.out.println("error message : " +
throwables.getMessage());
                }
            }
        } catch (ClassNotFoundException e) {
            e.printStackTrace();
        } catch (SQLException e) {
            e.printStackTrace();
        } finally {
            for (int i = 0; i < connections.length; i++) {
                connections[i].close();
            }
        }
        Thread.sleep(3000000);
        for (int i = 0; i < connections.length; i++) {
            System.out.println(connections[i]);
        }
    }
}
```

程序运行后，并没有在“Session Browser”界面中查看到未释放的 session 会话连接，说明在 finally 块中执行 connections[i].close()方法会释放连接。

## 3.16 实现Connection工厂类

因为每一次获得 Connection 都要重新创建一个 Connection 对象，所以代码中出现了大量重复，我们可以通过创建一个 Connection 工厂类封装创建 Connection 对象的过程来解决这一问题。Connection 工厂类 ConnectionFactory.java 的代码如下：

```
package testA;

import java.sql.Connection;
import java.sql.DriverManager;
import java.sql.ResultSet;
import java.sql.SQLException;
import java.sql.Statement;

public class ConnectionFactory {
    public  static  Connection  getConnection()  throws  SQLException,
ClassNotFoundException {
        String url = "jdbc:oracle:thin:@localhost:1521:orcl";
        String username = "y2";
        String password = "123";
        Class.forName("oracle.jdbc.OracleDriver");
        Connection connection = DriverManager.getConnection(url, username,
password);
        return connection;
    }

    public static void close(Connection connection) {
        try {
            if (connection != null) {
                connection.close();
                System.out.println("成功释放了 Connection 资源");
            }

        } catch (SQLException e) {
            e.printStackTrace();
        }
```

```
    }

    public static void close(Connection connection, Statement statement) {
        try {
            if (statement != null) {
                statement.close();
                System.out.println("成功释放了 statement 资源");
            }
            if (connection != null) {
                connection.close();
                System.out.println("成功释放了 connection 资源");
            }
        } catch (SQLException e) {
            e.printStackTrace();
        }
    }

    public static void close(Connection connection, Statement statement,
ResultSet resultSet) {
        try {
            if (resultSet != null) {
                resultSet.close();
                System.out.println("成功释放了 resultSet 资源");
            }
            if (statement != null) {
                statement.close();
                System.out.println("成功释放了 statement 资源");
            }
            if (connection != null) {
                connection.close();
                System.out.println("成功释放了 connection 资源");
            }
        } catch (SQLException e) {
            e.printStackTrace();
        }
    }
}
```

运行类的代码如下：

```
package testA;
```

```
import java.sql.SQLException;

public class Test19 {

    public static void main(String[] args) {
        try {
            System.out.println(ConnectionFactory.getConnection());
            System.out.println(ConnectionFactory.getConnection());
            System.out.println(ConnectionFactory.getConnection());
            System.out.println(ConnectionFactory.getConnection());
            System.out.println(ConnectionFactory.getConnection());
        } catch (ClassNotFoundException e) {
            e.printStackTrace();
        } catch (SQLException e) {
            e.printStackTrace();
        }
    }
}
```

程序运行结果如图 3-47 所示。

```
oracle.jdbc.driver.T4CConnection@4629104a
oracle.jdbc.driver.T4CConnection@4fccd51b
oracle.jdbc.driver.T4CConnection@44e81672
oracle.jdbc.driver.T4CConnection@60215eee
oracle.jdbc.driver.T4CConnection@4ca8195f
```

图 3-47

有了 Connection 连接工厂类就可以实现一个查询的示例，测试代码如下：

```
package testA;

import java.sql.Connection;
import java.sql.PreparedStatement;
import java.sql.ResultSet;
import java.sql.SQLException;

public class Test20 {

    public static void main(String[] args) {
        Connection connection = null;
```

```
        PreparedStatement ps = null;
        ResultSet rs = null;
        try {
            connection = ConnectionFactory.getConnection();
            String sql = "select * from userinfo order by id asc";
            ps = connection.prepareStatement(sql);
            rs = ps.executeQuery();
            while (rs.next()) {
                long id = rs.getLong("id");
                String username = rs.getString("username");
                String password = rs.getString("password");

                System.out.println(id + " " + username + " " + password);
            }
        } catch (ClassNotFoundException e) {
            e.printStackTrace();
        } catch (SQLException e) {
            e.printStackTrace();
        } finally {
            ConnectionFactory.close(connection, ps, rs);
        }
    }
}
```

程序运行结果如图 3-48 所示。

```
247 全部账号都一样的 账号AA
248 全部账号都一样的 账号AA
成功释放了ResultSet资源
成功释放了Statement资源
成功释放了Connection资源
```

图 3-48

## 3.17　多条件查询

查询条件存在多种情况，如果我们想只查询查询条件不为空的情况，那么程序代码应该如何实现呢？示例代码如下：

```
package test6;

import java.sql.Connection;
import java.sql.PreparedStatement;
import java.sql.ResultSet;
import java.sql.SQLException;
import java.util.ArrayList;
import java.util.List;
```

```
public class Test5 {
    public static void main(String[] args) {
        Connection connection = null;
        PreparedStatement ps = null;
        ResultSet rs = null;

        try {
            List queryValue = new ArrayList();

            long idValue = 105;
            String usernameValue = "a";
            String passwordValue = "a";

            String whereSQL = "";

            if (idValue != -1) {
                whereSQL = whereSQL + " and id = ?";
                queryValue.add(idValue);
            }

            if (usernameValue != null && !"".equals(usernameValue)) {
                whereSQL = whereSQL + " and username like ?";
                queryValue.add("%" + usernameValue + "%");
            }

            if (passwordValue != null && !"".equals(passwordValue)) {
                whereSQL = whereSQL + " and password like ?";
                queryValue.add("%" + passwordValue + "%");
            }

            String sql = "select * from userinfo where 1=1 ";
            sql = sql + " " + whereSQL + " order by id asc";
            System.out.println(sql);

            connection = ConnectionFactory.getConnection();
            ps = connection.prepareStatement(sql);
            for (int i = 0; i < queryValue.size(); i++) {
                ps.setObject(i + 1, queryValue.get(i));
            }
            rs = ps.executeQuery();
            while (rs.next()) {
```

```
                System.out.println(rs.getLong("id")      +      "      "      +
rs.getString("username") + " " + rs.getString("password"));
            }
        } catch (ClassNotFoundException e) {
            e.printStackTrace();
        } catch (SQLException e) {
            e.printStackTrace();
        } finally {
            try {
                ConnectionFactory.close(connection, ps, rs);
            } catch (SQLException e) {
                e.printStackTrace();
            }
        }
    }
}
```

## 3.18 DTO、ENTITY和DAO介绍

### 1. DTO

DTO（Data Transfer Object，数据传输对象）主要的作用是在不同类中进行数据的交换，将大量数据放入一个 DTO 中便于整体传输。比如，在 Service 类中可以先将数据封装进 DTO，再把 DTO 交给其他类进行处理，其他类还可以继续将 DTO 交给另外的类进行处理。DTO 就相当于接力赛跑中的接力棒，按照程序调用的顺序将数据送到正确的目的地，使用 DTO 的场景如图 3-49 所示。

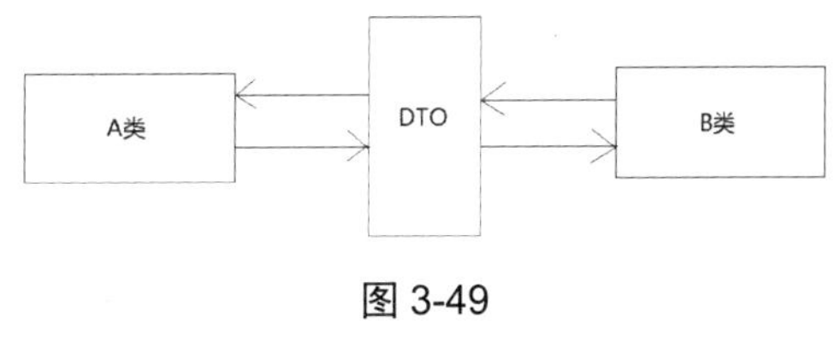


图 3-49

DTO 示例代码如下：

```
package dto;

public class Userinfo {

    private long id;
    private String username;
    private String password;
```

```
    public Userinfo() {
    }

    public Userinfo(long id, String username, String password) {
        super();
        this.id = id;
        this.username = username;
        this.password = password;
    }

    //set()和get()方法省略
}
```

### 2. ENTITY

ENTITY 实体类与 DTO 在结构和作用方面非常相似，但又有着本质上的不同，实体类一般是和数据表做映射，实体类中的属性对应表中的列，比如有结构如图 3-50 所示的数据表 Userinfo。

| ID | USERNAME | PASSWORD |
|---|---|---|
| 2 | 美国 | 美国人 |
| 3 | c | cc |
| 4 | da | dd |
| 5 | eaa | ee |
| 1145353 | 中国 | 中国人 |

图 3-50

那么，为了存储数据表 Userinfo 每行的数据就要创建一个对应的实体类，实体类的名称尽量与表名一样，但首字母大写；实体类中的属性名尽量与表中的列名一样。实体类 Userinfo.java 的代码如下：

```
package entity;

public class Userinfo {

    private long id;
    private String username;
    private String password;

    public Userinfo() {
    }

    public Userinfo(long id, String username, String password) {
        super();
        this.id = id;
        this.username = username;
        this.password = password;
```

```
    }

    //set()和get()方法省略
}
```

表与实体类的映射关系如图 3-51 所示。

在实体类 Userinfo.java 中，先创建与表中列名相同的属性名称，然后再创建针对这些属性的 set()和 get()方法，这样，Userinfo.java 类就可以封装表中的一条记录，将从数据表中查询的记录放入 Userinfo.java 的 ENTITY 实体类中，这是查询的阶段。另外，还可以将插入的数据先放入 ENTITY 实体类中，再从 ENTITY 实体类中取出数据并插入数据表中，这些场景都是 ENTITY 实体类的用武之地。

### 3. DAO

DAO（Data Access Object，数据访问对象）主要的作用就是封装对数据库操作的 JDBC 代码，把增删改查相关的 JDBC 代码放入 DAO 层中，对外提供操作数据库的 CURD 方法，封装的优点是便于后期代码的维护与排错。DAO 处理的数据来自实体类，如图 3-52 所示。

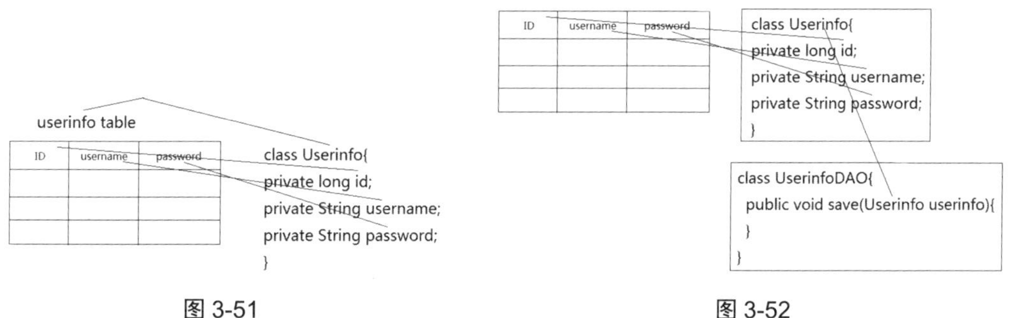


图 3-51　　　　图 3-52

项目中使用 DAO，ENTITY，DTO 的场景，如图 3-53 所示。

（1）一个 ENTITY 实体类需要有一个 DAO 类与之对应。

（2）DAO 需要使用 ENTITY 实体类与数据库进行交互。

（3）DTO 不是与数据表进行映射，而是在不同的类之间以面向对象的方式传递数据。虽然 DTO 和 ENTITY 都有一些属性，属性中也存储一些值，但 DTO 与 ENTITY 最本质的区别是不参与数据库的操作。

（4）DTO 适用的场景是 JSP 和 Servlet、Servlet 和 Service、Service 和 Service 进行传递数据，ENTITY 适用的场景是 ENTITY 和 DAO 传递数据。

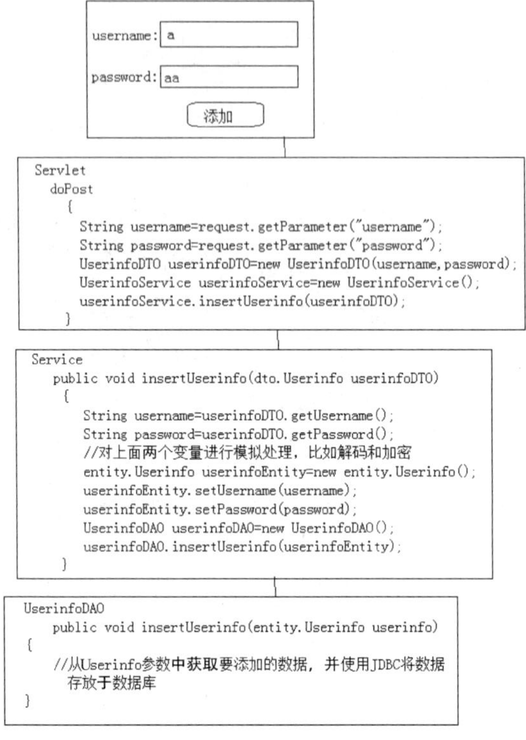


图 3-53

## 3.19 将JDBC操作封装为DAO数据访问对象

下面结合 ENTITY 实体类实现一个真实的 DAO 类，代码中使用 StringBuffer 组装 SQL 语句，并且支持针对 Connection，Statement 和 ResultSet 释放资源的示例。

创建映射数据表的实体类 Userinfo.java 代码如下：

```
public class Userinfo {

    private long id;
    private String username;
    private String password;

    public Userinfo() {
        super();
    }
```

```
    //省略 set()和 get()方法

}
```

数据访问层 DAO 的 UserinfoDAO.java 类代码如下：

```
public class UserinfoDAO {
    public void insert(Userinfo userinfo) {
        Connection conn = null;
        PreparedStatement ps = null;
        try {
            StringBuffer buffer = new StringBuffer();
            buffer.append(" insert into userinfo");
            buffer.append(" (id,username,password)");
            buffer.append(" values");
            buffer.append(" (idauto.nextval,?,?)");

            conn = ConnectionFactory.getConnection();
            ps = conn.prepareStatement(buffer.toString());
            ps.setString(1, userinfo.getUsername());
            ps.setString(2, userinfo.getPassword());
            ps.executeUpdate();
        } catch (SQLException e) {
            e.printStackTrace();
        } finally {
            try {
                ConnectionFactory.close(conn, ps);
            } catch (SQLException e) {
                e.printStackTrace();
            }
        }
    }

    public void delete(long userId) {
        Connection conn = null;
        PreparedStatement ps = null;
        try {
            StringBuffer buffer = new StringBuffer();
            buffer.append(" delete from userinfo");
            buffer.append(" where id=?");
```

```
            conn = ConnectionFactory.getConnection();
            ps = conn.prepareStatement(buffer.toString());
            ps.setLong(1, userId);
            ps.executeUpdate();
        } catch (SQLException e) {
            e.printStackTrace();
        } finally {
            try {
                ConnectionFactory.close(conn, ps);
            } catch (SQLException e) {
                e.printStackTrace();
            }
        }
    }

    public void update(Userinfo userinfo) {
        Connection conn = null;
        PreparedStatement ps = null;
        try {
            StringBuffer buffer = new StringBuffer();
            buffer.append(" update userinfo set");
            buffer.append(" username=?,");
            buffer.append(" password=?");
            buffer.append(" where id=?");

            conn = ConnectionFactory.getConnection();
            ps = conn.prepareStatement(buffer.toString());
            ps.setString(1, userinfo.getUsername());
            ps.setString(2, userinfo.getPassword());
            ps.setLong(3, userinfo.getId());
            ps.executeUpdate();
        } catch (SQLException e) {
            e.printStackTrace();
        } finally {
            try {
                ConnectionFactory.close(conn, ps);
            } catch (SQLException e) {
                e.printStackTrace();
            }
        }
```

```
    }

    public Userinfo selectById(long userId) {
        Userinfo userinfo = null;
        Connection conn = null;
        PreparedStatement ps = null;
        ResultSet rs = null;
        try {
            StringBuffer buffer = new StringBuffer();
            buffer.append(" select * from userinfo");
            buffer.append(" where id=?");

            conn = ConnectionFactory.getConnection();
            ps = conn.prepareStatement(buffer.toString());
            ps.setLong(1, userId);
            rs = ps.executeQuery();
            while (rs.next()) {
                long idDB = rs.getLong("id");
                String usernameDB = rs.getString("username");
                String passwordDB = rs.getString("password");

                userinfo = new Userinfo();
                userinfo.setId(idDB);
                userinfo.setUsername(usernameDB);
                userinfo.setPassword(passwordDB);
            }
        } catch (SQLException e) {
            e.printStackTrace();
        } finally {
            try {
                ConnectionFactory.close(conn, ps, rs);
            } catch (SQLException e) {
                e.printStackTrace();
            }
        }
        return userinfo;
    }

    public List<Userinfo> selectAll() {
        List<Userinfo> listUserinfo = new ArrayList<>();
        Connection conn = null;
```

```
        PreparedStatement ps = null;
        ResultSet rs = null;
        try {
            StringBuffer buffer = new StringBuffer();
            buffer.append("select * from userinfo order by id asc");

            conn = ConnectionFactory.getConnection();
            ps = conn.prepareStatement(buffer.toString());
            rs = ps.executeQuery();
            while (rs.next()) {
                long idDB = rs.getLong("id");
                String usernameDB = rs.getString("username");
                String passwordDB = rs.getString("password");

                Userinfo userinfo = new Userinfo();
                userinfo.setId(idDB);
                userinfo.setUsername(usernameDB);
                userinfo.setPassword(passwordDB);

                listUserinfo.add(userinfo);
            }
        } catch (SQLException e) {
            e.printStackTrace();
        } finally {
            try {
                ConnectionFactory.close(conn, ps, rs);
            } catch (SQLException e) {
                e.printStackTrace();
            }
        }
        return listUserinfo;
    }
}
```

从 DAO 层的代码来看，其大量使用了 ENTITY 实体类 Userinfo.java。如果不使用实体类，DAO 中方法的参数就会过多，不便于代码的维护与阅读，比如，下面程序代码方法的参数数量。

```
public void save(String username,String password,int age,String address,
String phone){
}
```

继续测试，设计执行 CURD 功能的代码。

执行插入功能的实现代码如下：

```
public class Test1_insert {
    public static void main(String[] args) {
        Userinfo userinfo = new Userinfo();
        userinfo.setUsername("中国");
        userinfo.setPassword("中国人");

        UserinfoDAO dao = new UserinfoDAO();
        dao.insert(userinfo);
    }
}
```

执行删除功能的实现代码如下：

```
public class Test1_delete {
    public static void main(String[] args) {
        UserinfoDAO dao = new UserinfoDAO();
        dao.delete(100049L);
    }
}
```

执行修改功能的实现代码如下：

```
public class Test1_update {
    public static void main(String[] args) {
        Userinfo userinfo = new Userinfo();
        userinfo.setId(100078L);
        userinfo.setUsername("中国地球人");
        userinfo.setPassword("中国人地球人");

        UserinfoDAO dao = new UserinfoDAO();
        dao.update(userinfo);
    }
}
```

执行查询全部记录功能的实现代码如下：

```
public class Test1_selectAll {
```

```
    public static void main(String[] args) {
        UserinfoDAO dao = new UserinfoDAO();
        List<Userinfo> listUserinfo = dao.selectAll();
        for (int i = 0; i < listUserinfo.size(); i++) {
            Userinfo userinfo = listUserinfo.get(i);
            System.out.println(userinfo.getId() + " " +
userinfo.getUsername() + " " + userinfo.getPassword());
        }
    }
}
```

执行查询单条记录功能的实现代码如下：

```
public class Test1_selectById {
    public static void main(String[] args) {
        UserinfoDAO dao = new UserinfoDAO();
        Userinfo userinfo = dao.selectById(100078L);
        System.out.println(userinfo.getId());
        System.out.println(userinfo.getUsername());
        System.out.println(userinfo.getPassword());
    }
}
```

## 3.20 允许MySQL被远程访问

在默认情况下，MySQL 是不允许远程连接到数据库的，连接时会出现如下异常信息：

```
"Host '某个 IP' is not allowed to connect to this MySQL server"
```

因此，要进行如下配置。

（1）在 CMD 中输入命令：mysql -uroot -p123，登录 MySQL 控制台。

（2）执行切换数据库命令：use mysql，切换到 MySQL 数据库。

（3）执行更新 SQL 语句：update user set host = '%' where user = 'root'。

（4）执持命令：flush privileges。

（5）重新启动 MySQL 服务后，就可以远程连接了，这时别忘了关闭操作系统的防火墙。

# 第 4 章 JDBC 实战技术

本章主要学习 JDBC 的必备技术，下面通过若干案例进行介绍。

## 4.1 元数据的获取

在 JDBC 中，可以获取与数据库有关的元数据，其代表数据库、数据表的真实物理信息。

先来看看 DatabaseMetaData 的元数据信息，示例代码如下：

```
package testA;

import java.sql.Connection;
import java.sql.DatabaseMetaData;
import java.sql.SQLException;

public class Test51 {
    public static void main(String[] args) {
        try {
            Connection connection = ConnectionFactory.getConnection();
            DatabaseMetaData metaData = connection.getMetaData();
            System.out.println("A=getDatabaseProductName=" +
metaData.getDatabaseProductName());
            System.out.println("B=getDatabaseProductVersion=" +
metaData.getDatabaseProductVersion());
            System.out.println("C=getDriverName=" +
metaData.getDriverName());
            System.out.println("D=getURL=" + metaData.getURL());
            System.out.println("E=getUserName=" + metaData.getUserName());
```

```
            connection.close();
        } catch (ClassNotFoundException e) {
            e.printStackTrace();
        } catch (SQLException e) {
            e.printStackTrace();
        }
    }
}
```

程序运行结果如图 4-1 所示。

```
获得Connection对象------: oracle.jdbc.driver.T4CConnection@1ed6993a
A=getDatabaseProductName=Oracle
B=getDatabaseProductVersion=Oracle Database 11g Enterprise Edition Release 11.2.0.1.0 - 64bit Production
With the Partitioning, OLAP, Data Mining and Real Application Testing options
C=getDriverName=Oracle JDBC driver
D=getURL=jdbc:oracle:thin:@localhost:1521:orcl
E=getUserName=Y2
```

图 4-1

继续获得 ParameterMetaData 有关的元数据信息。

注意：如果你使用 ojdbc6 或者 ojdbc7，则会出现异常，ojdbc8 在不支持获得某些元数据的情况下也会出现异常。当使用 MySQL 数据库时，如果想要获得正确的 ParameterMetaData 元数据，则需要在 URL 中添加如下参数：

```
"jdbc:mysql://localhost:3306/y2?generateSimpleParameterMetadata=true";
```

示例代码如下：

```
package test37;

import java.sql.Connection;
import java.sql.ParameterMetaData;
import java.sql.PreparedStatement;
import java.sql.SQLException;
import java.util.Date;

public class Test1 {
    public static void main(String[] args) {
        try {
            String sql = "insert into userinfo(id,username,insertdate)
values(idauto.nextval,?,?)";
            Connection connection = ConnectionFactory.getConnection();
            PreparedStatement ps = connection.prepareStatement(sql);
```

```
            ps.setString(1, "高洪岩");
            ps.setDate(2, new java.sql.Date(new Date().getTime()));
            ParameterMetaData data = ps.getParameterMetaData();
            int paramCount = data.getParameterCount();
            System.out.println("paramCount=" + paramCount);
            for (int i = 0; i < paramCount; i++) {
                System.out.println("getParameterClassName="                    +
data.getParameterClassName(i + 1) + " "
                        +              "getParameterTypeName="                 +
data.getParameterTypeName(i + 1));
            }
            ps.executeUpdate();
            ps.close();
            connection.close();
        } catch (ClassNotFoundException e) {
            e.printStackTrace();
        } catch (SQLException e) {
            e.printStackTrace();
        }
    }

}
```

程序运行结果如图 4-2 所示。

```
paramCount=2
getParameterClassName=java.lang.String getParameterTypeName=VARCHAR2
getParameterClassName=java.sql.Timestamp getParameterTypeName=DATE
```

图 4-2

ResultSet 接口也有对应的元数据，数据类型是 ResultSetMetaData。ResultSetMetaData 是查询的虚拟表的结构信息，并不是数据库中物理数据表的信息，示例代码如下：

```
package testA;

import java.sql.Connection;
import java.sql.PreparedStatement;
import java.sql.ResultSet;
import java.sql.ResultSetMetaData;
import java.sql.SQLException;

public class Test53 {
    public static void main(String[] args) {
```

```
        try {
            String sql = "select * from userinfo";
            Connection connection = ConnectionFactory.getConnection();
            PreparedStatement ps = connection.prepareStatement(sql);
            ResultSetMetaData metaData1 = ps.getMetaData();
            System.out.println("metaData1=" + metaData1);
            ResultSet rs = ps.executeQuery();
            ResultSetMetaData metaData2 = rs.getMetaData();
            System.out.println("metaData2=" + metaData2);
            int getColumnCount = metaData2.getColumnCount();
            System.out.println("getColumnCount=" + getColumnCount);
            System.out.println();
            for (int i = 0; i < getColumnCount; i++) {
                System.out.println("getColumnName="                          +
metaData2.getColumnName(i + 1)
                        + "                    getColumnClassName="    +
metaData2.getColumnClassName(i + 1));
            }
            rs.close();
            ps.close();
            connection.close();
        } catch (ClassNotFoundException e) {
            e.printStackTrace();
        } catch (SQLException e) {
            e.printStackTrace();
        }
    }
}
```

程序运行结果如图 4-3 所示。

```
metaData1=oracle.jdbc.driver.OracleResultSetMetaData@6069db50
metaData2=oracle.jdbc.driver.OracleResultSetMetaData@293a5bf6
getColumnCount=4

getColumnName=ID              getColumnClassName=java.math.BigDecimal
getColumnName=USERNAME               getColumnClassName=java.lang.String
getColumnName=PASSWORD               getColumnClassName=java.lang.String
getColumnName=INSERTDATE               getColumnClassName=java.sql.Timestamp
```

图 4-3

那么，这些元数据除了可用于打印，还有哪些使用上的意义呢？元数据对于软件设计工作具有非常重要的意义，它可以简化代码，使代码变得更加通用。

## 4.2　简化CURD的操作代码

前面我们创建了一个 UserinfoDAO.java 类，专门用来处理 userinfo 表中的数据，如果针对 bookinfo 表进行操作，那么要创建针对该表的 BookinfoDAO.java 类。一个表对应一个 DAO 类是正常的现象，但问题是这些 DAO 类中的 JDBC 代码基本都是重复的。程序员为了实现对每个表的 CURD 操作，不得不重复复制粘贴大部分都相同的 JDBC 代码，其中只是 SQL 代码不同，这大大降低了程序员的开发效率。那么，是否可以进行封装与简化，以减少重复的 JDBC 代码量呢？完全可以。封装与简化的原理就是借助于元数据，来看一下简化后的代码：

```
public class DBOperate {
    public List<Map> query(String sql, List valueList) throws SQLException,
ClassNotFoundException {
        List<Map> listMap = new ArrayList<>();
        Connection connection = ConnectionFactory.getConnection();
        PreparedStatement statement = connection.prepareStatement(sql);
        if (valueList != null && valueList.size() != 0) {
            for (int i = 0; i < valueList.size(); i++) {
                statement.setObject(i + 1, valueList.get(i));
            }
        }
        ResultSet resultSet = statement.executeQuery();
        int colCount = resultSet.getMetaData().getColumnCount();
        while (resultSet.next()) {
            Map rowMap = new HashMap();
            for (int i = 0; i < colCount; i++) {
                String colName = resultSet.getMetaData().getColumnName(i + 1);
                Object value = resultSet.getObject(i + 1);
                rowMap.put(colName, value);
            }
            listMap.add(rowMap);
        }
        ConnectionFactory.close(connection, statement, resultSet);
        return listMap;
    }

    public void update(String sql, List valueList) throws SQLException,
ClassNotFoundException {
```

```
        List<Map> listMap = new ArrayList<>();
        Connection connection = ConnectionFactory.getConnection();
        PreparedStatement statement = connection.prepareStatement(sql);
        if (valueList != null && valueList.size() != 0) {
            for (int i = 0; i < valueList.size(); i++) {
                statement.setObject(i + 1, valueList.get(i));
            }
        }
        statement.executeUpdate();
        ConnectionFactory.close(connection, statement);
    }
}
```

在实现 CURD 功能时，只需要在 DAO 类中调用 DBOperate.java 类中的两个方法即可实现增删改查操作，大大减少了 JDBC 代码量，示例代码如下：

```
public class UserinfoDAO {
    private DBOperate dbo = new DBOperate();
    public void insertUserinfo(Userinfo userinfo) throws SQLException {
        List paramValues = new ArrayList();
        paramValues.add(userinfo.getUsername());
        dbo.update("insert    into    userinfo(username)    values(?)",
paramValues);
    }
}
```

这时，DAO 类中不再出现 JDBC 代码了。

在查询功能中，将数据封装到 Map 的原理如图 4-4 所示。

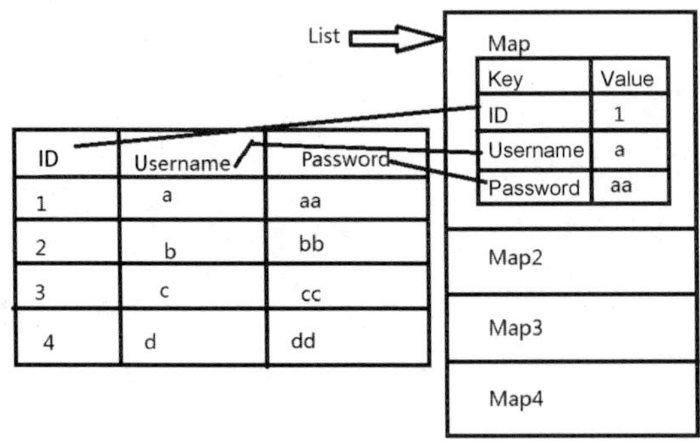


图 4-4

## 4.3　反射与泛型结合为泛型DAO

在与 JDBC 有关的框架中，如 Hibernate，MyBatis 等，都是把 JDBC 和泛型技术结合使用。由于 CURD 增删改查的方法是通用的，因此借助于泛型技术有利于规范 API 的写法。下面将 Hibernate 和 MyBatis 的原理进行展示。

本节使用 MySQL 数据库，反射与泛型技术结合使用后的泛型 DAO 类代码如下：

```
package dao;

import java.lang.reflect.Field;
import java.sql.Connection;
import java.sql.PreparedStatement;
import java.sql.ResultSet;
import java.sql.SQLException;
import java.util.ArrayList;
import java.util.List;

import dbtools.GetConnection;

public class BaseDAO {

    public <T> void save(T t)
            throws   IllegalArgumentException,    IllegalAccessException,
ClassNotFoundException, SQLException {
        String tableName = t.getClass().getSimpleName().toLowerCase();
        String sql = "insert into " + tableName;
        String colName = "";
        String colValue = "";
        List valueList = new ArrayList();
        Field[] fieldArray = t.getClass().getDeclaredFields();
        for (int i = 0; i < fieldArray.length; i++) {
            String eachFieldName = fieldArray[i].getName();
            if (!eachFieldName.equals("id")) {
                colName = colName + "," + eachFieldName;
                colValue = colValue + ",?";
                fieldArray[i].setAccessible(true);
                Object valueObject = fieldArray[i].get(t);
                valueList.add(valueObject);
            }
        }
```

```
        colName = colName.substring(1);
        colValue = colValue.substring(1);
        sql = sql + "(" + colName + ")" + " values" + "(" + colValue + ")";
        System.out.println(sql);

        Connection conn = GetConnection.getConnection();
        PreparedStatement ps = conn.prepareStatement(sql);
        for (int i = 0; i < valueList.size(); i++) {
            ps.setObject(i + 1, valueList.get(i));
        }
        ps.executeUpdate();
        ps.close();
        conn.close();
    }

    public <T> void delete1(Class<T> classObject, int deleteId) throws
ClassNotFoundException, SQLException {
        String tableName = classObject.getSimpleName();
        String sql = "delete from " + tableName + " where id=?";
        Connection conn = GetConnection.getConnection();
        PreparedStatement ps = conn.prepareStatement(sql);
        ps.setObject(1, deleteId);
        ps.executeUpdate();
        ps.close();
        conn.close();
    }

    public <T> void delete2(T t)
            throws    IllegalArgumentException,    IllegalAccessException,
ClassNotFoundException, SQLException {
        String tableName = t.getClass().getSimpleName();
        String sql = "delete from " + tableName + " where id=?";
        long deleteId = 0;
        Field[] fieldArray = t.getClass().getDeclaredFields();
        for (int i = 0; i < fieldArray.length; i++) {
            String eachFieldName = fieldArray[i].getName().toLowerCase();
            if (eachFieldName.equals("id")) {
                fieldArray[i].setAccessible(true);
                deleteId                                                  =
Long.parseLong(fieldArray[i].get(t).toString());
                break;
```

```
            }
        }

        Connection conn = GetConnection.getConnection();
        PreparedStatement ps = conn.prepareStatement(sql);
        ps.setObject(1, deleteId);
        ps.executeUpdate();
        ps.close();
        conn.close();
    }

    public <T> void delete3(T t) throws IllegalArgumentException,
IllegalAccessException, ClassNotFoundException,
            SQLException, NoSuchFieldException, SecurityException {
        String tableName = t.getClass().getSimpleName();
        String sql = "delete from " + tableName + " where id=?";

        Field idField = t.getClass().getDeclaredField("id");
        idField.setAccessible(true);
        long deleteId = Long.parseLong(idField.get(t).toString());

        Connection conn = GetConnection.getConnection();
        PreparedStatement ps = conn.prepareStatement(sql);
        ps.setObject(1, deleteId);
        ps.executeUpdate();
        ps.close();
        conn.close();
    }

    public <T> void update(T t) throws NumberFormatException,
IllegalArgumentException, IllegalAccessException,
            ClassNotFoundException, SQLException {
        String tableName = t.getClass().getSimpleName().toLowerCase();
        String sql = "update " + tableName;
        String whereSQL = "where id=?";
        String colName = "";
        List valueList = new ArrayList();
        Field[] fieldArray = t.getClass().getDeclaredFields();
        long updateUserId = 0;
        for (int i = 0; i < fieldArray.length; i++) {
            Field eachField = fieldArray[i];
```

```
            eachField.setAccessible(true);
            String eachColName = eachField.getName();
            if (eachColName.equals("id")) {
                updateUserId                                                     =
Long.parseLong(eachField.get(t).toString());
            } else {
                colName = colName + "," + eachColName + "=?";
                valueList.add(eachField.get(t));
            }
        }
        valueList.add(updateUserId);
        colName = colName.substring(1);
        sql = sql + " set " + colName + " " + whereSQL;

        Connection conn = GetConnection.getConnection();
        PreparedStatement ps = conn.prepareStatement(sql);
        for (int i = 0; i < valueList.size(); i++) {
            ps.setObject(i + 1, valueList.get(i));
        }
        ps.executeUpdate();
        ps.close();
        conn.close();
    }

    public <T> T get(Class<T> t, long id)
            throws    InstantiationException,     IllegalAccessException,
SQLException, ClassNotFoundException {
        T returnT = t.newInstance();
        String tableName = t.getSimpleName();
        String sql = "select * from " + tableName;
        String whereSQL = "where id=?";
        sql = sql + " " + whereSQL;

        Connection conn = GetConnection.getConnection();
        PreparedStatement ps = conn.prepareStatement(sql);
        ps.setObject(1, id);
        ResultSet rs = ps.executeQuery();
        while (rs.next()) {
            Field[] fieldArray = t.getDeclaredFields();
            for (int i = 0; i < fieldArray.length; i++) {
                Field eachField = fieldArray[i];
                eachField.setAccessible(true);
```

```
                String colName = fieldArray[i].getName();
                Object colValue = rs.getObject(colName);
                eachField.set(returnT, colValue);
            }
        }
        rs.close();
        ps.close();
        conn.close();

        return returnT;
    }

    public <T> List<T> getAll(Class<T> classObject)
            throws InstantiationException, IllegalAccessException,
SQLException, ClassNotFoundException {
        List<T> returnList = new ArrayList();
        String tableName = classObject.getSimpleName();
        String sql = "select * from " + tableName;

        Connection conn = GetConnection.getConnection();
        PreparedStatement ps = conn.prepareStatement(sql);
        ResultSet rs = ps.executeQuery();
        while (rs.next()) {
            T returnT = classObject.newInstance();
            Field[] fieldArray = classObject.getDeclaredFields();
            for (int i = 0; i < fieldArray.length; i++) {
                Field eachField = fieldArray[i];
                eachField.setAccessible(true);
                String colName = fieldArray[i].getName();
                Object colValue = rs.getObject(colName);
                eachField.set(returnT, colValue);
            }
            returnList.add(returnT);
        }
        rs.close();
        ps.close();
        conn.close();

        return returnList;
    }

}
```

DBOperate.java 类可以实现 CURD 功能，而泛型 DAO 类同样可以实现 CURD 功能。为什么框架技术的原理是基于泛型 DAO 类的呢？因为 DBOperate.java 类最明显的缺点是不能指定返回实体类的类型，List 只能存储 Map，而泛型 DAO 类可以由调用者指定实体类的类型，具有很好的通用性。

保存记录的运行代码如下：

```
public class Save {
    public static void main(String[] args)
            throws    IllegalArgumentException,    IllegalAccessException,
ClassNotFoundException, SQLException {
        Userinfo userinfo = new Userinfo();
        userinfo.setUsername("中国");
        userinfo.setPassword("中国人");

        BaseDAO dao = new BaseDAO();
        dao.save(userinfo);
    }
}
```

删除记录（写法 1）的运行代码如下：

```
public class Delete1 {
    public static void main(String[] args)
            throws    IllegalArgumentException,    IllegalAccessException,
ClassNotFoundException, SQLException {
        BaseDAO dao = new BaseDAO();
        dao.delete1(Userinfo.class, 408);
    }
}
```

删除记录（写法 2）的运行代码如下：

```
public class Delete2 {
    public static void main(String[] args)
            throws    IllegalArgumentException,    IllegalAccessException,
ClassNotFoundException, SQLException {
        Userinfo userinfo = new Userinfo();
        userinfo.setId(407);

        BaseDAO dao = new BaseDAO();
```

```
        dao.delete2(userinfo);
    }
}
```

删除记录（写法 3）的运行代码如下：

```
public class Delete3 {
    public static void main(String[] args)
            throws    IllegalArgumentException,    IllegalAccessException,
ClassNotFoundException,         SQLException,          NoSuchFieldException,
SecurityException {
        Userinfo userinfo = new Userinfo();
        userinfo.setId(406);

        BaseDAO dao = new BaseDAO();
        dao.delete3(userinfo);
    }
}
```

根据 ID 更新记录的运行代码如下：

```
public class Update {
    public static void main(String[] args)
            throws    IllegalArgumentException,    IllegalAccessException,
ClassNotFoundException, SQLException {
        Userinfo userinfo = new Userinfo();
        userinfo.setId(409);
        userinfo.setUsername("a");
        userinfo.setPassword("aa");

        BaseDAO dao = new BaseDAO();
        dao.update(userinfo);
    }
}
```

根据 ID 查询记录的运行代码如下：

```
public class Get {
    public static void main(String[] args) throws IllegalArgumentException,
IllegalAccessException,
            ClassNotFoundException, SQLException, InstantiationException {
        BaseDAO dao = new BaseDAO();
```

```
        Userinfo userinfo = dao.get(Userinfo.class, 409);
        System.out.println(userinfo.getId() + " " + userinfo.getUsername()
+ " " + userinfo.getPassword());
    }
}
```

查询全部记录的运行代码如下：

```
public class GetAll {
    public static void main(String[] args) throws IllegalArgumentException,
IllegalAccessException,
            ClassNotFoundException, SQLException, InstantiationException {
        BaseDAO dao = new BaseDAO();
        List<Userinfo> listUserinfo = dao.getAll(Userinfo.class);
        for (int i = 0; i < listUserinfo.size(); i++) {
            Userinfo userinfo = listUserinfo.get(i);
            System.out.println(userinfo.getId()      +      "      "      +
userinfo.getUsername() + " " + userinfo.getPassword());
        }
    }
}
```

使用泛型 DAO 类封装 JDBC 代码，这时 DAO 类中的方法可以复用，该写法也是 Hibernate 和 Mybatis 框架的原理，被称为 ORM（Object Relational Mapping，对象关系映射），原理如图 4-5 所示。

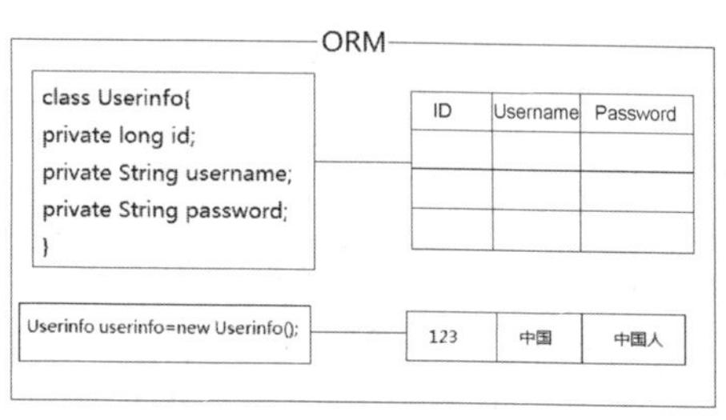


图 4-5

ORM 的原理如下。

（1）一个类对应一个表。

（2）一个类的一个实例对应一个表中的一行。

通过这种映射的机制可以将实体类中的数据转换成数据表中的一行，也可以将数据表中行的数据转换到实体类的属性中，全程都是以面向对象的方式进行开发，而且封装了 JDBC 代码，开发效率大大提升。

## 4.4 数据源和连接池的使用

当使用 JDBC 技术时，在大多数的情况下，获得 Connection 对象是通过调用 DriverManager

类的 getConnection()方法或 Driver 接口中的 getConnection()方法实现的。另外，在标准的 Java EE 技术中，DriverManager 类不推荐使用，而是推荐使用数据源接口来获得 Connection 对象，因为数据源接口是 Sun 公司提供的一个规范，面向规范开发，代码也能变得更加规范与易于维护。

数据源接口信息如图 4-6 所示。从接口信息可知，在 JDK 中并不存在数据源接口的实现类，该接口提供两个重载方法来获得 Connection 对象，如图 4-7 所示。

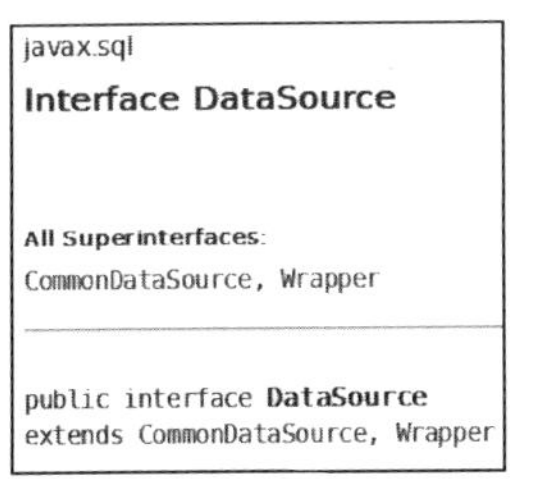


图 4-6

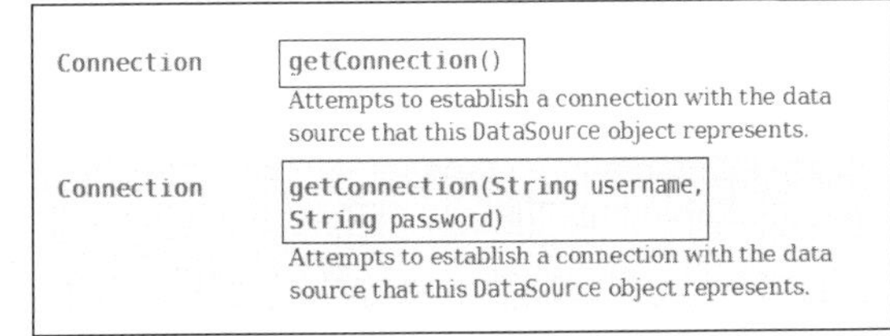


图 4-7

曾经使用 DriverManager.getConnection()方法获得 Connection 对象，现在要使用 DataSource.getConnection()方法了。

## 4.4.1　创建数据源接口的实现类

从数据源接口信息可知，该接口在 JDK 中并没有任何的实现类，那么如何使用数据源获得 Connection 对象呢？创建数据源接口的实现类即可，示例代码如下：

```
package mydatasource;

import java.io.PrintWriter;
import java.sql.Connection;
import java.sql.DriverManager;
import java.sql.SQLException;
import java.sql.SQLFeatureNotSupportedException;
import java.util.logging.Logger;

import javax.sql.DataSource;

public class MyDataSource implements DataSource {

    @Override
    public Connection getConnection() throws SQLException {
```

```
        Connection conn = null;
        try {
            String username = "y2";
            String password = "123";
            String url = "jdbc:oracle:thin:@localhost:1521:orcl";
            String driver = "oracle.jdbc.OracleDriver";
            Class.forName(driver);
            conn = DriverManager.getConnection(url, username, password);
        } catch (ClassNotFoundException e) {
            e.printStackTrace();
        }
        return conn;
    }

    @Override
    public Connection getConnection(String username, String password) throws
SQLException {
        Connection conn = null;
        try {
            String url = "jdbc:oracle:thin:@localhost:1521:orcl";
            String driver = "oracle.jdbc.OracleDriver";
            Class.forName(driver);
            conn = DriverManager.getConnection(url, username, password);
        } catch (ClassNotFoundException e) {
            e.printStackTrace();
        }
        return conn;
    }

    @Override
    public PrintWriter getLogWriter() throws SQLException {
        return null;
    }

    @Override
    public void setLogWriter(PrintWriter out) throws SQLException {
    }

    @Override
    public void setLoginTimeout(int seconds) throws SQLException {
    }
```

```
    @Override
    public int getLoginTimeout() throws SQLException {
        return 0;
    }

    @Override
    public Logger getParentLogger() throws SQLFeatureNotSupportedException
{
        return null;
    }

    @Override
    public <T> T unwrap(Class<T> iface) throws SQLException {
        return null;
    }

    @Override
    public boolean isWrapperFor(Class<?> iface) throws SQLException {
        return false;
    }

}
```

运行类代码如下：

```
package test;

import java.sql.Connection;
import java.sql.SQLException;
import javax.sql.DataSource;
import mydatasource.MyDataSource;

public class Test {
    public static void main(String[] args) throws SQLException {
        DataSource myDataSource = new MyDataSource();
        for (int i = 0; i < 5; i++) {
            Connection conn = myDataSource.getConnection();
            System.out.println(conn);
            conn.close();
        }
```

```
    }
}
```

程序运行后，控制台输出的信息如下：

```
oracle.jdbc.driver.T4CConnection@573fd745
oracle.jdbc.driver.T4CConnection@6aceb1a5
oracle.jdbc.driver.T4CConnection@2d6d8735
oracle.jdbc.driver.T4CConnection@ba4d54
oracle.jdbc.driver.T4CConnection@12bc6874
```

可见，已成功从数据源中获得了 Connection 对象，且每一次都是新的 Connection 对象。

### 4.4.2 使用驱动提供的 DataSource 接口的实现类

在一些 JDBC 驱动中已经提供了数据源接口的实现类，故不需要自己创建实现类，示例代码如下：

```
package test2;

import java.sql.Connection;
import java.sql.SQLException;

import oracle.jdbc.pool.OracleDataSource;

public class Test2 {

    public static void main(String[] args) throws SQLException {
        OracleDataSource source = new OracleDataSource();
        source.setUser("y2");
        source.setPassword("123");
        source.setURL("jdbc:oracle:thin:@localhost:1521:orcl");
        for (int i = 0; i < 5; i++) {
            Connection conn = source.getConnection();
            System.out.println(conn);
            conn.close();
        }
    }

}
```

oracle.jdbc.pool.OracleDataSource 类实现了 oracle.jdbc.datasource.OracleDataSource 接口，而此接口继承了 javax.sql.DataSource 接口，说明 oracle.jdbc.pool.OracleDataSource 类实现了 javax.sql.DataSource 接口。

程序运行后，控制台输出的信息如下：

```
oracle.jdbc.driver.T4CConnection@4f2410ac
oracle.jdbc.driver.T4CConnection@ba4d54
oracle.jdbc.driver.T4CConnection@12bc6874
oracle.jdbc.driver.T4CConnection@de0a01f
oracle.jdbc.driver.T4CConnection@4c75cab9
```

从控制台输出的信息来看，使用 OracleDataSource.java 类获得 Connection 对象时每一次都是新的连接。

使用 MySQL 驱动中的数据源来获得 Connection 对象的代码如下：

```
package test;

import java.sql.Connection;
import java.sql.SQLException;

import com.mysql.jdbc.jdbc2.optional.MysqlDataSource;

public class Test2 {
    public static void main(String[] args) throws SQLException {
        MysqlDataSource ds = new MysqlDataSource();
        ds.setUser("root");
        ds.setPassword("123123");
        ds.setURL("jdbc:mysql://localhost:3306/y2");
        for (int i = 0; i < 5; i++) {
            Connection conn = ds.getConnection();
            System.out.println(conn);
            conn.close();
        }
    }
}
```

### 4.4.3 DataSource 接口的弊端

不使用 DataSource 接口的情况，代码如下：

```
package test6;

import java.sql.Connection;

public class A {
    public Connection getConnection() {
        return null;// 模拟返回 Connection 对象
    }
}

package test6;

import java.sql.Connection;

public class B {
    public Connection getBBBBBBConnection() {
        return null;
    }
}

package test6;

import java.sql.Connection;

public class C {
    public Connection cccccccccccccCgetCCCCCCCCConnection() {
        return null;
    }
}
```

上面 A，B，C 三个类都是不同厂商提供获得 Connection 对象的工具类。那么，如何获得不同厂商的 Connection 对象呢？可以使用如下代码：

```
package test6;

import java.sql.Connection;
import java.sql.SQLException;

public class Test1 {
    public static Connection getConnection(A a) {
```

```
        return a.getConnection();
    }

    public static Connection getConnection(B b) {
        return b.getBBBBBBConnection();
    }

    public static Connection getConnection(C c) {
        return c.ccccccccccccccCgetCCCCCCCCConnection();
    }

    public static void main(String[] args) throws SQLException {
        getConnection(new A());
        getConnection(new B());
        getConnection(new C());
    }
}
```

每个厂商都需要写一个方法，如果有 30 个厂商，难道要写 30 个获得 Connection 对象的方法吗？数据源可以解决此类问题。使用数据源获得 Connection 对象的代码如下：

```
public class A implements DataSource {

public class B implements DataSource {

public class C implements DataSource {
```

从数据源获得 Connection 对象的代码如下：

```
package test7;

import java.sql.Connection;
import java.sql.SQLException;

import javax.sql.DataSource;

public class Test1 {
    public    static    Connection    getConnection(DataSource    ds)    throws
SQLException {
        return ds.getConnection();
    }
```

```
    public static void main(String[] args) throws SQLException {
        getConnection(new A());
        getConnection(new B());
        getConnection(new C());
    }
}
```

这样，不管有多少个厂商，获得 Connection 对象只需要一个方法即可，即从该方法的参数数据源中获得 Connection 对象。

### 4.4.4 连接池

获得 Connection 对象主要有两个步骤。

（1）连接数据库：在连接数据库的过程中，IP 寻址及使用用户名和密码登录 DBMS 都需要耗费大量的时间，其在总时间中占据非常大的比例。

（2）创建 Connection 对象并释放资源：在驱动内部先创建 Connection 接口的实现类，然后释放 Connection 对象资源，如图 4-8 所示。

在下一次操作数据库时，再执行上面两个同样的步骤。如果一个大型网站每天操作数据库的次数非常多，那么获得 Connection 对象需要耗费大量的时间，所以在使用数据源技术时，常常结合连接池（Connection Pool）技术。

连接池就是使用某一个集合来存放 Connection 对象，其优势是可以提升获得 Connection 对象的执行效率，具有连接池的项目结构如图 4-9 所示。

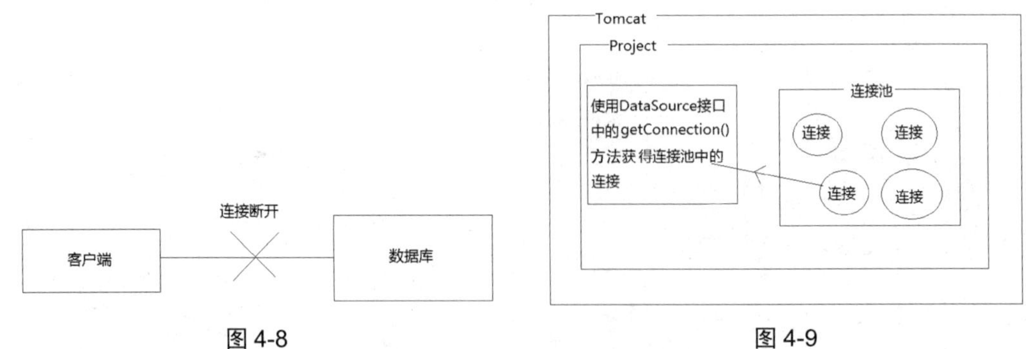


图 4-8　　图 4-9

连接池提升获得 Connection 对象效率的原因有两点。

（1）预先将很多 Connection 对象放入一个内存区中，由于内存区中的 Connection 对

象已经使用长连接的方式连接到数据库了，因此在操作数据库时只需要从内存区中获得已经创建好的 Connection 对象即可，不需要重复进行 IP 寻址和账号及密码的验证，也不需要重复创建新的 Connection 对象，这大大提升程序运行的效率。

（2）当显式执行 connection.close()方法时，关闭的是客户端与连接池中的 Connection 对象，如图 4-10 所示，这时连接池中的 Connection 对象和数据库还保持连接的状态。

将此 Connection 对象标注为空闲状态，以备后面的请求使用，这样就实现了 Connection 对象的复用，提升了程序运行的效率。而连接池中的 Connection 对象还是以长连接的方式连接到数据库中，并没有与数据库断开。使用长连接的优势是可以随时使用空闲的 Connection 对象来执行 SQL 语句，提升了与数据库通信的效率。

这里的“内存区”可以理解为“连接池”。

是否使用连接池的示例，如图 4-11 所示。

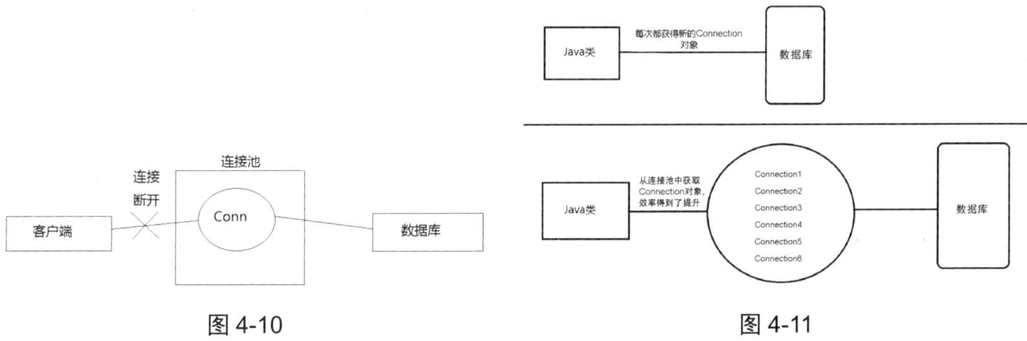


图 4-10　　　　图 4-11

## 4.4.5　不使用连接池与使用连接池的比较

不使用连接池获得 100 个 Connection 对象耗时的示例代码如下：

```
package testA;

import java.sql.Connection;
import java.sql.DriverManager;
import java.sql.SQLException;

public class Test35 {
    public static void main(String[] args) throws ClassNotFoundException,
SQLException {
        String url = "jdbc:oracle:thin:@localhost:1521:orcl";
```

```
        String username = "y2";
        String password = "123";
        Class.forName("oracle.jdbc.OracleDriver");
        long beginTime = System.currentTimeMillis();
        for (int i = 0; i < 100; i++) {
            Connection  connection  =  DriverManager.getConnection(url,
username, password);
            System.out.println(connection + " " + (i + 1));
            connection.close();
        }
        long endTime = System.currentTimeMillis();
        System.out.println(endTime - beginTime);
    }
}
```

程序运行结果如图 4-12 所示。

```
oracle.jdbc.driver.T4CConnection@2aafb23c 87
oracle.jdbc.driver.T4CConnection@2b80d80f 88
oracle.jdbc.driver.T4CConnection@3ab39c39 89
oracle.jdbc.driver.T4CConnection@2eee9593 90
oracle.jdbc.driver.T4CConnection@7907ec20 91
oracle.jdbc.driver.T4CConnection@546a03af 92
oracle.jdbc.driver.T4CConnection@721e0f4f 93
oracle.jdbc.driver.T4CConnection@28864e92 94
oracle.jdbc.driver.T4CConnection@6ea6d14e 95
oracle.jdbc.driver.T4CConnection@6ad5c04e 96
oracle.jdbc.driver.T4CConnection@6833ce2c 97
oracle.jdbc.driver.T4CConnection@725bef66 98
oracle.jdbc.driver.T4CConnection@2aaf7cc2 99
oracle.jdbc.driver.T4CConnection@6e3c1e69 100
2356
```

图 4-12

从控制台输出的信息来看，获得 100 个 Connection 对象需要耗时 2 秒多。如果使用连接池则会大大减少消耗的时间。

### 4.4.6 HikariCP 作为连接池

在此实验中使用 HikariCP 框架作为连接池，HikariCP 是第三方连接池框架，也是现在主流的连接池框架，在软件项目中使用比较广泛，其性能非常好，HikariCP 官方提供一张性能对比图，如图 4-13 所示。

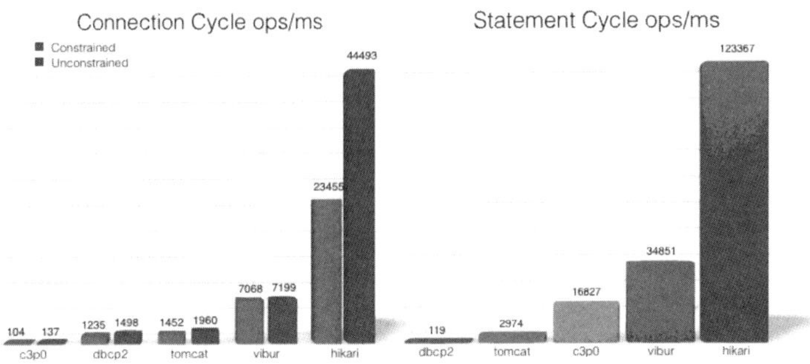


图 4-13

使用 HikariCP 获得 Connection 对象的示例代码如下：

```
package test;

import java.sql.Connection;
import java.sql.SQLException;

import com.zaxxer.hikari.HikariDataSource;

public class Test2 {

    public static void main(String[] args) throws SQLException {
        HikariDataSource ds = new HikariDataSource();
        ds.setDriverClassName("oracle.jdbc.OracleDriver");
        ds.setJdbcUrl("jdbc:oracle:thin:@localhost:1521:orcl");
        ds.setUsername("y2");
        ds.setPassword("123");
        ds.setMinimumIdle(10);
        ds.setMaximumPoolSize(10);

        long beginTime = System.currentTimeMillis();
        for (int i = 0; i < 100; i++) {
            Connection connection = ds.getConnection();
            System.out.println(connection + " " + (i + 1));
            connection.close();
        }
        long endTime = System.currentTimeMillis();
        System.out.println(endTime - beginTime);

    }

}
```

程序运行结果如图 4-14 所示。

```
HikariProxyConnection@1757676444 wrapping oracle.jdbc.driver.T4CConnection@3013d9e5 92
HikariProxyConnection@182738614 wrapping oracle.jdbc.driver.T4CConnection@3013d9e5 93
HikariProxyConnection@94345706 wrapping oracle.jdbc.driver.T4CConnection@3013d9e5 94
HikariProxyConnection@670035812 wrapping oracle.jdbc.driver.T4CConnection@3013d9e5 95
HikariProxyConnection@1870647526 wrapping oracle.jdbc.driver.T4CConnection@3013d9e5 96
HikariProxyConnection@1204167249 wrapping oracle.jdbc.driver.T4CConnection@3013d9e5 97
HikariProxyConnection@1047503754 wrapping oracle.jdbc.driver.T4CConnection@3013d9e5 98
HikariProxyConnection@1722023916 wrapping oracle.jdbc.driver.T4CConnection@3013d9e5 99
HikariProxyConnection@2009787198 wrapping oracle.jdbc.driver.T4CConnection@3013d9e5 100
438
```

图 4-14

控制台输出的信息显示，获得 100 个连接都使用同一个 Connection 对象，这个对象的标识是 3013d9e5，说明 Connection 对象被复用，而不是重新创建新的连接。从时间上来看，使用 HikariCP 后，获得 Connection 对象的效率得到了极大提升，获得 100 个 Connection 对象需要大约 438 毫秒。

注意：如果你分别在非连接池与连接池环境中创建 100 个新连接从时间上看是非常相近的，因为从连接池中获得 Connection 对象速度快的主要是 Connection 对象被复用了，而不是创建新的连接速度快。

## 4.5 在JDBC中处理数据库的事务

JDBC 支持处理数据库的事务，我们先来看看程序代码不报错的示例，插入两条记录的代码如下：

```
package test;

import java.sql.Connection;
import java.sql.DriverManager;
import java.sql.PreparedStatement;
import java.sql.SQLException;

public class Test10 {
    public static void main(String[] args) {
        Connection conn = null;
        PreparedStatement ps1 = null;
        PreparedStatement ps2 = null;
        try {
            String sql1 = "insert into userinfo(id,username,password)
values(idauto.nextval,'a','aa')";
            String sql2 = "insert into userinfo(id,username,password)
values(idauto.nextval,'b','bb')";
```

```
            String url = "jdbc:oracle:thin:@//localhost:1521/orcl";
            String usernameDB = "y2";
            String passwordDB = "123";
            Class.forName("oracle.jdbc.OracleDriver");
            conn = DriverManager.getConnection(url, usernameDB, passwordDB);
            ps1 = conn.prepareStatement(sql1);
            ps1.executeUpdate();
            ps2 = conn.prepareStatement(sql2);
            ps2.executeUpdate();
        } catch (ClassNotFoundException e) {
            e.printStackTrace();
        } catch (SQLException e) {
            e.printStackTrace();
        } finally {
            try {
                if (ps1 != null) {
                    ps1.close();
                }
                if (ps2 != null) {
                    ps2.close();
                }
                if (conn != null) {
                    conn.close();
                }
            } catch (SQLException e) {
                e.printStackTrace();
            }
        }
    }
}
```

在程序运行之前，首先要清空 Userinfo 数据表，再执行程序，成功插入两条记录，如图 4-15 所示。

| ID | USERNAME | PASSWORD |
|---|---|---|
| 100099 | a | aa |
| 100100 | b | bb |

图 4-15

当程序出现异常时，下面的代码会出现什么样的运行效果呢？代码如下：

```
package test;
```

```
import java.sql.Connection;
import java.sql.DriverManager;
import java.sql.PreparedStatement;
import java.sql.SQLException;

public class Test11 {
    public static void main(String[] args) {
        Connection conn = null;
        PreparedStatement ps1 = null;
        PreparedStatement ps2 = null;
        try {
            String sql1 = "insert into userinfo(id,username,password)
values(idauto.nextval,'a','aa')";
            String sql2 = "insert into userinfoNONONONONONONONONONONO
(id,username,password) values(idauto.nextval,'b','bb')";

            String url = "jdbc:oracle:thin:@//localhost:1521/orcl";
            String usernameDB = "y2";
            String passwordDB = "123";
            Class.forName("oracle.jdbc.OracleDriver");
            conn = DriverManager.getConnection(url, usernameDB, passwordDB);
            ps1 = conn.prepareStatement(sql1);
            ps1.executeUpdate();
            ps2 = conn.prepareStatement(sql2);
            ps2.executeUpdate();
        } catch (ClassNotFoundException e) {
            e.printStackTrace();
        } catch (SQLException e) {
            e.printStackTrace();
        } finally {
            try {
                if (ps1 != null) {
                    ps1.close();
                }
                if (ps2 != null) {
                    ps2.close();
                }
                if (conn != null) {
                    conn.close();
                }
            } catch (SQLException e) {
```

```
                    e.printStackTrace();
                }
            }
        }
}
```

程序运行前要先清空 userinfo 数据表，再执行程序。程序在第二次执行 insert 操作时出现了异常，因为在数据库中并不存在名称为 userinfoNONONONONONONONONONONONO 的数据表，但数据表 userinfo 中还是插入了一条记录，如图 4-16 所示。

| ID | USERNAME | PASSWORD |
|---|---|---|
| 100116 | a | aa |

图 4-16

按道理来说这是正确的，因为第一次 insert 没有出现异常，在默认的情况下会自动提交事务，第二次 insert 出现了异常，所以数据表就应该有一条记录，这是没有问题的，按正常的思维是可以理解的，但是如果这是一个转账的操作，在某一步出现异常，则应该整体回滚，而不是将数据表中的数据进行彻底更改。这样的程序代码在实现转账功能时会使财务系统发生混乱，这样的转账业务其实也不是理想状态下的转账流程，那么如何解决这一问题呢？

在 JDBC 中，处理事务回滚比较好的代码模式如下：

```
package test;

import java.sql.Connection;
import java.sql.DriverManager;
import java.sql.PreparedStatement;
import java.sql.SQLException;

public class Test11 {
    public static void main(String[] args) {
        Connection conn = null;
        PreparedStatement ps1 = null;
        PreparedStatement ps2 = null;
        try {
            String sql1 = "insert into userinfo(id,username,password)
values(idauto.nextval,'a','aa')";
            String sql2 = "insert into userinfoNO(id,username,password)
values(idauto.nextval,'b','bb')";

            String url = "jdbc:oracle:thin:@//localhost:1521/orcl";
            String usernameDB = "y2";
```

```
            String passwordDB = "123";
            Class.forName("oracle.jdbc.OracleDriver");
            conn = DriverManager.getConnection(url, usernameDB, passwordDB);
            conn.setAutoCommit(false);// （1）非自动提交事务
            ps1 = conn.prepareStatement(sql1);
            ps1.executeUpdate();
            ps2 = conn.prepareStatement(sql2);
            ps2.executeUpdate();
            conn.commit();// （2）提交事务
        } catch (ClassNotFoundException e) {
            e.printStackTrace();
        } catch (SQLException e) {
            e.printStackTrace();
            try {
                // （3）程序出现异常就回滚
                conn.rollback();
            } catch (SQLException e1) {
                e1.printStackTrace();
            }
        } finally {
            // 不管是否出现异常，资源都要释放
            try {
                if (ps1 != null) {
                    ps1.close();
                }
                if (ps2 != null) {
                    ps2.close();
                }
                if (conn != null) {
                    conn.close();
                }
            } catch (SQLException e) {
                e.printStackTrace();
            }
        }
    }
}
```

在执行程序前，首先要清空 userinfo 数据表，然后执行程序，这时程序运行出现了异常，但数据表 userinfo 中并没有出现新的记录，说明事务被回滚了，达到了预期的目的，效果如图 4-17 所示。

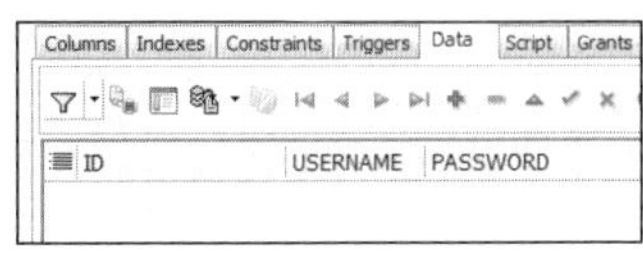


图 4-17

上面的代码也是在转账出现异常时数据库要回滚的代码模式。

## 4.6　多事务导致转账发生异常不回滚

当程序的代码写法不正确时，虽然你使用了 JDBC 中的事务回滚，但还是保证不了数据的一致性，来重现这个实验吧。

首先创建两张数据表 A 和 B，如图 4-18 所示。

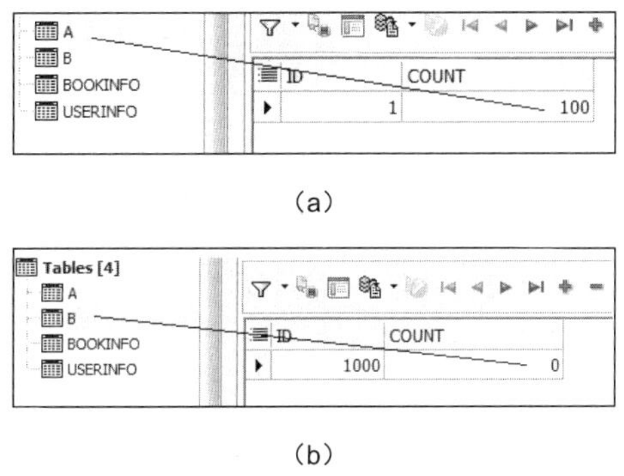


(a)

(b)

图 4-18

创建 ADAO1.java 类，代码如下：

```
public class ADAO1 {

    public void aService() {
        Connection connection = null;
        PreparedStatement ps = null;
        try {
            connection = ConnectionFactory.getConnection();
            connection.setAutoCommit(false);
            ps = connection.prepareStatement("update a set count=count-100");
            ps.executeUpdate();
            connection.commit();
```

```
        } catch (ClassNotFoundException e) {
            e.printStackTrace();
        } catch (SQLException e) {
            try {
                connection.rollback();
            } catch (SQLException e1) {
                e1.printStackTrace();
            }
            e.printStackTrace();
        } finally {
            ConnectionFactory.close(connection, ps);
        }

    }
}
```

创建 BDAO1.java 类，代码如下：

```
public class BDAO1 {

    public void aService() {
        Connection connection = null;
        PreparedStatement ps = null;
        try {
            connection = ConnectionFactory.getConnection();
            connection.setAutoCommit(false);
            ps = connection.prepareStatement("update b set count=
count+100");
            ps.executeUpdate();
            connection.commit();
        } catch (ClassNotFoundException e) {
            e.printStackTrace();
        } catch (SQLException e) {
            try {
                connection.rollback();
            } catch (SQLException e1) {
                e1.printStackTrace();
            }
            e.printStackTrace();
        } finally {
            ConnectionFactory.close(connection, ps);
```

```
        }

    }
}
```

运行类的代码如下：

```
public class Test37 {
    public static void main(String[] args) {
        ADAO1 adao = new ADAO1();
        adao.aService();

        BDAO1 bdao = new BDAO1();
        bdao.aService();
    }
}
```

程序运行后没有出现异常，因为代码都是正确的，转账也是成功的，效果如图 4-19 所示。

如果在这个过程中出现了异常，还会转账成功吗？继续实验。

首先将数据表 A 和 B 中的内容进行还原，效果如图 4-20 所示。

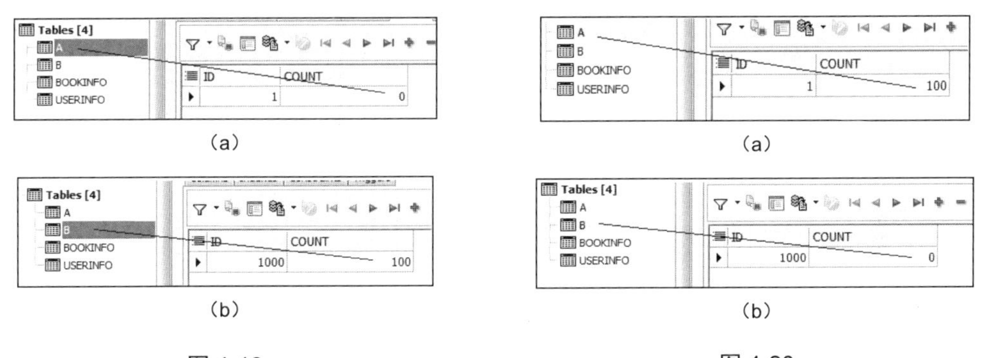


图 4-19　　图 4-20

创建新的 Java 类 BDAO2.java，故意制造一个运行异常，代码如下：

```
public class BDAO2 {

    public void aService() {
        Connection connection = null;
        PreparedStatement ps = null;
```

```
        try {
            connection = ConnectionFactory.getConnection();
            connection.setAutoCommit(false);
            // 没有 bbbbbbbbbbbb 表
            ps  =  connection.prepareStatement("update  bbbbbbbbbbbb  set
count=count+100");
            ps.executeUpdate();
            connection.commit();
        } catch (ClassNotFoundException e) {
            e.printStackTrace();
        } catch (SQLException e) {
            try {
                connection.rollback();
                System.out.println("---------------事务回滚了！");
            } catch (SQLException e1) {
                e1.printStackTrace();
            }
            e.printStackTrace();
        } finally {
            ConnectionFactory.close(connection, ps);
        }

    }
}
```

运行类的代码如下：

```
public class Test38 {
    public static void main(String[] args) {
        ADAO1 adao = new ADAO1();
        adao.aService();

        BDAO2 bdao = new BDAO2();
        bdao.aService();
    }
}
```

程序运行后，在控制台出现了异常，如图 4-21 所示。

```
成功释放了Statement资源
成功释放了Connection资源
---------------事务回滚了！
java.sql.SQLSyntaxErrorException: ORA-00942: 表或视图不存在

	at oracle.jdbc.driver.T4CTTIoer.processError(T4CTTIoer.java:440)
	at oracle.jdbc.driver.T4CTTIoer.processError(T4CTTIoer.java:396)
	at oracle.jdbc.driver.T4C8Oall.processError(T4C8Oall.java:837)
	at oracle.jdbc.driver.T4CTTIfun.receive(T4CTTIfun.java:445)
	at oracle.jdbc.driver.T4CTTIfun.doRPC(T4CTTIfun.java:191)
	at oracle.jdbc.driver.T4C8Oall.doOALL(T4C8Oall.java:523)
	at oracle.jdbc.driver.T4CPreparedStatement.doOall8(T4CPreparedStatement.java:207)
	at oracle.jdbc.driver.T4CPreparedStatement.executeForRows(T4CPreparedStatement.java:1010)
	at oracle.jdbc.driver.OracleStatement.doExecuteWithTimeout(OracleStatement.java:1315)
	at oracle.jdbc.driver.OraclePreparedStatement.executeInternal(OraclePreparedStatement.java:3576)
	at oracle.jdbc.driver.OraclePreparedStatement.executeUpdate(OraclePreparedStatement.java:3657)
	at oracle.jdbc.driver.OraclePreparedStatementWrapper.executeUpdate(OraclePreparedStatementWrapper.java:1350)
	at testA.BDAO2.aService(BDAO2.java:17)
	at testA.Test38.main(Test38.java:9)
成功释放了Statement资源
成功释放了Connection资源
```

图 4-21

数据表 A 和 B 的内容如图 4-22 所示。

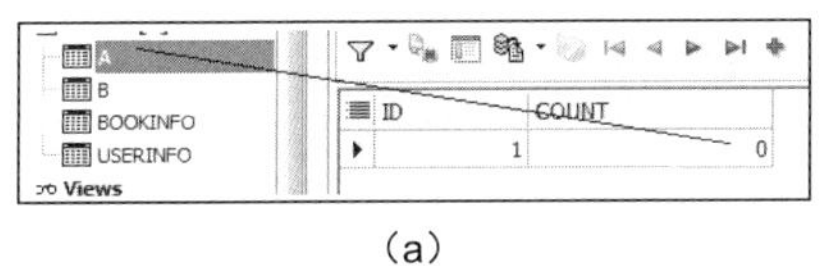


（a）

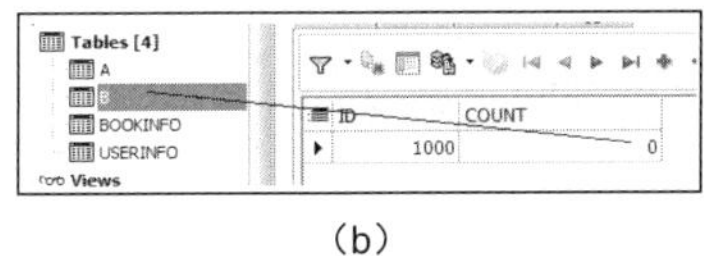


（b）

图 4-22

在程序代码中，出现的异常导致 100 元丢失了，这样的程序代码是不正确的，从运行的结果来分析，是由于事务没有回滚，但这么说也经不起推敲，因为在程序出现异常时，事务回滚的代码明明被执行了，那么为什么没有回滚呢？效果如图 4-23 所示。

```
public class BDAO2 {

    public void aService() {
        Connection connection = null;
        PreparedStatement ps = null;
        try {
            connection = ConnectionFactory.getConnection();
            connection.setAutoCommit(false);
            // 没有bbbbbbbbbbb这张表
            ps = connection.prepareStatement("update bbbbbbbbbbb set count=count+100");
            ps.executeUpdate();
            connection.commit();
        } catch (ClassNotFoundException e) {
            e.printStackTrace();
        } catch (SQLException e) {
            try {
                connection.rollback();
                System.out.println("---------------事务回滚了！");

Problems  Javadoc  Declaration  Console
<terminated> Test38 (4) [Java Application] C:\java1.8\bin\javaw.exe
成功释放了Statement资源
成功释放了Connection资源
---------------事务回滚了！
java.sql.SQLSyntaxErrorException: ORA-00942: 表或视图不存在
```

图 4-23

其最根本的原因是获得了两个 Connection 对象，我们可以将 ConnectionFactory.java 类的代码更改如下：

```
public class ConnectionFactory {
    public    static    Connection    getConnection()    throws    SQLException,
ClassNotFoundException {
        String url = "jdbc:oracle:thin:@localhost:1521:orcl";
        String username = "y2";
        String password = "123";
        Class.forName("oracle.jdbc.OracleDriver");
        Connection connection = DriverManager.getConnection(url, username,
password);
        System.out.println("获得 Connection 对象------: " + connection);//加
入打印 Connection 对象的程序代码
        return connection;
    }
```

还原数据表 A 和 B 中的数据，再次执行如下程序代码：

```
public class Test38 {
    public static void main(String[] args) {
        ADAO1 adao = new ADAO1();
        adao.aService();

        BDAO2 bdao = new BDAO2();
        bdao.aService();
    }
}
```

这时控制台出现了异常，从运行的结果来看，的确是获得了两个 Connection 对象，效果如图 4-24 所示。

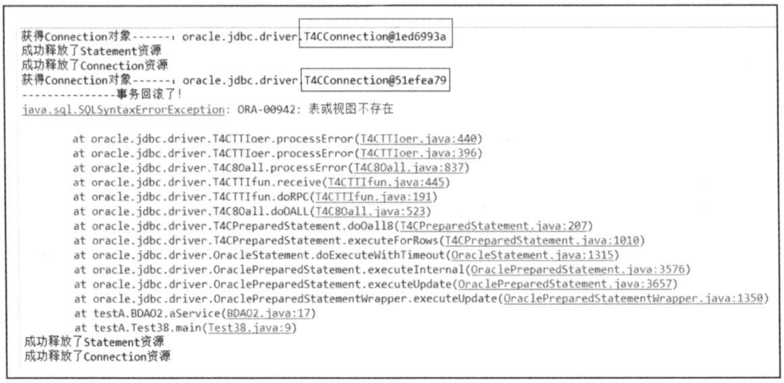

```
获得Connection对象------: oracle.jdbc.driver.T4CConnection@1ed6993a
成功释放了Statement资源
成功释放了Connection资源
获得Connection对象------: oracle.jdbc.driver.T4CConnection@51efea79
---------------事务回滚了!
java.sql.SQLSyntaxErrorException: ORA-00942: 表或视图不存在

	at oracle.jdbc.driver.T4CTTIoer.processError(T4CTTIoer.java:440)
	at oracle.jdbc.driver.T4CTTIoer.processError(T4CTTIoer.java:396)
	at oracle.jdbc.driver.T4C8Oall.processError(T4C8Oall.java:837)
	at oracle.jdbc.driver.T4CTTIfun.receive(T4CTTIfun.java:445)
	at oracle.jdbc.driver.T4CTTIfun.doRPC(T4CTTIfun.java:191)
	at oracle.jdbc.driver.T4C8Oall.doOALL(T4C8Oall.java:523)
	at oracle.jdbc.driver.T4CPreparedStatement.doOall8(T4CPreparedStatement.java:207)
	at oracle.jdbc.driver.T4CPreparedStatement.executeForRows(T4CPreparedStatement.java:1010)
	at oracle.jdbc.driver.OracleStatement.doExecuteWithTimeout(OracleStatement.java:1315)
	at oracle.jdbc.driver.OraclePreparedStatement.executeInternal(OraclePreparedStatement.java:3576)
	at oracle.jdbc.driver.OraclePreparedStatement.executeUpdate(OraclePreparedStatement.java:3657)
	at oracle.jdbc.driver.OraclePreparedStatementWrapper.executeUpdate(OraclePreparedStatementWrapper.java:1350)
	at testA.BDAO2.aService(BDAO2.java:17)
	at testA.Test38.main(Test38.java:9)
成功释放了Statement资源
成功释放了Connection资源
```

图 4-24

由于获得了两个 Connection 对象，且每个 Connection 对象都有自己的事务，因此第二个 Connection 对象在执行 SQL 语句时出现异常，自己回滚了，但并不影响第一个正常的 Connection 对象提交，即导致转账失败后，出现支出与收入不匹配的问题。

问题的根本原因已经找到了：获得了两个 Connection 对象，出现了两个事务。

那么，如何解决这个问题呢？这里可以使用 ThreadLocal 解决。

## 4.7　使用ThreadLocal解决问题

ThreadLocal 类的主要作用就是将值（value）绑定到当前的线程对象中，每个线程对象都拥有自己的值，其他的线程不能使用当前线程中的值。

注意：ThreadLocal 类不是 Map，也不存储任何值，其只是“中介”，即通过 ThreadLocal 类将值存储到当前线程的.ThreadLocalMap 对象中。

创建 GetConnection.java 类，代码如下：

```
public class GetConnection {
    public static ThreadLocal<Connection> tl = new ThreadLocal<>();
    public static Connection getConnection() throws SQLException,
ClassNotFoundException {
        Connection conn = tl.get();
        if (conn == null) {
            String url = "jdbc:oracle:thin:@localhost:1521:orcl";
            String username = "y2";
            String password = "123";
            Class.forName("oracle.jdbc.OracleDriver");
            conn = DriverManager.getConnection(url, username, password);
            tl.set(conn);
            conn.setAutoCommit(false);
        } else {
        }
        return conn;
    }

    public static void commit() {
        try {
            if (tl.get() != null) {
                tl.get().commit();
                tl.get().close();
```

```
                tl.set(null);
            }
        } catch (SQLException e) {
            e.printStackTrace();
        }
    }

    public static void rollback() {
        try {
            if (tl.get() != null) {
                tl.get().rollback();
                tl.get().close();
                tl.set(null);
            }
        } catch (SQLException e) {
            e.printStackTrace();
        }
    }

}
```

为了测试每个线程都拥有自己的 Connection 对象，创建运行类代码如下：

```
public class Test39 {
    public static void main(String[] args) {

        Thread t1 = new Thread() {
            public void run() {
                try {
                    for (int i = 0; i < 5; i++) {
                        System.out.println("t1                    "                    +
GetConnection.getConnection().hashCode());
                        Thread.sleep(100);
                    }
                } catch (ClassNotFoundException | SQLException e) {
                    e.printStackTrace();
                } catch (InterruptedException e) {
                    e.printStackTrace();
                }
            };
        };
```

```
        t1.start();

        Thread t2 = new Thread() {
            public void run() {
                try {
                    for (int i = 0; i < 5; i++) {
                        System.out.println("t2 " + GetConnection.
getConnection().hashCode());
                        Thread.sleep(100);
                    }
                } catch (ClassNotFoundException | SQLException e) {
                    e.printStackTrace();
                } catch (InterruptedException e) {
                    e.printStackTrace();
                }
            };
        };
        t2.start();
    }
}
```

程序运行后，你会发现每个线程都有自己的 Connection 对象，因为一共创建了两个 Connection 对象，如图 4-25 所示。

```
t2 1901917078
t1 1991745811
t2 1901917078
t1 1991745811
t1 1991745811
t2 1901917078
t2 1901917078
t1 1991745811
t1 1991745811
t2 1901917078
```

图 4-25

上面的程序代码满足了一个线程拥有一个 Connection 对象的条件，当一个线程执行不同 DAO 类中的代码操作数据库时，也就能实现有一步出现异常，数据库其他所有被更改的内容要全部回滚，这样在转账出现异常时，就能保证不会出现数值不匹配的问题了。下面开始完善转账业务代码。

创建 GetConnection.java 类的代码如下：

```
public class GetConnection {
    public static ThreadLocal<Connection> tl = new ThreadLocal<>();

    public  static  Connection  getConnection()  throws  SQLException,
ClassNotFoundException {
        Connection conn = tl.get();
        if (conn == null) {
            String url = "jdbc:oracle:thin:@localhost:1521:orcl";
            String username = "y2";
            String password = "123";
```

```
            Class.forName("oracle.jdbc.OracleDriver");
            conn = DriverManager.getConnection(url, username, password);

            tl.set(conn);
            conn.setAutoCommit(false);
        } else {
        }
        // 加入打印语句，查看线程与 Connection 对象的对应关系
        System.out.println(Thread.currentThread().getName()   +   "   拥 有
Connection:" + conn);
        return conn;
    }

    public static void commit() {
        try {
            if (tl.get() != null) {
                tl.get().commit();
                tl.get().close();
                tl.set(null);
                System.out.println("事务提交了");
            }
        } catch (SQLException e) {
            e.printStackTrace();
        }
    }

    public static void rollback() {
        try {
            if (tl.get() != null) {
                tl.get().rollback();
                tl.get().close();
                tl.set(null);
                System.out.println("事务回滚了");
            }
        } catch (SQLException e) {
            e.printStackTrace();
        }
    }

}
```

创建 ADAO_1.java 类的代码如下：

```
public class ADAO_1 {
    public void aService() throws ClassNotFoundException, SQLException {
        Connection connection = GetConnection.getConnection();
        PreparedStatement ps = connection.prepareStatement("update a set
count=count-100");
        ps.executeUpdate();
    }
}
```

创建 BDAO_1.java 类的代码如下：

```
public class BDAO_1 {
    public void aService() throws ClassNotFoundException, SQLException {
        Connection connection = GetConnection.getConnection();
        PreparedStatement ps = connection.prepareStatement("update b set
count=count+100");
        ps.executeUpdate();
    }
}
```

运行类的代码如下：

```
public class Test40 {
    public static void main(String[] args) {
        try {
            ADAO_1 adao = new ADAO_1();
            adao.aService();

            BDAO_1 bdao = new BDAO_1();
            bdao.aService();
        } catch (ClassNotFoundException e) {
            e.printStackTrace();
        } catch (SQLException e) {
            e.printStackTrace();
            GetConnection.rollback();
        } finally {
            GetConnection.commit();
}
    }
}
```

还原数据表 A 和 B 的数据如图 4-26 所示。

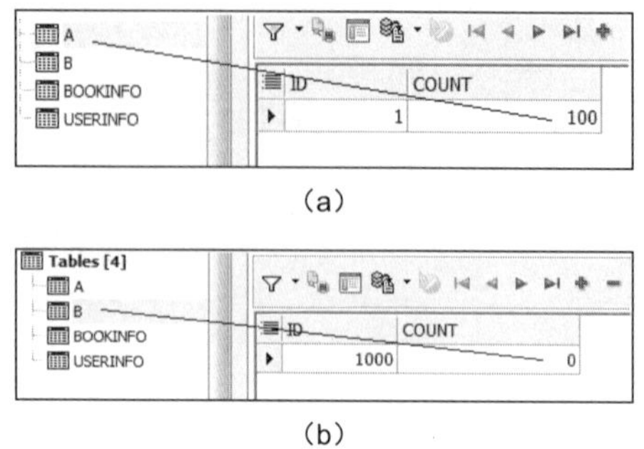


（a）

（b）

图 4-26

运行程序，结果并没有出现异常，转账成功，效果如图 4-27 所示。

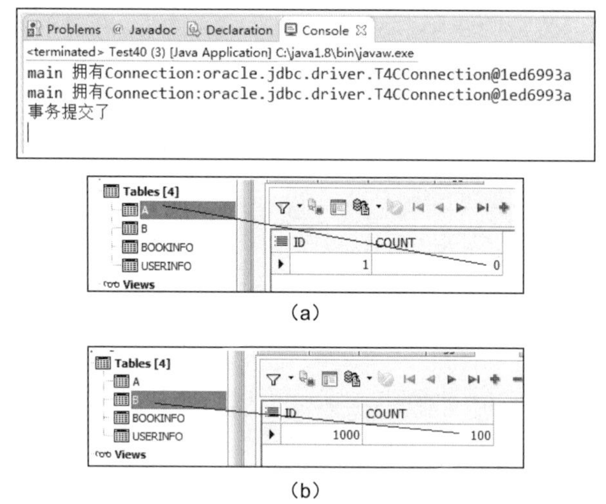


（a）

（b）

图 4-27

如果结果出现异常，也能正确回滚，更改 BDAO_1.java 类的代码如下：

```
public class BDAO_1 {
    public void aService() throws ClassNotFoundException, SQLException {
        Connection connection = GetConnection.getConnection();
        PreparedStatement   ps   =   connection.prepareStatement("update
zzzzzzzzzzzzzzzzzzzzz set count=count+100");
        ps.executeUpdate();
    }
}
```

还原数据表 A 和 B 的内容如图 4-28 所示。

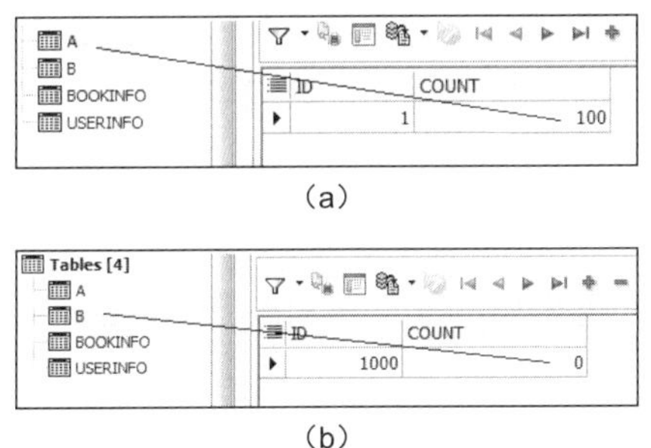


（a）

（b）

图 2-28

再次运行程序，出现异常，效果如图 4-29 所示。

```
main 拥有Connection:oracle.jdbc.driver.T4CConnection@1ed6993a
main 拥有Connection:oracle.jdbc.driver.T4CConnection@1ed6993a
java.sql.SQLSyntaxErrorException: ORA-00942: 表或视图不存在

	at oracle.jdbc.driver.T4CTTIoer.processError(T4CTTIoer.java:440)
	at oracle.jdbc.driver.T4CTTIoer.processError(T4CTTIoer.java:396)
	at oracle.jdbc.driver.T4C8Oall.processError(T4C8Oall.java:837)
	at oracle.jdbc.driver.T4CTTIfun.receive(T4CTTIfun.java:445)
	at oracle.jdbc.driver.T4CTTIfun.doRPC(T4CTTIfun.java:191)
	at oracle.jdbc.driver.T4C8Oall.doOALL(T4C8Oall.java:523)
	at oracle.jdbc.driver.T4CPreparedStatement.doOall8(T4CPreparedStatement.java:207)
	at oracle.jdbc.driver.T4CPreparedStatement.executeForRows(T4CPreparedStatement.java:1010)
	at oracle.jdbc.driver.OracleStatement.doExecuteWithTimeout(OracleStatement.java:1315)
	at oracle.jdbc.driver.OraclePreparedStatement.executeInternal(OraclePreparedStatement.java:3576)
	at oracle.jdbc.driver.OraclePreparedStatement.executeUpdate(OraclePreparedStatement.java:3657)
	at oracle.jdbc.driver.OraclePreparedStatementWrapper.executeUpdate(OraclePreparedStatementWrapper.java:1350)
	at testA.BDAO_1.aService(BDAO_1.java:13)
	at testA.Test40.main(Test40.java:14)
事务回滚了
```

图 4-29

数据表 A 和 B 的内容还是保持原状，如图 4-30 所示。这时，转账数值不匹配的问题彻底被解决。

在写代码时，大多数程序员是将使用了 ThreadLocal.java 类的 GetConnection.java 类与 DBOperate.java 类或泛型 DAO 类或普通 DAO 类结合使用，这既保证了 JDBC 代码结构的简洁性，又保证了事务的一致性，如图 4-31 所示。

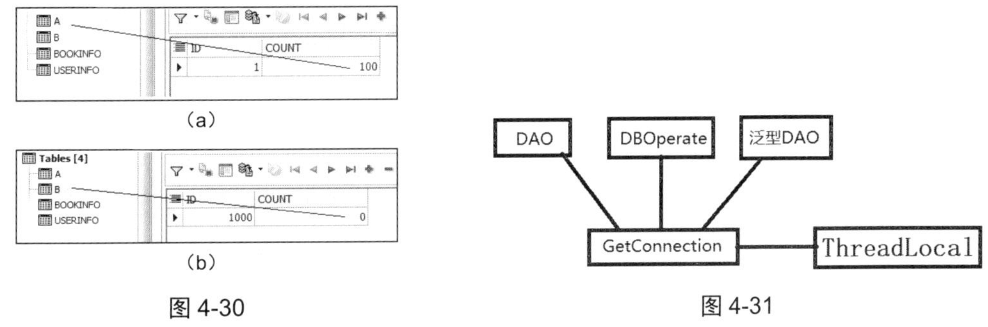


（a）

（b）

图 4-30

图 4-31

ThreadLocal 的问题总结如下。

（1）在转账时，为什么要使用 ThreadLocal 管理 Connection 对象呢？因为想实现每个线程拥有一个自己的 Connection 对象，自己回滚，不影响其他线程。

（2）ThreadLocal 类的作用是什么？在每个线程的“兜子”里存放自己的数据。

（3）ThreadLocal 自己存储数据吗？不存储，它扮演“中介”的角色，数据还是存放在 Thread 类中的 ThreadLocal.ThreadLocalMap 对象中。

（4）Thread 类中 ThreadLocal.ThreadLocalMap 对象的作用是什么？存储当前线程的私有数据。

（5）为什么要使用 ThreadLocal 向 Thread 类中的 ThreadLocal.ThreadLocalMap 对象存储数据？因为 ThreadLocal.ThreadLocalMap 的访问范围是同包 default 级别，所以要使用同包的 ThreadLocal 类来访问 Thread 类中的 ThreadLocal.ThreadLocalMap 对象。

（6）ThreadLocal.ThreadLocalMap 中的 key 和 value 分别指什么？key 就是 ThreadLocal 类的对象，value 值是任意的值，因为 set()方法的参数是 T。

（7）当通过 ThreadLocal 向 ThreadLocal.ThreadLocalMap 多次存储数据时，值能被覆盖吗？能被覆盖，因为它们使用相同的 key，key 就是 ThreadLocal 类的对象。

（8）ThreadLocal 类中 set()方法的参数 T 代表什么？泛型类型，目的是执行 set()方法之后的值在执行 get()方法时不需要强转。

## 4.8 使用JDBC操作CLOB

当 Oracle 数据库保存超大量的文本内容时，需要使用的字段类型为 CLOB（Character Large Object，字符大对象），JDBC 支持对 CLOB 字段类型进行操作。

MySQL 数据库使用 longtext 数据类型。

本节使用 Oracle 数据库作为测试环境。

### 4.8.1 添加 CLOB 类型的数据

添加 CLOB 类型数据的示例代码如下：

```
public class Test64 {
    public static void main(String[] args) throws ClassNotFoundException,
SQLException, FileNotFoundException {
```

```
        Reader reader2 = new StringReader("中国人 2");
        Reader reader3 = new FileReader("c:\\abc\\bigtext.txt");

        String sql = "insert into userinfo(id,bigtext) values(idauto.
nextval,?)";
        Connection conn = ConnectionFactory.getConnection();
        PreparedStatement ps = conn.prepareStatement(sql);

        Clob clob1 = conn.createClob();
        clob1.setString(1, "中国人 1");
        ps.setClob(1, clob1);
        ps.executeUpdate();

        ps.setClob(1, reader2);
        ps.executeUpdate();

        ps.setCharacterStream(1, reader3);
        ps.executeUpdate();

        ConnectionFactory.close(conn, ps);

    }
}
```

添加的大段文本还可以来自于一个自定义内容的.txt 文件，上面的程序代码就是从一个 bigtext.txt 文本文件中获得文本再插入到数据库中，该文件内容如图 4-32 所示。

程序运行后，成功添加了 3 条记录，并且在 bigtext 字段中都有对应的值。

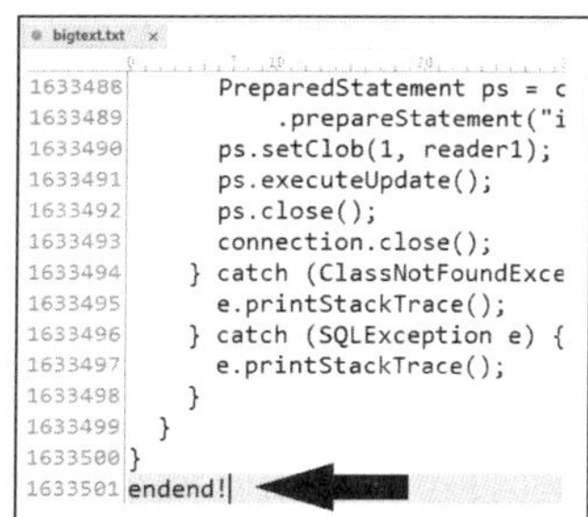


图 4-32

### 4.8.2　获取 CLOB 字段中的数据

获取 CLOB 字段数据的示例代码如下：

```
package testA;

import java.io.IOException;
import java.io.Reader;
import java.sql.Clob;
import java.sql.Connection;
```

```
import java.sql.PreparedStatement;
import java.sql.ResultSet;
import java.sql.SQLException;

public class Test42 {
    public static void main(String[] args) {
        try {
            Clob clobRef = null;

            Connection connection = ConnectionFactory.getConnection();
            PreparedStatement ps = connection.prepareStatement("select *
from userinfo where id=1145364");
            ResultSet rs = ps.executeQuery();
            while (rs.next()) {
                clobRef = rs.getClob(4);
            }
            Reader reader = clobRef.getCharacterStream();
            char[] charArray = new char[10000];
            int readLength = reader.read(charArray);
            while (readLength != -1) {
                String newString = new String(charArray, 0, readLength);
                System.out.println(newString);
                readLength = reader.read(charArray);
            }
            rs.close();
            ps.close();
            connection.close();
        } catch (ClassNotFoundException e) {
            e.printStackTrace();
        } catch (SQLException e) {
            e.printStackTrace();
        } catch (IOException e) {
            e.printStackTrace();
        }
    }
}
```

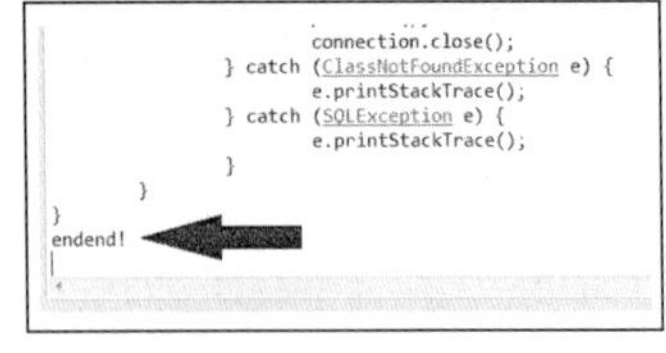


图 4-33

程序运行后的效果如图 4-33 所示。

另外，也可以使用 ResultSet 接口中的 getCharacterStream(int)方法从 CLOB 中获取大段文本，示例代码如下：

```
public class Test66 {
    public static void main(String[] args) throws ClassNotFoundException,
SQLException, IOException {
        // ID 为 1345400 对应的 bigtext 列中的值是 160 多万行的大文本
        String sql = "select bigtext from userinfo where id=1345400";
        Connection conn = ConnectionFactory.getConnection();
        PreparedStatement ps = conn.prepareStatement(sql);
        ResultSet rs = ps.executeQuery();
        Reader reader = null;
        while (rs.next()) {
            reader = rs.getCharacterStream(1);
        }
        char[] charArray = new char[10000];
        int readLength = reader.read(charArray);
        while (readLength != -1) {
            String newString = new String(charArray, 0, readLength);
            System.out.println(newString);
            readLength = reader.read(charArray);
        }
        reader.close();
        ConnectionFactory.close(conn, ps);

    }
}
```

## 4.9　使用JDBC操作BLOB

当 Oracle 数据库保存二进制内容（图片、视频、音频）时，需要使用的字段类型为 BLOB（Binary Large Object，二进制大对象），JDBC 支持对 BLOB 字段类型进行操作。

MySQL 数据库使用 longblob 数据类型。

本节使用 Oracle 数据库作为测试环境。

### 4.9.1　添加 BLOB 类型的数据

添加 BLOB 数据类型的代码如下：

```
public class Test67 {
    public static void main(String[] args) throws ClassNotFoundException,
SQLException, IOException {
```

```
        String sql = "insert into userinfo(id,bigbinary) values(idauto.
nextval,?)";
        Connection conn = ConnectionFactory.getConnection();
        PreparedStatement ps = conn.prepareStatement(sql);

        // 必须是 3 个 InputStream，不能使用 1 个 InputStream 进行多次插入操作
        InputStream picStream1 = new FileInputStream("c:\\abc\\myjpg.jpg");
        InputStream picStream2 = new FileInputStream("c:\\abc\\myjpg.jpg");
        InputStream picStream3 = new FileInputStream("c:\\abc\\myjpg.jpg");
        byte[] byteArray = new byte[picStream1.available()];
        picStream1.read(byteArray);

        Blob blob1 = conn.createBlob();
        blob1.setBytes(1, byteArray);
        ps.setBlob(1, blob1);
        ps.executeUpdate();

        ps.setBlob(1, picStream2);
        ps.executeUpdate();

        ps.setBinaryStream(1, picStream3);
        ps.executeUpdate();

        ConnectionFactory.close(conn, ps);

    }
}
```

程序运行后，成功添加了 3 条 BLOB 类型的数据。

当使用 MySQL 数据库时，如果执行程序后出现异常：

```
Packet for query is too large (1,274,435 > 1,048,576). You can change this
value on the server by setting the 'max_allowed_packet' variable.
```

则需要在 my.ini 文件的[mysqld]节点中先添加如下配置：

```
max_allowed_packet = 50M
```

然后重启 MySQL 服务。

## 4.9.2　获取 BLOB 字段的数据

获取 BLOB 字段的程序代码如下：

```
package testA;

import java.io.FileOutputStream;
import java.io.IOException;
import java.io.InputStream;
import java.sql.Blob;
import java.sql.Connection;
import java.sql.PreparedStatement;
import java.sql.ResultSet;
import java.sql.SQLException;

public class Test44 {
    public static void main(String[] args) {
        try {
            Blob blobRef = null;

            Connection connection = ConnectionFactory.getConnection();
            PreparedStatement ps = connection.prepareStatement("select *
from userinfo where id=1145369");
            ResultSet rs = ps.executeQuery();
            while (rs.next()) {
                blobRef = rs.getBlob(5);
            }
            InputStream inputStream = blobRef.getBinaryStream();

            FileOutputStream fosRef = new FileOutputStream("c:\\abc\\
newMyjpg.jpg");
            byte[] byteArray = new byte[10000];
            int readLength = inputStream.read(byteArray);
            while (readLength != -1) {
                fosRef.write(byteArray, 0, readLength);
                readLength = inputStream.read(byteArray);
            }
            inputStream.close();
            fosRef.close();

            rs.close();
```

```
            ps.close();
            connection.close();
        } catch (ClassNotFoundException e) {
            e.printStackTrace();
        } catch (SQLException e) {
            e.printStackTrace();
        } catch (IOException e) {
            e.printStackTrace();
        }
    }
}
```

程序运行后，在 abc 文件夹中成功创建新的.jpg 文件，并且可以成功打开该文件，效果如图 4-34 所示。

图 4-34

另外，使用 ResultSet 接口中的 getBinaryStream(int)方法也可以从 BLOB 字段中获得大型的二进制对象，示例代码如下：

```
public class Test69 {
    public static void main(String[] args) throws ClassNotFoundException,
SQLException, IOException {

        String sql = "select bigbinary from userinfo where id=1345414";

        Connection conn = ConnectionFactory.getConnection();
        PreparedStatement ps = conn.prepareStatement(sql);
```

```
        ResultSet rs = ps.executeQuery();
        InputStream inputStream = null;
        while (rs.next()) {
            inputStream = rs.getBinaryStream(1);
        }
        OutputStream   outputStream    =    new    FileOutputStream("c:\\abc\\
newNewMyjpg.jpg");
        byte[] byteArray = new byte[10000];
        int readLength = inputStream.read(byteArray);
        while (readLength != -1) {
            outputStream.write(byteArray, 0, readLength);
            readLength = inputStream.read(byteArray);
        }
        outputStream.close();
        inputStream.close();

        ConnectionFactory.close(conn, ps);

    }
}
```

CLOB 中存储的文件类型是文本，而 BLOB 中存储的是任何类型的文件。

在真实的软件项目中，建议不要使用 CLOB 和 BLOB 类型的字段，因为它们存储的数据会占用极大的空间，降低数据库运行的效率。替代 CLOB 的方式可以是先将大文本保存在.txt 文件中，然后使用 varchar 类型的字段存储路径，BLOB 也是如此。

## 4.10　实现Batch批处理

使用批处理可以减少数据库网络与 IO 操作的次数，提升程序执行的效率，它的运行原理就是一起执行多个 SQL 语句。

先来看看使用 PreparedStatement 接口连续插入 50000 条记录需要的时间，示例代码如下：

```
public class Test45 {
    public static void main(String[] args) {
        try {
            Connection connection = ConnectionFactory.getConnection();
            PreparedStatement ps = connection
                    .prepareStatement("insert  into  userinfo(id,username)
```

```
values(idauto.nextval,'abcdefg')");
            long beginTime = System.currentTimeMillis();
            for (int i = 0; i < 50000; i++) {
                ps.executeUpdate();
            }
            long endTime = System.currentTimeMillis();
            System.out.println(endTime - beginTime);
            ps.close();
            connection.close();
        } catch (ClassNotFoundException e) {
            e.printStackTrace();
        } catch (SQLException e) {
            e.printStackTrace();
        }
    }
}
```

程序从开始运行到结束需要的时间如图 4-35 所示。

```
获得Connection对象------: oracle.jdbc.driver.T4CConnection@1ed6993a
12791
```

图 4-35

下面使用批处理进行操作，示例代码如下：

```
public class Test46 {
    public static void main(String[] args) {
        try {
            Connection connection = ConnectionFactory.getConnection();
            PreparedStatement ps = connection
                    .prepareStatement("insert  into  userinfo(id,username)
values(idauto.nextval,'abcdefg')");
            long beginTime = System.currentTimeMillis();
            for (int i = 0; i < 50000; i++) {
                ps.addBatch();
            }
            ps.executeBatch();
            long endTime = System.currentTimeMillis();
            System.out.println(endTime - beginTime);
            ps.close();
            connection.close();
        } catch (ClassNotFoundException e) {
```

```
            e.printStackTrace();
        } catch (SQLException e) {
            e.printStackTrace();
        }
    }
}
```

程序从开始运行到结束需要的时间如图 4-36 所示。

```
获得Connection对象------: oracle.jdbc.driver.T4CConnection@1ed6993a
4207
|
```

图 4-36

从运行时间来看，使用批处理技术的确大大提升了运行速度。

MySQL 数据库开启批处理需要在 URL 中添加 rewriteBatchedStatements=true 参数，并且执行 connection.setAutoCommit(false)方法，代码禁止自动提交。

批处理技术大多数是在调用 insert 或 update 的 SQL 语句时使用的，而调用 insert SQL 语句的机会最多，其通常在向数据库导入海量数据时使用。

批处理技术还可以批量执行不同的 SQL 语句，示例代码如下：

```
public class Test47 {
    public static void main(String[] args) {
        try {
            Connection connection = ConnectionFactory.getConnection();
            Statement ps = connection.createStatement();

            ps.addBatch("insert          into          userinfo(id,username)
values(idauto.nextval,'a')");
            ps.addBatch("insert          into          userinfo(id,username)
values(idauto.nextval,'b')");

            ps.executeBatch();
            ps.close();
            connection.close();
        } catch (ClassNotFoundException e) {
            e.printStackTrace();
        } catch (SQLException e) {
            e.printStackTrace();
        }
    }
}
```

void addBatch(String SQL)方法不支持传入参数。在使用批处理技术实现同时插入两条记录时，插入的每条 SQL 语句都不一样。

## 4.11 插入Date数据类型并查询区间

在 Oracle 数据库的 userinfo 数据表中添加字段类型为 Date（日期）的 insertdate 字段，如果是 MySQL 数据库，则列类型为 datetime。

本节使用 Oracle 数据库进行测试。

添加两种日期格式的代码如下：

```
package test75;

import java.io.IOException;
import java.sql.Connection;
import java.sql.PreparedStatement;
import java.sql.SQLException;
import java.util.Calendar;
import java.util.Date;

public class Test1 {
    public  static  void  main(String[]  args)  throws  SQLException,
ClassNotFoundException, IOException {
        Connection connection = ConnectionFactory.getConnection();

        Calendar calendar = Calendar.getInstance();
        calendar.set(Calendar.YEAR, 2011);
        calendar.set(Calendar.MONTH, 3);
        calendar.set(Calendar.DAY_OF_MONTH, 13);
        calendar.set(Calendar.HOUR_OF_DAY, 0);
        calendar.set(Calendar.MINUTE, 0);
        calendar.set(Calendar.SECOND, 0);
        calendar.set(Calendar.MILLISECOND, 0);

        PreparedStatement ps = connection
                .prepareStatement("insert  into  userinfo(id,insertdate)
values(idauto.nextval,?)");
        ps.setDate(1, new java.sql.Date(calendar.getTime().getTime()));
        ps.executeUpdate();
```

```
            ps    =   connection.prepareStatement("insert   into   userinfo(id,
insertdate) values(idauto.nextval,?)");
            ps.setTimestamp(1, new java.sql.Timestamp(new Date().getTime()));
            ps.executeUpdate();

            ps.close();
            connection.close();
        }

}
```

程序执行后，userinfo 数据表有两条记录，如图 4-37 所示。

下面的测试将进行日期区间的查询，添加多条记录，如图 4-38 所示。

| ID | USERNAME | PASSWORD | BIGTEXT | BIGBINARY | INSERTDATE |
| --- | --- | --- | --- | --- | --- |
| 1345375 | | | (HugeClob) | (HugeBlob) | 2011/4/13 |
| 1345376 | | | (HugeClob) | (HugeBlob) | 2011/4/13 8:45:59 |

图 4-37

| ID | USERNAME | PASSWORD | BIGTEXT | BIGBINARY | INSERTDATE |
| --- | --- | --- | --- | --- | --- |
| 1 | | | (HugeClob) | (HugeBlob) | 2011/4/13 |
| 2 | | | (HugeClob) | (HugeBlob) | 2011/4/13 1:02:03 |
| 3 | | | (HugeClob) | (HugeBlob) | 2011/4/13 |
| 4 | | | (HugeClob) | (HugeBlob) | 2011/4/13 |
| 5 | | | (HugeClob) | (HugeBlob) | 2011/4/13 |
| 6 | | | (HugeClob) | (HugeBlob) | 2011/4/13 |
| 1345375 | | | (HugeClob) | (HugeBlob) | 2011/4/13 |
| 1345376 | | | (HugeClob) | (HugeBlob) | 2011/4/14 8:45:59 |

图 4-38

查询日期区间的程序代码如下：

```
public class Test49 {
    public static void main(String[] args) {
        try {

            String beginDateString = "2011-4-13 00:00:00";
            String endDateString = "2011-4-13 23:59:59";

            SimpleDateFormat  format  =  new  SimpleDateFormat("yyyy-MM-dd
hh:mm:ss");
            Date beginDate = format.parse(beginDateString);
            Date endDate = format.parse(endDateString);

            String sql = "select * from userinfo where insertdate >=? and
insertdate <=?";

            Connection connection = ConnectionFactory.getConnection();
            PreparedStatement ps = connection.prepareStatement(sql);
            ps.setTimestamp(1, new Timestamp(beginDate.getTime()));
            ps.setTimestamp(2, new Timestamp(endDate.getTime()));
```

```
            ResultSet rs = ps.executeQuery();
            while (rs.next()) {
                System.out.println(rs.getString("id")     +     "     "     +
rs.getTimestamp("insertdate"));
            }
            rs.close();
            ps.close();
            connection.close();
        } catch (ClassNotFoundException e) {
            e.printStackTrace();
        } catch (SQLException e) {
            e.printStackTrace();
        } catch (ParseException e) {
            // TODO Auto-generated catch block
            e.printStackTrace();
        }
    }
}
```

程序查询结果如图 4-39 所示。

```
1 2011-04-13 00:00:00.0
2 2011-04-13 01:02:03.0
3 2011-04-13 00:00:00.0
4 2011-04-13 00:00:00.0
5 2011-04-13 00:00:00.0
6 2011-04-13 00:00:00.0
1345375 2011-04-13 00:00:00.0
```

图 4-39

## 4.12 返回最新版的ID值

在实现 insert 功能时，使用 JDBC 代码可以获得 Oracle 数据库中刚刚记录的 ID 值，示例代码下：

```
package testA;

import java.sql.Connection;
import java.sql.PreparedStatement;
import java.sql.ResultSet;
import java.sql.SQLException;

public class Test50 {
    public static void main(String[] args) {
        try {
            int insertUserId = 0;
            String    sql    =    "insert    into    userinfo(id,username)
values(idauto.nextval,'a')";
            Connection connection = ConnectionFactory.getConnection();
```

```
            PreparedStatement ps = connection.prepareStatement(sql, new
String[] { "id" });
            ps.executeUpdate();
            ResultSet rs = ps.getGeneratedKeys();
            while (rs.next()) {
                insertUserId = rs.getInt(1);
            }
            rs.close();
            ps.close();
            connection.close();
            System.out.println(insertUserId);
        } catch (ClassNotFoundException e) {
            e.printStackTrace();
        } catch (SQLException e) {
            e.printStackTrace();
        }
    }
}
```

其中，connection.prepareStatement(sql, new String[] { "id" });的第二个参数 new String[] { "id" }的含义是获取刚刚记录的 ID 值。

程序运行结果如图 4-40 所示。

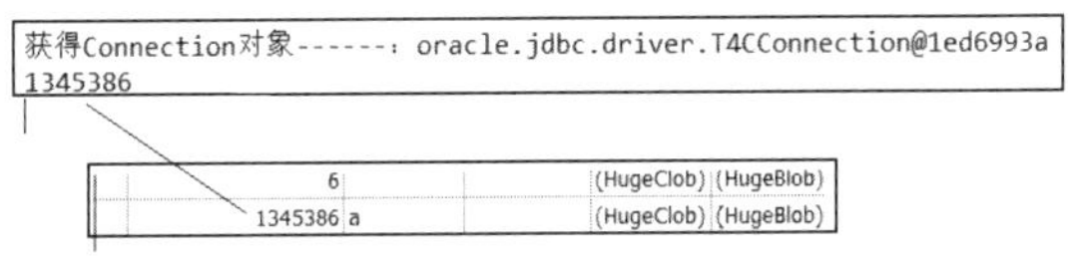


图 4-40

操作 MySQL 的代码如下：

```
package test;

import java.sql.Connection;
import java.sql.DriverManager;
import java.sql.PreparedStatement;
import java.sql.ResultSet;
import java.sql.SQLException;

public class Test2 {
```

```
    public static void main(String[] args) {
        try {
            String url = "jdbc:mysql://localhost:3306/y2";
            String username = "root";
            String password = "123123";
            Class.forName("com.mysql.cj.jdbc.Driver");
            Connection   connection   =   DriverManager.getConnection(url,
username, password);

            int insertUserId = 0;
            String sql = "insert into userinfo(username) values('a')";
            PreparedStatement   ps   =   connection.prepareStatement(sql,
PreparedStatement.RETURN_GENERATED_KEYS);
            ps.executeUpdate();
            ResultSet rs = ps.getGeneratedKeys();
            while (rs.next()) {
                insertUserId = rs.getInt(1);
            }
            rs.close();
            ps.close();
            connection.close();
            System.out.println(insertUserId);
        } catch (ClassNotFoundException e) {
            e.printStackTrace();
        } catch (SQLException e) {
            e.printStackTrace();
        }

    }

}
```

## 4.13 事务的ACID特性

现代数据库必须具备 ACID 特性，具体包括以下方面。

（1）原子性（Atomicity）。原子性是指事务包含的所有操作要么全部成功进行提交，要么全部失败进行回滚。

（2）一致性（Consistency）。一致性是指事务必须使数据库从一个一致性状态变换到另一个一致性状态，也就是说，一个事务执行之前和执行之后都必须处于一致性状态。比如转账操作，假设用户 A 和用户 B 两者的钱加起来一共是 5000 元，那么不管 A 和 B 之间如何转账，转几次，事务结束后两个用户的钱相加还是 5000 元，这就是事务的一致性。需要注意的是，确保一致性是用户的责任，而不是 DBMS 的责任。

（3）隔离性（Isolation）。隔离性是指当多个用户并发访问数据库时，比如操作同一张表，数据库为每个用户开启的事务不能被其他事务干扰，多个并发事务之间要相互隔离。

（4）持久性（Durability）。持久性是指一个事务一旦被提交，那么对数据库的改变就是永久性的，即使在数据库系统遇到故障的情况下，也不会丢失已提交事务的数据。

## 4.14　数据库事务的类型

数据库事务的类型主要分为 3 种。

（1）读取类事务：select。

（2）更新类事务：update。

（2）添加类事务：insert。

在操作同一条记录的前提下，3 种不同种类的事务两两结合会有如下 9 种方式。

（1）select-select：不会出现问题。

（2）select-update：会出现问题***。

（3）select-insert：会出现问题***。

（4）update-select：会出现问题***。

（5）update-update：会出现问题***。

（6）update-insert：不会出现问题。

（7）inserst-select：不会出现问题。

（8）insert-update：不会出现问题。

（9）insert-insert：不会出现问题。

其中，

（1）select-update：会出现问题***。当一个读取类事务读取一条数据时，如果另一个更新类事务修改了这条数据，则会出现“不可重复读（Non Repeatable Read）”，也就是读取到了最新的数据。

（2）select-insert：会出现问题***。当一个读取类事务读取一条数据时，如果另一个添加类事务插入了一条新数据，则可能会多读一条数据，出现“幻读（Phantom Read）”。

（3）update-select：会出现问题***。当一个更新类事务更新一条数据时，如果另一个读取类事务读取了还没提交的更新，则会出现“脏读（Dirty Read）”。

（4）update-update：会出现问题***。当两个更新类事务同时修改一条数据时，会造成更新数据的丢失，因为前一个事务做的修改会被后一个事务所覆盖，这是非常严重的情况。

（1）（3）（4）组合方式是对同一条记录的并发操作，可能会对程序的运行结果产生致命的影响，尤其是对准确性要求极高的金融系统，绝不允许出现这 3 种情况。组合方式（2）适用于社交网站、论坛等场景。

## 4.15 脏读、可重复读、不可重复读及幻读的解释

（1）脏读。脏读是指在一个事务处理过程中读取了另一个未提交事务中的数据。比如，用户 A 向用户 B 转账 100 元，执行如下两条 SQL 语句：

```
update account set money=money+100 where name='B'; //A 向 B 转账 100 元
update account set money=money-100 where name='A'; //A 自减 100 元
```

当执行第一条 update 语句时，A 通知 B 查看账户，B 发现钱已到账（此时即发生了脏读，因为事务并没有提交）。在执行第二条 update 语句时，系统出现异常，导致事务回滚，那么 B 其实并没有收到 100 元。

（2）可重复读。可重复读是指对于数据库中的某个数据，一个事务范围内多次查询返回了相同的数据值。

（3）不可重复读。不可重复读是指对于数据库中的某个数据，一个事务范围内多次查询返回了不同的数据值，这是由于在查询间隔被另一个事务修改并提交了。比如，如果事务 T1 正在读取某一数据记录，而事务 T2 修改了这个数据记录并且提交事务，则当事务 T1 再次读取该数据记录时，就会得到不同的结果，发生了不可重复读。

在某些情况下，不可重复读并不是问题，比如，多次查询某个数据记录，当然是以最后查询的结果为主。

（4）幻读（虚读）。例如事务 T1 将一个表中所有行的某个列值“1”修改为“2”，这时事务 T2 又在表中插入了列值为“1”的一条新的数据记录，并且提交了事务。而如果事务 T1 查看刚刚修改的数据，则会发现还有一行没有修改，其实这行是事务 T2 添加的，就好像产生幻觉一样，发生了幻读。

幻读和不可重复读都是读取了另一条已经提交的事务（这点和脏读不同），所不同的是：①不可重复读查询的是同一个数据项；②幻读针对的是一批数据。

## 4.16 事务隔离级别

事务隔离级别定义了并发事务在修改数据时是如何相互影响的，4 种事务隔离级别的解释如下。

（1）ISOLATION_READ_UNCOMMITTED：读取未提交的数据，这是最低的隔离级别。在并发的事务中，它允许一个事务可以读到另一个事务未提交的更新数据（会出现脏读，不可重复读和幻读）。

（2）ISOLATION_READ_COMMITTED：读取已提交的数据。在并发的事务中，一个事务修改数据提交后才能被另外一个事务读取（会出现不可重复读和幻读）。

（3）ISOLATION_REPEATABLE_READ：可重复读，可以防止脏读、不可重复读，但可能会出现幻读。

（4）ISOLATION_SERIALIZABLE：事务采用顺序执行，不能并行执行，花费时间比较多，但也是最可靠的事务隔离级别，能有效地避免脏读、不可重复读及幻读。

每种数据库的默认隔离级别是不同的，例如 SQL Server 和 Oracle 默认是 READ COMMITED，MySQL 默认是 REPEATABLE READ。

我们可以把脏读、可重复读、不可重复读、幻读理解成权限明细（比如增删改查功能），把 ISOLATION_READ_UNCOMMITTED，ISOLATION_READ_COMMITTED，ISOLATION_REPEATABLE_READ，ISOLATION_SERIALIZABLE 理解成角色（普通用户、管理员、超级管理员），每个角色有不同的权限明细，来规定这个角色可以做什么，不可以做什么。

### 4.16.1 TRANSACTION_READ_UNCOMMITTED

下面测试 TRANSACTION_READ_UNCOMMITTED 隔离级别。

#### 1. 测试脏读

还原数据表 a 的内容，如图 4-41 所示，示例代码如下：

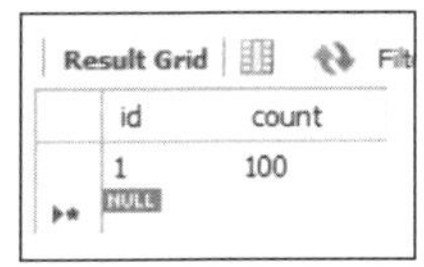


图 4-41

```
package com.ghy.www.transaction_read_uncommitted;

import java.sql.*;
```

```
public class DirtyRead {
    public static void main(String[] args) throws InterruptedException {

        Thread a = new Thread() {
            @Override
            public void run() {
                try {
                    String username = "root";
                    String password = "123123";
                    String                             url                             =
"jdbc:mysql://localhost:3306/y2?serverTimezone=Asia/Shanghai";
                    Class.forName("com.mysql.cj.jdbc.Driver");
                    Connection    conn    =    DriverManager.getConnection(url,
username, password);

conn.setTransactionIsolation(Connection.TRANSACTION_READ_UNCOMMITTED);
                    conn.setAutoCommit(false);

                    String sql = "select * from a where id=1";
                    PreparedStatement ps = null;
                    {
                        ps = conn.prepareStatement(sql);
                        ResultSet rs = ps.executeQuery();
                        while (rs.next()) {
                            System.out.println("id=" + rs.getString("id") + "
count=" + rs.getString("count"));
                        }
                    }
                    Thread.sleep(3000);
                    {
                        ResultSet rs = ps.executeQuery();
                        while (rs.next()) {
                            System.out.println("id=" + rs.getString("id") + "
count=" + rs.getString("count"));
                        }
                    }
                } catch (SQLException throwables) {
                    throwables.printStackTrace();
                } catch (ClassNotFoundException e) {
                    e.printStackTrace();
```

```
            } catch (InterruptedException e) {
                e.printStackTrace();
            }
        }
    };

    Thread b = new Thread() {
        @Override
        public void run() {
            try {
                String username = "root";
                String password = "123123";
                String                              url                              =
"jdbc:mysql://localhost:3306/y2?serverTimezone=Asia/Shanghai";
                Class.forName("com.mysql.cj.jdbc.Driver");
                Connection    conn    =    DriverManager.getConnection(url,
username, password);

conn.setTransactionIsolation(Connection.TRANSACTION_REPEATABLE_READ);
                conn.setAutoCommit(false);

                String sql = "update a set count=count+100 where id=1";
                PreparedStatement ps = conn.prepareStatement(sql);
                ps.executeUpdate();
                Thread.sleep(10000);
            } catch (SQLException throwables) {
                throwables.printStackTrace();
            } catch (ClassNotFoundException e) {
                e.printStackTrace();
            } catch (InterruptedException e) {
                e.printStackTrace();
            }
        }
    };
    a.start();
    Thread.sleep(1000);
    b.start();
  }
}
```

运行结果如下：

```
id=1 count=100
id=1 count=200
```

结论：出现脏读。

### 2. 测试不可重复读

还原数据表 a 的内容，示例代码如下：

```
package com.ghy.www.transaction_read_uncommitted;

import java.sql.*;

public class NonRepeatableRead {
    public static void main(String[] args) throws InterruptedException {

        Thread a = new Thread() {
            @Override
            public void run() {
                try {
                    String username = "root";
                    String password = "123123";
                    String url = "jdbc:mysql://localhost:3306/
y2?serverTimezone=Asia/Shanghai";
                    Class.forName("com.mysql.cj.jdbc.Driver");
                    Connection    conn    =    DriverManager.getConnection(url,
username, password);

conn.setTransactionIsolation(Connection.TRANSACTION_READ_UNCOMMITTED);
                    conn.setAutoCommit(false);

                    String sql = "select * from a where id=1";
                    PreparedStatement ps = null;
                    {
                        ps = conn.prepareStatement(sql);
                        ResultSet rs = ps.executeQuery();
                        while (rs.next()) {
                            System.out.println("id=" + rs.getString("id") + "
count=" + rs.getString("count"));
                        }
                    }
                    Thread.sleep(3000);
```

```
                {
                    ResultSet rs = ps.executeQuery();
                    while (rs.next()) {
                        System.out.println("id=" + rs.getString("id") + "
count=" + rs.getString("count"));
                    }
                }
            } catch (SQLException throwables) {
                throwables.printStackTrace();
            } catch (ClassNotFoundException e) {
                e.printStackTrace();
            } catch (InterruptedException e) {
                e.printStackTrace();
            }
        }
    };

    Thread b = new Thread() {
        @Override
        public void run() {
            try {
                String username = "root";
                String password = "123123";
                String url = "jdbc:mysql://localhost:3306/
y2?serverTimezone=Asia/Shanghai";
                Class.forName("com.mysql.cj.jdbc.Driver");
                Connection    conn    =    DriverManager.getConnection(url,
username, password);

conn.setTransactionIsolation(Connection.TRANSACTION_REPEATABLE_READ);
                conn.setAutoCommit(false);

                String sql = "update a set count=count+100 where id=1";
                PreparedStatement ps = conn.prepareStatement(sql);
                ps.executeUpdate();
                ps.close();
                conn.commit();
                conn.close();
            } catch (SQLException throwables) {
                throwables.printStackTrace();
            } catch (ClassNotFoundException e) {
```

```
                e.printStackTrace();
            }
        }
    };
    a.start();
    Thread.sleep(1000);
    b.start();
  }
}
```

运行结果如下：

```
id=1 count=100
id=1 count=200
```

结论：出现不可重复读。

### 3. 测试幻读

还原数据表 a 的内容，示例代码如下：

```
package com.ghy.www.transaction_read_uncommitted;

import java.sql.*;

public class PhantomRead {
    public static void main(String[] args) throws InterruptedException {
        Thread a = new Thread() {
            @Override
            public void run() {
                try {
                    String username = "root";
                    String password = "123123";
                    String url = "jdbc:mysql://localhost:3306/
y2?serverTimezone=Asia/Shanghai";
                    Class.forName("com.mysql.cj.jdbc.Driver");
                    Connection conn = DriverManager.getConnection(url,
username, password);

conn.setTransactionIsolation(Connection.TRANSACTION_READ_UNCOMMITTED);
                    conn.setAutoCommit(false);

                    String sql = "select * from a order by id asc";
```

```
                PreparedStatement ps = null;
                {
                    ps = conn.prepareStatement(sql);
                    ResultSet rs = ps.executeQuery();
                    while (rs.next()) {
                        System.out.println("id=" + rs.getString("id") + "
count=" + rs.getString("count"));
                    }
                }
                System.out.println("---------");
                Thread.sleep(4000);
                {
                    ResultSet rs = ps.executeQuery();
                    while (rs.next()) {
                        System.out.println("id=" + rs.getString("id") + "
count=" + rs.getString("count"));
                    }
                }
            } catch (SQLException throwables) {
                throwables.printStackTrace();
            } catch (ClassNotFoundException e) {
                e.printStackTrace();
            } catch (InterruptedException e) {
                e.printStackTrace();
            }
        }
    };

    Thread b = new Thread() {
        @Override
        public void run() {
            try {
                String username = "root";
                String password = "123123";
                String url = "jdbc:mysql://localhost:3306/
y2?serverTimezone=Asia/Shanghai";
                Class.forName("com.mysql.cj.jdbc.Driver");
                Connection    conn    =    DriverManager.getConnection(url,
username, password);

conn.setTransactionIsolation(Connection.TRANSACTION_REPEATABLE_READ);
                conn.setAutoCommit(false);
```

```
                String sql = "insert into a(count) values(200)";
                PreparedStatement ps = conn.prepareStatement(sql);
                ps.executeUpdate();
                ps.close();
                conn.commit();
                conn.close();
            } catch (SQLException throwables) {
                throwables.printStackTrace();
            } catch (ClassNotFoundException e) {
                e.printStackTrace();
            }
        }
    };
    a.start();
    Thread.sleep(2000);
    b.start();
  }
}
```

运行结果如下：

```
id=1 count=100
---------
id=1 count=100
id=5 count=200
```

结论：出现幻读。

## 4.16.2 TRANSACTION_READ_COMMITTED

下面测试 TRANSACTION_READ_COMMITTED 隔离级别。

### 1. 测试脏读

还原数据表 a 的内容，示例代码如下：

```
package com.ghy.www.transaction_read_committed;

import java.sql.*;

public class DirtyRead {
   public static void main(String[] args) throws InterruptedException {
      Thread a = new Thread() {
```

```
            @Override
            public void run() {
                try {
                    String username = "root";
                    String password = "123123";
                    String url = "jdbc:mysql://localhost:3306/
y2?serverTimezone=Asia/Shanghai";
                    Class.forName("com.mysql.cj.jdbc.Driver");
                    Connection   conn   =   DriverManager.getConnection(url,
username, password);

conn.setTransactionIsolation(Connection.TRANSACTION_READ_COMMITTED);
                    conn.setAutoCommit(false);

                    String sql = "select * from a where id=1";
                    PreparedStatement ps = null;
                    {
                        ps = conn.prepareStatement(sql);
                        ResultSet rs = ps.executeQuery();
                        while (rs.next()) {
                            System.out.println("id=" + rs.getString("id") + "
count=" + rs.getString("count"));
                        }
                    }
                    Thread.sleep(3000);
                    {
                        ResultSet rs = ps.executeQuery();
                        while (rs.next()) {
                            System.out.println("id=" + rs.getString("id") + "
count=" + rs.getString("count"));
                        }
                    }
                } catch (SQLException throwables) {
                    throwables.printStackTrace();
                } catch (ClassNotFoundException e) {
                    e.printStackTrace();
                } catch (InterruptedException e) {
                    e.printStackTrace();
                }
            }
        };

        Thread b = new Thread() {
```

```
            @Override
            public void run() {
                try {
                    String username = "root";
                    String password = "123123";
                    String url = "jdbc:mysql://localhost:3306/
y2?serverTimezone=Asia/Shanghai";
                    Class.forName("com.mysql.cj.jdbc.Driver");
                    Connection    conn    =    DriverManager.getConnection(url,
username, password);

conn.setTransactionIsolation(Connection.TRANSACTION_REPEATABLE_READ);
                    conn.setAutoCommit(false);

                    String sql = "update a set count=count+100 where id=1";
                    PreparedStatement ps = conn.prepareStatement(sql);
                    ps.executeUpdate();
                    Thread.sleep(10000);
                } catch (SQLException throwables) {
                    throwables.printStackTrace();
                } catch (ClassNotFoundException e) {
                    e.printStackTrace();
                } catch (InterruptedException e) {
                    e.printStackTrace();
                }
            }
        };
        a.start();
        Thread.sleep(1000);
        b.start();
    }
}
```

运行结果如下：

```
id=1 count=100
id=1 count=100
```

结论：未出现脏读。

### 2. 测试不可重复读

还原数据表 a 的内容，示例代码如下：

```
package com.ghy.www.transaction_read_committed;

import java.sql.*;

public class NonRepeatableRead {
    public static void main(String[] args) throws InterruptedException {

        Thread a = new Thread() {
            @Override
            public void run() {
                try {
                    String username = "root";
                    String password = "123123";
                    String url = "jdbc:mysql://localhost:3306/
y2?serverTimezone=Asia/Shanghai";
                    Class.forName("com.mysql.cj.jdbc.Driver");
                    Connection    conn    =    DriverManager.getConnection(url,
username, password);

conn.setTransactionIsolation(Connection.TRANSACTION_READ_COMMITTED);
                    conn.setAutoCommit(false);

                    String sql = "select * from a where id=1";
                    PreparedStatement ps = null;
                    {
                        ps = conn.prepareStatement(sql);
                        ResultSet rs = ps.executeQuery();
                        while (rs.next()) {
                            System.out.println("id=" + rs.getString("id") + "
count=" + rs.getString("count"));
                        }
                    }
                    Thread.sleep(3000);
                    {
                        ResultSet rs = ps.executeQuery();
                        while (rs.next()) {
                            System.out.println("id=" + rs.getString("id") + "
count=" + rs.getString("count"));
                        }
                    }
                } catch (SQLException throwables) {
                    throwables.printStackTrace();
                } catch (ClassNotFoundException e) {
```

```
                e.printStackTrace();
            } catch (InterruptedException e) {
                e.printStackTrace();
            }
        }
    };

    Thread b = new Thread() {
        @Override
        public void run() {
            try {
                String username = "root";
                String password = "123123";
                String url = "jdbc:mysql://localhost:3306/
y2?serverTimezone=Asia/Shanghai";
                Class.forName("com.mysql.cj.jdbc.Driver");
                Connection   conn   =   DriverManager.getConnection(url,
username, password);

conn.setTransactionIsolation(Connection.TRANSACTION_REPEATABLE_READ);
                conn.setAutoCommit(false);

                String sql = "update a set count=count+100 where id=1";
                PreparedStatement ps = conn.prepareStatement(sql);
                ps.executeUpdate();
                ps.close();
                conn.commit();
                conn.close();
            } catch (SQLException throwables) {
                throwables.printStackTrace();
            } catch (ClassNotFoundException e) {
                e.printStackTrace();
            }
        }
    };
    a.start();
    Thread.sleep(1000);
    b.start();
  }
}
```

运行结果如下：

```
id=1 count=100
id=1 count=200
```

结论：出现不可重复读。

### 3. 测试幻读

还原数据表 a 的内容，示例代码如下：

```
package com.ghy.www.transaction_read_committed;

import java.sql.*;

public class PhantomRead {
    public static void main(String[] args) throws InterruptedException {
        Thread a = new Thread() {
            @Override
            public void run() {
                try {
                    String username = "root";
                    String password = "123123";
                    String url = "jdbc:mysql://localhost:3306/
y2?serverTimezone=Asia/Shanghai";
                    Class.forName("com.mysql.cj.jdbc.Driver");
                    Connection    conn    =    DriverManager.getConnection(url,
username, password);

conn.setTransactionIsolation(Connection.TRANSACTION_READ_COMMITTED);
                    conn.setAutoCommit(false);

                    String sql = "select * from a order by id asc";
                    PreparedStatement ps = null;
                    {
                        ps = conn.prepareStatement(sql);
                        ResultSet rs = ps.executeQuery();
                        while (rs.next()) {
                            System.out.println("id=" + rs.getString("id") + "
count=" + rs.getString("count"));
                        }
                    }
                    System.out.println("---------");
                    Thread.sleep(4000);
                    {
```

```
                    ResultSet rs = ps.executeQuery();
                    while (rs.next()) {
                        System.out.println("id=" + rs.getString("id") + "
count=" + rs.getString("count"));
                    }
                }
            } catch (SQLException throwables) {
                throwables.printStackTrace();
            } catch (ClassNotFoundException e) {
                e.printStackTrace();
            } catch (InterruptedException e) {
                e.printStackTrace();
            }
        }
    };

    Thread b = new Thread() {
        @Override
        public void run() {
            try {
                String username = "root";
                String password = "123123";
                String url = "jdbc:mysql://localhost:3306/
y2?serverTimezone=Asia/Shanghai";
                Class.forName("com.mysql.cj.jdbc.Driver");
                Connection conn = DriverManager.getConnection(url,
username, password);

conn.setTransactionIsolation(Connection.TRANSACTION_REPEATABLE_READ);
                conn.setAutoCommit(false);

                String sql = "insert into a(count) values(200)";
                PreparedStatement ps = conn.prepareStatement(sql);
                ps.executeUpdate();
                ps.close();
                conn.commit();
                conn.close();
            } catch (SQLException throwables) {
                throwables.printStackTrace();
            } catch (ClassNotFoundException e) {
                e.printStackTrace();
            }
        }
```

```
        };
        a.start();
        Thread.sleep(2000);
        b.start();
    }
}
```

运行结果如下：

```
id=1 count=100
---------
id=1 count=100
id=5 count=200
```

结论：出现幻读。

## 4.16.3 TRANSACTION_REPEATABLE_READ

下面测试 TRANSACTION_REPEATABLE_READ 隔离级别。

**1. 测试脏读**

还原数据表 a 的内容，示例代码如下：

```
package com.ghy.www.transaction_repeatable_read;

import java.sql.*;

public class DirtyRead {
    public static void main(String[] args) throws InterruptedException {
        Thread a = new Thread() {
            @Override
            public void run() {
                try {
                    String username = "root";
                    String password = "123123";
                    String url = "jdbc:mysql://localhost:3306/
y2?serverTimezone=Asia/Shanghai";
                    Class.forName("com.mysql.cj.jdbc.Driver");
                    Connection    conn    =    DriverManager.getConnection(url,
username, password);
```

```
conn.setTransactionIsolation(Connection.TRANSACTION_REPEATABLE_READ);
                conn.setAutoCommit(false);

                String sql = "select * from a where id=1";
                PreparedStatement ps = null;
                {
                    ps = conn.prepareStatement(sql);
                    ResultSet rs = ps.executeQuery();
                    while (rs.next()) {
                        System.out.println("id=" + rs.getString("id") + "
count=" + rs.getString("count"));
                    }
                }
                Thread.sleep(3000);
                {
                    ResultSet rs = ps.executeQuery();
                    while (rs.next()) {
                        System.out.println("id=" + rs.getString("id") + "
count=" + rs.getString("count"));
                    }
                }
            } catch (SQLException throwables) {
                throwables.printStackTrace();
            } catch (ClassNotFoundException e) {
                e.printStackTrace();
            } catch (InterruptedException e) {
                e.printStackTrace();
            }
        }
    };

    Thread b = new Thread() {
        @Override
        public void run() {
            try {
                String username = "root";
                String password = "123123";
                String url = "jdbc:mysql://localhost:3306/
y2?serverTimezone=Asia/Shanghai";
                Class.forName("com.mysql.cj.jdbc.Driver");
                Connection   conn   =   DriverManager.getConnection(url,
username, password);
```

```
conn.setTransactionIsolation(Connection.TRANSACTION_REPEATABLE_READ);
                conn.setAutoCommit(false);

                String sql = "update a set count=count+100 where id=1";
                PreparedStatement ps = conn.prepareStatement(sql);
                ps.executeUpdate();
                Thread.sleep(10000);
            } catch (SQLException throwables) {
                throwables.printStackTrace();
            } catch (ClassNotFoundException e) {
                e.printStackTrace();
            } catch (InterruptedException e) {
                e.printStackTrace();
            }
        }
    };
    a.start();
    Thread.sleep(1000);
    b.start();
  }
}
```

运行结果如下：

```
id=1 count=100
id=1 count=100
```

结论：未出现脏读。

### 2. 测试不可重复读

还原数据表 a 的内容，示例代码如下：

```
package com.ghy.www.transaction_repeatable_read;

import java.sql.*;

public class NonRepeatableRead {
    public static void main(String[] args) throws InterruptedException {

        Thread a = new Thread() {
            @Override
            public void run() {
```

```
            try {
                String username = "root";
                String password = "123123";
                String url = "jdbc:mysql://localhost:3306/
y2?serverTimezone=Asia/Shanghai";
                Class.forName("com.mysql.cj.jdbc.Driver");
                Connection    conn    =    DriverManager.getConnection(url,
username, password);

conn.setTransactionIsolation(Connection.TRANSACTION_REPEATABLE_READ);
                conn.setAutoCommit(false);

                String sql = "select * from a where id=1";
                PreparedStatement ps = null;
                {
                    ps = conn.prepareStatement(sql);
                    ResultSet rs = ps.executeQuery();
                    while (rs.next()) {
                        System.out.println("id=" + rs.getString("id") + "
count=" + rs.getString("count"));
                    }
                }
                Thread.sleep(3000);
                {
                    ResultSet rs = ps.executeQuery();
                    while (rs.next()) {
                        System.out.println("id=" + rs.getString("id") + "
count=" + rs.getString("count"));
                    }
                }
            } catch (SQLException throwables) {
                throwables.printStackTrace();
            } catch (ClassNotFoundException e) {
                e.printStackTrace();
            } catch (InterruptedException e) {
                e.printStackTrace();
            }
        }
    };

    Thread b = new Thread() {
        @Override
        public void run() {
```

```
                try {
                    String username = "root";
                    String password = "123123";
                    String url = "jdbc:mysql://localhost:3306/
y2?serverTimezone=Asia/Shanghai";
                    Class.forName("com.mysql.cj.jdbc.Driver");
                    Connection    conn    =    DriverManager.getConnection(url,
username, password);

conn.setTransactionIsolation(Connection.TRANSACTION_REPEATABLE_READ);
                    conn.setAutoCommit(false);

                    String sql = "update a set count=count+100 where id=1";
                    PreparedStatement ps = conn.prepareStatement(sql);
                    ps.executeUpdate();
                    ps.close();
                    conn.commit();
                    conn.close();
                } catch (SQLException throwables) {
                    throwables.printStackTrace();
                } catch (ClassNotFoundException e) {
                    e.printStackTrace();
                }
            }
        };
        a.start();
        Thread.sleep(1000);
        b.start();
    }
}
```

运行结果如下：

```
id=1 count=100
id=1 count=100
```

结论：未出现不可重复读。

### 3. 测试幻读

还原数据表 a 的内容，示例代码如下：

```
package com.ghy.www.transaction_repeatable_read;
```

```
import java.sql.*;

public class PhantomRead {
    public static void main(String[] args) throws InterruptedException {
        Thread a = new Thread() {
            @Override
            public void run() {
                try {
                    String username = "root";
                    String password = "123123";
                    String url = "jdbc:mysql://localhost:3306/
y2?serverTimezone=Asia/Shanghai";
                    Class.forName("com.mysql.cj.jdbc.Driver");
                    Connection    conn    =    DriverManager.getConnection(url,
username, password);

conn.setTransactionIsolation(Connection.TRANSACTION_REPEATABLE_READ);
                    conn.setAutoCommit(false);

                    String sql = "select * from a order by id asc";
                    PreparedStatement ps = null;
                    {
                        ps = conn.prepareStatement(sql);
                        ResultSet rs = ps.executeQuery();
                        while (rs.next()) {
                            System.out.println("id=" + rs.getString("id") + "
count=" + rs.getString("count"));
                        }
                    }
                    System.out.println("---------");
                    Thread.sleep(4000);
                    {
                        ResultSet rs = ps.executeQuery();
                        while (rs.next()) {
                            System.out.println("id=" + rs.getString("id") + "
count=" + rs.getString("count"));
                        }
                    }
                } catch (SQLException throwables) {
                    throwables.printStackTrace();
                } catch (ClassNotFoundException e) {
                    e.printStackTrace();
                } catch (InterruptedException e) {
```

```
                e.printStackTrace();
            }
        }
    };

    Thread b = new Thread() {
        @Override
        public void run() {
            try {
                String username = "root";
                String password = "123123";
                String url = "jdbc:mysql://localhost:3306/
y2?serverTimezone=Asia/Shanghai";
                Class.forName("com.mysql.cj.jdbc.Driver");
                Connection    conn    =    DriverManager.getConnection(url,
username, password);

conn.setTransactionIsolation(Connection.TRANSACTION_REPEATABLE_READ);
                conn.setAutoCommit(false);

                String sql = "insert into a(count) values(200)";
                PreparedStatement ps = conn.prepareStatement(sql);
                ps.executeUpdate();
                ps.close();
                conn.commit();
                conn.close();
            } catch (SQLException throwables) {
                throwables.printStackTrace();
            } catch (ClassNotFoundException e) {
                e.printStackTrace();
            }
        }
    };
    a.start();
    Thread.sleep(2000);
    b.start();
  }
}
```

运行结果如下：

```
id=1 count=100
---------
```

```
id=1 count=100
```

结论：未出现幻读，原因是 MySQL 解决了这个问题。

## 4.16.4 TRANSACTION_SERIALIZABLE

下面测试 TRANSACTION_SERIALIZABLE 隔离级别。

### 1. 测试脏读

还原数据表 a 的内容，示例代码如下：

```
package com.ghy.www.transaction_serializable;

import java.sql.*;

public class DirtyRead {
    public static void main(String[] args) throws InterruptedException {
        Thread a = new Thread() {
            @Override
            public void run() {
                try {
                    String username = "root";
                    String password = "123123";
                    String url = "jdbc:mysql://localhost:3306/
y2?serverTimezone=Asia/Shanghai";
                    Class.forName("com.mysql.cj.jdbc.Driver");
                    Connection    conn    =    DriverManager.getConnection(url,
username, password);

conn.setTransactionIsolation(Connection.TRANSACTION_SERIALIZABLE);
                    conn.setAutoCommit(false);

                    String sql = "select * from a where id=1";
                    PreparedStatement ps = null;
                    {
                        ps = conn.prepareStatement(sql);
                        ResultSet rs = ps.executeQuery();
                        while (rs.next()) {
                            System.out.println("id=" + rs.getString("id") + "
count=" + rs.getString("count"));
                        }
                    }
```

```
                Thread.sleep(3000);
                {
                    ResultSet rs = ps.executeQuery();
                    while (rs.next()) {
                        System.out.println("id=" + rs.getString("id") + "
count=" + rs.getString("count"));
                    }
                }
            } catch (SQLException throwables) {
                throwables.printStackTrace();
            } catch (ClassNotFoundException e) {
                e.printStackTrace();
            } catch (InterruptedException e) {
                e.printStackTrace();
            }
        }
    };

    Thread b = new Thread() {
        @Override
        public void run() {
            try {
                String username = "root";
                String password = "123123";
                String url = "jdbc:mysql://localhost:3306/
y2?serverTimezone=Asia/Shanghai";
                Class.forName("com.mysql.cj.jdbc.Driver");
                Connection conn = DriverManager.getConnection(url,
username, password);
                conn.setTransactionIsolation(Connection.
TRANSACTION_REPEATABLE_READ);
                conn.setAutoCommit(false);

                String sql = "update a set count=count+100 where id=1";
                PreparedStatement ps = conn.prepareStatement(sql);
                ps.executeUpdate();
                Thread.sleep(10000);
            } catch (SQLException throwables) {
                throwables.printStackTrace();
            } catch (ClassNotFoundException e) {
                e.printStackTrace();
            } catch (InterruptedException e) {
                e.printStackTrace();
```

```
                }
            }
        };
        a.start();
        Thread.sleep(1000);
        b.start();
    }
}
```

运行结果如下：

```
id=1 count=100
id=1 count=100
```

结论：未出现脏读。

### 2. 测试不可重复读

还原数据表 a 的内容，示例代码如下：

```
package com.ghy.www.transaction_serializable;

import java.sql.*;

public class NonRepeatableRead {
    public static void main(String[] args) throws InterruptedException {

        Thread a = new Thread() {
            @Override
            public void run() {
                try {
                    String username = "root";
                    String password = "123123";
                    String url = "jdbc:mysql://localhost:3306/
y2?serverTimezone=Asia/Shanghai";
                    Class.forName("com.mysql.cj.jdbc.Driver");
                    Connection    conn    =    DriverManager.getConnection(url,
username, password);

conn.setTransactionIsolation(Connection.TRANSACTION_SERIALIZABLE);
                    conn.setAutoCommit(false);

                    String sql = "select * from a where id=1";
```

```
                PreparedStatement ps = null;
                {
                    ps = conn.prepareStatement(sql);
                    ResultSet rs = ps.executeQuery();
                    while (rs.next()) {
                        System.out.println("id=" + rs.getString("id") + "
count=" + rs.getString("count"));
                    }
                }
                Thread.sleep(3000);
                {
                    ResultSet rs = ps.executeQuery();
                    while (rs.next()) {
                        System.out.println("id=" + rs.getString("id") + "
count=" + rs.getString("count"));
                    }
                }
            } catch (SQLException throwables) {
                throwables.printStackTrace();
            } catch (ClassNotFoundException e) {
                e.printStackTrace();
            } catch (InterruptedException e) {
                e.printStackTrace();
            }
        }
    };

    Thread b = new Thread() {
        @Override
        public void run() {
            try {
                String username = "root";
                String password = "123123";
                String url = "jdbc:mysql://localhost:3306/
y2?serverTimezone=Asia/Shanghai";
                Class.forName("com.mysql.cj.jdbc.Driver");
                Connection    conn    =    DriverManager.getConnection(url,
username, password);

conn.setTransactionIsolation(Connection.TRANSACTION_REPEATABLE_READ);
                conn.setAutoCommit(false);

                String sql = "update a set count=count+100 where id=1";
```

```
                PreparedStatement ps = conn.prepareStatement(sql);
                ps.executeUpdate();
                ps.close();
                conn.commit();
                conn.close();
            } catch (SQLException throwables) {
                throwables.printStackTrace();
            } catch (ClassNotFoundException e) {
                e.printStackTrace();
            }
        }
    };
    a.start();
    Thread.sleep(1000);
    b.start();
  }
}
```

运行结果如下：

```
id=1 count=100
id=1 count=100
```

结论：未出现不可重复读。

### 3. 测试幻读

还原数据表 a 的内容，示例代码如下：

```
package com.ghy.www.transaction_serializable;

import java.sql.*;

public class PhantomRead {
    public static void main(String[] args) throws InterruptedException {
        Thread a = new Thread() {
            @Override
            public void run() {
                try {
                    String username = "root";
                    String password = "123123";
                    String url = "jdbc:mysql://localhost:3306/
y2?serverTimezone=Asia/Shanghai";
```

```
                Class.forName("com.mysql.cj.jdbc.Driver");
                Connection   conn   =   DriverManager.getConnection(url,
username, password);

conn.setTransactionIsolation(Connection.TRANSACTION_SERIALIZABLE);
                conn.setAutoCommit(false);

                String sql = "select * from a order by id asc";
                PreparedStatement ps = null;
                {
                    ps = conn.prepareStatement(sql);
                    ResultSet rs = ps.executeQuery();
                    while (rs.next()) {
                        System.out.println("id=" + rs.getString("id") + "
count=" + rs.getString("count"));
                    }
                }
                System.out.println("---------");
                Thread.sleep(4000);
                {
                    ResultSet rs = ps.executeQuery();
                    while (rs.next()) {
                        System.out.println("id=" + rs.getString("id") + "
count=" + rs.getString("count"));
                    }
                }
            } catch (SQLException throwables) {
                throwables.printStackTrace();
            } catch (ClassNotFoundException e) {
                e.printStackTrace();
            } catch (InterruptedException e) {
                e.printStackTrace();
            }
        }
    };

    Thread b = new Thread() {
        @Override
        public void run() {
            try {
                String username = "root";
                String password = "123123";
                String                           url                        =
```

```
"jdbc:mysql://localhost:3306/y2?serverTimezone=Asia/Shanghai";
                Class.forName("com.mysql.cj.jdbc.Driver");
                Connection   conn   =   DriverManager.getConnection(url,
username, password);
                conn.setAutoCommit(false);

                String sql = "insert into a(count) values(200)";
                PreparedStatement ps = conn.prepareStatement(sql);
                ps.executeUpdate();
                ps.close();
                conn.commit();
                conn.close();
            } catch (SQLException throwables) {
                throwables.printStackTrace();
            } catch (ClassNotFoundException e) {
                e.printStackTrace();
            }
        }
    };
    a.start();
    Thread.sleep(2000);
    b.start();
  }
}
```

运行结果如下：

```
id=1 count=100
---------
id=1 count=100
```

结论：未出现幻读。

# 第 5 章 Servlet 核心技术

## 5.1 Servlet简介

什么是 Servlet? 先来看看百度百科的解释，如图 5-1 所示。根据百度百科的解释可以总结 Servlet 的知识点如下。

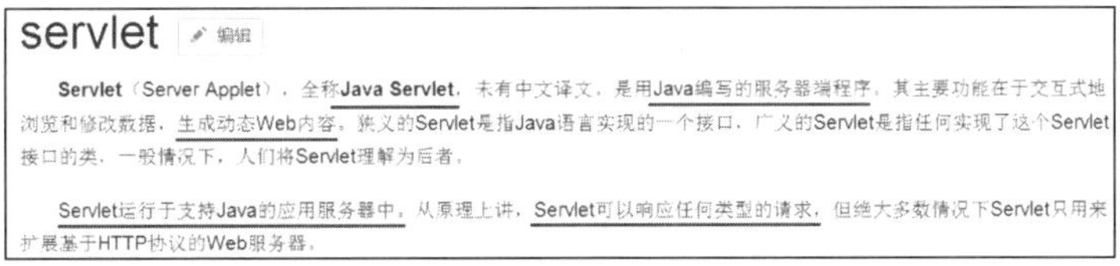

servlet 编辑

**Servlet**（Server Applet），全称**Java Servlet**，未有中文译文，是用Java编写的服务器端程序，其主要功能在于交互式地浏览和修改数据，生成动态Web内容。狭义的Servlet是指Java语言实现的一个接口，广义的Servlet是指任何实现了这个Servlet接口的类，一般情况下，人们将Servlet理解为后者。

Servlet运行于支持Java的应用服务器中，从原理上讲，Servlet可以响应任何类型的请求，但绝大多数情况下Servlet只用来扩展基于HTTP协议的Web服务器。

图 5-1

（1）Servlet 是基于 Java 语言的，用 Java 编写的技术。

（2）Servlet 是基于 Web 的。

（3）Servlet 是 Java 语言与 Web 进行交互的一个接口，支持 HTTP 协议。

（4）Servlet 主要用于 Web 项目的后台处理。

（5）使用 Servlet 可以实现动态的网页，这个动态并不是指视觉上很花哨的特效，而是展示的数据可以不像 HTML 文件那样固定，显示的数据是可以通过程序来定制的，具有数据展示动态性，也就是具有后期维护性。

这 5 点已基本将 Servlet 所提供的功能进行了概括，本章通过对 Servlet 的学习，可以掌握基于 Java 语言整合 Web 技术的程序设计，并不局限于在 IntelliJ IDEA 的 Console 控制台面板中进行数据的展示与处理，完全可以将数据在浏览器中进行显示，软件的后台程序还可以与前端界面的操作者进行交互。

Servlet 是从 Web 开发到 Web 框架使用的必学技术，它是 Java Web 框架底层的实现。

在 Java Web 领域中，处理 HTTP 协议最快的组件就是 Servlet，所以掌握 Servlet 技术是非常重要的。

Servlet 在项目中的位置及作用如图 5-2 所示。

（1）Web 项目：初学 Java 时创建的都是 Java 项目，运行的环境都是在控制台中，而 Web 项目以 HTML 为界面和客户进行交互，而不使用 Java 项目的控制台。

（2）Tomcat：Tomcat 是运行 Java Web 项目的一个软件，Java Web 项目想要被外界所访问，必须要放入 Tomcat 中。Tomcat 里面部署 Java Web 项目，而 Java Web 项目里面包含 Servlet 代码。Tomcat 实现了 JavaEE 规范和标准。

（3）增删改查：使用 JDBC 技术来操作数据库，实现对数据库的更改。

当在浏览器中输入 URL 后在内部会发生 9 步流程，如图 5-3 所示。

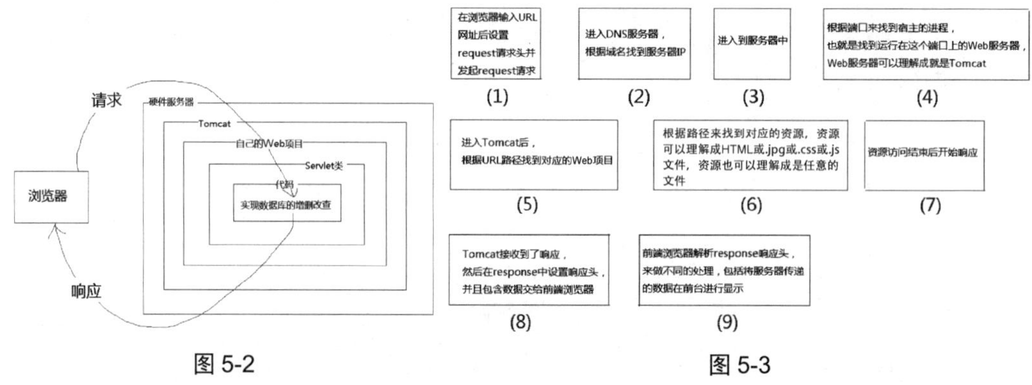


图 5-2　　　　图 5-3

在计算机行业中，“服务器”这个术语同时代表硬件的服务器与软件的服务器，硬件的服务器就是一台高性能的电脑，这台电脑主要提供存储数据、处理数据、传输数据的功能，常见的硬件服务器品牌有 IBM、惠普。软件服务器大多数是指处理 HTTP 协议的软件，它可以将 B 浏览器端与 S 服务器端进行通信连接，在 S 端处理从 B 端传输过来的数据，对 HTTP 协议标准进行实现，所以 Tomcat，Resin，Weblogic，JBoss，Websphere，Glassfish 等软件可以被称为“Web 容器”或“Web 服务器”。

总结一下 Servlet 及相关的知识。

（1）Servlet 是 Sun 公司制定的一种支持 Web 开发的组件规范，以程序员的角度来看，Servlet 就是一个具有处理 Web 功能的 Java 类。

（2）Servlet 基于 Web 完成一定的业务功能，Servlet 的代码不能随意设计，要遵守一定的继承与实现关系，因为 Servlet 是一个标准和规范。

（3）Servlet 组件不能独立运行，需要依赖 Servlet 容器才能运行，说明拥有了一个 Servlet 类是没有用的，因为没有运行的环境，常见的运行环境就是 Tomcat，Tomcat 是 Servlet 容器，也称 Java Web 容器。Java Web 容器提供 Servlet 组件的运行环境，可以管理 Servlet 组件的生命周期，包括创建 Servlet 以及销毁 Servlet。

（4）Tomcat 能执行动态的 Java 代码，这些 Java 代码就包括 Servlet，所以使用 Tomcat 后就可以使用 Java 代码处理动态资源的请求，比如可以在 Servlet 中操作数据库，执行一些业务性的代码等功能。

## 5.2　更改访问Tomcat的端口号

如果在启动 Tomcat 后，在控制台出现异常提示信息是有关端口被占用的，可以将 Tomcat 默认的端口 8080 改成其他的，因为 8080 端口在 Tomcat 启动之前已经被其他进程所占用，在 IntelliJ IDEA 中更改 Tomcat 端口如图 5-4 所示。

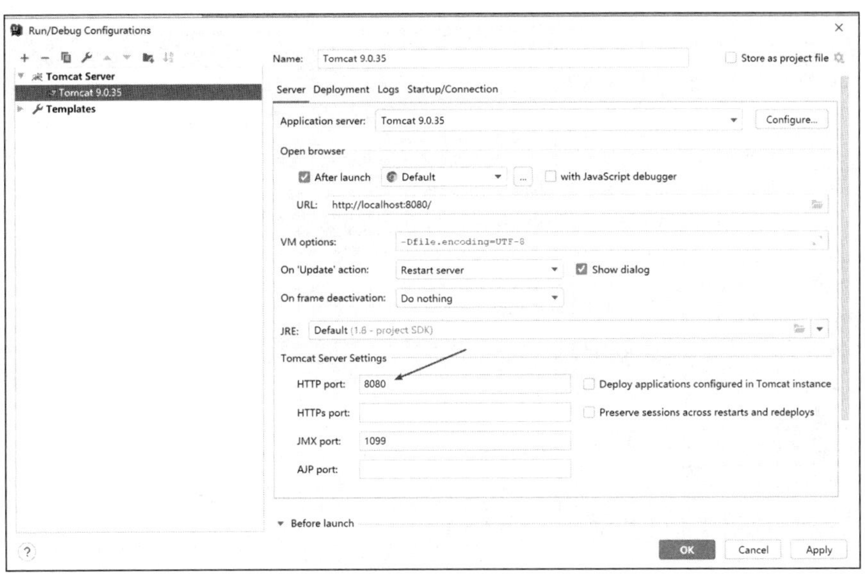


图 5-4

如果不是通过 IntelliJ IDEA 部署 Web 项目时，则可以直接在硬盘中编辑 server.xml 配置文件，步骤如下。

（1）打开 apache-tomcat\conf 文件夹。

（2）将 server.xml 文件中的配置：

```
<Connector port="8080" protocol="HTTP/1.1"
               connectionTimeout="20000"
               redirectPort="8443" />
```

改成

```
<Connector port="8081" protocol="HTTP/1.1"
               connectionTimeout="20000"
               redirectPort="8443" />
```

（3）保存文件。

（4）在 CMD 中执行命令：apache-tomcat\bin\startup.bat 重新启动 Tomcat 即可，如果 8081 没有被占用，则 Tomcat 顺利启动。

查看端口占用情况使用如下命令：

```
C:\>netstat -ano|findstr "8080"
  TCP    0.0.0.0:8080         0.0.0.0:0          LISTENING        13224
  TCP    127.0.0.1:14731      127.0.0.1:8080     TIME_WAIT        0
  TCP    127.0.0.1:14733      127.0.0.1:8080     TIME_WAIT        0
  TCP    [::]:8080            [::]:0             LISTENING        13224
  TCP    [::1]:8080           [::1]:14736        ESTABLISHED      13224
  TCP    [::1]:8080           [::1]:14737        ESTABLISHED      13224
  TCP    [::1]:14736          [::1]:8080         ESTABLISHED      4304
  TCP    [::1]:14737          [::1]:8080         ESTABLISHED      4304
C:\>tasklist |findstr         "13224"
java.exe                 13224 Console           4                251,160 K
C:\>
```

通过以上命令可以分析出，端口 8080 被 java.exe 进程所占用。

## 5.3 Servlet技术开发

本节主要介绍 Servlet 技术细节与使用方法，快速使用该技术构造基于 Web 的软件项目，在学习中着重掌握并积累 Servlet 在出现异常时的解决办法，增加 Web 开发与调试的经验。

### 5.3.1　Servlet 的继承与实现关系

前面章节已经使用过 Servlet 进行字符串信息的打印，大体了解 Servlet 的创建与使用的过程，下面开始学习 Servlet 类的继承结构与接口实现关系。

Servlet 类的继承结构如图 5-5 所示。从图中可以发现，自定义的 Servlet 类 Test.java 的父类是 HttpServlet，继续向上的父类是 GenericServlet，顶级父类就是 Object 了，学习 GenericServlet 类的结构有助于从宏观上了解 Servlet 的类继承与接口的实现关系，因为它是 Servlet 最主要的父类。类 GenericServlet.java 的信息如图 5-6 所示。

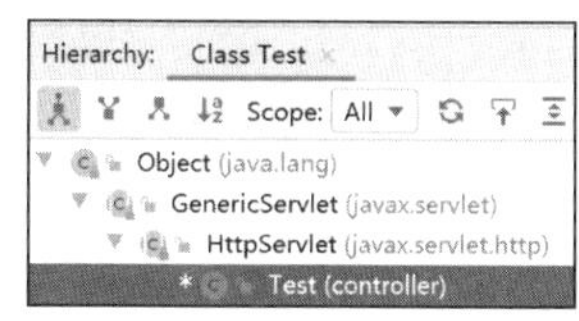


图 5-5

```
public abstract class GenericServlet implements Servlet, ServletConfig,
        java.io.Serializable {
```

图 5-6

从源代码中可以发现，类 GenericServlet.java 是一个抽象类，实现了 Servlet 和 ServletConfig 接口。接口 Servlet 的信息如图 5-7 所示。

接口 Servlet 主要提供了 Servlet 生命周期的过程，比如初始化 init()和销毁 destroy()方法。接口 ServletConfig 的信息如图 5-8 所示。

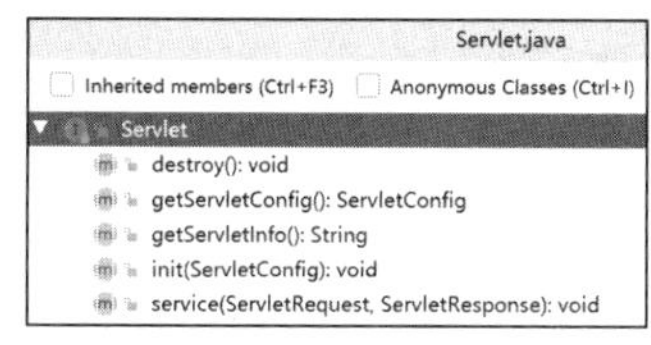


图 5-7

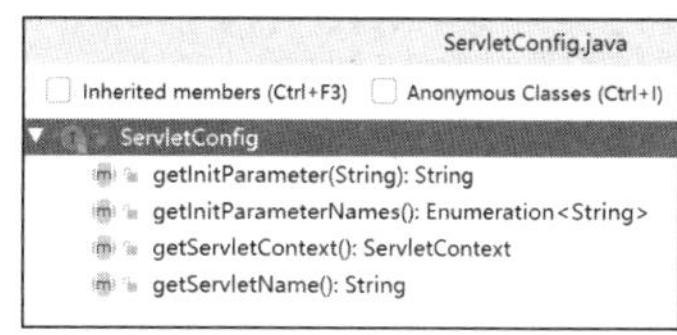


图 5-8

接口 ServletConfig 提供了获得配置信息数据的功能。

两个接口一共有 9 种方法，说明只要实现了这两个接口的实现类也就拥有这 9 种方法，这 9 种方法是可以在子类中进行调用的，类结构如图 5-9 所示。

在 Outline 大纲中带向上箭头的方法代表是重写或实现的操作。

GenericServlet.java 类不光实现了 Servlet 和 ServletConfig 接口中全部的 9 种方法，还加入了自己的扩展方法，比如方法 log()。

再来看看子类 HttpServlet.java，源代码信息如图 5-10 所示。

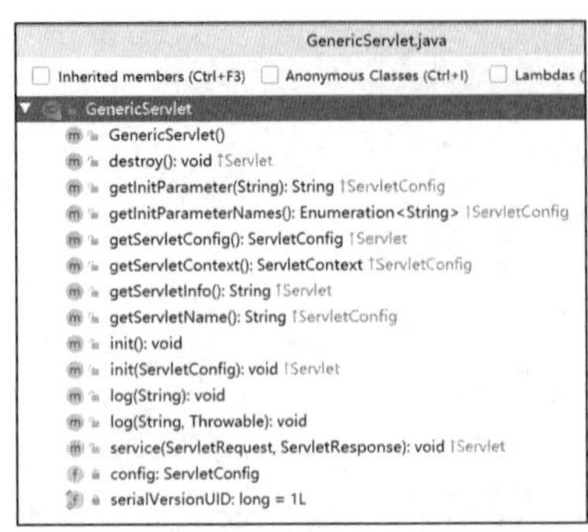


图 5-9

```
public abstract class HttpServlet extends GenericServlet {
```

图 5-10

从源代码中可以发现，类 HttpServlet.java 也是一个抽象类，父类是 GenericServlet.java。类 HttpServlet.java 没有实现其他的接口，方法列表如图 5-11 所示。

在 Outline 中可以发现，类 HttpServlet.java 只将父类的 public void service（ServletRequest req, ServletResponse res）方法进行了重写，其他的方法都是被保护的，并且是新添加的方法，在子类中可以重写并使用这些方法，比如 doGet()或 doPost()等。

经过上面的分析可以总结出 Servlet 继承与实现关系图如图 5-12 所示。

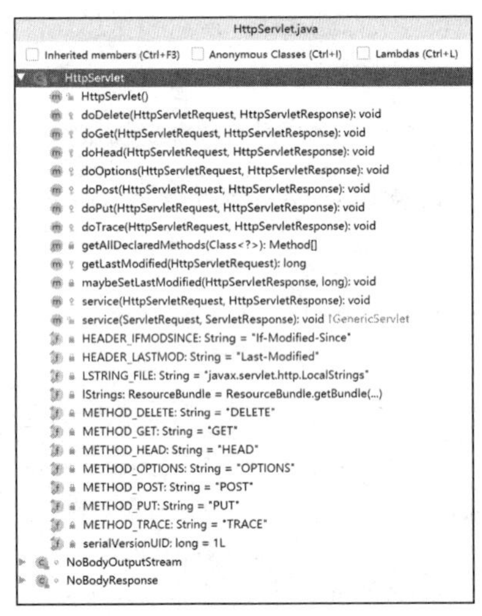


图 5-11

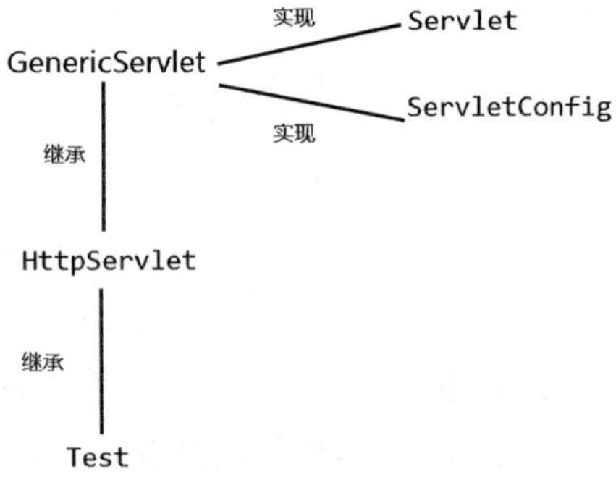

GenericServlet
实现
Servlet
实现
ServletConfig
继承
HttpServlet
继承
Test

图 5-12

从继承与实现关系图中可以发现，Servlet 的关系结构并不复杂，学习 Servlet 其实就是在学习 Servlet 与 ServletConfig 接口。

### 5.3.2　创建基于 xml 的 Servlet 案例

前面章节使用的是 IntelliJ IDEA 工具以向导式的方式创建 Servlet 并运行，其实并不能从根本上了解 Servlet 的创建与配置，本节就来丰富这个知识点。

创建全新的 Web 项目，项目中包含 web.xml 配置文件，如图 5-13 所示。创建新的 Java 类，效果如图 5-14 所示。

图 5-13

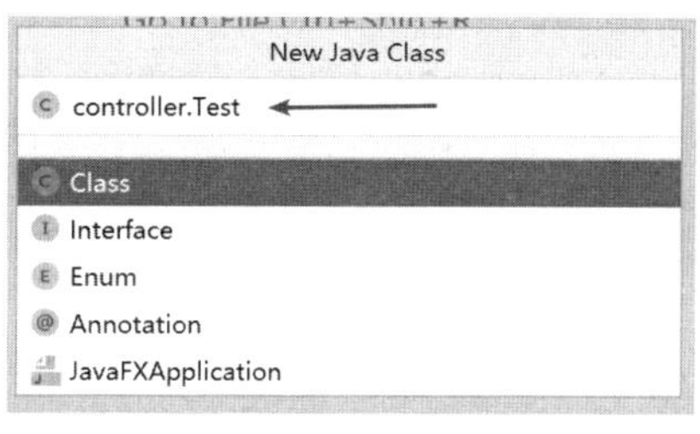


图 5-14

类 Test 继承自 HttpServlet，并重写 doGet()方法，完整的示例代码如下：

```
package controller.controller;

import javax.servlet.ServletException;
import javax.servlet.http.HttpServlet;
import javax.servlet.http.HttpServletRequest;
import javax.servlet.http.HttpServletResponse;
import java.io.IOException;

public class Test extends HttpServlet {
    @Override
    protected void doGet(HttpServletRequest req, HttpServletResponse resp)
throws ServletException, IOException {
        System.out.println("进入了 doGet()方法");
    }
}
```

一个普通的 Java 类虽然被定义成了一个 Servlet，但依然并不能处理 Web 的请求与响应，这时就要把这个 Java 类注册到系统中，变成一个可以真正处理 HTTP 请求的 Servlet，在哪里注册呢？在 web.xml 文件中。

编辑 web.xml 文件，代码如下：

```
<?xml version="1.0" encoding="UTF-8"?>
<web-app xmlns:xsi="http://www.w3.org/2001/XMLSchema-instance"
    xmlns="http://java.sun.com/xml/ns/javaee"
    xsi:schemaLocation="http://java.sun.com/xml/ns/javaee
http://java.sun.com/xml/ns/javaee/web-app_2_5.xsd"
    id="WebApp_ID" version="2.5">
    <display-name>web2</display-name>

    <servlet>
        <servlet-name>xxx</servlet-name>
        <servlet-class>controller.Test</servlet-class>
    </servlet>

    <servlet-mapping>
        <servlet-name>xxx</servlet-name>
        <url-pattern>/abcabc123</url-pattern>
    </servlet-mapping>

    <welcome-file-list>
        <welcome-file>index.html</welcome-file>
        <welcome-file>index.htm</welcome-file>
        <welcome-file>index.jsp</welcome-file>
        <welcome-file>default.html</welcome-file>
        <welcome-file>default.htm</welcome-file>
        <welcome-file>default.jsp</welcome-file>
    </welcome-file-list>
</web-app>
```

在 web.xml 文件中添加如下的核心代码：

```
<servlet>
    <servlet-name>xxx</servlet-name>
    <servlet-class>controller.Test</servlet-class>
</servlet>
<servlet-mapping>
    <servlet-name>xxx</servlet-name>
    <url-pattern>/abcabc123</url-pattern>
</servlet-mapping>
```

此代码的作用是配置 Servlet，可以使 Servlet 的 Java 类接收 HTTP 请求，标签的具体解释如下。

（1）标签<servlet-mapping>的主要作用是配置 Servlet 的访问路径信息。

（2）标签<url-pattern>的主要作用是配置具体的 Servlet 访问路径信息，也就是用什么具体的 URL 路径来访问指定的 Servlet，URL 路径不允许重复。

（3）标签<servlet>的主要作用是配置 Servlet 的 Java 类信息。

（4）标签<servlet-class>的主要作用是配置具体的 Servlet 类信息，包括包名及 Servlet 类名称。

（5）为什么输入一个网址就能执行对应的 Servlet 类呢？也就是<servlet-mapping>标签和<servlet>标签是如何合作工作的呢？关键的中介就是使用<servlet-name>标签设置别名，别名相同即可成功进行 URL 映射。

Servlet 的执行流程如下。

（1）浏览器中执行地址 abcabc123。

（2）请求进入 Tomcat，Tomcat 根据浏览器上的 URL 路径找到对应的 Web 项目。

（3）在指定 Web 项目中的 web.xml 配置文件寻找有没有标签<url-pattern>值为 abcabc123 的配置。

（4）如果存在，则获取同级<servlet-name>xxx</servlet-name>标签中的 xxx 配置别名。

（5）再根据这个 xxx 别名到 web.xml 文件中寻找有没有父标签是<servlet>，而子标签<servlet-name>中的值恰恰也正是 xxx 的配置。

（6）如果存在，则通过反射技术创建<servlet>的同级子标签<servlet-class>中的 Servlet 类的实例，进而执行 doGet()方法。

程序运行效果如图 5-15 所示。

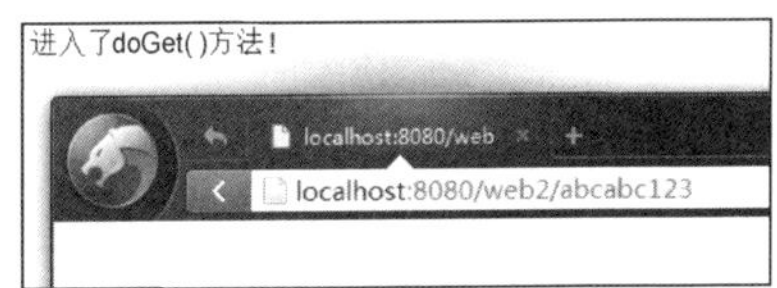


图 5-15

通过这个案例从零基础到掌握了 Servlet 的手动创建与 web.xml 配置，对 Servlet 内部的配置与处理更加熟悉。掌握查看 web.xml 中的配置即可正确执行 Servlet，是本小节的学习重点，也就是手动配置 web.xml 中的代码。

下面再来看一下 Servlet 执行的具体过程。

（1）在浏览器中输入网址 http://localhost:8080/test1web/servletName。

（2）发起访问服务器的客户端请求。

（3）正确进入 Tomcat。

（4）根据 URL 中的地址“test1web/servletName”，Tomcat 就能知道要到达 test1web 项目中名称为 servletName 的 Servlet。

（5）在 web.xml 配置文件中寻找有没有这个 Servlet 路径配置。

（6）如果存在这个 Servlet 路径配置，就将该 Servlet 通过反射技术进行实例化，再执行 doGet()或者 doPost()方法。

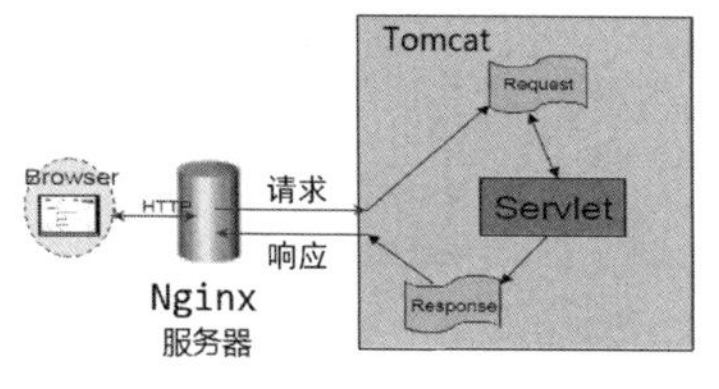


图 5-16

（7）Servlet 执行完毕后发出响应，响应对象中存储的内容就是客户端要显示的 HTML 代码。

（8）浏览器接收到响应，将 HTML 代码在浏览器中显示出来。

执行过程如图 5-16 所示。

### 5.3.3 正确与错误配置 Servlet 的不同情况

在 web.xml 文件中，错误地配置 Servlet 会出现若干异常，而正确地配置 Servlet 也会出现多种情况，下面就来演示一下。

创建新的 Web 项目，并在 controller 包中创建自定义 Servlet 类代码如下：

```
package controller;

import java.io.IOException;

import javax.servlet.ServletException;
import javax.servlet.http.HttpServlet;
import javax.servlet.http.HttpServletRequest;
import javax.servlet.http.HttpServletResponse;

public class MyServlet extends HttpServlet {
    @Override
    protected void doGet(HttpServletRequest req, HttpServletResponse resp)
throws ServletException, IOException {
        System.out.println("MyServlet go !");
    }
}
```

（1）当别名不一样时，web.xml 配置文件代码如下：

```
<servlet>
    <servlet-name>别名不一样</servlet-name>
    <servlet-class>controller.MyServlet</servlet-class>
</servlet>

<servlet-mapping>
    <servlet-name>myServlet</servlet-name>
    <url-pattern>/myServlet</url-pattern>
</servlet-mapping>
```

启动 Tomcat 后控制台出现异常信息如下：

```
java.lang.IllegalArgumentException: Servlet mapping specifies an unknown
servlet name [myServlet]
```

说明别名必须相同。

（2）当包名错误时，web.xml 配置文件代码如下：

```
<servlet>
<servlet-name>myServlet</servlet-name>
<servlet-class>错误的包名.MyServlet</servlet-class>
</servlet>

<servlet-mapping>
    <servlet-name>myServlet</servlet-name>
    <url-pattern>/myServlet</url-pattern>
</servlet-mapping>
```

启动 Tomcat 后，控制台并没有出现异常信息，但在使用路径 http://localhost:8080/web3/myServlet 访问这个 Servlet 时却出现异常信息如下：

```
java.lang.ClassNotFoundException: 错误的包名.MyServlet
```

由于包名是错误的，访问 Servlet 就会出现异常。

（3）当映射的 Servlet 路径 URL 出现重复时，web.xml 配置文件代码如下：

```
<servlet>
    <servlet-name>myServletA</servlet-name>
```

```
    <servlet-class>controller.MyServlet</servlet-class>
</servlet>
<servlet-mapping>
    <servlet-name>myServletA</servlet-name>
    <url-pattern>/myServlet</url-pattern>
</servlet-mapping>

<servlet>
    <servlet-name>myServletB</servlet-name>
    <servlet-class>controller.MyServlet</servlet-class>
</servlet>
<servlet-mapping>
    <servlet-name>myServletB</servlet-name>
    <url-pattern>/myServlet</url-pattern>
</servlet-mapping>
```

启动 Tomcat 后，控制台出现异常信息如下：

```
java.lang.IllegalArgumentException: The servlets named [myServletA] and [myServletB] are both mapped to the url-pattern [/myServlet] which is not permitted
```

映射的 Servlet 路径 URL 不能出现重复。

（4）当使用不同路径访问同一个 Servlet 时，web.xml 配置文件代码如下：

```
<servlet>
    <servlet-name>myServletA</servlet-name>
    <servlet-class>controller.MyServlet</servlet-class>
</servlet>
<servlet-mapping>
    <servlet-name>myServletA</servlet-name>
    <url-pattern>/myServletB</url-pattern>
</servlet-mapping>

<servlet>
    <servlet-name>myServletB</servlet-name>
    <servlet-class>controller.MyServlet</servlet-class>
</servlet>
<servlet-mapping>
    <servlet-name>myServletB</servlet-name>
    <url-pattern>/myServletA</url-pattern>
```

```
</servlet-mapping>
```

启动 Tomcat 后，控制台并没有出现异常，并且能使用下面两个 URL 访问同一个 Servlet：

http://localhost:8080/web3/myServletA

http://localhost:8080/web3/myServletB

说明不同的 Servlet 路径可以对应到同一个包中同一个 Servlet 类。

注意：此种写法创建了两个 Servlet 类的对象（可以通过添加 Servlet 构造方法对打印信息进行验证）。

（5）当配置代码使用相同的别名时，web.xml 配置文件代码如下：

```
<servlet>
    <servlet-name>zzzzzzzzzz</servlet-name>
    <servlet-class>controller.MyServlet</servlet-class>
</servlet>
<servlet-mapping>
    <servlet-name>zzzzzzzzzz</servlet-name>
    <url-pattern>/myServletA</url-pattern>
</servlet-mapping>

<servlet>
    <servlet-name>zzzzzzzzzz</servlet-name>
    <servlet-class>controller.MyServlet</servlet-class>
</servlet>
<servlet-mapping>
    <servlet-name>zzzzzzzzzz</servlet-name>
    <url-pattern>/myServletB</url-pattern>
</servlet-mapping>
```

上面的配置代码使用相同的别名，保存后，IntelliJ IDEA 出现 XML 校验异常，但在 Tomcat 启动时没有出现异常，执行 http://localhost:8080/web3/myServletA 和 http://localhost:8080/web3/myServletB 后只创建一个 Servlet 对象。

### 5.3.4 创建基于注解的 Servlet 案例

创建全新的 Web 项目，项目中不包含 web.xml 配置文件，如图 5-17 所示。创建新的 Java 类，效果如图 5-18 所示。

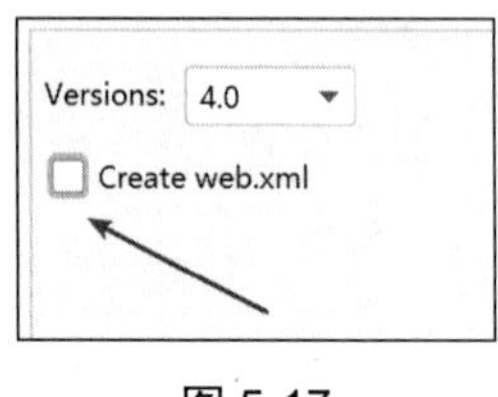


图 5-17

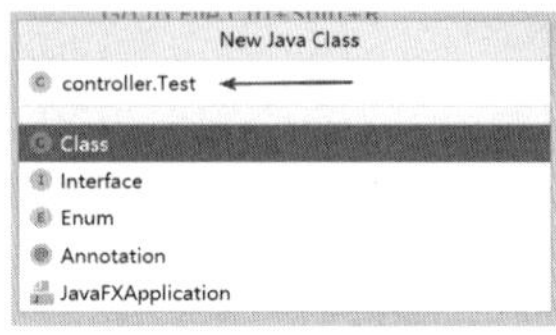


图 5-18

类 Test 继承自 HttpServlet，重写 doGet()方法，并添加如下注解：

```
@WebServlet(name = "Test", urlPatterns = "/Test")
```

完整的示例代码如下：

```
package controller;

import javax.servlet.ServletException;
import javax.servlet.annotation.WebServlet;
import javax.servlet.http.HttpServlet;
import javax.servlet.http.HttpServletRequest;
import javax.servlet.http.HttpServletResponse;
import java.io.IOException;

@WebServlet(name = "Test", urlPatterns = "/Test")
public class Test extends HttpServlet {
   @Override
   protected void doGet(HttpServletRequest req, HttpServletResponse resp)
throws ServletException, IOException {
      System.out.println("Test run !");
   }
}
```

运行 Servlet 后，控制台输出结果如图 5-19 所示。

图 5-19

使用注解后不再需要 web.xml 文件，减少了 XML 配置文件的代码量。

## 5.3.5　接口 Servlet

接口 Servlet 中的 API 的主要功能就是处理与 Servlet 生命周期相关的工作，如初始化、服务、销毁等。再来查看一下接口 Servlet 的方法声明，如图 5-20 所示。

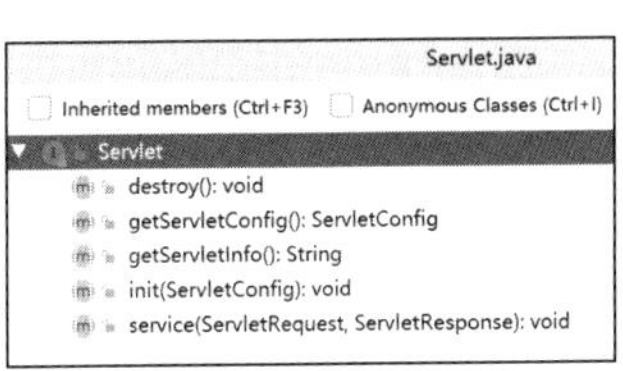


图 5-20

在本节就把这 5 个方法通过案例进行介绍，从而细化对 Servlet 接口的理解。

### 1. public void init(ServletConfig config)方法的使用

public void init(ServletConfig config)方法的主要作用就是在第一次访问 Servlet 时，Servlet 对象被 Tomcat 实例化，并且只执行一次 init()方法以实现对 Servlet 做一些初始化的工作，但通常都是使用参数 ServletConfig 来获得与 Servlet 有关的初始化配置信息。

每创建一个 Servlet 实例时，当前创建的 Servlet 的 public void init(ServletConfig config)方法只执行一次。

来测试一下 public void init(ServletConfig config)方法只执行一次的特性，示例代码如下：

```
public class MyServlet extends HttpServlet {

    public MyServlet() {
        System.out.println("public MyServlet() hashCode:" + this.hashCode());
    }

    @Override
    public void init(ServletConfig config) throws ServletException {
        System.out.println("public void init(ServletConfig config)");
    }

    @Override
    protected void doGet(HttpServletRequest req, HttpServletResponse resp)
throws ServletException, IOException {
        System.out.println("进入了doGet()方法!");
    }
}
```

程序运行后的效果如图 5-21 所示。

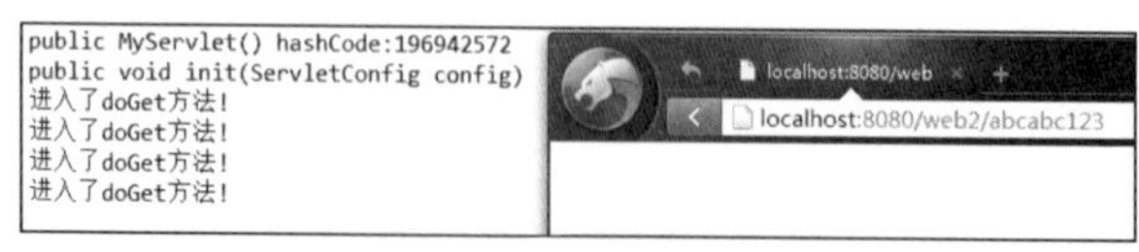


图 5-21

在本示例中只验证 public void init(ServletConfig config)方法仅执行一次的特性，此方法可以在创建 Servlet 时对自定义 XML 文件进行解析，仅解析一次即可，不需要重复进行解析，因为解析后的信息通常是放入一个共享变量中，所以 init()方法最适合实现此需求，此知识点在框架技术中有所涉及。

Servlet 的构造方法和 public void init（ServletConfig config）方法都是执行一次的效果，这两个方法的使用场景如下。

（1）使用 public void init（ServletConfig config）方法大多数是通过 ServletConfig config 参数获得此 Servlet 有关配置信息的。

（2）构造方法适合初始化 Servlet 自身的属性值。

public void init（ServletConfig config）方法的参数是 ServletConfig 类型，关于接口 ServletConfig 的使用请参看后面的章节。

### 2. public ServletConfig getServletConfig()方法的使用

关于 public ServletConfig getServletConfig()方法的使用请参看 ServletConfig 有关的章节。

### 3. public String getServletInfo()方法的使用

public String getServletInfo()方法的主要作用就是返回此 Servlet 的相关注释信息，比如 Servlet 创建的日期、作者、版本等，在默认的情况下，此方法返回一个空的字符串，可以重写此方法达到获得 Servlet 信息的目的。

先来看看返回空字符串的示例，代码如下：

```
public class MyServlet extends HttpServlet {
    @Override
    protected void doGet(HttpServletRequest req, HttpServletResponse resp)
throws ServletException, IOException {
        System.out.println("进入了 doGet()方法 getServletInfo=" + "|" +
this.getServletInfo() + "|");
    }
}
```

程序运行结果如图 5-22 所示。

图 5-22

重写此方法可以达到获得信息的目的，代码如下：

```
public class MyServlet extends HttpServlet {

    @Override
    public String getServletInfo() {
        return "介绍 Servlet 的相关信息";
    }

    @Override
    protected void doGet(HttpServletRequest req, HttpServletResponse resp)
throws ServletException, IOException {
        System.out.println("进入了 doGet() 方法 getServletInfo=" + "|" +
this.getServletInfo() + "|");
    }
}
```

程序运行结果如图 5-23 所示。

图 5-23

#### 4. public void destroy()方法的使用

public void destroy()方法被调用的时机主要有两种情况。

（1）在执行 Redeploy 重部署时被调用。

（2）在使用“stop”按钮停止 Tomcat 服务时被调用。

实验代码如下：

```
package controller;

import javax.servlet.ServletException;
import javax.servlet.annotation.WebServlet;
```

```
import javax.servlet.http.HttpServlet;
import javax.servlet.http.HttpServletRequest;
import javax.servlet.http.HttpServletResponse;
import java.io.IOException;

@WebServlet(name = "Test", urlPatterns = "/Test")
public class Test extends HttpServlet {

    public Test() {
        System.out.println("public Test()");
    }

    @Override
    protected void doGet(HttpServletRequest req, HttpServletResponse resp)
throws ServletException, IOException {
        System.out.println("doGet run !");
    }

    @Override
    public void destroy() {
        System.out.println("public void destroy()");
    }
}
```

运行这个 Servlet 后，控制台信息如图 5-24 所示。

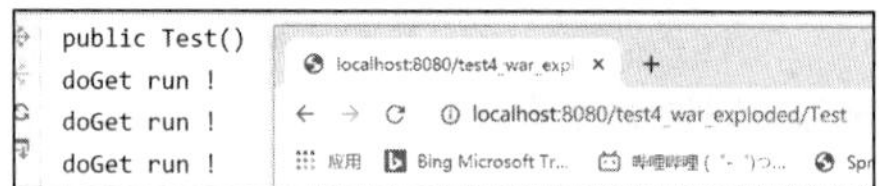


图 5-24

（1）执行 Redeploy 重部署时被调用。

单击“Tomcat”按钮，如图 5-25 所示，选择“Redeploy”选项，如图 5-26 所示。

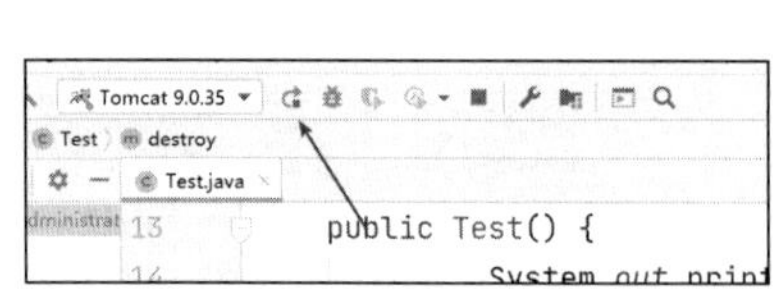


图 5-25

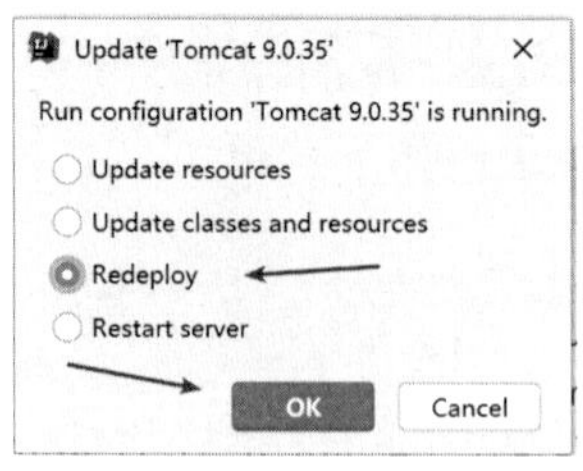


图 5-26

控制台输出结果如图 5-27 所示。

（2）单击“stop”按钮停止 Tomcat 服务被调用。public void destroy()方法被调用的情形就是单击“stop”按钮时 destroy()方法也会被调用，效果如图 5-28 所示。

图 5-27　　　　图 5-28

**5. service(ServletRequest req, ServletResponse res)方法的使用**

所有 Servlet 的父类都是 GenericServlet.java，它提供了 Servlet 功能最基本的抽象，包括 service()方法：

```
@Override
public abstract void service(ServletRequest req, ServletResponse res)
    throws ServletException, IOException;
```

该方法的主要作用是使前台发送过来的 HTTP 请求与后台自定义 Servlet 类进行对接，执行后台业务功能，也就是前台发送来的 HTTP 请求最终到达 doGet()方法时，中途需要经过 service(request,response)方法进行中转，由 service(request,response)方法来决定到底是执行 doGet()方法还是 doPost()方法。

service(request,response)方法接收 Tomcat 封装过的请求及响应对象，通过请求对象可以获得客户端的数据，而使用响应对象可以向客户端传递一些数据。

service()方法在类 GenericServlet.java 中是抽象的，效果如图 5-29 所示。

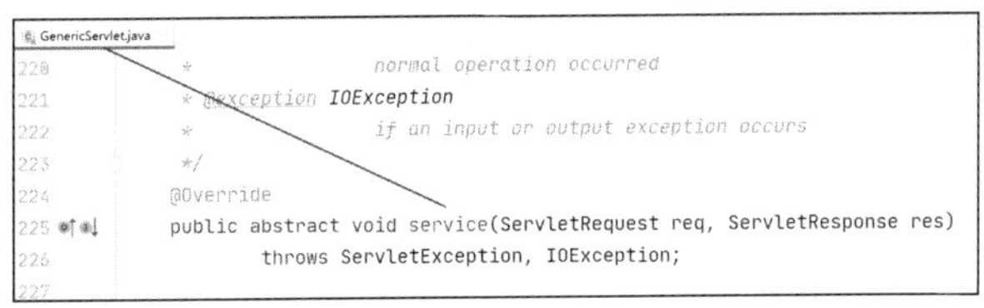


图 5-29

说明在 GenericServlet.java 类的子类中必须重写这个方法才可以实现具体的功能，

下面来做一个实验。

创建一个 Servlet，父类不是 HttpServlet.java，而是 GenericServlet.java，代码如下：

```
package controller;

import javax.servlet.GenericServlet;
import javax.servlet.ServletException;
import javax.servlet.ServletRequest;
import javax.servlet.ServletResponse;
import javax.servlet.annotation.WebServlet;
import java.io.IOException;

@WebServlet(name = "MyServlet", urlPatterns = "/MyServlet")
public class MyServlet extends GenericServlet {
    @Override
    public void service(ServletRequest req, ServletResponse res) throws
ServletException, IOException {
        System.out.println(req);
        System.out.println(res);
    }
}
```

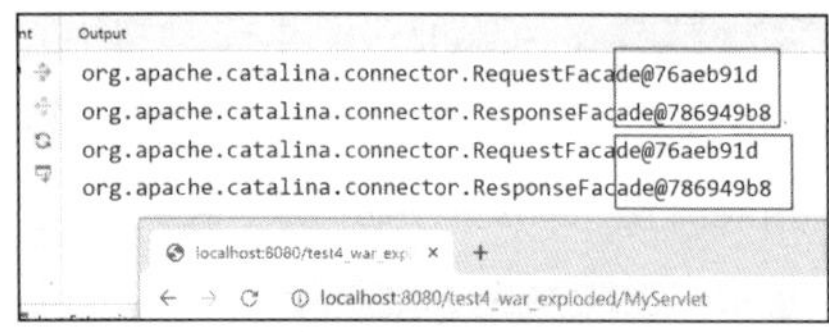


图 5-30

部署到 Tomcat 的执行效果如图 5-30 所示。

从控制台输出的结果来看，浏览器每发起一个新的 HTTP 请求去执行 Servlet 时，不会创建新的请求和响应对象，而是复用内存中的这两个对象，浏览器每发起一个新的 HTTP 请求就是创建一个新的 Socket 处理过程。

注意：有的时候创建新的请求和响应对象，但多次运行还是会看到复用这两个对象了。

在 GenericServlet.java 的子类 MyServlet.java 中的 service()方法里执行具体的业务就可以实现前台与后台的数据交互了。

但在真实的软件项目中，并没有多少人愿意直接继承自 GenericServlet.java 类来进行软件设计，因为它的封装比较原始，所以就要使用 HttpServlet.java 类作为自定义 Servlet 的父类进行设计，因为 HttpServlet.java 类重写了父类 GenericServlet.java 的 service(request, response)方法，源代码如下：

```
@Override
public void service(ServletRequest req, ServletResponse res)
    throws ServletException, IOException {

    HttpServletRequest  request;
    HttpServletResponse response;

    try {
        request = (HttpServletRequest) req;
        response = (HttpServletResponse) res;
    } catch (ClassCastException e) {
        throw new ServletException("non-HTTP request or response");
    }
    service(request, response);
}
```

HttpServlet.java 类中的 service(ServletRequest req, ServletResponse res)方法要调用 service(HttpServletRequest req, HttpServletResponse resp)方法来作为具体执行某个方法的参考，因为根据前台提交数据的方式不同，会调用不同的方法，比如使用 get 提交类型要调用 doGet()方法，post 提交类型要调用 doPost()方法，这个功能在 HttpServlet.java 类中的 service(HttpServletRequest req, HttpServletResponse resp)方法得到了封装，源代码如下：

```
protected void service(HttpServletRequest req, HttpServletResponse resp)
    throws ServletException, IOException {

    String method = req.getMethod();

    if (method.equals(METHOD_GET)) {
        long lastModified = getLastModified(req);
        if (lastModified == -1) {
            // servlet doesn't support if-modified-since, no reason
            // to go through further expensive logic
            doGet(req, resp);
        } else {
            long ifModifiedSince;
            try {
                ifModifiedSince = req.getDateHeader(HEADER_IFMODSINCE);
            } catch (IllegalArgumentException iae) {
                // Invalid date header - proceed as if none was set
                ifModifiedSince = -1;
```

```
            }
            if (ifModifiedSince < (lastModified / 1000 * 1000)) {
                // If the servlet mod time is later, call doGet()
                // Round down to the nearest second for a proper compare
                // A ifModifiedSince of -1 will always be less
                maybeSetLastModified(resp, lastModified);
                doGet(req, resp);
            } else {
                resp.setStatus(HttpServletResponse.SC_NOT_MODIFIED);
            }
        }

    } else if (method.equals(METHOD_HEAD)) {
        long lastModified = getLastModified(req);
        maybeSetLastModified(resp, lastModified);
        doHead(req, resp);

    } else if (method.equals(METHOD_POST)) {
        doPost(req, resp);

    } else if (method.equals(METHOD_PUT)) {
        doPut(req, resp);

    } else if (method.equals(METHOD_DELETE)) {
        doDelete(req, resp);

    } else if (method.equals(METHOD_OPTIONS)) {
        doOptions(req,resp);

    } else if (method.equals(METHOD_TRACE)) {
        doTrace(req,resp);

    } else {
        //
        // Note that this means NO servlet supports whatever
        // method was requested, anywhere on this server.
        //

        String errMsg = lStrings.getString("http.method_not_implemented");
        Object[] errArgs = new Object[1];
        errArgs[0] = method;
```

```
        errMsg = MessageFormat.format(errMsg, errArgs);

        resp.sendError(HttpServletResponse.SC_NOT_IMPLEMENTED, errMsg);
    }
}
```

通过源代码发现，service(HttpServletRequest req, HttpServletResponse resp)方法主要的作用就是根据变量 String method = req.getMethod()值来决定下一步调用哪些方法，起到承上启下的作用。那么在开发时，为了方便地使用自定义 MyServlet，通常的情况下 MyServlet 的父类是 HttpServlet.java 类。

执行请求对象的 getMethod()方法时返回请求的类型，常见的就是 get 和 post，这两种类型可以使用 HTML 语言中的<form>标签来进行决定，示例代码如下：

```
<form action="myServlet" method="get">
    <input type="submit" value="get 提交"/>
</form>
<br/>
<form action="myServlet" method="post">
    <input type="submit" value="post 提交"/>
</form>
```

一个 Servlet 的生命周期如下。

（1）实例化 Servlet 对象。

（2）执行 init(ServletConfig)方法。

（3）执行 service(ServletRequest,ServletResponse)方法。

（4）执行 doXXX()方法。

（5）执行 destroy()销毁方法。

Servlet 生命周期可以总结成 4 个过程：实例化、初始化、服务和销毁。

### 5.3.6　接口 ServletConfig

在学习接口 Servlet 时，方法 public void init (ServletConfig config)使用到了接口 ServletConfig，所以在本节将把该接口中全部方法的功能进行实验。

接口 ServletConfig 的方法声明如图 5-31 所示。

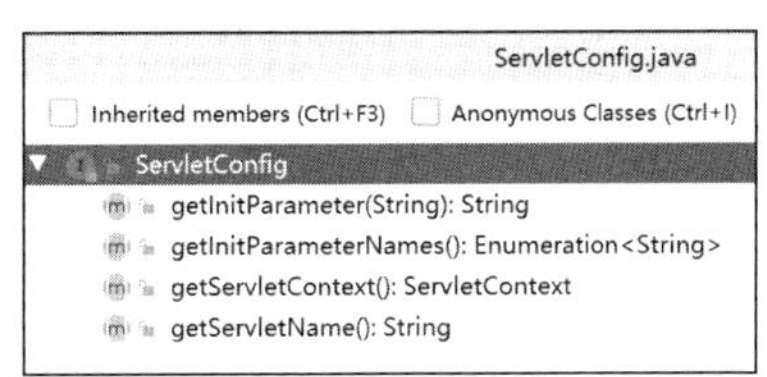


图 5-31

### 1. public String getServletName()方法的使用

getServletName()方法的主要作用是获取 web.xml 配置文件中<servlet-name>文本</servlet-name>的配置文本值，其实就是获得配置 Servlet 的别名。

Servlet 代码如下：

```
public class MyServlet extends HttpServlet {
    @Override
    protected void doGet(HttpServletRequest req, HttpServletResponse resp)
throws ServletException, IOException {
        System.out.println("servletName=" + this.getServletName());
    }
}
```

执行 this.getServletName()代码其实是调用 GenericServlet 类中的 public String getServletName()方法，其源代码如下：

```
@Override
public String getServletName() {
   return config.getServletName();
}
```

通过源代码可以发现，getServletName()方法在内部要依赖于 ServletConfig config 对象来取得 ServletName，如果 config 对象为 NULL，也就是并没有对其进行赋值，运行程序就会出现 java.lang.NullPointerException 空指针异常。赋值的源代码如下：

```
@Override
public void init(ServletConfig config) throws ServletException {
   this.config = config;
   this.init();
}
```

如果继承自 GenericServlet.java 的子类重写 public void init(ServletConfig config)方法，并没有在 public void init(ServletConfig config)方法中执行：

```
super.init(config);
```

则父类 GenericServlet.java 中的 config 对象没有机会赋值，对象就是 NULL，运行程序就会出现 java.lang.NullPointerException 空指针异常。此情况在下一节中有详细介绍。

配置文件 web.xml 中的配置代码如下：

```
<servlet>
    <servlet-name>zzz</servlet-name>
    <servlet-class>web2.MyServlet</servlet-class>
</servlet>
<servlet-mapping>
    <servlet-name>zzz</servlet-name>
    <url-pattern>/MyServlet</url-pattern>
</servlet-mapping>
```

打开 http://localhost:8080/web2/MyServlet，输出的结果如图 5-32 所示。

```
信息: Server startup in 443 ms
servletName=zzz
```

图 5-32

**2. 在重写public void init(ServletConfig config)方法中执行super.init(config)代码的必要性**

注意，在自定义 Servlet 类中的 doGet()方法里运行代码：

```
System.out.println(this.getServletName())
```

并且还重写了 public void init(ServletConfig config)方法，则需要在 public void init(ServletConfig config)方法中执行 super.init(config)代码，示例代码如下：

```
public class MyServlet extends HttpServlet {
    @Override
    public void init(ServletConfig config) throws ServletException {
        super.init(config);
    }
    @Override
    protected void doGet(HttpServletRequest req, HttpServletResponse resp)
throws ServletException, IOException {
        System.out.println(this.getServletName());
    }
}
```

执行 super.init(config)代码的目的是执行父类 GenericServlet.java 的 public void init(ServletConfig config)方法来对 ServletConfig config 实例变量进行初始化赋值，赋值的源代码如图 5-33 所示。

```
GenericServlet.java
    @Override
    public void init(ServletConfig config) throws ServletException {
        this.config = config;
        this.init();
    }
```

图 5-33

这样当在 doGet()方法中调用 System.out.println(this.getServletName())代码时就不会出现 java.lang.NullPointerException 空指针异常，因为 getServletName()方法所依赖的 ServletConfig config 对象已经被赋值，不再为 NULL。

反之，如果不调用 super.init(config)方法，也就是 ServletConfig config 对象的值一直是 NULL，当调用 getServletName()方法时则会出现 java.lang.NullPointerException 空指针异常。

在重写的 public void init(ServletConfig config)方法中调用 super.init(config)方法的优势是对 GenericServlet.java 类中的 ServletConfig config 对象进行初始化赋值，从而可以在类 GenericServlet.java 的任何子类的任何方法中来访问非空的 ServletConfig config 对象实现业务上的需求。

### 3. public String getInitParameter(String name)方法结合web.xml使用

在 web.xml 文件中配置 Servlet 时可以加入一些额外的信息，public String getInitParameter(String name)方法的作用就是获取这些信息，示例代码如下：

```
public class MyServlet extends HttpServlet {
    @Override
    public void init(ServletConfig config) throws ServletException {
        super.init(config);
        System.out.println("public void init=" + config.getInitParameter
("xmlFileName1"));
        System.out.println("public void init=" + config.getInitParameter
("xmlFileName2"));
    }

    @Override
    protected void doGet(HttpServletRequest req, HttpServletResponse resp)
throws ServletException, IOException {
        System.out.println("protected void doGet run !");
        System.out.println("protected void doGet=" + this.getInitParameter
("xmlFileName1"));
        System.out.println("protected void doGet=" + this.getInitParameter
```

```
("xmlFileName2"));
    }
}
```

配置文件 web.xml 中的代码如下：

```
<servlet>
    <servlet-name>myServlet</servlet-name>
    <servlet-class>controller.MyServlet</servlet-class>
    <init-param>
        <param-name>xmlFileName1</param-name>
        <param-value>a.xml</param-value>
    </init-param>
    <init-param>
        <param-name>xmlFileName2</param-name>
        <param-value>b.xml</param-value>
    </init-param>
</servlet>
<servlet-mapping>
    <servlet-name>myServlet</servlet-name>
    <url-pattern>/myServlet</url-pattern>
</servlet-mapping>
```

执行 Servlet 的路径后，控制台输出结果如下：

```
public void init=a.xml
public void init=b.xml
protected void doGet run !
protected void doGet=a.xml
protected void doGet=b.xml
```

public void init(ServletConfig config)方法和 getInitParameter("xmlFileName1")方法结合使用的场景是当 Servlet 在第一次被实例化时要解析一些 XML 配置文件，但这些 XML 配置文件的解析次数仅一次就够，故可以在 Servlet 的 public void init(ServletConfig config)方法中进行解析。

#### 4. public String getInitParameter(String name)方法结合注解使用

创建 Servlet 示例代码如下：

```
package controller;
```

```
import java.io.IOException;

import javax.servlet.ServletConfig;
import javax.servlet.ServletException;
import javax.servlet.annotation.WebInitParam;
import javax.servlet.annotation.WebServlet;
import javax.servlet.http.HttpServlet;
import javax.servlet.http.HttpServletRequest;
import javax.servlet.http.HttpServletResponse;

@WebServlet(urlPatterns = "/Test", initParams = { @WebInitParam(name =
"xmlFile1", value = "a.xml"),
        @WebInitParam(name = "xmlFile2", value = "b.xml") })
public class Test extends HttpServlet {
    @Override
    public void init(ServletConfig config) throws ServletException {
        super.init(config);
        System.out.println("init " + config.getInitParameter("xmlFile1"));
        System.out.println("init " + config.getInitParameter("xmlFile2"));
    }

    protected void doGet(HttpServletRequest request, HttpServletResponse
response)
            throws ServletException, IOException {
        System.out.println("doGet " + this.getInitParameter("xmlFile1"));
        System.out.println("doGet " + this.getInitParameter("xmlFile2"));
    }
}
```

**5. public Enumeration<String> getInitParameterNames()方法的使用**

当有多个初始化配置值时，可以使用 getInitParameterNames()方法来循环获得每个配置的<param-name>值。

还是使用前面章节配置文件 web.xml 中的配置代码如下：

```
public class MyServlet extends HttpServlet {
    @Override
    public void init(ServletConfig config) throws ServletException {
        super.init(config);
        Enumeration<String> enumObject = config.getInitParameterNames();
```

```
        while (enumObject.hasMoreElements()) {
            String key = enumObject.nextElement();
            String value = config.getInitParameter(key);
            System.out.println(key + " " + value);
        }
    }

    @Override
    protected void doGet(HttpServletRequest req, HttpServletResponse resp)
throws ServletException, IOException {
        System.out.println("protected void doGet run !");
    }
}
```

运行 Servlet 后，控制台输出结果如下：

```
xmlFileName2 b.xml
xmlFileName1 a.xml
protected void doGet run !
```

请大家自行尝试在注解版中的使用。

### 6. public void init(config)方法与public void init()方法的区别

类 GenericServlet.java 中 public void init(ServletConfig config)方法的源代码如下：

```
@Override
public void init(ServletConfig config) throws ServletException {
    this.config = config;
    this.init();
}
```

从上面的源代码可以分析出：当创建自定义的 MyServlet 时，在 MyServlet 中重写 public void init(ServletConfig config)方法并且结合 super.init(config)方法会调用父类 GenericServlet.java 的 public void init(ServletConfig config) 方法，而父类 GenericServlet.java 中的有参 public void init(ServletConfig config)方法会调用无参 public void init()方法。

通过源代码可以发现，类 GenericServlet.java 中不仅存在有参的方法 public void init(ServletConfig config)，还存在无参的方法 public void init()，这两个方法在使用上有什么区别呢？在实现 init 初始化的过程中，可以将业务分成两种：一种是依赖于

ServletConfig 对象，另外一种是不依赖 ServletConfig 对象，依赖 ServletConfig 对象的初始化代码可以放入 public void init(ServletConfig config)方法中，不依赖于 ServletConfig 对象的初始化代码可以放入 public void init()方法，以进行 init 初始化代码的分工，示例代码如下：

```
package controller;

import javax.servlet.ServletConfig;
import javax.servlet.ServletException;
import javax.servlet.http.HttpServlet;

public class MyServlet extends HttpServlet {
    @Override
    public void init() throws ServletException {
        String username = "anyValue";
        String password = "anyValue";
        String url = "anyValue";
        String driverName = "anyValue";
        System.out.println("public void init() 访问数据库");
    }

    @Override
    public void init(ServletConfig config) throws ServletException {
        super.init(config);
        System.out.println("public void init(ServletConfig config)解析 XML
文件");
        // 可以将 xml 文件名存放在 web.xml 中，获得 web.xml 中的 xml 文件名
        // 是需要依赖于 ServletConfig 接口的，此案例在后面的实验有介绍
    }
}
```

### 7. public void init(config)方法和public void init()方法的不同运行情况

在使用 public void init(ServletConfig config)和 public void init()方法时，有以下 4 种不同的使用方式。

（1）在 MyServlet 中。

<A>不重写 public void init(ServletConfig config)方法；

<B>不重写 public void init()方法。

Tomcat容器在内部会调用GenericServlet类中的public void init(ServletConfig config)方法，而 public void init(ServletConfig config)方法会调用 public void init()空方法。

（2）在 MyServlet 中。

<A>重写 public void init(ServletConfig config)方法，并且不调用 super.init(config)代码；

<B>重写或不重写 public void init()方法。

Tomcat 容器在内部会调用 MyServlet 类中的 public void init(ServletConfig config)方法，并不会调用 GenericServlet 类中的 public void init(ServletConfig config)和 public void init()空方法或 MyServlet 类中的重写 public void init()方法。

（3）在 MyServlet 中。

<A>重写 public void init(ServletConfig config)方法，并且调用 super.init(config)代码；

<B>不重写 public void init()方法。

Tomcat 容器在内部会调用 MyServlet 类中的 public void init(ServletConfig config)方法，继续调用 GenericServlet 类中的 public void init(ServletConfig config)和 public void init()空方法。

（4）在 MyServlet 中。

<A>重写 public void init(ServletConfig config)方法，并且调用 super.init(config)代码；

<B>重写 public void init()方法。

Tomcat 容器在内部会调用 MyServlet 类中的 public void init(ServletConfig config)方法，继续调用 GenericServlet 类中的 public void init(ServletConfig config)方法，然后调用 MyServlet 类中的 public void init()空方法。

以上 4 种方式可以通过调试源代码的方式进行分析。

#### 8. public ServletContext getServletContext()方法的使用

关于 public ServletContext getServletContext()方法的使用请参看与 ServletContext 有关章节的介绍。

### 5.3.7 使用<load-on-startup>配置 Servlet

在 Tomcat 容器启动时，可以先自动对 Servlet 进行实例化，再进行初始化，这样的功能是需要在 web.xml 配置文件中使用<load-on-startup>标记完成的，实验代码如下：

```
<servlet>
```

```
    <servlet-name>MyServlet</servlet-name>
    <servlet-class>web2.MyServlet</servlet-class>
    <load-on-startup>0</load-on-startup>
</servlet>
<servlet>
    <servlet-name>MyServlet2</servlet-name>
    <servlet-class>web2.MyServlet2</servlet-class>
    <load-on-startup>2</load-on-startup>
</servlet>
<servlet>
    <servlet-name>MyServlet3</servlet-name>
    <servlet-class>web2.MyServlet3</servlet-class>
    <load-on-startup>1</load-on-startup>
</servlet>
```

```
public MyServlet()
MyServlet类的public void init(ServletConfig config)方法也会被自动调用!
public MyServlet3()
public MyServlet2()
```

图 5-34

Tomcat 启动后，3 个 Servlet 的构造方法执行顺序的效果如图 5-34 所示。

配置<load-on-startup>总结如下。

（1）<load-on-startup>表示是否在容器启动时就实例化并初始化这个 Servlet。

（2）<load-on-startup>的值必须是一个整数，表示 Servlet 被实例化的顺序。

（3）当<load-on-startup>值为 0 或大于 0 时，表示在应用启动时就自动加载并初始化这个 Servlet。

（4）<load-on-startup>正数的值越小，该 Servlet 的优先级越高。

（5）当<load-on-startup>值小于 0 或没有指定时，则表示容器在该 Servlet 被初次访问时才会去实例化。

（6）如果<load-on-startup>值一样，则 Servlet 初始化的顺序不确定，因为在 Tomcat 源代码中将 Servlet 放入 HashMap 中，而 HashMap 中是无序的，是根据 Hash 码算出来的。

### 5.3.8 使用注解实现<load-on-startup>的功能

创建 3 个 Servlet 的示例代码如下：

```
@WebServlet(urlPatterns = "/Test1", loadOnStartup = 3)
public class Test1 extends HttpServlet {
    public Test1() {
        System.out.println("public Test1()");
```

```
    }

    @Override
    public void init(ServletConfig config) throws ServletException {
        super.init(config);
        System.out.println("public Test1   init");
    }

    protected void doGet(HttpServletRequest request, HttpServletResponse
response)
            throws ServletException, IOException {
    }
}

@WebServlet(urlPatterns = "/Test2", loadOnStartup = 2)
public class Test2 extends HttpServlet {
    public Test2() {
        System.out.println("public Test2()");
    }

    @Override
    public void init(ServletConfig config) throws ServletException {
        super.init(config);
        System.out.println("public Test2   init");
    }

    protected void doGet(HttpServletRequest request, HttpServletResponse
response)
            throws ServletException, IOException {
    }
}

@WebServlet(urlPatterns = "/Test3", loadOnStartup = 1)
public class Test3 extends HttpServlet {
    public Test3() {
        System.out.println("public Test3()");
    }

    @Override
    public void init(ServletConfig config) throws ServletException {
        super.init(config);
        System.out.println("public Test3   init");
    }
```

```
    protected void doGet(HttpServletRequest request, HttpServletResponse
response)
            throws ServletException, IOException {
    }
}
```

启动 Tomcat 后控制台输出结果如下：

```
public Test3()
public Test3  init
public Test2()
public Test2  init
public Test1()
public Test1  init
```

### 5.3.9 执行 doGet()方法或 doPost()方法的方式

创建 HTML 的代码如下：

```
<body>
    <a href="MyServlet">get</a>
    <br />

    <form action="MyServlet">
        <input type="submit" value="get">
    </form>
    <br />

    <form action="MyServlet" method="get">
        <input type="submit" value="get">
    </form>
    <br />

    <form action="MyServlet" method="post">
        <input type="submit" value="post">
    </form>
</body>
```

依次单击 4 个按钮后，使用浏览器的“F12”开发工具进行抓包，抓包内容如图 5-35 所示。

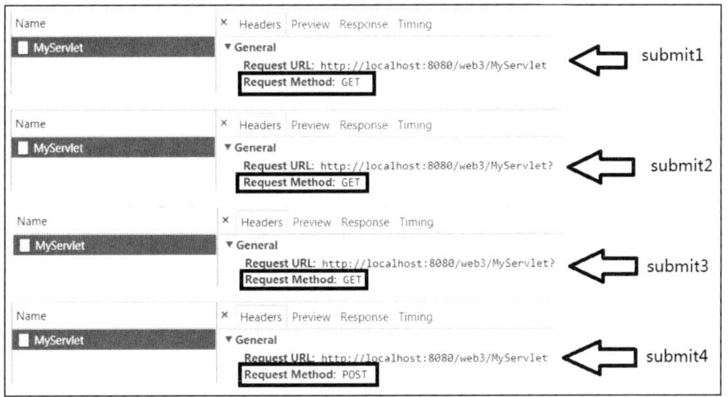


图 5-35

在服务器端，你可以使用如下代码来获得客户端提交的方式：

```
public class MyServlet extends HttpServlet {
    @Override
    protected void doGet(HttpServletRequest req, HttpServletResponse resp)
throws ServletException, IOException {
        System.out.println("protected void doGet " + req.getMethod());
    }

    @Override
    protected void doPost(HttpServletRequest req, HttpServletResponse resp)
throws ServletException, IOException {
        System.out.println("protected void doPost " + req.getMethod());
    }
}
```

依次点击 4 个按钮后，控制台输出结果如下：

```
protected void doGet GET
protected void doGet GET
protected void doGet GET
protected void doPost POST
```

说明有 3 种方式可以执行 doGet()方法，而只有一种方式执行 doPost()方法，就是<form>标签的 method 属性值必须是 post。

服务器端可以获取提交的类型，根据这个类型值就可以调用指定的 doGet()或 doPost()方法，调用的逻辑就是在 HttpServlet 类中的 protected void service(HttpServletRequest req,

HttpServletResponse resp)方法里，前面章节已经看过此代码，简化的部分核心源代码如下：

```
protected void service(HttpServletRequest req, HttpServletResponse resp)
{
    String method = req.getMethod();
    if (method.equals(METHOD_GET)) {
    doGet(req, resp);
    }
    else if (method.equals(METHOD_POST)) {
        doPost(req, resp);
    }
}
```

### 5.3.10 doGet()方法与 doPost()方法的区别

提交方式为 get 和 post 有什么区别呢？先创建一个 HTML 文件作为实验的环境，然后再来研究，HTML 文件代码如下：

```
<!DOCTYPE html>
<html>
    <head>
        <meta charset="UTF-8">
        <title>Insert title here</title>
    </head>
    <body>
        <form action="myServlet1" method="get">
            username:<input type="text" name="username" />
            <br/>
            password:<input type="text" name="password" />
            <br/>
            <input type="submit" value="submit1" />
        </form>
        <br/>
        <form action="myServlet1" method="post">
            username:<input type="text" name="username" />
            <br/>
            password:<input type="text" name="password" />
            <br/>
            <input type="submit" value="submit2" />
        </form>
    </body>
</html>
```

使用<form method="get">方式提交的表单数据在 URL 中作为参数传递到服务器端，在<form action="myServlet1" method="get">中的 username 文本域输入字符 a，在 password 文本域输入字符 b，再单击“submit1”按钮后，浏览器地址栏中出现了 username 和 password 以及它们的值 a 和 b，完整的 URL 地址如下：

http://localhost:8080/web3/myServlet1?username=a&password=b

访问路径和参数之间要使用“?”问号进行分隔，参数名和参数值使用“=”等号进行搭配，多个参数之间使用“&”进行分隔。

而使用<form method="post">方式提交的表单数据不在 URL 中作为参数显示出来，在<form action="myServlet1" method="post">中的 username 文本域输入字符 aa，在 password 文本域输入字符 bb，再单击“submit2”按钮后，浏览器地址栏中并没有出现 username 和 password 以及它们的值 aa 和 bb，完整的 URL 地址如下：

http://localhost:8080/web3/myServlet1

文本域 username 和 password 的值 aa 和 bb 是在请求体中进行保存的，然后再传递给服务器端，使用浏览器的“F12”快捷键抓包可以看到这两个值在 Form Data 中，如图 5-36 所示。

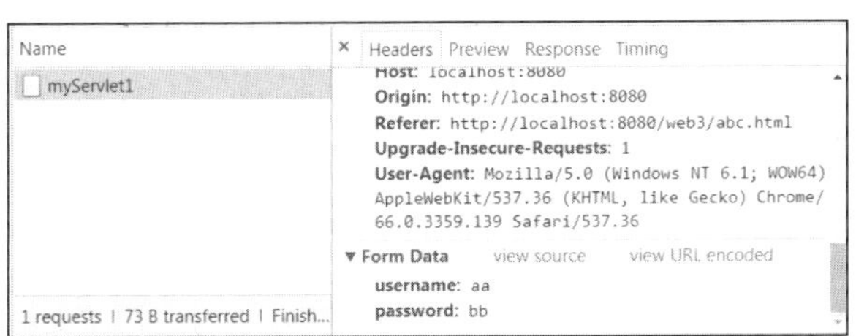


图 5-36

不能从 post 提交时地址栏不显示提交的数据就认定 post 提交是安全的，因为使用工具完全可以分析出 post 提交的数据内容而进行数据截获，所以在安全性上 get 和 post 基本都是不安全的。

另外一个区别是 get 传输数据时，受限于 URL 网址的最大长度，在传输数据量上没有 post 大，因为上传文件时只可以使用 post 提交，根据这个观点可以说明 get 传输数据量小，而 post 传输数据量大。

通过以上讲解，可以做以下两方面总结。

（1）get 提交方式：数据在浏览器地址栏显示出来，可以通过工具分析出数据，传输的数据量小。

（2）post 提交方式：数据不在浏览器地址栏显示出来，也可以通过工具分析出数据，传输的数据量大。

### 5.3.11 Application context 选项的作用

访问项目的上下文路径可以通过选项 Application context 进行设置，如图 5-37 所示。

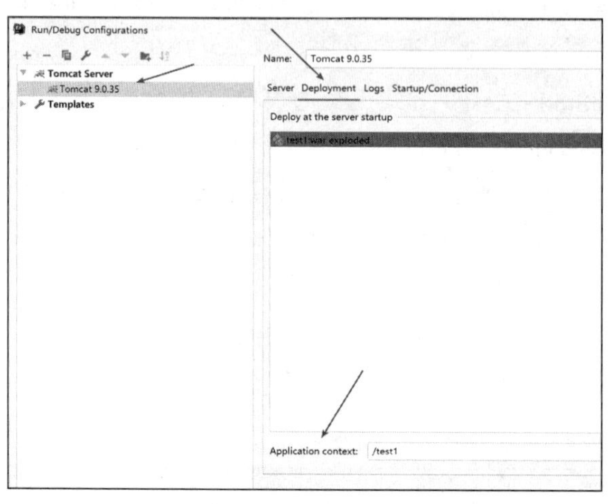


图 5-37

在 Tomcat 访问 Web 项目时就可以使用这个路径进行访问。

### 5.3.12 HttpServletRequest 和 HttpServletResponse 接口的使用

HttpServletRequest 接口主要用来封装客户端的请求信息，里面包含客户端的若干数据信息，将这些信息交给服务器端。HttpServletResponse 接口主要用来封装服务器端的响应信息，里面包含服务器端的若干数据信息，将这些信息交给浏览器。

#### 1. 装饰者设计模式简介

装饰者设计模式可以动态地给一个对象添加一些额外的职责，就增加功能来说，装饰者设计模式相比生成子类更为灵活。先来看看不使用装饰者设计模式来实现一些功能，在代码上有什么缺点。

创建 a 公司的业务接口，代码如下：

```
package a;
```

```
public interface IMyServiceA {
    public void saveMethod(String data);
}
```

创建 a 公司业务接口的实现类，代码如下：

```
package a;

public class MyServiceA implements IMyServiceA {
    @Override
    public void saveMethod(String data) {
        System.out.println("将数据" + data + "保存到数据库");
    }
}
```

创建 b 公司的业务类，代码如下：

```
package b;

import a.IMyServiceA;

public class MyServiceB {
    private IMyServiceA myServiceA;

    public MyServiceB(IMyServiceA myServiceA) {
        super();
        this.myServiceA = myServiceA;
    }

    public void runSaveService(String bData) {
        myServiceA.saveMethod(bData);
    }
}
```

创建 b 公司的运行类，代码如下：

```
package b;

import a.IMyServiceA;
import a.MyServiceA;

public class Test1 {
```

```
    public static void main(String[] args) {
        IMyServiceA a = new MyServiceA();
        MyServiceB b = new MyServiceB(a);
        b.runSaveService("-B 公司的数据-");
    }
}
```

程序运行后，控制台输出结果如下：

```
将数据-b 公司的数据-保存到数据库
```

b 公司调用 a 公司的公共服务来实现保存数据的功能，但 a 公司并不清楚当前公共服务的执行效率，所以 a 公司更改了业务类的代码如下：

```
package a;

public class MyServiceA implements IMyServiceA {
    @Override
    public void saveMethod(String data) {
        System.out.println("begin " + System.currentTimeMillis());
        System.out.println("将数据" + data + "保存到数据库");
        System.out.println("end " + System.currentTimeMillis());
        System.out.println("将耗时的数据通过 socket 网络传输到 a 公司的性能采集服务
器中");
    }
}
```

执行 B 公司的运行类后，控制台输出结果如下：

```
begin 1525944305772
将数据-B 公司的数据-保存到数据库
  end 1525944305773
将耗时的数据通过 socket 网络传输到 A 公司的性能采集服务器中
```

这时，A 公司 MyServiceA.java 类最新版本的代码写法出现“紧耦合”，也就是保存功能的代码和计时功能的代码混杂在一起，不利于软件代码的维护，因为并没有分隔开来。

### 2. 使用装饰者设计模式解决问题

上面“紧耦合”的问题可以使用装饰者设计模式来解决，使用装饰者设计模式可以动态地给一个对象添加额外的功能。下面是完整使用装饰者设计模式的程序代码。

创建 A 公司的业务接口，代码如下：

```
package a;

public interface IMyServiceA {
    public void saveMethod(String data);
}
```

创建 A 公司业务接口的实现类，代码如下：

```
package a;

public class MyServiceA implements IMyServiceA {
    @Override
    public void saveMethod(String data) {
        System.out.println("将数据" + data + "保存到数据库");
    }
}
```

创建装饰者类 MyServiceAWrapper.java，代码如下：

```
package a;

public class MyServiceAWrapper implements IMyServiceA {

    private IMyServiceA myServiceA;

    public MyServiceAWrapper(IMyServiceA myServiceA) {
        this.myServiceA = myServiceA;
    };

    @Override
    public void saveMethod(String data) {
        System.out.println("begin " + System.currentTimeMillis());
        myServiceA.saveMethod(data);
        System.out.println(" end " + System.currentTimeMillis());
        System.out.println("将耗时的数据通过 socket 网络传输到 A 公司的性能采集服务
器中");
    }

}
```

创建 B 公司的业务类，代码如下：

```
package b;

import a.IMyServiceA;

public class MyServiceB {
    private IMyServiceA myServiceA;

    public MyServiceB(IMyServiceA myServiceA) {
        super();
        this.myServiceA = myServiceA;
    }

    public void runSaveService(String bData) {
        myServiceA.saveMethod(bData);
    }
}
```

创建 B 公司的运行类，代码如下：

```
package b;

import a.IMyServiceA;
import a.MyServiceA;
import a.MyServiceAWrapper;

public class Test1 {
    public static void main(String[] args) {
        IMyServiceA a = new MyServiceAWrapper(new MyServiceA());
        MyServiceB b = new MyServiceB(a);
        b.runSaveService("-B 公司的数据-");
    }
}
```

程序运行后控制台输出结果如下：

```
begin 1525944565648
将数据-B 公司的数据-保存到数据库
 end 1525944565648
将耗时的数据通过 socket 网络传输到 A 公司的性能采集服务器中
```

使用装饰者设计模式后，保存数据的功能代码和计时的功能代码完全分开，便于后期维护。

装饰者设计模式的本质是对原有对象在功能上进行增加和扩展，而又不更改原有对象的程序代码，主要的使用方式就是添加更多的功能性方法，也包括针对某些方法进行功能上的升级，比如刚刚实现的计时监控功能。

### 3. Servlet中的装饰者设计模式

在 Servlet 中也存在使用装饰者设计模式，ServletRequest 和 ServletResponse 两个接口的继承关系如图 5-38 所示。

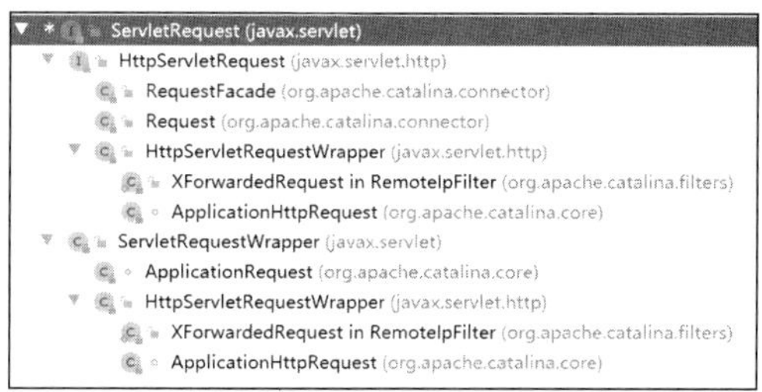


（a）

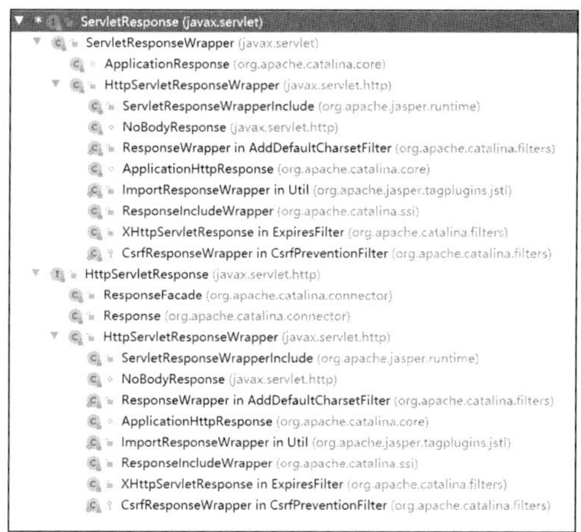


（b）

图 5-38

接口 HttpServletRequest 的父接口是 ServletRequest，接口 HttpServletResponse 的父接口是 ServletResponse。

类 ServletRequestWrapper 使用装饰者设计模式对 ServletRequest 对象进行包装，简化后的核心源代码如下：

```
public class ServletRequestWrapper implements ServletRequest {
    public ServletRequestWrapper(ServletRequest request) {
    }
```

类 ServletRequestWrapper 的子类 HttpServletRequestWrapper 也具有同样的功能，只是对 HttpServletRequest 对象进行包装，源代码如下：

```
public class HttpServletRequestWrapper extends ServletRequestWrapper
implements HttpServletRequest {
    public HttpServletRequestWrapper(HttpServletRequest request) {
        super(request);
    }
```

类 ServletResponseWrapper 使用装饰者设计模式对 ServletResponse 对象进行包装，源代码如下：

```
public class ServletResponseWrapper implements ServletResponse {
    public ServletResponseWrapper(ServletResponse response) {
    }
```

类 ServletResponseWrapper 的子类 HttpServletResponseWrapper 也具有同样的功能，只是对 HttpServletResponse 对象进行包装，源代码如下：

```
public class HttpServletResponseWrapper extends ServletResponseWrapper
implements HttpServletResponse {
    public HttpServletResponseWrapper(HttpServletResponse response) {
        super(response);
    }
```

子类 HttpServletRequestWrapper 和 HttpServletResponseWrapper 提供的功能是最多的，关于如何使用这两个子类来实现包装的效果在下面的章节有介绍。

类 ServletRequestWrapper 和 HttpServletRequestWrapper 类的装饰范围如图 5-39 所示。

类 ServletRequestWrapper 装饰范围是外框区域；类 HttpServletRequestWrapper 装饰范围是内框区域。

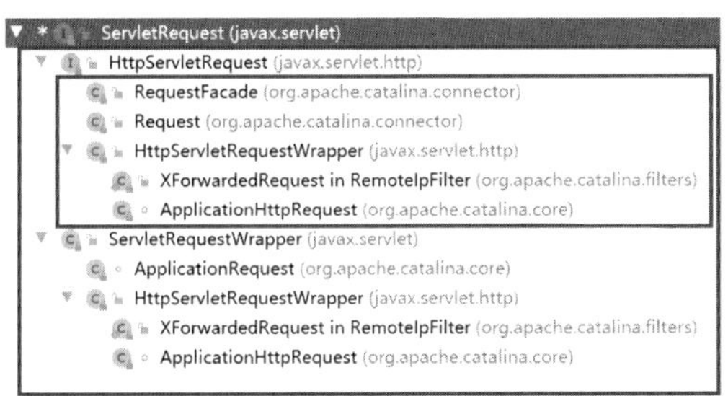


图 5-39

### 4. 使用HttpServletRequestWrapper达到装饰的效果

如果想实现对 getMethod()方法进行包装，达到查看执行时间的效果，就需要使用包装类进行功能的二次扩展。那么包装类对谁包装呢？当然是请求对象了，因为 getMethod()方法就是在请求对象中。那么请求对象具体的类型是什么呢？可以使用如下代码进行获得：

```
public class test1 extends HttpServlet {
    protected void doGet(HttpServletRequest request, HttpServletResponse
response)
            throws ServletException, IOException {
        System.out.println(request);
    }
}
```

程序运行结果如下：

```
org.apache.catalina.connector.RequestFacade@62e9a60b
```

请求对象具体的类型是 org.apache.catalina.connector.RequestFacade，getMethod()方法是在 RequestFacade 类的对象中进行调用的，说明要对 org.apache.catalina.connector.RequestFacade 类的对象进行功能的包装。在 Web 项目中创建针对 org.apache.catalina.connector.RequestFacade 类的包装器，代码如下：

```
package controller;

import org.apache.catalina.connector.Request;
import org.apache.catalina.connector.RequestFacade;

public class MyRequestWrapper extends RequestFacade {
```

```
    public MyRequestWrapper(Request request) {
        super(request);
    }
}
```

设计出这样的代码后，MyRequestWrapper 类永远依赖于 RequestFacade 类，证明 MyRequestWrapper 类永远依赖于 Tomcat 环境，因为 RequestFacade 类是 Tomcat 提供的，如果换成其他的 Web 容器，则此代码无法进行移植，其他的 Web 容器中使用的类并不是 org.apache.catalina.connector.RequestFacade，所以要避免 MyRequestWrapper 类依赖于第三方厂商提供的类。

本节的初衷是对 javax.servlet.http.HttpServletRequest 接口中的 getMethod()方法做扩展，那么来看看 org.apache.catalina.connector.RequestFacade 类实现的接口信息是什么，是不是也是 javax.servlet.http.HttpServletRequest 接口呢？代码如下：

```
public class test1 extends HttpServlet {
    protected void doGet(HttpServletRequest request, HttpServletResponse
response)
            throws ServletException, IOException {
        System.out.println(request);
        System.out.println(request.getClass().getSuperclass().getName());

    System.out.println(request.getClass().getInterfaces()[0].getName());
    }
}
```

控制台输出的结果如下：

```
org.apache.catalina.connector.RequestFacade@171aa7ca
java.lang.Object
javax.servlet.http.HttpServletRequest
```

从输出的结果来看，类 org.apache.catalina.connector.RequestFacade 的父类是 Object 类，并且 org.apache.catalina.connector.RequestFacade 类也实现了 javax.servlet.http.HttpServletRequest 接口。

既然 MyRequestWrapper 类不能依赖于第三方的 org.apache.catalina.connector.RequestFacade 类，那么 MyRequestWrapper 类可以实现 javax.servlet.http.HttpServletRequest 接口吗？测试的部分代码如下：

```
public class MyRequestWrapper implements HttpServletRequest {

    @Override
    public Object getAttribute(String name) {
        // TODO Auto-generated method stub
        return null;
    }

    @Override
    public Enumeration<String> getAttributeNames() {
        // TODO Auto-generated method stub
        return null;
    }

    @Override
    public String getCharacterEncoding() {
        // TODO Auto-generated method stub
        return null;
    }

    @Override
    public void setCharacterEncoding(String env) throws
UnsupportedEncodingException {
        // TODO Auto-generated method stub

    }

    @Override
    public int getContentLength() {
        // TODO Auto-generated method stub
        return 0;
    }

    @Override
    public long getContentLengthLong() {
        // TODO Auto-generated method stub
        return 0;
    }

    @Override
    public String getContentType() {
        // TODO Auto-generated method stub
        return null;
```

```
    }

    @Override
    public ServletInputStream getInputStream() throws IOException {
        // TODO Auto-generated method stub
        return null;
    }
```

由于 javax.servlet.http.HttpServletRequest 是接口，所以 MyRequestWrapper 并不是二次扩展，而是首次的功能实现，没有达到包装、扩展功能的目的，所以实现 javax.servlet.http.HttpServletRequest 接口并不是正确的解决办法，还要使用继承类的方式，并且还不能继承第三方厂商的类，比如 org.apache.catalina.connector.RequestFacade，也不能实现 javax.servlet.http.HttpServletRequest 接口，那 JDK 中有没有提供一个实现 javax.servlet.http.HttpServletRequest 接口的类，并且该类还可以提供二次包装的效果呢？答案是：有！该类就是 HttpServletRequestWrapper，从类的名称来看是装饰者，是对 HttpServletRequest 对象进行装饰的，所以类 MyRequestWrapper 可以继承 HttpServletRequestWrapper 类进行二次包装，实现功能的二次扩展。MyRequestWrapper 类依赖的是 JDK 中的类，并不是第三方厂商提供的类，代码具有移植性。代码移植性是指相同的代码可以在不同的环境中运行，不需要改代码。类 HttpServletRequestWrapper 的继承与实现关系如图 5-40 所示。

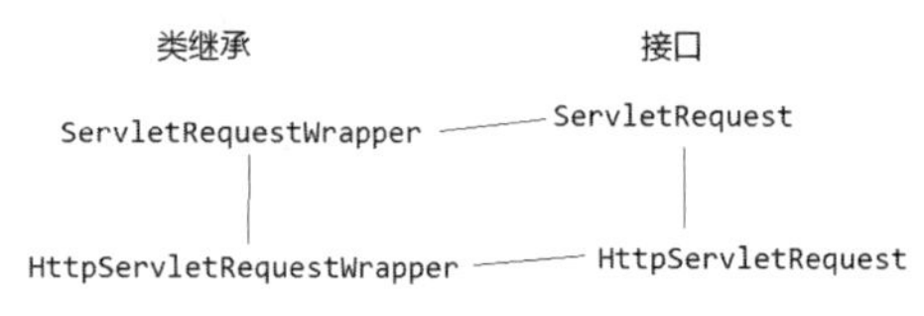


图 5-40

类 HttpServletRequestWrapper 源代码如下：

```
public class HttpServletRequestWrapper extends ServletRequestWrapper implements
        HttpServletRequest {

    public HttpServletRequestWrapper(HttpServletRequest request) {
        super(request);
    }

    @Override
    public String getMethod() {
        return this._getHttpServletRequest().getMethod();
}

//其他方法省略
}
```

从 HttpServletRequestWrapper 类的源代码来看，可以对 HttpServletRequest 对象进行装饰。创建 HttpServletRequestWrapper 类的子类，代码如下：

```
package myrequestwrapper;

import javax.servlet.http.HttpServletRequest;
import javax.servlet.http.HttpServletRequestWrapper;

public class MyHttpServletRequestWrapper extends HttpServletRequestWrapper
{

    public MyHttpServletRequestWrapper(HttpServletRequest request) {
        super(request);
    }

    @Override
    public String getMethod() {
        String methodName = "";
        try {
            System.out.println("begin run " + System.currentTimeMillis());
            Thread.sleep(3000);
            methodName = super.getMethod();
            System.out.println("  end run " + System.currentTimeMillis());
        } catch (InterruptedException e) {
            e.printStackTrace();
        }
        return methodName;
    }

}
```

装饰类 MyHttpServletRequestWrapper 的主要作用是计算执行 getMethod()方法所需要的时间，增加“计时”这个额外的功能。

创建自定义的 Servlet 代码如下：

```
package controller;

import java.io.IOException;

import javax.servlet.ServletException;
```

```
import javax.servlet.http.HttpServlet;
import javax.servlet.http.HttpServletRequest;
import javax.servlet.http.HttpServletResponse;

import myrequestwrapper.MyHttpServletRequestWrapper;

public class MyServlet extends HttpServlet {
    @Override
    protected void doGet(HttpServletRequest req, HttpServletResponse resp)
throws ServletException, IOException {
        printMethodName(new MyHttpServletRequestWrapper(req));
    }

    // 方法 printMethodName()的作用就是获取请求的方式
    // 调用 MyHttpServletRequestWrapper 装饰类中的 getMethod()方法
    public void printMethodName(HttpServletRequest req) {
        System.out.println(req.getMethod());
    }
}
```

执行 Servlet 后，控制台输出结果如下：

```
begin run 1525942351526
 end run 1525942354527
GET
```

### 5. 使用getParameter()方法和getParameterValues()方法获得客户端表单的值

使用 HttpServletRequest 接口中的 public String getParameter(String name)方法和 public String[] getParameterValues(String name)方法可以获取客户端表单提交的值。public String getParameter(String name)方法获取单值；public String[] getParameterValues(String name)方法获取多值。

前台 my.html 代码如下：

```
<body>
    <form action="MyServlet" method="post">
        username:<input type="text" name="username" value="账号"/><br>
        password:<input type="password" name="password" value="密码"/><br>
        hiddenValue:<input type="hidden" name="hiddenValue" value="隐藏域的
值"/><br>
```

```
        bigText:<textarea name="bigText">超大文本</textarea><br/>
        mycheckbox:<br/>
        <input type="checkbox" name="mycheckbox" value="A"/><br>
        <input type="checkbox" name="mycheckbox" value="AA"/><br>
        <input type="checkbox" name="mycheckbox" value="AAA"/><br>
        myradio:<br/>
        <input type="radio" name="myradio" value="B"/><br>
        <input type="radio" name="myradio" value="BB"/><br>
        select1<br/>
        <select name="select1">
          <option value="a1">aa1</option>
          <option value="a2">aa2</option>
          <option value="a3">aa3</option>
        </select>
        <br/>
        select2<br/>
        <select name="select2" size="4">
          <option value="a1">aa1</option>
          <option value="a2">aa2</option>
          <option value="a3">aa3</option>
        </select>
        <br/>
        select3<br/>
        <select name="select3" size="4" multiple="multiple">
          <option value="a1">aa1</option>
          <option value="a2">aa2</option>
          <option value="a3">aa3</option>
        </select>
        <br/>
        <input type="submit" value="提交"/><br>
    </form>
</body>
```

后台 Servlet 示例代码如下：

```
public class MyServlet extends HttpServlet {

    @Override
    protected void doPost(HttpServletRequest req, HttpServletResponse resp)
throws ServletException, IOException {
        System.out.println(req.getParameter("username"));
        System.out.println(req.getParameter("password"));
```

```
        System.out.println(req.getParameter("hiddenValue"));
        System.out.println(req.getParameter("bigText"));
        System.out.println();
        System.out.println();
        String[] mycheckboxArray = req.getParameterValues("mycheckbox");
        for (int i = 0; i < mycheckboxArray.length; i++) {
            System.out.println(mycheckboxArray[i] + " ");
        }
        System.out.println();
        System.out.println();
        System.out.println(req.getParameter("myradio"));
        System.out.println();
        System.out.println();
        System.out.println(req.getParameter("select1"));
        System.out.println(req.getParameter("select2"));
        System.out.println();
        System.out.println();
        String[] select3Array = req.getParameterValues("select3");
        for (int i = 0; i < select3Array.length; i++) {
            System.out.println(select3Array[i] + " ");
        }
    }
}
```

设置界面内容如图 5-41 所示。

单击“提交”按钮后，成功进入 Servlet，但打印的中文却是乱码，如图 5-42 所示。

图 5-41

图 5-42

遇到这种情况可以使用 HttpServletRequest 接口的 setCharacterEncoding()方法解决，在 doPost()方法中的第一行进行调用，如图 5-43 所示。

```
@Override
protected void doPost(HttpServletRequest
    req.setCharacterEncoding("utf-8");
```

图 5-43

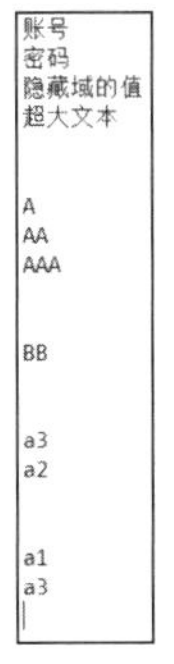


图 5-44

在浏览器单击“F5”键刷新界面，控制台正确显示出了中文，如图 5-44 所示。

**6. 使用getParameterMap ()方法和getParameterNames ()方法获取客户端表单的值**

如果知道前台表单 name，可以使用 public String getParameter (String name)和 public String[] getParameterValues(String name)方法获取表单值；如果不知道表单 name 则要使用 public Map<String, String[]> getParameterMap() 方 法 和 public Enumeration<String> getParameterNames()方法。

public Map<String, String[]> getParameterMap()方法的作用是将 name 和 value 封装到 Map 中，public Enumeration<String> getParameterNames()方法返回 Enumeration 对象的方法中包含所有 name。

前台 HTML 代码如下：

```
<body>
   <form action="MyServlet" method="post">
      <input type="username" name="usernameParam" value="账号"/>
      <br/>
      <input type="checkbox" name="mycheckbox" value="a" />
      <br/>
      <input type="checkbox" name="mycheckbox" value="b" />
      <br/>
      <input type="checkbox" name="mycheckbox" value="c" />
      <br/>
```

```
        <input type="checkbox" name="mycheckbox" value="d" />
        <br/>
        <input type="submit" value="submit">
    </form>
</body>
```

后台 Servlet 代码如下：

```
public class MyServlet extends HttpServlet {
    @Override
    protected void doPost(HttpServletRequest request, HttpServletResponse 
response)
            throws ServletException, IOException {
        request.setCharacterEncoding("utf-8");

        Map<String, String[]> map = request.getParameterMap();
        Iterator iterator = map.keySet().iterator();
        while (iterator.hasNext()) {
            String key = "" + iterator.next();
            System.out.print(key + "  ");
            String[] values = map.get(key);
            for (int i = 0; i < values.length; i++) {
                System.out.print(values[i]);
            }
            System.out.println();
        }
        System.out.println();
        map = request.getParameterMap();
        iterator = map.keySet().iterator();
        while (iterator.hasNext()) {
            String name = "" + iterator.next();
            System.out.print(name + "   ");
            String[] values = request.getParameterValues(name);
            for (int i = 0; i < values.length; i++) {
                System.out.print(values[i]);
            }
            System.out.println();
        }
        System.out.println();
        Enumeration<String> enumObject = request.getParameterNames();
        while (enumObject.hasMoreElements()) {
```

```
            String name = enumObject.nextElement();
            System.out.print(name + "   ");
            String[] values = request.getParameterValues(name);
            for (int i = 0; i < values.length; i++) {
                System.out.print(values[i]);
            }
            System.out.println();
        }

    }
}
```

程序运行结果如下：

```
usernameParam 账号
mycheckbox  acd

usernameParam  账号
mycheckbox  acd

usernameParam  账号
mycheckbox  acd
```

### 7. 非线程安全与Servlet单例性

Tomcat 为了高性能地处理后台业务，减少服务器内存占用率，将 Servlet 设计成单例的，创建测试代码如下：

```
public class MyServlet extends HttpServlet {
    public MyServlet() {
        System.out.println("public MyServlet() " + this.hashCode());
    }

    @Override
    protected void doGet(HttpServletRequest request, HttpServletResponse
response)
            throws ServletException, IOException {
        System.out.println("protected void doGet " + this.hashCode());
    }
}
```

程序运行结果如下：

```
public MyServlet() 399531991
protected void doGet 399531991
protected void doGet 399531991
protected void doGet 399531991
protected void doGet 399531991
```

从程序运行结果来看，Servlet 的确是单例的。如果有多个线程访问同一个 Servlet 对象相同的实例变量，则会出现线程安全问题，示例代码如下：

```
package controller;

import java.io.IOException;

import javax.servlet.ServletException;
import javax.servlet.http.HttpServlet;
import javax.servlet.http.HttpServletRequest;
import javax.servlet.http.HttpServletResponse;

public class Test extends HttpServlet {

    public Test() {
        System.out.println("public Test() " + this.hashCode());
    }

    private String username;
    private String password;

    protected void doGet(HttpServletRequest request, HttpServletResponse
response)
            throws ServletException, IOException {

        try {
            username = request.getParameter("username");
            if (username.equals("a")) {
                Thread.sleep(10000);
            }
            password = request.getParameter("password");
            System.out.println(Thread.currentThread().getName()     +     "
username=" + username + " password=" + password);
```

```
        } catch (InterruptedException e) {
            e.printStackTrace();
        }

    }

}
```

首先在浏览器窗口 1 中输入网址：

http://localhost:8080/web1/Test?username=a&password=aa

并按下“Enter”键，然后在 10 秒钟内快速在浏览器窗口 2 中输入网址：

http://localhost:8080/web1/Test?username=b&password=bb

并按下“Enter”键，10 秒钟过后，username 和 password 的值会出现混乱，效果如图 5-45 所示。

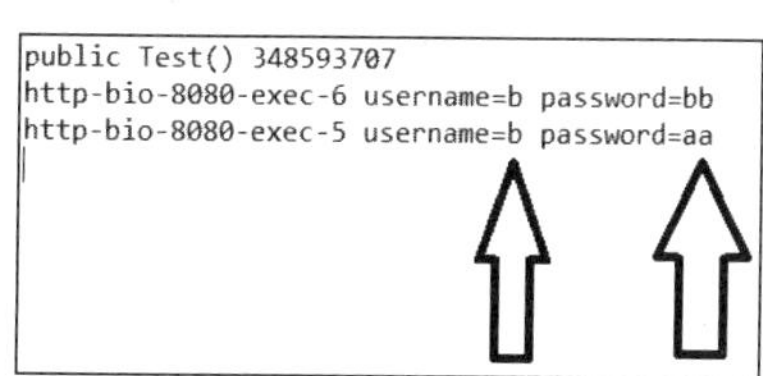


图 5-45

出现这种情况的根本原因就是 Servlet 是单例的，多个线程访问同一个 Servlet 对象中相同的实例变量，出现值与预期的结果不一样的情况，发生了“非线程安全”问题，所以在 Servlet 中尽量不要使用带有“写”操作的实例变量，极易发生值被覆盖的情况，也就是出现非线程安全问题，解决的办法就是将 username 和 password 放入 doGet()方法中，变成绑定各个线程对象的私有变量，示例代码如下：

```
package controller;

import java.io.IOException;

import javax.servlet.ServletException;
import javax.servlet.http.HttpServlet;
import javax.servlet.http.HttpServletRequest;
import javax.servlet.http.HttpServletResponse;

public class Test extends HttpServlet {

    public Test() {
        System.out.println("public Test() " + this.hashCode());
    }
```

```
    protected void doGet(HttpServletRequest request, HttpServletResponse
response)
            throws ServletException, IOException {
        try {
            String username = request.getParameter("username");
            if (username.equals("a")) {
                Thread.sleep(10000);
            }
            String password = request.getParameter("password");
            System.out.println(Thread.currentThread().getName()     +     "
username=" + username + " password=" + password);
        } catch (InterruptedException e) {
            e.printStackTrace();
        }
    }

}
```

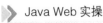

```
public Test() 348593707
http-bio-8080-exec-6 username=b password=bb
http-bio-8080-exec-5 username=a password=aa
```

图 5-46

当再执行上面的两个网址时，运算的结果是正确的，如图 5-46 所示。

注意：不要在 Servlet 中声明带有“写”操作的实例变量，“写”操作就是对变量赋值，也就是执行 set()方法；“读”操作就是对变量取值，也就是执行 get()方法。

**8. 使用response.setContentType("text/html;charset=GBK or UTF-8")方法解决乱码问题**

接口 HttpServletResponse 的主要作用就是对 HTTP 的响应处理，从服务器端向客户端传递一些数据，比如在浏览器上生成显示的 HTML 代码，示例代码如下：

```
public class MyServlet extends HttpServlet {
    @Override
    protected void doGet(HttpServletRequest request, HttpServletResponse
response)
            throws ServletException, IOException {
        PrintWriter out = response.getWriter();
        out.println("中国");
        out.close();
    }
}
```

程序运行后的效果如图 5-47 所示。在浏览器中显示的并不是中文，而是乱码“??”信息，更改代码如下：

图 5-47

```
public class MyServlet extends HttpServlet {
    @Override
    protected void doGet(HttpServletRequest request, HttpServletResponse
response)
            throws ServletException, IOException {
        response.setContentType("text/html;charset=GBK");
        PrintWriter out = response.getWriter();
        out.println("中国");
        out.close();
    }
}
```

程序运行后乱码问题得到解决，如图 5-48 所示。

图 5-48

为什么要使用代码 response.setContentType("text/html;charset=GBK")来解决乱码问题呢？

如果不执行 response.setContentType("text/html;charset=GBK")代码，在执行 out.println("中国")代码时，进入 Tomcat9 的内部流程，会将“中国”字符串转换成“iso-8859-1”编码的字节数组，字节数组转换成 String 正是乱码“??”，Tomcat 将乱码“??”字符串交给浏览器，浏览器将乱码“??”处理成浏览器默认的编码 GBK 编码后还是乱码“??”，所以在前端浏览器里显示了乱码“??”，验证代码如下：

```
public class Test1 {
    public static void main(String[] args) throws
UnsupportedEncodingException {
        // 提供数据
        String username = "中国";
        // 进入 Tomcat
        byte[] byteArray = username.getBytes("iso-8859-1");
        // 交给浏览器并显示
```

```
        System.out.println(new String(byteArray, "GBK"));
    }
}
```

程序运行结果还是“??”。

如果执行了 response.setContentType("text/html;charset=GBK")代码，则在执行 out.println("中国")代码时，进入 Tomcat9 的内部流程，会将“中国”字符串转换成 GBK 编码的字节数组，Tomcat 将这个 GBK 编码的字节数组交给浏览器，浏览器将 GBK 编码的字节数组转换成响应头中的编码格式 GBK 编码后还是“中国”，所以在前端浏览器里不显示乱码。验证代码如下：

```
public class Test2 {
    public static void main(String[] args) throws
UnsupportedEncodingException {
        // 提供数据
        String username = "中国";
        // 进入 Tomcat
        byte[] byteArray = username.getBytes("GBK");
        // 交给浏览器并显示
        System.out.println(new String(byteArray, "GBK"));
    }
}
```

程序运行结果是“中国”，不再是乱码“??”。

如果 MyServlet 代码更改如下：

```
public class MyServlet extends HttpServlet {
    @Override
    protected void doGet(HttpServletRequest request, HttpServletResponse
response)
            throws ServletException, IOException {
        response.setContentType("text/html;charset=utf-8");
        PrintWriter out = response.getWriter();
        out.println("中国");
        out.close();
    }
}
```

则在浏览器中也不会出现乱码，出现这样的情况是由于执行了 response.setContentType

("text/html;charset=utf-8")代码。在执行 out.println("中国")代码时，进入 Tomcat9 的内部流程，会将“中国”字符串转换成 utf-8 编码的字节数组，Tomcat 将 utf-8 编码的字节数组交给浏览器，浏览器将“utf-8”编码的字节数组转换成响应头中的编码格式 utf-8 编码后还是“中国”，所以在前端浏览器里不显示乱码。验证代码如下：

```
public class Test3 {
    public static void main(String[] args) throws
UnsupportedEncodingException {
        // 提供数据
        String username = "中国";
        // 进入 Tomcat
        byte[] byteArray = username.getBytes("utf-8");
        // 交给浏览器并显示
        System.out.println(new String(byteArray, "utf-8"));
    }
}
```

建议使用：

```
response.setContentType("text/html;charset=utf-8")
```

不要使用：

```
response.setContentType("text/html;charset=GBK")
```

因为 utf-8 编码的范围比 GBK 要大。

代码 response.setContentType("text/html;charset=XXXXX")有两个作用：

（1）告诉 Tomcat 使用什么编码对数据进行编码；

（2）告诉浏览器使用什么编码对数据进行解码并显示数据。

### 9. 使用response.setCharacterEncoding()方法解决乱码问题

在 Servlet 中，解决乱码还可以使用另外一种方式，就是执行 response.setCharacterEncoding()方法。方法 response.setCharacterEncoding("GBK")的作用是设置 Tomcat 传给浏览器的数据的编码格式。

创建 MyServlet 代码如下：

```
package controller;
```

```
import java.io.IOException;
import java.io.PrintWriter;

import javax.servlet.ServletException;
import javax.servlet.http.HttpServlet;
import javax.servlet.http.HttpServletRequest;
import javax.servlet.http.HttpServletResponse;

public class MyServlet extends HttpServlet {
    @Override
    protected void doGet(HttpServletRequest request, HttpServletResponse
response)
            throws ServletException, IOException {
        // 执行代码 response.setCharacterEncoding
        // 不在响应头中添加编码信息
        // response.setCharacterEncoding()方法的作用是对服务器端进行设置，使返回
数据的编码与客户端默认编码一样，这样就不会在前台显示乱码了
        response.setCharacterEncoding("GBK");

        // username 是 GBK 编码
        String username = "我是中国人";
        // utf8String 是 utf-8 编码
        String utf8String = new String(username.getBytes("utf-8"), "utf-8");
        PrintWriter out = response.getWriter();
        out.println(utf8String);
        out.close();
    }
}
```

程序运行后显示结果“我是中国人”，没有发生乱码的情况。

utf-8 编码格式的字符串变量 String utf8String 的作用就是实现服务器端和客户端编码不同的效果。因为上面执行了 response.setCharacterEncoding("GBK")方法，所以在 Tomcat 内部又将临时的 utf-8 编码格式的字符串变量 String utf8String 转换成了 GBK 编码，验证代码：

```
public class Test4 {
    public static void main(String[] args) throws
UnsupportedEncodingException {
        // 提供数据
        String usernameGBK = "中国";
```

```
        // 查看 GBK 字节数组中的内容
        byte[] gbkArray = usernameGBK.getBytes("GBK");
        // 查看 utf-8 字节数组中的内容
        // 下面的代码模拟是：
        //  String  utf8String  =  new  String(username.getBytes("utf-8"),
"utf-8") 中的 username.getBytes("utf-8")
        byte[] utf8Array = usernameGBK.getBytes("utf-8");
        // 下面两个 for 打印不同编码字节数组中的信息
        for (int i = 0; i < gbkArray.length; i++) {
            System.out.print(gbkArray[i] + " ");
        }
        System.out.println();
        for (int i = 0; i < utf8Array.length; i++) {
            System.out.print(utf8Array[i] + " ");
        }
        System.out.println();
        // 将 utf8Array 中的字节数组信息转换为 utf-8 编码
        String usernameUTF8 = new String(utf8Array, "utf-8");
        // 打印结果：中国
        System.out.println("usernameUTF8=" + usernameUTF8);
        // 字符串变量 usernameUTF8 中存储的数据并不是乱码
        // 所以 usernameUTF8.getBytes("GBK")模拟的就是响应
setCharacterEncoding("GBK")的执行将"中国"字符串转换成 GBK 编码
        gbkArray = usernameUTF8.getBytes("GBK");
        for (int i = 0; i < gbkArray.length; i++) {
            System.out.print(gbkArray[i] + " ");
        }
        System.out.println();
        // 模拟前台浏览器将 gbkArray 字节数组转换为 GBK
        // 在前台显示“中国”
        System.out.println(new String(gbkArray, "GBK"));
    }
}
```

程序运行后，浏览器并没有出现乱码，效果如图 5-49 所示。

另外，在执行 response.setCharacterEncoding("GBK")方法后，并不在响应头中添加编码信息，只是在服务器端将数据转换成指定的编码，它主要的目的就是使服务器端和浏览器端使用的编码统一，如果不统一，则会出现乱码，效果如图 5-50 所示。

```
-42 -48 -71 -6
-28 -72 -83 -27 -101 -67
usernameUTF8=中国
-42 -48 -71 -6
中国
```

图 5-49

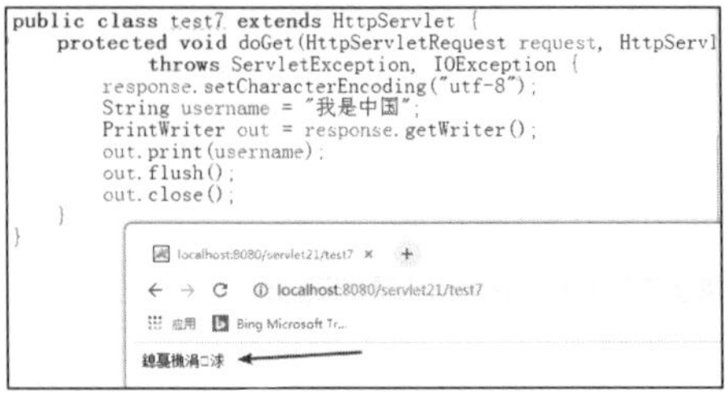

```
public class test7 extends HttpServlet {
    protected void doGet(HttpServletRequest request, HttpServl
            throws ServletException, IOException {
        response.setCharacterEncoding("utf-8");
        String username = "我是中国";
        PrintWriter out = response.getWriter();
        out.print(username);
        out.flush();
        out.close();
    }
}
```

图 5-50

代码 response.setCharacterEncoding("utf-8")只是在服务器端将中文转换成 utf-8 编码，但前台却使用 GBK 进行解码，前后端编码不统一，导致出现乱码。

再细化来说，不出现乱码的原因是执行了 response.setCharacterEncoding("GBK")方法后，将 utf-8 编码格式的字符串变量 String utf8String 在 Tomcat 内部又转回了 GBK 编码的字节数组，Tomcat 将这个 GBK 编码的字节数组交给浏览器，浏览器将 GBK 编码的字节数组处理成其默认的编码格式 GBK 编码后，还是“我是中国人”，所以在前端浏览器中不显示乱码。

总结一下：

（1）如果在代码中单独使用方法 response.setContentType()会控制服务器端与浏览器端的编码，在响应头中添加编码信息，告诉浏览器用指定的编码进行解析。

（2）如果在代码中单独使用 response.setCharacterEncoding(charsetName)方法，则会将服务器端返回的数据编码转换成目的 charsetName 编码，浏览器如果默认使用 charsetName 就不会出现中文乱码，而且不在 response 响应头中添加编码信息。

（3）当 response.setCharacterEncoding()方法和 response.setContentType()方法同时使用时，最后执行哪个方法，返回数据的编码就是最后执行方法所设置的编码，并且在响应头中添加编码信息并传给客户端。在功能上，response.setContentType()方法更加完善，因为可以设置响应头，在项目中可以只使用 response.setContentType()方法处理中文乱码。

（4）response.setCharacterEncoding()方法只控制服务器端编码，response.setContentType()方法控制服务器端与客户端编码。

### 10. 使用RequestDispatcher接口实现forward转发

接口 RequestDispatcher 负责请求的分发工作，可以将请求转交给其他资源，也就是让这个请求继续访问其他资源。获得 RequestDispatcher 对象的代码如下：

```
RequestDispatcher requestDispatcher = request.getRequestDispatcher("资源名
```

```
称");
```

实现转发操作使用 public void forward(ServletRequest request, ServletResponse response)方法，代码如下：

```
requestDispatcher.forward(request, response);
```

在分工协作的场景下，比如某一个 Servlet 处理 A 功能，其他某一个 Servlet 处理 B 功能，示例代码如图 5-51 所示。

```
MyServlet.java
package web2;

import java.io.IOException;

public class MyServlet extends HttpServlet {

    @Override
    protected void doGet(HttpServletRequest req, HttpServletResponse resp) throws ServletException, IOException {
        System.out.println("向数据库插入一条记录");
    }

}
```

```
MyServlet2.java
package web2;

import java.io.IOException;

public class MyServlet2 extends HttpServlet {

    protected void doGet(HttpServletRequest request, HttpServletResponse response)
            throws ServletException, IOException {
        System.out.println("从数据库中取出数据并列表");
    }

}
```

图 5-51

那如何实现这两个 Servlet 进行跳转交互呢，也就是执行完 MyServlet 的业务后，程序自动跳转到 MyServlet2 中，添加数据并列表，使用 RequestDispatcher 接口的 forward() 方法即可，示例代码如下：

```
public class MyServlet extends HttpServlet {
    @Override
    protected void doGet(HttpServletRequest req, HttpServletResponse resp)
throws ServletException, IOException {
        System.out.println("向数据库插入一条记录");
        req.getRequestDispatcher("MyServlet2").forward(req, resp);
    }
}
```

程序运行后控制台输出结果如图 5-52 所示。

图 5-52

那么，如何实现传递数据呢？更改 MyServlet 代码如下：

```
public class MyServlet extends HttpServlet {
    @Override
    protected void doGet(HttpServletRequest req, HttpServletResponse resp)
throws ServletException, IOException {
        System.out.println("向数据库插入一条记录");
        req.setAttribute("myServletKey", "myServletValue");
        req.getRequestDispatcher("MyServlet2").forward(req, resp);
    }

}
```

ServletRequest 接口中的 public void setAttribute(String name, Object o)方法可以向请求作用域中存储任何类型的数据值。更改 MyServlet2 代码如下：

```
public class MyServlet2 extends HttpServlet {
    protected void doGet(HttpServletRequest request, HttpServletResponse
response)
            throws ServletException, IOException {
        System.out.println(" 从 数 据 库 中 取 出 数 据 并 列 表  " +
request.getAttribute("myServletKey"));
    }

}
```

程序运行结果如图 5-53 所示。

图 5-53

能获取到值的原因是 MyServlet 类将数据使用 setAttribute()方法存放到一个变量中，当执行 MyServlet2 时，使用 getAttribute()方法从这个变量中取值，就实现了两个 Servlet 之间传输数据。

在“F12”中进行调试可以发现，转发是服务器端的行为，仅发起了一个请求，但却执行了两个 Servlet，说明第二个 Servlet 是 Tomcat 调用执行的，并不是浏览器直接执行的，而是间接执行的，如图 5-54 所示。

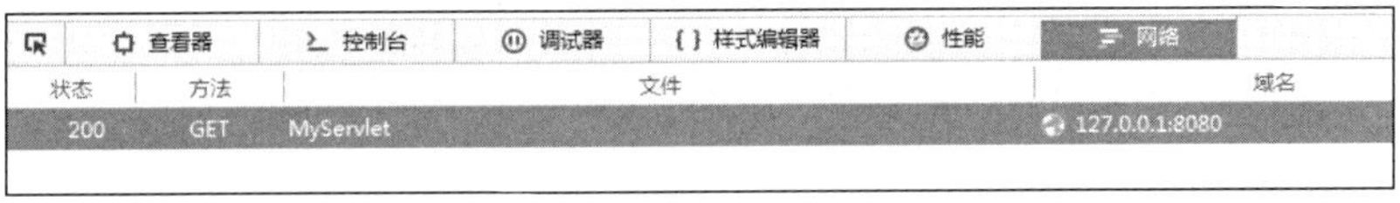

图 5-54

如果反复在浏览器上按下“F5”键会发现，重复执行了插入一条记录的 Servlet，导致数据有可能被重复添加，如图 5-55 所示。

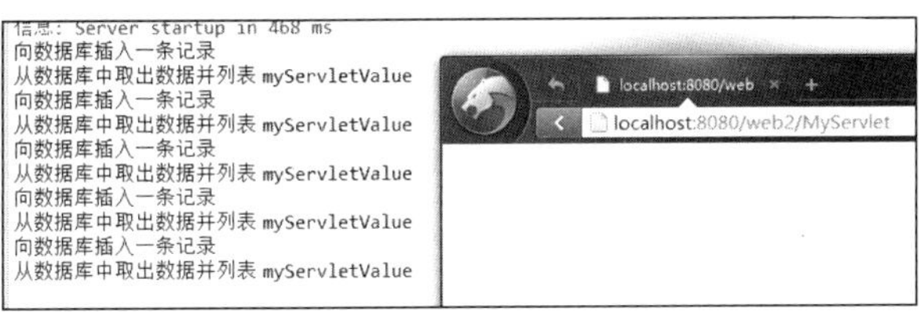


图 5-55

因为发生重复添加记录是浏览器的地址依然还是 MyServlet 所致的，所以每次按下“F5”键都要执行一次 MyServlet 添加操作。那么，如何避免这种情况的发生呢？使用重定向即可。

### 11. 使用HttpServletResponse接口处理重定向

重定向的操作是由 HttpServletResponse 接口提供的。

更改 MyServlet 代码如下：

```
public class MyServlet extends HttpServlet {
    @Override
    protected void doGet(HttpServletRequest req, HttpServletResponse resp)
throws ServletException, IOException {
        System.out.println("向数据库插入一条记录");
        req.setAttribute("myServletKey", "myServletValue");
        // req.getRequestDispatcher("MyServlet2").forward(req, resp);
        resp.sendRedirect("MyServlet2");
    }
}
```

先关闭当前浏览器，再执行 MyServlet，程序运行并重复按下“F5”键后不再重复添加多条记录，如图 5-56 所示。

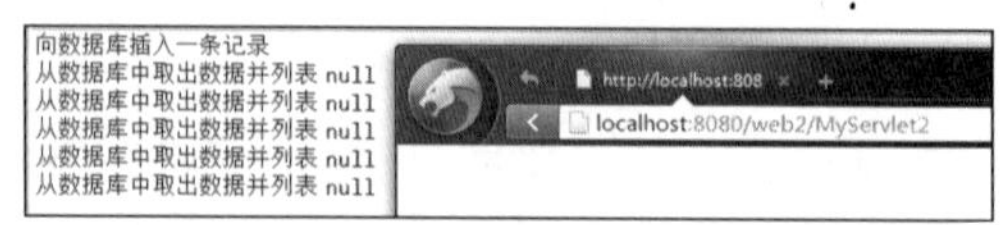


图 5-56

没有重复执行插入的关键点在于 URL 的地址被更改成 MyServlet2 了，说明重定向后 URL 地址发生变化，但数据值并没有得到传递，因为重定向发起新的请求，数据存放在旧的请求对象中，新的请求对象中没有数据，所以打印数据值为 NULL，在“F12”中进行调试，可以发现重定向真的发起了两个请求，如图 5-57 所示。

图 5-57

重定向是客户端的行为，响应状态码值为 302，再结合地址，浏览器就会向新的 URL 重新发起请求了，最终一共发起两次请求。

如果在使用重定向时想传递数据值，则不能使用 request.setAttribute(key,value)方法，可以将值作为重定向地址的参数传递过去，示例代码如下：

```
response.sendRedirect("test9?gotoPage=200&password=123");//URL 网址只能传递字符类型
```

接收值时使用如下代码：

```
String gotoPage = "" + request.getParameter("gotoPage");
String gotoPage = "" + request.getParameter("password");
```

下面总结一下转发与重定向。

（1）转发：服务器端行为，请求的 Attribute 中的数据可以获取到，浏览器地址不变，执行了一次请求，可以传递任意类型的数据。

（2）重定向：客户端行为，请求的 Attribute 中的数据不可以获取到，传输数据要使用 URL 传参?paramName=paramValue 的方式，参数值只能是字符串，浏览器地址发生变

化，执行了两次请求，传递的数据类型只能是字符。

### 12. 转发与重定向图示

转发能取到请求中 Attributes 数据的原因如图 5-58 所示。重定向取不到请求中 Attributes 数据的原因如图 5-59 所示。

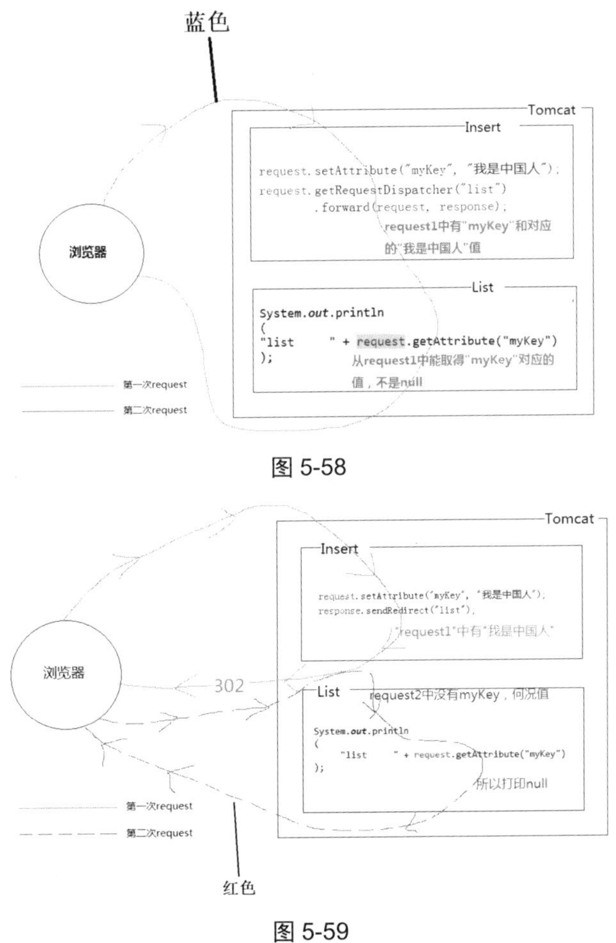


图 5-58

图 5-59

前面的图示仅仅是从“F12”开发者工具发起请求的次数来分析的，下面从 Tomcat9 的源代码级别来分析。

（1）转发不丢失数据是因为发起第一个 insert 请求时，将数据放入 request1 对象中，转发到 List 时创建新的 request2，但在创建 request2 时将 request1 作为构造方法的参数传递给 request2，所以 List 使用的 request2 就能取到值了，值并没有丢失。

（2）重定向丢失数据的原因是发起第二次请求后进入到 Tomcat9 中，Tomcat 会将请求池中以前请求对象中的 Attributes 数据全部清除，这时进入到 List 就取不到数据了。

### 13. 使用RequestDispatcher接口实现include()方法

接口 RequestDispatcher 的 include()方法主要就是用来包含请求信息，它可以将一个 Servlet 中输出的内容与其他 Servlet 输出的内容进行整合，进行整体信息的显示。

创建名称为 MyServlet 的代码如下：

```
public class MyServlet extends HttpServlet {
    @Override
    protected void doGet(HttpServletRequest req, HttpServletResponse resp)
throws ServletException, IOException {
        resp.setCharacterEncoding("utf-8");
        resp.setContentType("text/html");
        PrintWriter out = resp.getWriter();
        out.println("我是 MyServlet");
        out.println("<br/>");
        out.flush();
        out.close();
        req.setAttribute("fromKey", "fromValue");
        req.getRequestDispatcher("MyServlet2").include(req, resp);
    }
}
```

创建名称为 MyServlet2 的代码如下：

```
public class MyServlet2 extends HttpServlet {
    protected void doGet(HttpServletRequest request, HttpServletResponse
response)
            throws ServletException, IOException {
        response.setCharacterEncoding("utf-8");
        response.setContentType("text/html");
        PrintWriter out = response.getWriter();
        out.println("我是 MyServlet2 " + request.getAttribute("fromKey"));
        out.flush();
        out.close();
    }
}
```

实际运行结果如图 5-60 所示。

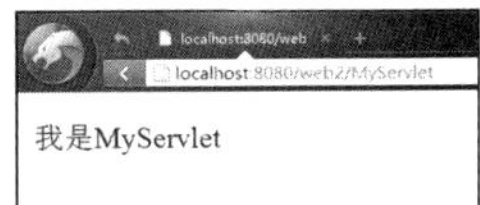

图 5-60

出现这样的情况是因为 MyServlet 将 out 对象调用了 close() 方法，更改 MyServlet 代码如下：

```
public class MyServlet extends HttpServlet {
    @Override
    protected void doGet(HttpServletRequest req, HttpServletResponse resp)
throws ServletException, IOException {
        resp.setCharacterEncoding("utf-8");
        resp.setContentType("text/html");
        PrintWriter out = resp.getWriter();
        out.println("我是 MyServlet");
        out.println("<br/>");
        // out.flush();
        // out.close(); 不要关闭输出流
        req.setAttribute("fromKey", "fromValue");
        req.getRequestDispatcher("MyServlet2").include(req, resp);
    }
}
```

图 5-61

程序运行效果如图 5-61 所示。

### 14. 如何在get和post中处理中文提交

在接收有中文的参数值时，get 和 post 的处理方式是不一样的。

前台 HTML 代码如下：

```
<!DOCTYPE html>
<html>
   <head>
      <meta charset="UTF-8">
      <title>Insert title here</title>
   </head>
   <body>
      <form action="login" method="get">
         username:<input type="text" name="username">
         <br/>
         <input type="submit" value="submit">
      </form>
```

```
        <br/>
        <form action="login" method="post">
            username:<input type="text" name="username">
            <br/>
            <input type="submit" value="submit">
        </form>
    </body>
</html>
```

后台 Servlet 代码如下：

```
public class login extends HttpServlet {
    protected void doGet(HttpServletRequest request, HttpServletResponse
response)
            throws ServletException, IOException {
        String username = request.getParameter("username");
        username = new String(username.getBytes("iso-8859-1"), "utf-8");
        System.out.println(username);
    }

    protected void doPost(HttpServletRequest request, HttpServletResponse
response)
            throws ServletException, IOException {
        request.setCharacterEncoding("utf-8");
        String username = request.getParameter("username");
        System.out.println(username);
    }

}
```

在 Tomcat9 中使用<form method=get>以 get 方式提交时，对网址使用的默认编码为 utf-8，说明在 Tomcat9 运行时不需要如下代码：

```
username = new String(username.getBytes("iso-8859-1"), "utf-8");
```

在使用 Tomcat 旧版本时需要上面的代码，但不管使用 Tomcat 的哪种版本，在使用 post 提交时，还是需要调用如下代码：

```
request.setCharacterEncoding("utf-8");
```

或

```
@Override
protected void doPost(HttpServletRequest request, HttpServletResponse
response)
        throws ServletException, IOException {
    String username = request.getParameter("username");
    username = new String(username.getBytes("iso-8859-1"), "utf-8");
    System.out.println("post " + username);
}
```

来解决 post 提交值是乱码的问题，因为任何版本的 Tomcat 对 post 提交的编码默认是 iso-8859-1。

如果前台 HTML 文件配置如下：

```
<head>
<meta charset="UTF-8">
<title>Insert title here</title>
</head>
```

后台 Servlet 代码如下：

```
protected void doPost(HttpServletRequest request, HttpServletResponse
response)
        throws ServletException, IOException {
    request.setCharacterEncoding("GBK");
    String username = request.getParameter("username");
    System.out.println(username);
}
```

那么程序运行后，Servlet 输出的结果是乱码，这是因为前后台编码不统一。

**15. request.getParameter()方法和request.getAttribute()方法的区别**

request.getParameter()方法用于接收从容器外部传递的数据；request.getAttribute()方法用于接收从容器内部传递的数据。

### 5.3.13 配置 Servlet 具有后缀

Servlet 路径可以有后缀。创建 Servlet 类代码如下：

```
package controller;

import java.io.IOException;
```

```
import javax.servlet.ServletException;
import javax.servlet.http.HttpServlet;
import javax.servlet.http.HttpServletRequest;
import javax.servlet.http.HttpServletResponse;

public class test1 extends HttpServlet {
    protected void doGet(HttpServletRequest request, HttpServletResponse
response)
            throws ServletException, IOException {
        System.out.println("test1 doGet run !");
    }
}
```

在 web.xml 文件中配置 Servlet 代码如下：

```
<servlet>
    <servlet-name>test1</servlet-name>
    <servlet-class>controller.test1</servlet-class>
</servlet>
<servlet-mapping>
    <servlet-name>test1</servlet-name>
    <url-pattern>/test1.action</url-pattern>
</servlet-mapping>
```

当访问 controller 包中的 test1.java 时，可以设置访问路径的后缀，本测试后缀名称为 action。使用如下完整 URL 访问这个 Servlet：http://localhost:8080/webX/ test1.action，控制台输出结果如下：

```
test1 doGet run !
```

请自行尝试使用注释版本配置访问路径的后缀。

## 5.4 请求与响应

本节将使用工具来研究请求与响应模型在数据传递时的数据格式及具体过程，还要学习 get 与 post 提交方式上的差异。

### 5.4.1 请求/响应模型

在 Web 开发中，从客户端发起请求到服务器端获取数据，再把获取的数据进行加工

返回客户端的模式称为请求/响应模型，访问网站的过程如图 5-62 所示。

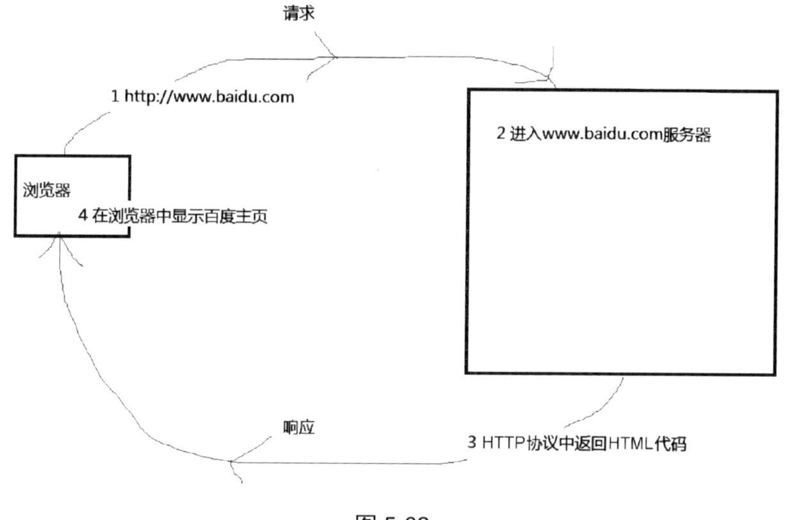


图 5-62

访问网站的过程包括如下 5 步。

（1）在浏览器中输入网址。

（2）在 DNS（Domain Name Server）服务器中找到百度服务器的 IP 地址。

（3）进入百度服务器。

（4）百度服务器返回主页的 HTML 代码。

（5）浏览器接收这些代码在界面上显示出来。

这 5 步其实就是 Web 从开始到结束的运行过程，该过程称为请求/响应模型。请求就是从客户端到服务器；响应就是从服务器到客户端。

## 5.4.2　请求与响应的数据格式

在请求与响应的过程中，传输与返回数据是要遵循 HTTP 协议格式的。在发起请求时，客户端要向服务器端传递 3 部分内容。

（1）请求行：请求方式（get/post）和请求资源的路径。

（2）请求头：浏览器与服务器之间通信的一些约定。比如浏览器可以告诉服务器是什么浏览器，版本是什么等客户端相关的信息。

（3）请求的数据内容：如果是 post 提交方式，则客户端的数据会存放到请求数据体里面，在 Form Data 节点下显示；如果是 get 提交方式，则客户端的数据会放到请求行中的请求资源路径之后。

在进行响应时，服务器端要向客户端传递 3 部分内容。

（1）状态行：也称状态码，代表服务器执行后的不同结果，状态码值为 200 代表响应正常、404 代表找不到服务器端资源、500 代表服务器系统内部错误。

（2）响应头：存储服务器向客户端传递数据的数据类型、数据量大小、执行时间、服务器产品名称等相关信息，还可以保存服务器向浏览器发送的 Cookie 数据。

（3）响应的数据内容：服务器处理完成后的结果，大多数就是要求浏览器显示的数据，这些数据的类型很大概率是 HTML 代码，也有可能是二进制数据流，比如在实现下载文件功能时。

下面开始创建实验用的环境，创建一个 Web 项目，创建 login.html 文件，代码如下：

```
<!DOCTYPE html>
<html>
    <head>
        <meta charset="UTF-8">
    </head>
    <body>
        <form action="login" method="post">
            username:<input type="text" name="username">
            <br/>
            password:<input type="text" name="password">
            <br/>
            <input type="submit" value="submit">
            <br/>
        </form>
        <br/>
        <form action="login" method="get">
            username:<input type="text" name="username">
            <br/>
            password:<input type="text" name="password">
            <br/>
            <input type="submit" value="submit">
            <br/>
        </form>
    </body>
</html>
```

创建一个名称为 login 的 Servlet，代码如下：

```
public class login extends HttpServlet {
    protected void doGet(HttpServletRequest request, HttpServletResponse
response)
            throws ServletException, IOException {
        this.doPost(request, response);
    }

    protected void doPost(HttpServletRequest request, HttpServletResponse
response)
            throws ServletException, IOException {
        System.out.println("goto login servlet!");
        request.getRequestDispatcher("ok.html").forward(request,
response);
    }
}
```

创建 ok.html 文件，代码如下：

```
<!DOCTYPE html>
<html>
   <head>
      <meta charset="UTF-8">
   </head>
   <body>
      this is ok page!
   </body>
</html>
```

将项目部署到 Tomcat 中，并在浏览器中打开 login.html 文件，然后按下“F12”键调出 HTTP 分析工具，显示界面如图 5-63 所示。

在 username 和 password 文本框中分别输入“a”和“aa”字符后再单击最上面的“submit”（提交）按钮，将数据提交到 Servlet 中，浏览器 HTTP 分析工具就捕获到请求数据格式了，效果如图 5-64 所示。响应数据格式如图 5-65 所示。

图 5-63

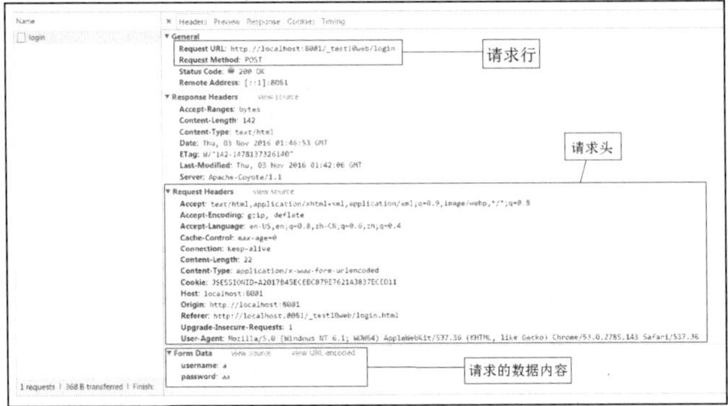


图 5-64

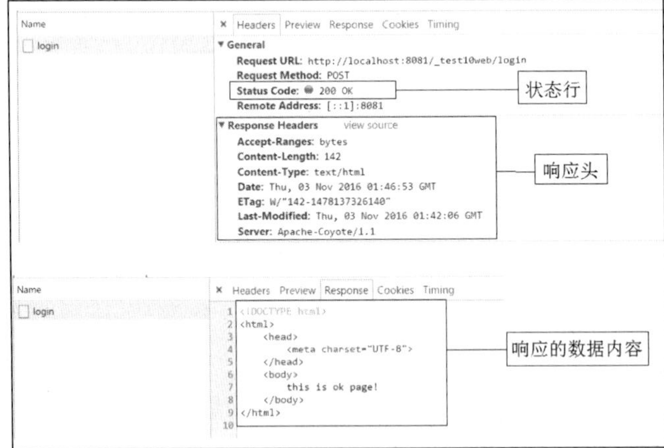


图 5-65

## 5.5　使用Servlet+JDBC实现基于Web的CURD增删改查

下面使用 Servlet 类实现增删改查的示例，目的是掌握如下 9 个知识点。

（1）掌握 JDBC 代码。

（2）掌握 HTML 代码。

（3）掌握调用关系与顺序。

（4）理解使用重定向的目的。

（5）会写 DAO 代码。

（6）理解实体类的作用。

（7）理解 HTML 前台和后台 Servlet 交互。

（8）掌握 Servlet 及其交互。

（9）掌握所有环节的报错 DEBUG 调试与分析。

创建实体类 Userinfo 代码如下：

```
package com.ghy.www.entity;

import java.util.Date;

public class Userinfo {
    private int id;
    private String username;
    private String password;
    private int age;
    private Date insertdate;

    public Userinfo() {
    }

    public int getId() {
        return id;
    }

    public void setId(int id) {
        this.id = id;
    }
```

```
    public String getUsername() {
        return username;
    }

    public void setUsername(String username) {
        this.username = username;
    }

    public String getPassword() {
        return password;
    }

    public void setPassword(String password) {
        this.password = password;
    }

    public int getAge() {
        return age;
    }

    public void setAge(int age) {
        this.age = age;
    }

    public Date getInsertdate() {
        return insertdate;
    }

    public void setInsertdate(Date insertdate) {
        this.insertdate = insertdate;
    }
}
```

创建获得 Connection 对象的 ConnectionTools 类代码如下：

```
package com.ghy.www.dbtools;

import java.sql.*;

public class ConnectionTools {
    public static Connection getConnection() throws ClassNotFoundException,
SQLException {
```

```
        String          url          =          "jdbc:mysql://localhost:3306/
y2?serverTimezone=Asia/Shanghai";
        String drvierName = "com.mysql.cj.jdbc.Driver";
        String username = "root";
        String password = "123123";
        Class.forName(drvierName);
        return DriverManager.getConnection(url, username, password);
    }

    public static void close(Connection conn) {
        try {
            if (conn != null && conn.isClosed() == false) {
                conn.close();
            }
        } catch (SQLException throwables) {
            throwables.printStackTrace();
        }
    }

    public static void close(Connection conn, Statement stat) {
        try {
            if (conn != null && conn.isClosed() == false) {
                conn.close();
            }
            if (stat != null && stat.isClosed() == false) {
                stat.close();
            }
        } catch (SQLException throwables) {
            throwables.printStackTrace();
        }
    }

    public static void close(Connection conn, Statement stat, ResultSet rs)
{
        try {
            if (conn != null && conn.isClosed() == false) {
                conn.close();
            }
            if (stat != null && stat.isClosed() == false) {
                stat.close();
            }
            if (rs != null && rs.isClosed() == false) {
```

```
                rs.close();
            }
        } catch (SQLException throwables) {
            throwables.printStackTrace();
        }
    }
}
```

创建数据访问对象的 DAO 代码如下：

```
package com.ghy.www.dao;

import com.ghy.www.dbtools.ConnectionTools;
import com.ghy.www.entity.Userinfo;

import java.sql.*;
import java.util.ArrayList;
import java.util.List;

public class UserinfoDAO {
    public void insertUserinfo(Userinfo userinfo) throws SQLException,
ClassNotFoundException {
        StringBuffer buffer = new StringBuffer();
        buffer.append("insert into userinfo ");
        buffer.append("(username,password,age,insertdate) ");
        buffer.append("values(?,?,?,?)");

        Connection conn = ConnectionTools.getConnection();
        PreparedStatement ps = conn.prepareStatement(buffer.toString());
        ps.setString(1, userinfo.getUsername());
        ps.setString(2, userinfo.getPassword());
        ps.setInt(3, userinfo.getAge());
        ps.setTimestamp(4,                                            new
Timestamp(userinfo.getInsertdate().getTime()));
        ps.executeUpdate();
        ConnectionTools.close(conn, ps);
    }

    public List<Userinfo> selectAllUserinfo() throws SQLException,
ClassNotFoundException {
        List<Userinfo> listUserinfo = new ArrayList<>();
```

```
        StringBuffer buffer = new StringBuffer();
        buffer.append("select * from userinfo ");
        buffer.append("order by id asc");

        Connection conn = ConnectionTools.getConnection();
        PreparedStatement ps = conn.prepareStatement(buffer.toString());
        ResultSet rs = ps.executeQuery();
        while (rs.next()) {
            int id = rs.getInt("id");
            String username = rs.getString("username");
            String password = rs.getString("password");
            int age = rs.getInt("age");
            Date insertDate = rs.getDate("insertdate");

            Userinfo userinfo = new Userinfo();
            userinfo.setId(id);
            userinfo.setUsername(username);
            userinfo.setPassword(password);
            userinfo.setAge(age);
            userinfo.setInsertdate(insertDate);

            listUserinfo.add(userinfo);
        }
        ConnectionTools.close(conn, ps, rs);
        return listUserinfo;
    }

    public void deleteUserinfoById(int userId) throws SQLException,
ClassNotFoundException {
        StringBuffer buffer = new StringBuffer();
        buffer.append("delete from userinfo where id=?");

        Connection conn = ConnectionTools.getConnection();
        PreparedStatement ps = conn.prepareStatement(buffer.toString());
        ps.setInt(1, userId);
        ps.executeUpdate();
        ConnectionTools.close(conn, ps);
    }

    public Userinfo selectUserinfoById(int userId) throws SQLException,
ClassNotFoundException {
        Userinfo userinfo = null;
```

```
        StringBuffer buffer = new StringBuffer();
        buffer.append("select * from userinfo ");
        buffer.append("where id=?");

        Connection conn = ConnectionTools.getConnection();
        PreparedStatement ps = conn.prepareStatement(buffer.toString());
        ps.setInt(1, userId);
        ResultSet rs = ps.executeQuery();
        while (rs.next()) {
            int id = rs.getInt("id");
            String username = rs.getString("username");
            String password = rs.getString("password");
            int age = rs.getInt("age");
            Date insertDate = rs.getDate("insertdate");

            userinfo = new Userinfo();
            userinfo.setId(id);
            userinfo.setUsername(username);
            userinfo.setPassword(password);
            userinfo.setAge(age);
            userinfo.setInsertdate(insertDate);

        }
        ConnectionTools.close(conn, ps, rs);
        return userinfo;
    }

    public void updateUserinfoById(Userinfo userinfo) throws SQLException,
ClassNotFoundException {
        StringBuffer buffer = new StringBuffer();
        buffer.append("update userinfo set ");
        buffer.append("username=?, password=?, age=?, insertdate=? ");
        buffer.append("where id=?");

        Connection conn = ConnectionTools.getConnection();
        PreparedStatement ps = conn.prepareStatement(buffer.toString());
        ps.setString(1, userinfo.getUsername());
        ps.setString(2, userinfo.getPassword());
        ps.setInt(3, userinfo.getAge());
        ps.setTimestamp(4,                                              new
Timestamp(userinfo.getInsertdate().getTime()));
```

```
        ps.setInt(5, userinfo.getId());
        ps.executeUpdate();
        ConnectionTools.close(conn, ps);
    }
}
```

创建业务类代码如下：

```
package com.ghy.www.service;

import com.ghy.www.dao.UserinfoDAO;
import com.ghy.www.entity.Userinfo;

import java.sql.SQLException;
import java.util.List;

public class UserinfoService {
    private UserinfoDAO userinfoDAO = new UserinfoDAO();

    public void insert(Userinfo userinfo) throws SQLException, 
ClassNotFoundException {
        userinfoDAO.insertUserinfo(userinfo);
    }

    public List<Userinfo> selectAllUserinfo() throws SQLException, 
ClassNotFoundException {
        return userinfoDAO.selectAllUserinfo();
    }

    public void deleteUserinfoById(int userId) throws SQLException, 
ClassNotFoundException {
        userinfoDAO.deleteUserinfoById(userId);
    }

    public Userinfo selectUserinfoById(int userId) throws SQLException, 
ClassNotFoundException {
        return userinfoDAO.selectUserinfoById(userId);
    }

    public void updateUserinfoById(Userinfo userinfo) throws SQLException, 
ClassNotFoundException {
        userinfoDAO.updateUserinfoById(userinfo);
```

```
    }

    public void insertUserinfo(Userinfo userinfo) throws SQLException,
ClassNotFoundException {
        userinfoDAO.insertUserinfo(userinfo);
    }
}
```

创建数据列表 list 及添加 insert 功能界面的 Servlet 代码如下：

```
package com.ghy.www.controller;

import com.ghy.www.entity.Userinfo;
import com.ghy.www.service.UserinfoService;

import javax.servlet.ServletException;
import javax.servlet.annotation.WebServlet;
import javax.servlet.http.HttpServlet;
import javax.servlet.http.HttpServletRequest;
import javax.servlet.http.HttpServletResponse;
import java.io.IOException;
import java.io.PrintWriter;
import java.sql.SQLException;
import java.text.SimpleDateFormat;
import java.util.Date;
import java.util.List;

@WebServlet(name = "SelectAllUserinfo", urlPatterns = "/SelectAllUserinfo")
public class SelectAllUserinfo extends HttpServlet {
    protected void doGet(HttpServletRequest request, HttpServletResponse
response) throws ServletException, IOException {
        try {
            UserinfoService userinfoService = new UserinfoService();
            List<Userinfo>                    listUserinfo                    =
userinfoService.selectAllUserinfo();

            String formBegin = "<form action='InsertUserinfo' method='post'>";
            String formEnd = "</form>";
            String tableBegin = "<table border='1'>";
            String tableEnd = "</table>";
            String trEnd = "</tr>";
            String trBegin = "<tr>";
```

```
        String tdBegin = "<td>";
        String tdEnd = "</td>";

        SimpleDateFormat format = new SimpleDateFormat("yyyy-MM-dd");

        StringBuffer finalBuffer = new StringBuffer();
        finalBuffer.append(trBegin);
        finalBuffer.append(tdBegin + "ID" + tdEnd);
        finalBuffer.append(tdBegin + "USERNAME" + tdEnd);
        finalBuffer.append(tdBegin + "AGE" + tdEnd);
        finalBuffer.append(tdBegin + "PASSWORD" + tdEnd);
        finalBuffer.append(tdBegin + "INSERTDATE" + tdEnd);
        finalBuffer.append(tdBegin + "OPERATE" + tdEnd);
        finalBuffer.append(trEnd);

        StringBuffer dataTR = new StringBuffer();
        for (int i = 0; i < listUserinfo.size(); i++) {
            Userinfo userinfo = listUserinfo.get(i);

            int id = userinfo.getId();
            String username = userinfo.getUsername();
            String password = userinfo.getPassword();
            int age = userinfo.getAge();
            Date insertDate = userinfo.getInsertdate();
            String insertDateString = format.format(insertDate);


            String deleteA = "<a href='DeleteUserinfoById?id=" + id + "'>
删除</a>";
            String updateA = "<a href='EditUserinfoById?id=" + id + "'>
编辑</a>";

            StringBuffer eachTR = new StringBuffer();
            eachTR.append(trBegin);
            eachTR.append(tdBegin + id + tdEnd);
            eachTR.append(tdBegin + username + tdEnd);
            eachTR.append(tdBegin + password + tdEnd);
            eachTR.append(tdBegin + age + tdEnd);
            eachTR.append(tdBegin + insertDateString + tdEnd);
            eachTR.append(tdBegin + deleteA + "  " + updateA +
tdEnd);
            eachTR.append(trEnd);
```

```
            dataTR = dataTR.append(eachTR);
        }

        finalBuffer.append(dataTR.toString());
        finalBuffer.insert(0, tableBegin);
        finalBuffer.insert(finalBuffer.length(), tableEnd);

        StringBuffer insertTR = new StringBuffer();
        insertTR.append(trBegin);
        insertTR.append(tdBegin + "USERNAME" + tdEnd);
        insertTR.append(tdBegin + "<input type='text' name='username'
value=''>" + tdEnd);
        insertTR.append(trEnd);

        insertTR.append(trBegin);
        insertTR.append(tdBegin + "PASSWORD" + tdEnd);
        insertTR.append(tdBegin + "<input type='text' name='password'
value=''>" + tdEnd);
        insertTR.append(trEnd);

        insertTR.append(trBegin);
        insertTR.append(tdBegin + "AGE" + tdEnd);
        insertTR.append(tdBegin + "<input type='text' name='age'
value=''>" + tdEnd);
        insertTR.append(trEnd);

        insertTR.append(trBegin);
        insertTR.append(tdBegin + "INSERTDATE" + tdEnd);
        insertTR.append(tdBegin + "<input type='text' name='insertdate'
value=''>" + tdEnd);
        insertTR.append(trEnd);

        insertTR.append(trBegin);
        insertTR.append("<td colspan='2'>");
        insertTR.append("<input type='submit' value='提交'/>" + tdEnd);
        insertTR.append(trEnd);

        insertTR.insert(0, tableBegin);
        insertTR.insert(insertTR.length(), tableEnd);

        insertTR.insert(0, formBegin);
```

```
            insertTR.insert(insertTR.length(), formEnd);

            finalBuffer.append("<br/><br/>");
            finalBuffer.append(insertTR.toString());

            response.setContentType("text/html;charset=utf-8");
            PrintWriter out = response.getWriter();
            out.print(finalBuffer.toString());
            out.flush();
            out.close();
        } catch (SQLException throwables) {
            throwables.printStackTrace();
        } catch (ClassNotFoundException e) {
            e.printStackTrace();
        } finally {
        }
    }
}
```

创建执行删除 delete 功能的 Servlet 类代码如下：

```
package com.ghy.www.controller;

import com.ghy.www.service.UserinfoService;

import javax.servlet.ServletException;
import javax.servlet.annotation.WebServlet;
import javax.servlet.http.HttpServlet;
import javax.servlet.http.HttpServletRequest;
import javax.servlet.http.HttpServletResponse;
import java.io.IOException;
import java.sql.SQLException;

@WebServlet(name    =    "DeleteUserinfoById",    urlPatterns    =
"/DeleteUserinfoById")
public class DeleteUserinfoById extends HttpServlet {
    protected void doGet(HttpServletRequest request, HttpServletResponse
response) throws ServletException, IOException {
        try {
            String userId = request.getParameter("id");
            int userIdInt = Integer.parseInt(userId);
```

```
            UserinfoService userinfoService = new UserinfoService();
            userinfoService.deleteUserinfoById(userIdInt);

            response.sendRedirect("SelectAllUserinfo");
        } catch (SQLException throwables) {
            throwables.printStackTrace();
        } catch (ClassNotFoundException e) {
            e.printStackTrace();
        } finally {
        }
    }
}
```

创建执行添加 insert 功能的 Servlet 类代码如下：

```
package com.ghy.www.controller;

import com.ghy.www.entity.Userinfo;
import com.ghy.www.service.UserinfoService;

import javax.servlet.ServletException;
import javax.servlet.annotation.WebServlet;
import javax.servlet.http.HttpServlet;
import javax.servlet.http.HttpServletRequest;
import javax.servlet.http.HttpServletResponse;
import java.io.IOException;
import java.sql.SQLException;
import java.text.ParseException;
import java.text.SimpleDateFormat;
import java.util.Date;

@WebServlet(name = "InsertUserinfo", urlPatterns = "/InsertUserinfo")
public class InsertUserinfo extends HttpServlet {
    protected void doPost(HttpServletRequest request, HttpServletResponse
response) throws ServletException, IOException {
        try {
            request.setCharacterEncoding("utf-8");

            String username = request.getParameter("username");
            String password = request.getParameter("password");
            String age = request.getParameter("age");
            String insertdate = request.getParameter("insertdate");
```

```
        int ageInt = Integer.parseInt(age);
        SimpleDateFormat format = new SimpleDateFormat("yyyy-MM-dd");
        Date dateObj = format.parse(insertdate);

        Userinfo userinfo = new Userinfo();
        userinfo.setUsername(username);
        userinfo.setPassword(password);
        userinfo.setAge(ageInt);
        userinfo.setInsertdate(dateObj);

        UserinfoService userinfoService = new UserinfoService();
        userinfoService.insertUserinfo(userinfo);

        response.sendRedirect("SelectAllUserinfo");
      } catch (SQLException throwables) {
        throwables.printStackTrace();
      } catch (ClassNotFoundException e) {
        e.printStackTrace();
      } catch (ParseException e) {
        e.printStackTrace();
      } finally {
      }

   }
}
```

创建显示旧数据的 Servlet 类代码如下：

```
package com.ghy.www.controller;

import com.ghy.www.entity.Userinfo;
import com.ghy.www.service.UserinfoService;

import javax.servlet.ServletException;
import javax.servlet.annotation.WebServlet;
import javax.servlet.http.HttpServlet;
import javax.servlet.http.HttpServletRequest;
import javax.servlet.http.HttpServletResponse;
import java.io.IOException;
import java.io.PrintWriter;
import java.sql.SQLException;
```

```
import java.text.SimpleDateFormat;

@WebServlet(name = "EditUserinfoById", urlPatterns = "/EditUserinfoById")
public class EditUserinfoById extends HttpServlet {
   protected void doGet(HttpServletRequest request, HttpServletResponse
response) throws ServletException, IOException {

      try {
         String userId = request.getParameter("id");
         int userIdInt = Integer.parseInt(userId);

         UserinfoService userinfoService = new UserinfoService();
         Userinfo                         userinfo                        =
userinfoService.selectUserinfoById(userIdInt);

         String    formBegin    =    "<form    action='UpdateUserinfoById'
method='post'>";
         String formEnd = "</form>";
         String tableBegin = "<table border='1'>";
         String tableEnd = "</table>";
         String trBegin = "<tr>";
         String trEnd = "</tr>";
         String tdBegin = "<td>";
         String tdEnd = "</td>";

         SimpleDateFormat format = new SimpleDateFormat("yyyy-MM-dd");

         StringBuffer finalBuffer = new StringBuffer();
         finalBuffer.append(trBegin);

         finalBuffer.append(tdBegin + "USERNAME" + tdEnd);
         finalBuffer.append(tdBegin  +  "<input  type='hidden'  name='id'
value='" + userinfo.getId() + "'/>" + "<input type='text' name='username'
value='" + userinfo.getUsername() + "'/>" + tdEnd);
         finalBuffer.append(trEnd);

         finalBuffer.append(trBegin);
         finalBuffer.append(tdBegin + "PASSWORD" + tdEnd);
         finalBuffer.append(tdBegin + "<input type='text' name='password'
value='" + userinfo.getPassword() + "'/>" + tdEnd);
         finalBuffer.append(trEnd);
```

```
            finalBuffer.append(trBegin);
            finalBuffer.append(tdBegin + "AGE" + tdEnd);
            finalBuffer.append(tdBegin   +   "<input  type='text'  name='age'
value='" + userinfo.getAge() + "'/>" + tdEnd);
            finalBuffer.append(trEnd);

            finalBuffer.append(trBegin);
            finalBuffer.append(tdBegin + "INSERTDATE" + tdEnd);
            finalBuffer.append(tdBegin      +      "<input      type='text'
name='insertdate'  value='"  +  format.format(userinfo.getInsertdate())  +
"'/>" + tdEnd);
            finalBuffer.append(trEnd);

            finalBuffer.append(trBegin);
            finalBuffer.append("<td colspan='2'>");
            finalBuffer.append("<input  type='submit'  name='submit'  value='
提交'/>" + tdEnd);

            finalBuffer.append(trEnd);

            finalBuffer.insert(0, tableBegin);
            finalBuffer.insert(finalBuffer.length(), tableEnd);

            finalBuffer.insert(0, formBegin);
            finalBuffer.insert(finalBuffer.length(), formEnd);

            response.setContentType("text/html;charset=utf-8");
            PrintWriter out = response.getWriter();
            out.print(finalBuffer.toString());
            out.flush();
            out.close();
        } catch (SQLException throwables) {
            throwables.printStackTrace();
        } catch (ClassNotFoundException e) {
            e.printStackTrace();
        } finally {
        }

    }
}
```

创建在数据库中 update 新数据的 Servlet 类代码如下：

```
package com.ghy.www.controller;

import com.ghy.www.entity.Userinfo;
import com.ghy.www.service.UserinfoService;

import javax.servlet.ServletException;
import javax.servlet.annotation.WebServlet;
import javax.servlet.http.HttpServlet;
import javax.servlet.http.HttpServletRequest;
import javax.servlet.http.HttpServletResponse;
import java.io.IOException;
import java.sql.SQLException;
import java.text.ParseException;
import java.text.SimpleDateFormat;
import java.util.Date;

@WebServlet(name        =        "UpdateUserinfoById",        urlPatterns        =
"/UpdateUserinfoById")
public class UpdateUserinfoById extends HttpServlet {
    protected void doPost(HttpServletRequest request, HttpServletResponse
response) throws ServletException, IOException {
        try {
            request.setCharacterEncoding("utf-8");

            String userId = request.getParameter("id");
            String username = request.getParameter("username");
            String password = request.getParameter("password");
            String age = request.getParameter("age");
            String insertdate = request.getParameter("insertdate");

            int userIdInt = Integer.parseInt(userId);
            int ageInt = Integer.parseInt(age);
            SimpleDateFormat format = new SimpleDateFormat("yyyy-MM-dd");
            Date dateObj = format.parse(insertdate);

            Userinfo userinfo = new Userinfo();
            userinfo.setId(userIdInt);
            userinfo.setUsername(username);
            userinfo.setPassword(password);
            userinfo.setAge(ageInt);
            userinfo.setInsertdate(dateObj);
```

```
            UserinfoService userinfoService = new UserinfoService();
            userinfoService.updateUserinfoById(userinfo);

            response.sendRedirect("SelectAllUserinfo");
        } catch (SQLException throwables) {
            throwables.printStackTrace();
        } catch (ClassNotFoundException e) {
            e.printStackTrace();
        } catch (ParseException e) {
            e.printStackTrace();
        } finally {
        }
    }
}
```

这些 Servlet 类之间的调用关系如图 5-66 所示。

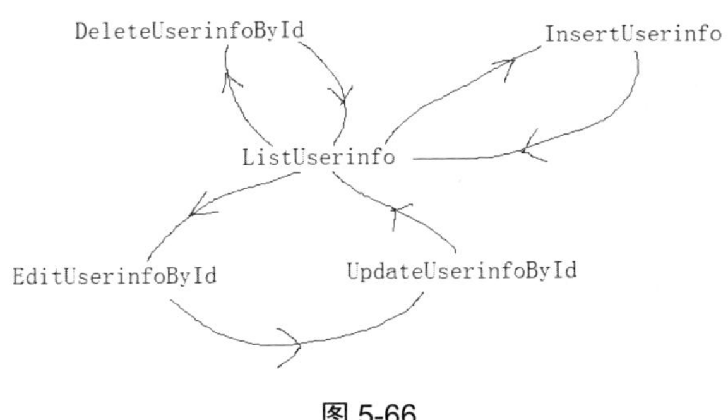


图 5-66

以上就是只使用 Servlet 结合 JDBC 代码实现 CURD 增删改查操作的全部流程。

# 第 6 章 Cookie 对象

什么是 Cookie？Cookie 是在浏览器客户端保存数据的一种方案。Cookie 中的数据保存在用户的个人电脑上而并不保存在服务器端。

为什么要使用 Cookie？如果一个大型网站访问的人数很多，每个人需要在服务端保存自己的数据，则会给服务器的内存占用造成很大的压力，为了缓解服务器的内存压力，可以将用户的私有数据以分布式的方式保存到本地，这种情况就是使用 Cookie 的时机。

创建 Cookie 的方式分为以下两种。

（1）服务器端创建：在 Server 服务器端执行 Servlet 时运行相关的 Java 代码将 Cookie 添加到响应对象中，然后 Client 客户端浏览器接收到服务器端响应发送过来在本地存储 Cookie 的命令，浏览器就开始在本地保存 Cookie。

（2）客户端创建：在浏览器客户端使用 JavaScript 语言直接在客户端创建 Cookie。

Server 服务器端能获取客户端 Cookie 的原因是当浏览器发起请求时，请求中携带客户端的 Cookie 并传给服务器端，服务器端就能获取客户端的 Cookie 值了。

每个浏览器都拥有自己的 Cookie，浏览器之间的 Cookie 并不能实现共享，但可以使用导入工具进行导入，比如在安装火狐浏览器时，就有一个选项来询问是否将 IE 浏览器中的 Cookie 导入火狐浏览器中。

那如何用 Java 代码操作 Cookie 呢？来看下面的介绍。

## 6.1 创建Cookie

创建 Servlet 代码如下：

```
package controller;
```

```
import java.io.IOException;

import javax.servlet.ServletException;
import javax.servlet.http.Cookie;
import javax.servlet.http.HttpServlet;
import javax.servlet.http.HttpServletRequest;
import javax.servlet.http.HttpServletResponse;

public class test1 extends HttpServlet {
    protected void doGet(HttpServletRequest request, HttpServletResponse
response)
            throws ServletException, IOException {
        Cookie cookie = new Cookie("myname", "myvalue");
        // setMaxAge()方法是设置 Cookie 的有效时间，超时后 Cookie 无效，无效的效果就
是在服务器端取不到客户端的 Cookie，也就是客户端发起请求时，不将超时过期的 Cookie 传送给
服务器
        cookie.setMaxAge(36000); // 时间单位是秒
        response.addCookie(cookie);
    }

}
```

cookie.setMaxAge(36000)代码的作用是设置 Cookie 的有效时间，Cookie 有效时间的算法是：new Date(System.currentTimeMillis()+maxAge*1000L)。方法 System.currentTimeMillis()获得的是 1970 年 1 月 1 日到当前时间所流逝的毫秒数，在此基础上与 maxAge*1000L 进行加法操作，结果就是未来的时间，若 Cookie 的生命周期超过此时间，则此 Cookie 就是无效的。

执行 Servlet 后，就在客户端保存了一个名称为 myname、值为 myvalue 的 Cookie，为什么程序代码在服务器端执行，Cookie 却保存到客户端了呢？是什么机制造成这样的效果？答案就是响应对象。在执行完 Servlet 后，响应对象中保存刚刚在服务器端使用 Java 代码创建的 Cookie，然后交给浏览器，浏览器接收到创建 Cookie 的命令，这时客户端也就有了 Cookie 数据，响应对象中保存 Cookie 的证据如图 6-1 所示。

在如图 6-2 所示的响应头 Set-Cookie 命令中保存的是 Cookie 在客户端的过期时间。

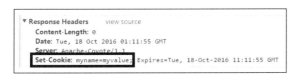

```
▼ Response Headers    view source
   Content-Length: 0
   Date: Tue, 18 Oct 2016 01:11:55 GMT
   Server: Apache-Coyote/1.1
   Set-Cookie: myname=myvalue; Expires=Tue, 18-Oct-2016 11:11:55 GMT
```

图 6-1

```
Expires=Tue, 18-Oct    11:11:55 GMT
```

图 6-2

超过这个时间，Cookie 即失效。如果一个 Cookie 由于超时造成失效，那么客户端发起请求时是不会将失效的 Cookie 发送给服务器端的。

## 6.2　查询Cookie

创建 Servlet 的代码如下：

```
public class test2 extends HttpServlet {
    protected void doGet(HttpServletRequest request, HttpServletResponse
response)
            throws ServletException, IOException {
        Cookie[] cookieArray = request.getCookies();
        // 如果请求中没有从客户端传递任何的 Cookie，则 cookieArray 即为 null
        if (cookieArray != null) {
            for (int i = 0; i < cookieArray.length; i++) {
                if (cookieArray[i].getName().equals("myname")) {
                    System.out.println(cookieArray[i].getValue());
                    break;
                }
            }
        }
    }

}
```

运行 Servlet 后，在控制台输出 Cookie 值，如图 6-3 所示。

前面介绍过，Cookie 保存在客户端，服务器端 Servlet 怎么能获取到 Cookie 值呢？难道是请求对象造成的？没错，的确是请求对象。每次发起请求时，请求自动携带着客户端的 Cookie 对象并传递给服务器端，这时服务器端就能获取客户端的 Cookie 值了，请求携带 Cookie 对象并传递给服务器端的证据如图 6-4 所示。

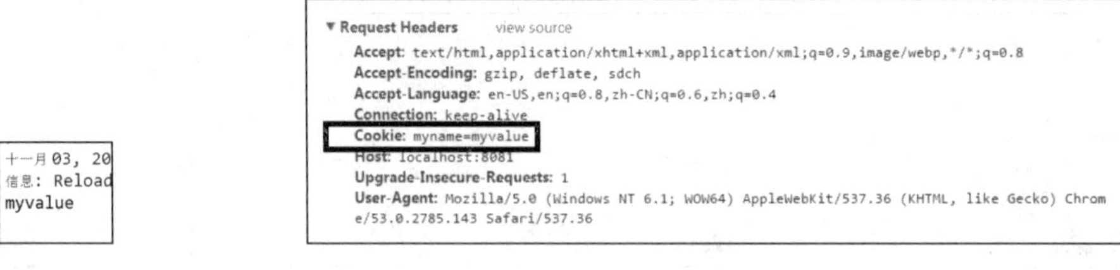


图 6-3　　　　图 6-4

## 6.3　修改Cookie

创建 Servlet 的代码如下：

```
public class test3 extends HttpServlet {

    protected void doGet(HttpServletRequest request, HttpServletResponse
response)
            throws ServletException, IOException {
        Cookie cookie = new Cookie("myname", "myvalueNEW");
        cookie.setMaxAge(36000);
        response.addCookie(cookie);
    }

}
```

修改 Cookie 和创建 Cookie 的代码基本一样，唯一的区别就是值是最新的，Cookie 的名称是一样的，造成的结果就是将 Cookie 中的旧值覆盖掉。

## 6.4　删除Cookie

创建 Servlet 的代码如下：

```
public class test4 extends HttpServlet {

    protected void doGet(HttpServletRequest request, HttpServletResponse
response)
            throws ServletException, IOException {
        Cookie cookie = new Cookie("myname", null);
        cookie.setMaxAge(0);
        response.addCookie(cookie);
    }

}
```

删除 Cookie 的原理是当执行 Servlet 代码后，在响应对象中设置 Cookie 的过期时间为“01-Jan-1970 00:00:10 GMT”，如图 6-5 所示。

```
▼ Response Headers    view source
    Content-Length: 0
    Date:
    Server: Apache-Coyote/1.1
    Set-Cookie: myname=""; Expires=Thu, 01-Jan-1970 00:00:10 GMT
```

图 6-5

## 6.5 设置setMaxAge()值为负数

调用 setMaxAge()方法传入负数的含义是 Cookie 在浏览器的内存中存在，如果浏览器关闭，则 Cookie 失效。创建 Servlet 代码如下：

```
public class test5 extends HttpServlet {

    protected void doGet(HttpServletRequest request, HttpServletResponse
response)
            throws ServletException, IOException {
        Cookie cookie = new Cookie("myname", "myvalue");
        cookie.setMaxAge(-1);
        response.addCookie(cookie);
    }

}
```

打开 http://localhost:8081/_test10web/test5 后，浏览器内存中就保存了一个 Cookie，使用查询 Cookie 的 Servlet 是可以找到 Cookie 值的，如图 6-6 所示。如果关闭浏览器再重新启动浏览器，执行查询 Cookie 的 Servlet 代码后，控制台并没有获取 Cookie 值，说明 Cookie 随着浏览器的关闭而失效。执行 setMaxAge(-1)代码，在传入参数为负数的情况下，浏览器不是将 Cookie 放入硬盘中，而是放入内存中，抓包的内容如图 6-7 所示。

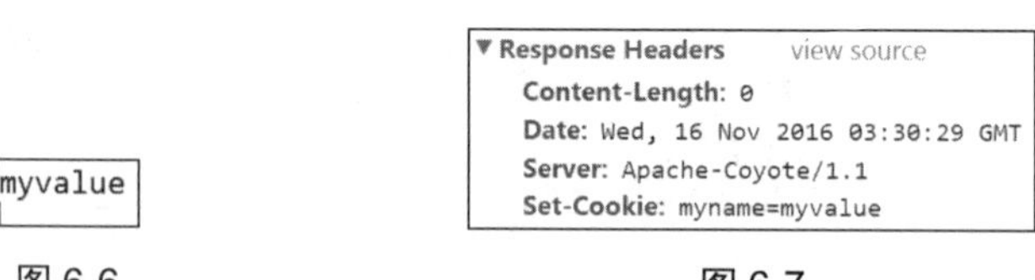


图 6-6　　图 6-7

## 6.6 使用Cookie存储中文或空格

如果使用某些旧版本的浏览器，这些浏览器不能直接存储中文，需要进行额外处理。创建 Cookie 保存中文的 Servlet 代码如下：

```
public class test6 extends HttpServlet {
    protected void doGet(HttpServletRequest request, HttpServletResponse
response)
            throws ServletException, IOException {
        String cookieValue = java.net.URLEncoder.encode("大中国", "utf-8");
```

```
        Cookie cookie = new Cookie("myname", cookieValue);
        cookie.setMaxAge(36000);
        response.addCookie(cookie);
    }
}
```

获取 Cookie 中文值的 Servlet 代码如下：

```
public class test7 extends HttpServlet {
    protected void doGet(HttpServletRequest request, HttpServletResponse
response)
            throws ServletException, IOException {
        Cookie[] cookieArray = request.getCookies();
        if (cookieArray != null) {
            for (int i = 0; i < cookieArray.length; i++) {
                if (cookieArray[i].getName().equals("myname")) {
                    String cookieValue = cookieArray[i].getValue();
                    cookieValue = java.net.URLDecoder.decode(cookieValue,
"utf-8");
                    System.out.println(cookieValue);
                    break;
                }
            }
        }
    }
}
```

保存 Cookie 后再查询，成功显示中文值，效果如图 6-8 所示。

在较新版本的 IE、Chrome 和 Firefox 浏览器中，Cookie 值默认支持双字节，不需要编码和解码的过程。如果使用旧版浏览器则需要编码和解码的过程。使用新版本的 Tomcat 时，如果 Cookie 值有空格，则需要进行编/解码，不然会出现异常。

大中国

图 6-8

## 6.7　为什么找不到Cookie

创建 Servlet 代码如下：

```
public class test20 extends HttpServlet {

    protected void doGet(HttpServletRequest request, HttpServletResponse
response)
```

```
            throws ServletException, IOException {
        Cookie cookie = new Cookie("myKey", "myValue");
        cookie.setMaxAge(36000);
        response.addCookie(cookie);
        Cookie[] getCookie = request.getCookies();
        if (getCookie != null) {
            for (int i = 0; i < getCookie.length; i++) {
                System.out.println(getCookie[i].getName()    +    "    "    +
getCookie[i].getValue());
            }
        } else {
            System.out.println("无Cookie");
        }
    }

}
```

```
无Cookie
```

图 6-9

执行此 Servlet 在控制台输出的信息如图 6-9 所示。当再次按“F5”键刷新浏览器时，控制台输出的信息如图 6-10 所示。

在第一次执行 Servlet 时，找不到 Cookie 的原因是 request1 中并没有 myKey 这个 Cookie，所以结果当然取不到 Cookie 了。但是当第一次执行完此 Servlet 后，浏览器保存了 response1 中携带着的 Cookie，当下一次发起新的 request2 请求访问此 Servlet 时，request2 带着 Cookie 到达 Servlet，这时就可以获取到 Cookie 中的值了，因为 request2 带着 response1 中的 Cookie 传给服务器了。

```
无Cookie
myKey myValue
```

图 6-10

注意：测试此实验之前需要初始化运行环境，步骤为首先重启 Tomcat，然后删除 Cookie，关闭浏览器，再打开浏览器，输入网址执行 Servlet。

## 6.8 创建工具类封装Cookie操作

创建 Cookie 工具类的代码如下：

```
public class CookieToolsOld {

    private HttpServletRequest request;
    private HttpServletResponse response;

    public CookieToolsOld(HttpServletRequest request, HttpServletResponse
```

```
response) {
        super();
        this.request = request;
        this.response = response;
    }

    public void save(String cookieName, String cookieValue, int maxAge) {
        try {
            cookieValue = java.net.URLEncoder.encode(cookieValue, "utf-8");
            Cookie cookie = new Cookie(cookieName, cookieValue);
            cookie.setMaxAge(maxAge);
            response.addCookie(cookie);
        } catch (UnsupportedEncodingException e) {
            e.printStackTrace();
        }
    }

    public String getValue(String cookieName) {
        String cookieValue = null;
        try {
            Cookie[] cookieArray = request.getCookies();
            if (cookieArray != null) {
                for (int i = 0; i < cookieArray.length; i++) {
                    if (cookieArray[i].getName().equals(cookieName)) {
                        cookieValue = cookieArray[i].getValue();
                        cookieValue                                    =
java.net.URLDecoder.decode(cookieValue, "utf-8");
                        break;
                    }


                }
            }
        } catch (UnsupportedEncodingException e) {
            e.printStackTrace();
        }
        return cookieValue;
    }

    public void delete(String cookieName) {
        Cookie cookie = new Cookie(cookieName, "");
        cookie.setMaxAge(0);
        response.addCookie(cookie);
```

```
    }
}
```

在操作 Cookie 时使用此工具类即可。

## 6.9 使用Cookie实现免登录

免登录的效果是曾经登录成功过，则不用登录，直接显示内容；如果以前从未登录成功过，则需进行登录操作，其实就是“记住密码”功能。

创建显示登录界面的 Servlet，源代码如下：

```
package controller;

import java.io.IOException;
import java.io.PrintWriter;

import javax.servlet.ServletException;
import javax.servlet.http.HttpServlet;
import javax.servlet.http.HttpServletRequest;
import javax.servlet.http.HttpServletResponse;

public class ShowLoginUI extends HttpServlet {
    protected void doGet(HttpServletRequest request, HttpServletResponse 
response)
            throws ServletException, IOException {

        response.setContentType("text/html;charset=utf-8");
        PrintWriter out = response.getWriter();

        String loginResult = request.getParameter("loginResult");
        if (loginResult != null && loginResult.equals("0")) {
            out.print("登录失败！");
        }

        out.println("<form action='LoginValidate' method='post'>");
        out.println("username:<input type='text' name='username'><br/>");
        out.println("password:<input type='text' name='password'><br/>");
        out.println("<input type='submit' value='提交'><br/>");
        out.println("</form>");
        out.flush();
```

```
        out.close();
    }
}
```

创建登录验证的 Servlet，源代码如下：

```
package controller;

import java.io.IOException;

import javax.servlet.ServletException;
import javax.servlet.http.Cookie;
import javax.servlet.http.HttpServlet;
import javax.servlet.http.HttpServletRequest;
import javax.servlet.http.HttpServletResponse;

public class LoginValidate extends HttpServlet {
    protected void doPost(HttpServletRequest request, HttpServletResponse
response)
            throws ServletException, IOException {
        String username = request.getParameter("username");
        String password = request.getParameter("password");
        if (username.equals("a") && password.equals("aa")) {
            Cookie cookie = new Cookie("currentLoginUser", username);
            cookie.setMaxAge(3600000);
            response.addCookie(cookie);
            response.sendRedirect("ListBookList");
        } else {
            response.sendRedirect("ShowLoginUI?loginResult=0");
        }
    }

}
```

创建显示图书列表的 Servlet，源代码如下：

```
package controller;

import java.io.IOException;
import java.io.PrintWriter;

import javax.servlet.ServletException;
import javax.servlet.http.Cookie;
```

```
import javax.servlet.http.HttpServlet;
import javax.servlet.http.HttpServletRequest;
import javax.servlet.http.HttpServletResponse;

public class ListBookList extends HttpServlet {
    protected void doGet(HttpServletRequest request, HttpServletResponse
response)
            throws ServletException, IOException {
        boolean isLogined = false;
        Cookie[] cookieArray = request.getCookies();
        if (cookieArray != null) {
            for (int i = 0; i < cookieArray.length; i++) {
                Cookie eachCookie = cookieArray[i];
                if (eachCookie.getName().equals("currentLoginUser")) {
                    isLogined = true;
                    break;
                }
            }
        }
        if (isLogined == true) {
            response.setContentType("text/html;charset=utf-8");
            PrintWriter out = response.getWriter();
            out.println("<a href='ShowBookinfo?id=1'>book1</a><br/>");
            out.println("<a href='ShowBookinfo?id=2'>book2</a><br/>");
            out.println("<a href='ShowBookinfo?id=3'>book3</a><br/>");
            out.println("<a href='ShowBookinfo?id=4'>book4</a><br/>");
            out.println("<a href='ShowBookinfo?id=5'>book5</a><br/>");
            out.flush();
            out.close();
        } else {
            response.sendRedirect("ShowLoginUI");
        }
    }

}
```

# 第 7 章
# HttpSession 接口

在客户端保存用户私有数据可以使用 Cookie 技术，而在服务器端保存用户私有数据则可以使用 HttpSession 技术，这两种技术的区别可以从 3 个角度进行说明。

（1）安全性：相对于 Cookie，HttpSession 在数据保存方面更安全，因为 HttpSession 中的数据存放在服务器端，而 Cookie 中的数据保存在客户端，在浏览器中可以随意查看 Cookie 中的值。

（2）存取的数据类型：HttpSession 存取的数据类型可以是任意的对象，存取的数据量也更大，而 Cookie 存取的数据类型只能是 String，存取的数据量要比 HttpSession 小。

（3）内存占用率：HttpSession 的缺点是所有的数据都存放在服务器端，对服务器内存的压力会增大，但 HttpSession 会采用激活（从数据库或文件中取出对象并放入内存）、钝化（将内存中的对象保存到文件或数据库中）机制将 HttpSession 中的数据临时保存到文件或者数据库中，也就是将内存中的数据保存到硬盘，或将硬盘中的数据重新加载回内存，以降低 HttpSession 对内存的高占用率。

钝化类似于序列化，将内存中的数据序列化到硬盘中；激活类似于反序列化，将硬盘中的数据重新加载回内存。总结一下这两种技术的优缺点。

（1）Cookie。Cookie 的优点是简单、方便，在客户端存储用户信息，降低了服务器端的内存占用率；缺点是不太安全，因为数据放在客户端本地，存储数据类型只能是 String。

（2）HttpSession。HttpSession 的优点是由于数据存放在服务器端，数据安全性较高，存储的数据类型是 Object，存储的数据量多一些；缺点是由于数据都存在服务器端，造成内存占用率较高。

# 7.1 HttpSession接口的使用

创建向 HttpSession 保存数据的 Servlet，代码如下：

```
package controller;

import java.io.IOException;

import javax.servlet.ServletException;
import javax.servlet.http.HttpServlet;
import javax.servlet.http.HttpServletRequest;
import javax.servlet.http.HttpServletResponse;
import javax.servlet.http.HttpSession;

public class test10 extends HttpServlet {
    protected void doGet(HttpServletRequest request, HttpServletResponse
response)
            throws ServletException, IOException {
        HttpSession session = request.getSession();
        session.setAttribute("mySessionKey", "mySessionValue");
    }
}
```

创建从 HttpSession 获取数据的 Servlet，代码如下：

```
package controller;

import java.io.IOException;

import javax.servlet.ServletException;
import javax.servlet.http.HttpServlet;
import javax.servlet.http.HttpServletRequest;
import javax.servlet.http.HttpServletResponse;

public class test11 extends HttpServlet {
    protected void doGet(HttpServletRequest request, HttpServletResponse
response)
            throws ServletException, IOException {
        Object                          sessionValue                          =
request.getSession().getAttribute("mySessionKey");
        System.out.println(sessionValue);
```

```
    }
}
```

程序运行后，成功从 HttpSession 中获取曾经保存过的数据，如图 7-1 所示。

```
mySessionValue
```

图 7-1

## 7.2　HttpServletRequest接口与HttpSession的区别

前面章节使用 HttpSession 存储用户的私有数据，示例代码如下：

```
public class test13 extends HttpServlet {
    protected void doGet(HttpServletRequest request, HttpServletResponse
response)
            throws ServletException, IOException {
        HttpSession session = request.getSession();
        session.setAttribute("mySessionNewKey", "mySessionNewValue");
        String sessionValue = (String) session.getAttribute("mySessionNewKey");
        System.out.println(sessionValue);
    }
}
```

```
mySessionNewValue
```

图 7-2

程序运行结果如图 7-2 所示。

使用 HttpSession 与使用 HttpServletRequest 存取数据的效果一样，示例代码如下：

```
public class test12 extends HttpServlet {
    protected void doGet(HttpServletRequest request, HttpServletResponse
response)
            throws ServletException, IOException {
        request.setAttribute("myRequestKey", "myRequestValue");
        String requestValue = (String) request.getAttribute("myRequestKey");
        System.out.println(requestValue);
    }
}
```

程序运行后也可以获取曾经放入 HttpServletRequest 中的私有数据，如图 7-3 所示。

```
信息: Reloading Co
myRequestValue
```

图 7-3

实验进行到这，向请求中存取数据与向 session 中存取数据这两种方式到底有什么区别呢？继续向下学习。

创建向请求中存数据的 Servlet，代码如下：

```
public class test14 extends HttpServlet {
    protected void doGet(HttpServletRequest request, HttpServletResponse
response)
            throws ServletException, IOException {
        request.setAttribute("myRequestKey", "myRequestValue");
    }
}
```

创建从请求中取数据的 Servlet，代码如下：

```
public class test15 extends HttpServlet {
    protected void doGet(HttpServletRequest request, HttpServletResponse
response)
            throws ServletException, IOException {
        String requestValue = (String) request.getAttribute("myRequestKey");
        System.out.println(requestValue);
    }
}
```

NULL

图 7-4

首先打开 http://localhost:8081/_test10web/test14，再打开 http://localhost:8081/_test10web/test15，控制台输出了 NULL 值，如图 7-4 所示。

出现 NULL 值的原因是客户端发起了两次请求，在第二次请求中并没有存储过数据，获取数据时当然返回值是 NULL。

继续实验。创建向 session 中存数据的 Servlet，代码如下：

```
public class test16 extends HttpServlet {
    protected void doGet(HttpServletRequest request, HttpServletResponse
response)
            throws ServletException, IOException {
        HttpSession session = request.getSession();
        session.setAttribute("mySessionNewKey", "mySessionNewValue");
    }
}
```

创建从 session 中取数据的 Servlet 代码如下：

```
public class test17 extends HttpServlet {
    protected void doGet(HttpServletRequest request, HttpServletResponse
response)
            throws ServletException, IOException {
        HttpSession session = request.getSession();
        String sessionValue = (String) session.getAttribute("mySessionNewKey");
        System.out.println(sessionValue);
    }
}
```

首先打开 http://localhost:8081/_test10web/test16，再打开 http://localhost:8081/_test10web/test17，控制台输出了字符串值，如图 7-5 所示。

```
mySessionNewValue
```

图 7-5

前面实验分别在两个 Servlet 中对请求和 session 进行数据的存取，都是发起了两次请求，为什么从请求中获取不到值，而从 session 中可以呢？

从请求作用域中取不出来值的原因是客户端一共发起了两个请求，值存放在 request1 中，从 request2 中尝试取值当然取不到了，但 session 却可以取到值，session 是怎么做到的呢？从 HttpSession 中能取到自己值的原因是在 Cookie 里面存储服务器端产生的 sessionId，然后每次请求时将客户端 Cookie 中的 sessionId 与服务器端的 sessionId 进行匹配，如果相同，则找到对应的值。这一段话就是 HttpSession 能找到自己数据的原理，下面用实验来验证这个原理。

## 7.3　Session与Cookie的运行机制

下面开始研究为什么能在 session 中获取自己的数据，也就是为什么在不同的 Servlet 中可以操作 session 中自己的数据。研究分为两个阶段。

（1）创建 Cookie 的阶段：session 的运行机制主要通过 Cookie 对象，IE 首次访问服务器端时，在服务器端生成一个 Map 对象，Map 的 key 就是唯一的 sessionId。sessionId 在服务器端生成，Map 的值就是向 HttpSession 存放的 value 值。服务器端代码执行结束后以响应对象的方式进行返回，其中返回的数据就带有符合 HTTP 协议的 Cookie 数据。IE 接收到响应返回来的数据，再根据数据中的格式就能识别出要在客户端创建一个 Cookie 的要求，所以在客户端就创建了一个 Cookie。Cookie 中的值就是刚刚在服务器端生成的那个唯一的 sessionId 值，这是生成 Cookie 阶段。

（2）匹配自己数据的阶段：IE 第二次访问服务器端，在请求中携带着刚刚创建的

Cookie。Cookie 中的值是 sessionId，到达服务器端时，根据 Cookie 中的 sessionId 找到旧的 session 对象，并不创建新的 session，再从 Map 对象 key-value 映射中取出值并进行后期处理。这是 Cookie 在服务器端匹配的阶段使用 HttpSession 对象能找到自己数据的原因，这就是原理的执行过程。

根据以上过程得知，HttpSession 的原理就是 Cookie。看来 sessionId 就是原理中的原理啊，如何获取呢？创建 Servlet 代码如下：

```
public class test18 extends HttpServlet {
    protected void doGet(HttpServletRequest request, HttpServletResponse
response)
            throws ServletException, IOException {
        HttpSession session = request.getSession();
        System.out.println(session.getId());
    }
}
```

```
8517CE7EDEC3D2D0492F18B794A316B4
```

图 7-6

程序运行后，在控制台输出 sessionId 值，如图 7-6 所示。一定要记住，sessionId 在服务器端被创建。

当创建了 session 对象后，sessionId 是保存在 Cookie 中的。响应将 Cookie 交给客户端，客户端再次发起请求将 Cookie 提交到服务器，然后在服务器端根据 Cookie 中保存的 sessionId 到内存中获取与此 sessionId 有关联的值。这也是为什么获取 HttpSession 要使用请求对象的原因，过程如下：

request→Cookie→sessionId→Tomcat→request→Cookie→sessionId→HttpSession→找到值

下面用示例的方式来重现原理文字所讲的执行过程。

创建一个 Servlet 对象，核心代码如下：

```
public class test101 extends HttpServlet {
    public void doGet(HttpServletRequest request, HttpServletResponse
response) throws ServletException, IOException {
        System.out.println("testServlet");
    }
}
```

将此项目部署到 Tocmat 并打开 URL：http://localhost:8081/_test10web/test101。

请求与响应的过程成功被浏览器中的 HTTP 工具捕获，如图 7-7 所示。在请求头节点中显示出请求头的内容，从显示的信息来看并没有 Cookie 对象发送给服务器。在请求头节点中显示出响应头的内容，从显示的信息来看也没有将 Cookie 对象发送给浏览器。说明在此过程中并没有产生任何的 Cookie 对象，而且多次刷新浏览器界面也没有捕获到 Cookie 对象的传输过程。

```
▼ 响应头 (73 字节)
    Content-Length: "0"
    Date: "Mon, 28 Aug 2017 03:21:19 GMT"
▼ 请求头 (359 字节)
    Host: "localhost:8080"
    User-Agent: "Mozilla/5.0 (Windows NT 6.1; ... Gecko/20100101 Firefox/54.0"
    Accept: "text/html,application/xhtml+x...lication/xml;q=0.9,*/*;q=0.8"
    Accept-Language: "zh-CN,zh;q=0.8,en-US;q=0.5,en;q=0.3"
    Accept-Encoding: "gzip, deflate"
    Connection: "keep-alive"
    Upgrade-Insecure-Requests: "1"
```

图 7-7

更改 Servlet 代码如下：

```
public class test101 extends HttpServlet {
    public void doGet(HttpServletRequest request, HttpServletResponse
response) throws ServletException, IOException {
        System.out.println("testServlet");
        request.getSession().setAttribute("username", "usernameValue");
    }
}
```

重启 Tomcat，按“F5”键刷新浏览器界面，HTTP 工具捕获到的信息如图 7-8 所示。从捕获的信息中可以看到请求头中并没有传输 Cookie 对象给服务器端，但在响应头中可以看到服务器端返回了一个 Cookie 给浏览器，如图 7-9 所示。

```
× Headers  Preview  Response  Cookies  Timing
▼ General
    Request URL: http://localhost:8081/_test10web/test101
    Request Method: GET
    Status Code: 200 OK
    Remote Address: [::1]:8081
▼ Response Headers    view source
    Content-Length: 0
    Date: Thu, 03 Nov 2016 04:25:15 GMT
    Server: Apache-Coyote/1.1
    Set-Cookie: JSESSIONID=AA08791BCE7B95DB427906EA112A9CDB; Path=/_test10web/; HttpOnly
▼ Request Headers    view source
    Accept: text/html,application/xhtml+xml,application/xml;q=0.9,image/webp,*/*;q=0.8
    Accept-Encoding: gzip, deflate, sdch
    Accept-Language: en-US,en;q=0.8,zh-CN;q=0.6,zh;q=0.4
    Cache-Control: max-age=0
    Connection: keep-alive
    Host: localhost:8081
    Upgrade-Insecure-Requests: 1
    User-Agent: Mozilla/5.0 (Windows NT 6.1; WOW64) AppleWebKit/537.36 (KHTML, like Gecko) Chrome/53.0.2785.143 Safari/537.36
```

图 7-8

```
Set-Cookie: JSESSIONID=AA08791BCE7B95DB427906EA112A9CDB; Path=/_test10web/; HttpOnly
```

图 7-9

Cookie 中的内容为：

```
JSESSIONID=AA08791BCE7B95DB427906EA112A9CDB; Path=/_test10web/; HttpOnly
```

记住 sessionId 的后 4 位值 9CDB，这时浏览器中已经有一个 Cookie 了。再次刷新浏览器，HTTP 分析工具再次捕获 HTTP 请求，内容如图 7-10 所示。从 HTTP 分析工具中可以看到，浏览器通过请求对象把 Cookie 传递给服务器端，Cookie 的值正是 9CDB，这时，Tomcat 就可以在 Map 中根据 9CDB 的 key 找到对应的 value 处理后面的业务了，流程示例图如图 7-11 所示。

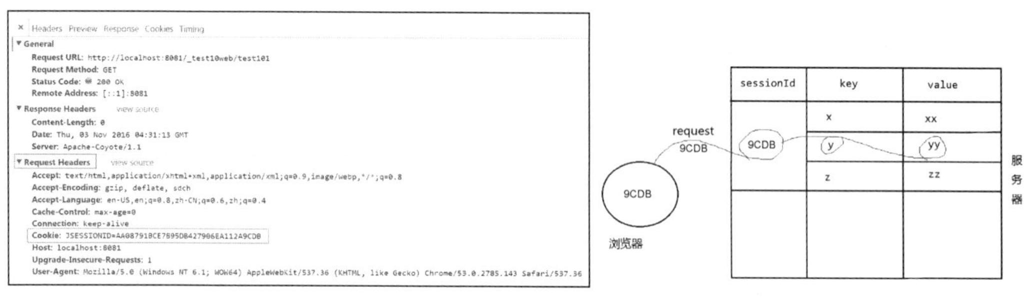


图 7-10　　　　图 7-11

是否创建新的 HttpSession 对象取决于如下两种情况。

（1）如果请求中带着 sessionId，并且请求对象中的 session 属性不为空，则使用原来的 sessionId。

（2）如果请求中不带 sessionId，说明请求对象中的 session 属性为空，则创建新的 session 对象并关联新的 sessionId。

注意：请求对象负责创建 HttpSession 对象，而响应对象并不负责创建 HttpSession 对象，响应对象只是将 sessionId 放入响应体中交给浏览器，让浏览器在客户端进行保存。响应对象相当于 Cookie 的搬运工。

本节主要介绍的是 Cookie 与 session 之间的关系，用户能根据 Cookie 在 session 中找到自己的数据，因为 Cookie 中保存的是 sessionId，那么如果 Cookie 在浏览器中被屏蔽，也就是浏览器不允许存储 Cookie 的效果是怎样的呢？继续来学习这个实验吧！

## 7.4　HttpSession接口与URL重写

先不要屏蔽浏览器保存 Cookie 的功能，创建 Servlet 代码如下：

```
public class test19 extends HttpServlet {
    protected void doGet(HttpServletRequest request, HttpServletResponse
response)
            throws ServletException, IOException {
        System.out.println(request.getSession().getId());
    }
}
```

在浏览器中打开网址：http://localhost:8081/_test10web/test19，该 Servlet 被多次运行后的效果如图 7-12 所示。

在浏览器客户端保存了 sessionId 值为 AA08791BCE7B95DB427906EA112A9CDB 的 Cookie，从第二次请求时将此 sessionId 发送给服务器。服务器监测到客户端有 sessionId 就不再重新创建新的 sessionId 了。那如果浏览器屏蔽 Cookie 会是什么样的效果呢？

```
AA08791BCE7B95DB427906EA112A9CDB
AA08791BCE7B95DB427906EA112A9CDB
AA08791BCE7B95DB427906EA112A9CDB
AA08791BCE7B95DB427906EA112A9CDB
AA08791BCE7B95DB427906EA112A9CDB
AA08791BCE7B95DB427906EA112A9CDB
```

图 7-12

谷歌的 Chrome 浏览器屏蔽 Cookie 的设置如图 7-13 所示。

图 7-13

关闭浏览器再重新打开浏览器，并在浏览器中打开 http://localhost:8081/_test10web/test19，该 Servlet 被多次运行后的效果如图 7-14 所示。

```
8CC4714891B1E9C04C792388E9C979A1
804E07DD826258779C214C2663E678F1
9A25B9D676CEB2995EBA3BAF97881F70
20C6BB165B24117B1394FBCF2A3452A0
496D276D88CF89930AE680A6B4BFDFB7
0103DE687AC3A6B94C89FA6154A62A87
```

图 7-14

当浏览器屏蔽了 Cookie 之后，说明在浏览器客户端从来不保存 Cookie，也就是客户端根本不存在 sessionId。如果客户端发起请求时服务器监测不到客户端有 sessionId，这时就会创建出新的 sessionId 给客户端，但客户端却依然不保存 Cookie，这就造成了在服务器端创建多个 sessionId 的情况。

通过上面的过程可以分析出，如果客户端真的屏蔽了 Cookie 的保存功能，则用户不能从服务器的 HttpSession 中获取自己的私有数据，造成数据的丢失，如果想要重获 HttpSession 中的数据可以使用“URL 重写”的方式，下面开始进行实验。

创建 Servlet 代码如下：

```
public class test25 extends HttpServlet {
    protected void doGet(HttpServletRequest request, HttpServletResponse 
response)
            throws ServletException, IOException {
        String usernameValue = request.getParameter("username");

        HttpSession session = request.getSession();
        String sessionId = session.getId();

        Object sessionValue = session.getAttribute("sessionKey");
        if (sessionValue == null) {
            System.out.println(sessionId + "从未关联数据,将" + usernameValue 
+ "存放到 session 中");
            session.setAttribute("sessionKey", usernameValue);
        } else {
            System.out.println(sessionId + "关联数据，值是：" + sessionValue);
        }
    }

}
```

在浏览器进程 1 中输入如下网址执行 Servlet：http://localhost:8080/web8/test25?username=aaa，运行效果如图 7-15 所示。

打开新的浏览器进程 2 后，输入 http://localhost:8080/ web8/test25?username=bbb，再次调用该 Servlet，运行效果如图 7-16 所示。

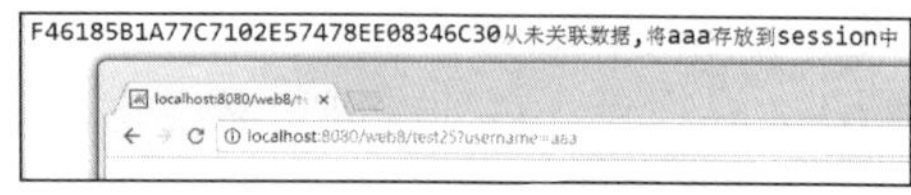


图 7-15

图 7-16

出现两个 sessionId 是正确的运行结果，因为打开了两个浏览器进程，每次在新的浏览器进程中执行 Servlet 都会创建新的 sessionId，永远也不会找到自己旧的数据，比如在浏览器进程 1 中按下“F5”键刷新浏览器，执行的效果如图 7-17 所示。

session 中旧的数据永远也找不到了，那如果想要找到 session 中旧的数据该如何处理呢？使用“URL 重写”即可，更改 URL 地址如下：

http://localhost:8080/web8/test25;jsessionid=F46185B1A77C7102E57478EE08346C30

执行后，在控制台成功找到自己旧的数据，如图 7-18 所示。

图 7-17

图 7-18

此实验证明可以通过 URL 重写的方式来获取自己或别人 session 中的数据，但这样有损软件的安全性。使用表格的方式来模拟在 HttpSession 中保存数据的结构如图 7-19 所示。

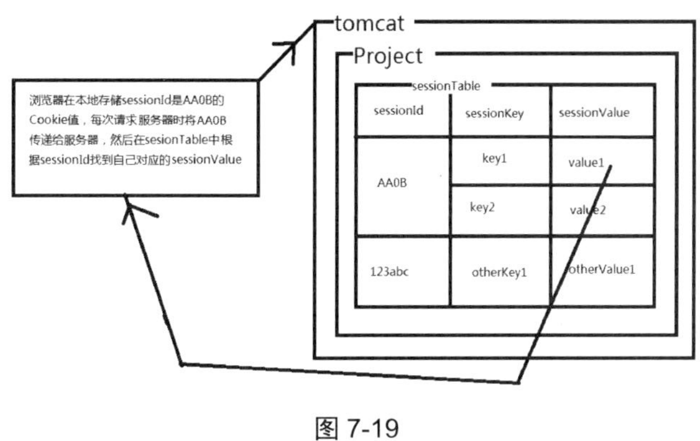


图 7-19

再来总结一下在 HttpSession 中如何能找到自己数据的过程。

### 1. 在服务器端向HttpSession中保存数据

（1）客户端发起请求，然后到达 Tomcat 执行 Servlet。

（2）Servlet 要执行的 Java 代码如下：

```
HttpSession session = request.getSession();
session.setAttribute("mySessionNewKey", "mySessionNewValue");
```

（3）当第一次执行本 Servlet 时，Tomcat 监测到请求中并没有 sessionId 传递过来，所以当执行 request.getSession()代码时在内部创建出新的 HttpSession 对象，再创建出唯一的 sessionId 标识，假设 sessionId 值为 AA0B，然后在 sessionTable 中根据 sessionId、sessionKey 和 sessionValue 进行存储。

（4）存储完毕后开始响应，响应对象中携带着 sessionId 是 AA0B 的 Cookie，浏览器接收响应后在客户端保存 name 为 jsessionId，value 是 AA0B 的 Cookie。

以上就是在服务器端向 HttpSession 保存自己数据的过程。

#### 2. 从HttpSession中获取保存的数据

（1）浏览器开始发起新的请求，目的就是获取自己 session 中的数据，请求对象中携带着 name 为 jsessionId，value 是 AA0B 的 Cookie。

（2）request 请求到达服务器端进入 Tomcat，在 Tomcat 中获取请求对象中的 Cookie 对象，Cookie 对象的 name 为 jsessionId，value 是 AA0B，然后执行 Servlet。

（3）Servlet 运行如下代码：

```
HttpSession session = request.getSession();
String sessionValue = (String)session.getAttribute("mySessionNewKey");
System.out.println(sessionValue);
```

当运行代码 HttpSession session = request.getSession()后，Tomcat 监测到请求中有 sesssionId 传递过来，所以不再创建新的 HttpSession 对象，使用旧的 HttpSession 对象并根据传递过来的 sessionId（AA0B）来获取对应的值，这样就可以根据客户端传递过来的 AA0B 到 sessionTable 中找到曾经保存的值。

Tomcat 中真实的类结构如图 7-20 所示。

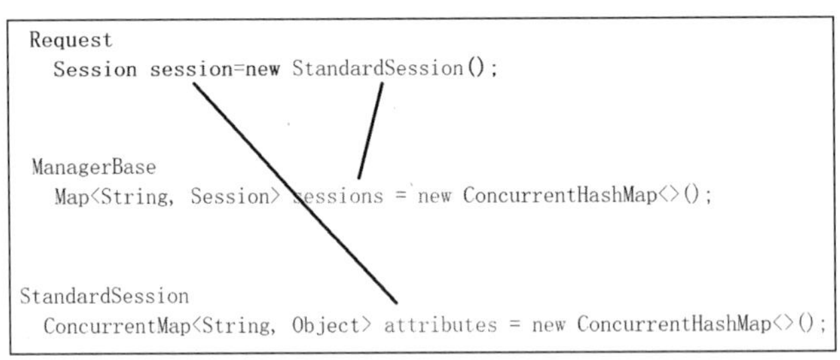


图 7-20

（1）请求中 session 对象的由来：根据 sessionId 到 ManagerBase 中的 Map<String, Session> sessions 进行查询，如果根据 sessionId 找到对应的 session 对象，则返回，否则创建新的 StandardSession 类的对象。

（2）运行 session 的 setAttribute()方法后数据存储的位置：向 session 中存储的数据是保存在 StandardSession 类中的 ConcurrentMap<String, Object> attributes 对象里。

此实验进行完毕，可以恢复浏览器能保存 Cookie 的功能了。学习技术的同时一定要研究技术的原理，这对理解技术如何使用有很大的帮助。

## 7.5 使用HttpSession实现免登录功能

当没有成功登录时不允许查看列表中的数据，重定向到登录界面，如果曾经已经登录成功过，则显示列表中的数据，不需要重复进行登录。此实验在前面章节中曾经使用 Cookie 进行实现，本章节使用 HttpSession 进行实现。

创建显示登录界面的 Servlet，源代码如下：

```
package controller;

import java.io.IOException;
import java.io.PrintWriter;

import javax.servlet.ServletException;
import javax.servlet.http.HttpServlet;
import javax.servlet.http.HttpServletRequest;
import javax.servlet.http.HttpServletResponse;

public class ShowLoginUI extends HttpServlet {
    protected void doGet(HttpServletRequest request, HttpServletResponse
response)
            throws ServletException, IOException {
        response.setContentType("text/html;charset=utf-8");
        PrintWriter out = response.getWriter();

        String loginResult = request.getParameter("loginResult");
        if (loginResult != null && loginResult.equals("0")) {
            out.print("登录失败！");
        }

        out.println("<form action='LoginValidate' method='post'>");
        out.println("username:<input type='text' name='username'><br/>");
        out.println("password:<input type='text' name='password'><br/>");
        out.println("<input type='submit' value='提交'><br/>");
        out.println("</form>");
        out.flush();
        out.close();
    }
}
```

创建登录验证的 Servlet，源代码如下：

```
package controller;

import java.io.IOException;

import javax.servlet.ServletException;
import javax.servlet.http.HttpServlet;
import javax.servlet.http.HttpServletRequest;
import javax.servlet.http.HttpServletResponse;

public class LoginValidate extends HttpServlet {
    protected void doPost(HttpServletRequest request, HttpServletResponse
response)
            throws ServletException, IOException {
        String username = request.getParameter("username");
        String password = request.getParameter("password");
        if (username.equals("a") && password.equals("aa")) {
            request.getSession().setAttribute("currentLoginUser",
username);
            response.sendRedirect("ListBookList");
        } else {
            response.sendRedirect("ShowLoginUI?loginResult=0");
        }
    }

}
```

创建显示图书列表的 Servlet，源代码如下：

```
package controller;

import java.io.IOException;
import java.io.PrintWriter;

import javax.servlet.ServletException;
import javax.servlet.http.HttpServlet;
import javax.servlet.http.HttpServletRequest;
import javax.servlet.http.HttpServletResponse;

public class ListBookList extends HttpServlet {
    protected void doGet(HttpServletRequest request, HttpServletResponse
response)
            throws ServletException, IOException {
```

```
        boolean isLogined = false;
        if (request.getSession().getAttribute("currentLoginUser") != null)
{
            isLogined = true;
        }

        if (isLogined == true) {
            response.setContentType("text/html;charset=utf-8");
            PrintWriter out = response.getWriter();
            out.println("<a href='ShowBookinfo?id=1'>book1</a><br/>");
            out.println("<a href='ShowBookinfo?id=2'>book2</a><br/>");
            out.println("<a href='ShowBookinfo?id=3'>book3</a><br/>");
            out.println("<a href='ShowBookinfo?id=4'>book4</a><br/>");
            out.println("<a href='ShowBookinfo?id=5'>book5</a><br/>");
            out.flush();
            out.close();
        } else {
            response.sendRedirect("ShowLoginUI");
        }
    }

}
```

## 7.6　使用HttpSession实现简易购物车功能

实现简易购物车功能的完整项目代码在本章中已经提供，本实验使用 MVC 开发方式，MVC 的主要作用就是分层开发、分工协作，每个程序员负责自己最擅长模块的开发任务，使用此开发方式后便于后期代码维护，哪个层出现了问题就到哪个层中进行修改。

MVC 中的 M 代表模型，模型有两个体现，一个是传递的数据是数据模型，另外一个是业务模型，也就是软件的核心功能。在本节中，M 代表 Model；V 代表 View，使用 HTML 或 JSP 文件来在客户端显示用户界面；C 代表 Controller，控制器的作用是控制软件大方向的执行流程，比如执行哪些 Service 服务层，控制层使用 Servlet 来进行处理。每层的作用总结如下。

M（Model，模型）：业务模型层，作用是实现软件的核心功能，使用自创建 Service.java 服务业务类进行实现。

V（View，视图）：视图层，作用是显示用户界面，使用 HTML 或 JSP 文件进行实现。

C（Controller，控制器）：控制层，作用是控制软件大方向的执行流程，使用 Servlet 进行实现。

MVC 各层之间的调用关系如下。

（1）实现登录功能，过程如图 7-21 所示。

（2）显示数据列表，过程如图 7-22 所示。

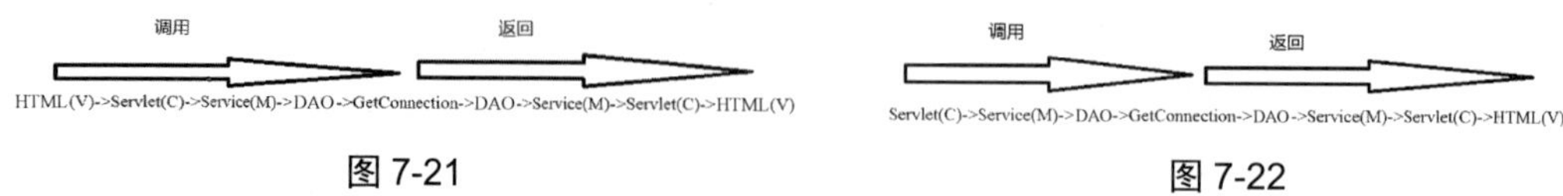


图 7-21　　图 7-22

## 7.6.1　创建一个 V 层

创建登录页面 login.html 文件，此文件就是 V 层，代码如下：

```
<!DOCTYPE html>
<html>
   <head>
      <meta charset="utf-8">
      <title>Insert title here</title>
   </head>
   <body>
      <form action="Login" method="post">
         username:<input type="text" name="username">
         <br/>
         password:<input type="text" name="password">
         <br/>
         <input type="submit" value="登录">
      </form>
   </body>
</html>
```

## 7.6.2　创建三个 C 层

创建图书列表 Servlet，此 Servlet 就是 C 层，代码如下：

```
public class ListBook extends HttpServlet {
   protected void doGet(HttpServletRequest request, HttpServletResponse
response)
           throws ServletException, IOException {
```

```
        response.setContentType("text/html;charset=utf-8");

        String tableBegin = "<table border='1'>";
        String tableEnd = "</table>";
        String trBegin = "<tr>";
        String trEnd = "</tr>";
        String tdBegin = "<td>";
        String tdEnd = "</td>";
        String allTRHTML = "";
        String finalHTML = "";

        BookinfoService bookinfoService = new BookinfoService();
        List<Bookinfo> listBookinfo = bookinfoService.getAllBookinfo();
        for (int i = 0; i < listBookinfo.size(); i++) {
            Bookinfo bookinfo = listBookinfo.get(i);
            long id = bookinfo.getId();
            String bookname = bookinfo.getBookname();
            String author = bookinfo.getAuthor();

            String putCartA = "<a href='PutCart?id=" + id + "'>放入购物车</a>";

            String eachTRHTML = "";
            eachTRHTML = trBegin + tdBegin + id + tdEnd + tdBegin + bookname
+ tdEnd + tdBegin + author + tdEnd
                    + tdBegin + putCartA + tdEnd + trEnd;
            allTRHTML = allTRHTML + eachTRHTML;
        }
        finalHTML = tableBegin + allTRHTML + tableEnd;
        PrintWriter out = response.getWriter();
        out.print(finalHTML);
        out.flush();
        out.close();
    }

}
```

创建登录验证功能的 Servlet，此 Servlet 也是 C 层，代码如下：

```
public class Login extends HttpServlet {
    protected void doPost(HttpServletRequest request, HttpServletResponse
response)
            throws ServletException, IOException {
```

```
        String username = request.getParameter("username");
        String password = request.getParameter("password");
        UserinfoService userinfoService = new UserinfoService();
        boolean loginResult = userinfoService.login(username, password);
        if (loginResult == false) {
            response.sendRedirect("login.html");
        } else {
            response.sendRedirect("ListBook");
        }

    }
}
```

创建放入购物车的 Servlet，此层还是 C 层，代码如下：

```
public class PutCart extends HttpServlet {
    protected void doGet(HttpServletRequest request, HttpServletResponse
response)
            throws ServletException, IOException {
        String bookId = request.getParameter("id");

        // 从 session 中获取自己的购物车
        Map cartMap = (Map) request.getSession().getAttribute("cartMap");

        CartService cartService = new CartService();
        cartMap = cartService.putCart(bookId, cartMap);// 处理购物车 Map
        request.getSession().setAttribute("cartMap", cartMap);// 再放入
session
        response.sendRedirect("ListBook");
    }

}
```

### 7.6.3 创建两个 entity 实体类

创建实体类 Bookinfo.java，代码如下：

```
public class Bookinfo {
    private long id;
    private String bookname;
    private String author;
```

```
    public Bookinfo() {
    }

    public Bookinfo(long id, String bookname, String author) {
        super();
        this.id = id;
        this.bookname = bookname;
        this.author = author;
    }

    //省略 get()方法和 set()方法
}
```

创建实体类 Userinfo.java，代码如下：

```
public class Userinfo {

}
```

## 7.6.4　创建两个 DAO 数据访问层

创建 BookinfoDAO.java 类，代码如下：

```
public class BookinfoDAO {
    public List<Bookinfo> getAllBookinfo() {
        List returnList = new ArrayList();
        returnList.add(new Bookinfo(1, "书籍 1", "A1"));
        returnList.add(new Bookinfo(2, "书籍 2", "A2"));
        returnList.add(new Bookinfo(3, "书籍 3", "A3"));
        returnList.add(new Bookinfo(4, "书籍 4", "A4"));
        returnList.add(new Bookinfo(5, "书籍 5", "A5"));
        returnList.add(new Bookinfo(6, "书籍 6", "A6"));
        return returnList;
    }
}
```

创建 UserinfoDAO.java 类，代码如下：

```
public class UserinfoDAO {
    public Userinfo getUserinfo(String username, String password) {
        Random random = new Random();
        if (random.nextInt() % 2 == 0) {
            return new Userinfo();
```

```
        } else {
            return null;
        }
    }
}
```

## 7.6.5 创建三个 Model 业务逻辑层

创建 UserinfoService.java 业务类，该类属于 M 层，代码如下：

```
public class UserinfoService {
    private UserinfoDAO userinfoDAO = new UserinfoDAO();

    public boolean login(String username, String password) {
        if (username == null || "".equals(username)) {
            return false;
        }
        if (password == null || "".equals(password)) {
            return false;
        }
        Userinfo userinfo = userinfoDAO.getUserinfo(username, password);
        if (userinfo == null) {
            return false;
        } else {
            return true;
        }
    }
}
```

创建书籍服务层 BookinfoService.java，代码如下：

```
public class BookinfoService {
    private BookinfoDAO bookinfoDAO = new BookinfoDAO();

    public List<Bookinfo> getAllBookinfo() {
        return bookinfoDAO.getAllBookinfo();
    }
}
```

创建购物车服务层 CartService.java，代码如下：

```
public class CartService {
```

```
    public Map putCart(String bookId, Map<String, Integer> cartMap) {
        if (cartMap == null) {
            // 第一次操作购物车
            cartMap = new HashMap<String, Integer>();
            cartMap.put(bookId, 1);
        } else {
            // 不是第一次操作购物车，session 中有 Map
            if (cartMap.containsKey(bookId) == true) {
                // 多次添加相同 bookId 的书籍到车中，数量加 1
                Integer newBookNum = cartMap.get(bookId) + 1;
                cartMap.put(bookId, newBookNum);
            } else {
                // 有车的存在，但 bookId 是第一次添加
                cartMap.put(bookId, 1);
            }
        }
        Iterator<String> iterator = cartMap.keySet().iterator();
        while (iterator.hasNext()) {
            String bookIdMap = iterator.next();
            Integer bookNumMap = cartMap.get(bookIdMap);
            System.out.println(bookIdMap + " " + bookNumMap);
        }
        return cartMap;
    }
}
```

# 第 8 章 ServletContext 接口

当使用请求对象的 setAttribute()方法存储数据时，如果发生响应，则请求中的数据即失效，也就是在离开 Tomcat 时，请求中的数据即失效。请求中数据的生命周期是 B 阶段，如图 8-1 所示。

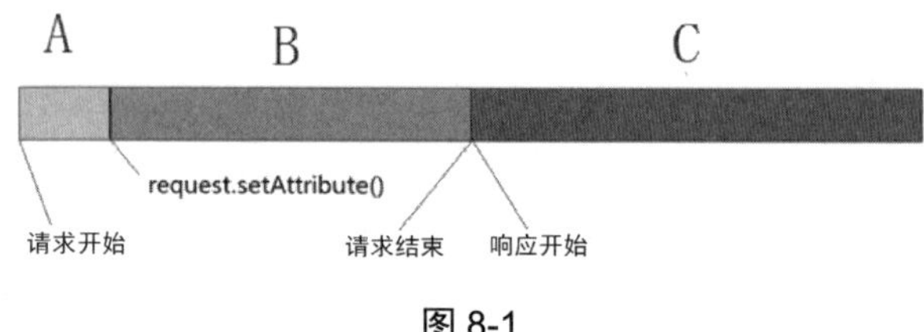


图 8-1

使用 session 对象的 setAttribute()方法存储数据的生命周期取决于以下两点。

（1）Web 容器的 session 超时时间：如果在指定的时间内客户端并没有访问服务器端自己的数据，则 Web 容器自动删除 session 中超时过期的数据，释放内存，如图 8-2 所示。Tomcat 启动后，在后台创建一个线程，以轮询的方式扫描这张表，确定哪些数据已经超时，如果超时就删除这些超时的数据。能找到 session 中自己数据的原理是客户端的 Cookie 中保存了服务器端的 sessionId。不同厂商的 Web 容器对 session 超时的时间设置也不同，Tomcat 根据不同版本，设置 session 超时时间为 30 ~ 45 分钟，具体时间要查看 tomcatpath/conf/web.xml 文件中的<session-timeout>配置。

| sessionId | key | value | lastAccessTime |
|---|---|---|---|
| 09AB | A | AA | 2000-1-1 02:02:02 |
| | B | BB | 2000-1-1 02:02:02 |
| | C | CC | 2000-1-1 02:02:04 |
| | | | |

图 8-2

（2）手动写 session.invalidate()或 session.removeAttribute(name)代码进行删除。

注意：不要以为关闭浏览器后，服务器端 session 中的数据就清除了，这个理解是错误的。关闭浏览器后，服务器端 session 中的数据还是处于保留的状态，只不过客户端的 sessionId 随着浏览器的关闭而销毁，再次发起请求的时候并没有传递 sessionId 给服务器端，造成找不到自己的数据，而数据还是在服务器端保留，直到超时才销毁。

请求中的私有数据只可以在当前一次请求中被访问，session 中的私有数据会根据 sessionId 在多次会话中被访问，但请求和 session 中的数据只能被自己访问，那么在 Java Web 技术中，如果想实现让所有人都可以访问相同的共享数据，则可以使用 ServletContext 接口。在演示 ServletContext 核心功能之前，先来验证一下 session 中的数据并不能被所有人使用，只能被自己使用的效果。

## 8.1　session中的数据是私有的

创建向 session 存放数据的 Servlet 类，代码如下：

```
public class sessionPutValue extends HttpServlet {
    protected void doGet(HttpServletRequest request, HttpServletResponse
response)
            throws ServletException, IOException {
        request.getSession().setAttribute("mySessionKey",
"mySessionValue");
        System.out.println("已经向 session 中存放数据");
    }
}
```

创建从 session 获取数据的 Servlet 类，代码如下：

```
public class sessionGetValue extends HttpServlet {
    protected void doGet(HttpServletRequest request, HttpServletResponse
response)
            throws ServletException, IOException {
        System.out.println(" 从  session  中 获 取 数 据 ： " +
request.getSession().getAttribute("mySessionKey"));
    }
}
```

首先输入网址 http://localhost:8081/_test14web/sessionPutValue，向 session 中存放数据，如图 8-3 所示；然后在相同的浏览器进程中输入网址 http://localhost:8081/_test14web/sessionGetValue，从 session 中获取数据，如图 8-4 所示。关闭当前的浏览器进程，再打开新的浏览器进程，在地址栏中输入网址 http://localhost:8081/_test14web/ sessionGetValue

就不会再获取 session 中的数据了，如图 8-5 所示。

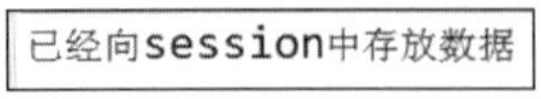

已经向session中存放数据

图 8-3

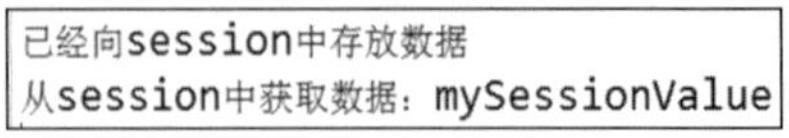

已经向session中存放数据
从session中获取数据：mySessionValue

图 8-4

从session中获取数据：null

图 8-5

出现这样的情况是因为浏览器关闭后，Cookie 中的 sessionId 也一同被销毁，再发起新的请求时并没有将 sessionId 重新传递到服务器端，造成找不到 session 中自己数据的情况。此实验说明 session 中自己的私有数据是有可能找不到的。

回顾完 session 技术的使用，再来学习新的技术：ServletContext，该技术可以实现数据的完全共享，每个人都可以访问 ServletContext 中的数据。

## 8.2 ServletContext中的数据是公共的

创建向 ServletContext 存放数据的 Servlet 类，代码如下：

```
public class test1 extends HttpServlet {
    protected void doGet(HttpServletRequest request, HttpServletResponse
response)
            throws ServletException, IOException {
        ServletContext servletContext = request.getServletContext();
        servletContext.setAttribute("publicValue", "共享值");
    }
}
```

创建从 ServletContext 获取数据的 Servlet 类，代码如下：

```
public class test2 extends HttpServlet {
    protected void doGet(HttpServletRequest request, HttpServletResponse
response)
            throws ServletException, IOException {
    System.out.println(request.getServletContext().getAttribute("publicVa
lue"));
    }
}
```

首先打开网址 http://localhost:8081/_test14web/test1，向 ServletContext 中存放数据；然后在相同的浏览器进程中输入网址 http://localhost:8081/_test14web/test2 从 ServletContext 中获取数据，如图 8-6 所示。关闭当前的浏览器进程，再打开新

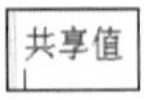

共享值

图 8-6

的浏览器进程，在地址栏中输入网址 http://localhost:8081/_test14web/test2 还会获取 ServletContext 中的数据，如图 8-7 所示。

此实验说明 ServletContext 中的数据呈共享的状态，每个人都可以进行读取与写入，要考虑非线程安全的情况，所以在大多数的情况下，ServletContext 中的数据都是只读的，只写入一次，很少进行重复写入。

共享值
共享值

图 8-7

使用 ServletContext 保存数据时，数据的生命周期直到 Tomcat 服务关闭时才销毁。作用域 HttpServletRequest，HttpSession，ServletContext 保存的数据存储在服务器端中，客户端保存的数据有 Cookie 以及通过响应返回的 HTML 代码。请求作用域只获取自己私有数据的原理是当前执行的线程拥有自己私有的 Request.java 对象。session 作用域只获取自己私有数据的原理是浏览器拥有自己私有的 sessionId，根据这个私有的 sessionId 到服务器找到自己的私有数据。ServletContext 作用域获取共享数据的原理是所有的请求都访问同一个 ApplicationContext 对象中的 Map attribute 属性。

## 8.3　ServletConfig.getInitParameter()方法的弊端

创建获得初始化参数的 Servlet 代码如下：

```
package controller;

import java.io.IOException;

import javax.servlet.ServletException;
import javax.servlet.http.HttpServlet;
import javax.servlet.http.HttpServletRequest;
import javax.servlet.http.HttpServletResponse;

public class test1 extends HttpServlet {
    protected void doGet(HttpServletRequest request, HttpServletResponse
response)
            throws ServletException, IOException {

    System.out.println(this.getServletConfig().getInitParameter("test1Par
am"));
    }
}
```

创建获得初始化参数的 Servlet 代码如下：

```
package controller;

import java.io.IOException;

import javax.servlet.ServletException;
import javax.servlet.http.HttpServlet;
import javax.servlet.http.HttpServletRequest;
import javax.servlet.http.HttpServletResponse;

public class test2 extends HttpServlet {
    protected void doGet(HttpServletRequest request, HttpServletResponse
response)
            throws ServletException, IOException {

    System.out.println(this.getServletConfig().getInitParameter("test1Par
am"));
    }

}
```

配置文件 web.xml 代码如下：

```
<servlet>
    <servlet-name>test1</servlet-name>
    <servlet-class>controller.test1</servlet-class>
    <init-param>
        <param-name>test1Param</param-name>
        <param-value>test1Value</param-value>
    </init-param>
</servlet>
<servlet-mapping>
    <servlet-name>test1</servlet-name>
    <url-pattern>/test1</url-pattern>
</servlet-mapping>
<servlet>
    <servlet-name>test2</servlet-name>
    <servlet-class>controller.test2</servlet-class>
</servlet>
<servlet-mapping>
    <servlet-name>test2</servlet-name>
    <url-pattern>/test2</url-pattern>
```

```
</servlet-mapping>
```

运行网址 http://localhost:8080/web6/test1，在控制台输出结果为 test1Value；运行网址 http://localhost:8080/web6/test2，在控制台输出结果为 NULL。说明在<servlet>中的<init-param>参数只能被当前的 Servlet 对象访问，其他 Servlet 并不能获取这个初始化参数值，如果想实现所有的 Servlet 都可以访问这个初始化参数，可以在 web.xml 文件中使用<context-param></context-param>标签来进行配置，并结合 ServletContext.getInitParameter()方法。

## 8.4　使用ServletContext.getInitParameter()方法解决问题

我们可以在 web.xml 文件中使用<context-param>标签来配置 ServletContext 的公共初始化参数达到创建公共信息的作用。所有的 Servlet 都可以访问<context-param>标签中的配置。

创建获得初始化参数的 Servlet 类代码如下：

```
package controller;

import java.io.IOException;

import javax.servlet.ServletException;
import javax.servlet.http.HttpServlet;
import javax.servlet.http.HttpServletRequest;
import javax.servlet.http.HttpServletResponse;

public class test1 extends HttpServlet {
    protected void doGet(HttpServletRequest request, HttpServletResponse
response)
            throws ServletException, IOException {
        System.out.println("test1="                                        +
this.getServletContext().getInitParameter("publicKey"));
    }
}
```

创建获得初始化参数的 Servlet 类代码如下：

```
package controller;
```

```
import java.io.IOException;

import javax.servlet.ServletException;
import javax.servlet.http.HttpServlet;
import javax.servlet.http.HttpServletRequest;
import javax.servlet.http.HttpServletResponse;

public class test2 extends HttpServlet {
    protected void doGet(HttpServletRequest request, HttpServletResponse
response)
            throws ServletException, IOException {
        System.out.println("test2="                                        +
this.getServletContext().getInitParameter("publicKey"));
    }
}
```

在 web.xml 文件中添加如下配置代码：

```
<servlet>
    <servlet-name>test1</servlet-name>
    <servlet-class>controller.test1</servlet-class>
</servlet>
<servlet-mapping>
    <servlet-name>test1</servlet-name>
    <url-pattern>/test1</url-pattern>
</servlet-mapping>
<servlet>
    <servlet-name>test2</servlet-name>
    <servlet-class>controller.test2</servlet-class>
</servlet>
<servlet-mapping>
    <servlet-name>test2</servlet-name>
    <url-pattern>/test2</url-pattern>
</servlet-mapping>

<context-param>
    <param-name>publicKey</param-name>
    <param-value>publicValue</param-value>
</context-param>
```

打开网址 http://localhost:8080/web6/test1，在控制台输出结果为 test1=publicValue；

打开网址 http://localhost:8080/web6/test2，在控制台输出结果为 test2=publicValue。

## 8.5　实现charset编码可配置

在<context-param>标签中配置的参数值在所有的 Servlet 中都可以被访问到，在框架技术中主要使用<context-param>标签配置默认加载的 XML 文件，以进行后期的 DOM 解析；也可以应用<context-param>标签实现 Filter（过滤器）过滤的编码可配置化，web.xml 文件中添加配置，示例代码如下：

```
<context-param>
    <param-name>myCharSet</param-name>
    <param-value>utf-8</param-value>
</context-param>
```

Servlet 类代码如下：

```
public class test3 extends HttpServlet {

    private String charSet = "";

    @Override
    public void init(ServletConfig config) throws ServletException {
        super.init(config);
        charSet = this.getServletContext().getInitParameter("myCharSet");
        System.out.println("init method charSet = " + charSet);
    }

    protected void doPost(HttpServletRequest request, HttpServletResponse
response)
            throws ServletException, IOException {
        request.setCharacterEncoding(charSet);
        System.out.println("post charSet = " + charSet);
    }

}
```

前台使用<form method="post">进行提交，程序运行结果如下：

```
init method charSet = utf-8
get charSet = utf-8
```

## 8.6 使用getRealPath("/")方法获取项目的运行路径

可以使用代码：

```
request.getSession().getServletContext().getRealPath("/");
```

获取当前项目的运行路径，ServletContext.getRealPath("/")方法在上传文件时经常使用，示例 Servlet 代码如下：

```
public class test5 extends HttpServlet {
    protected void doGet(HttpServletRequest request, HttpServletResponse
response)
            throws ServletException, IOException {
        String                              path                              =
request.getSession().getServletContext().getRealPath("/");
        System.out.println(path);
    }
}
```

# 第 9 章 Filter 接口

Filter 是在 Java Servlet 规范 2.3 版本中定义的，它能够对 Servlet 容器的请求和响应对象进行检查和修改。Filter 本身并不产生请求对象和响应对象，它只能起到过滤作用。Filter 能够在 Servlet 被调用之前检查请求对象，获取请求头和请求中的内容，可以在 Servlet 被调用之后检查响应对象，修改响应头和响应中的内容，常用在编码统一处理、权限验证等场景。

Filter 执行的时机是在调用 Web 组件之前和之后，这些 Web 组件可以是 Servlet、HTML 文件、JSP 文件、CSS 或 JS 文件，还可以是 JPG 文件等，访问这些文件都可以被 Filter 拦截，调用时机如图 9-1 所示。

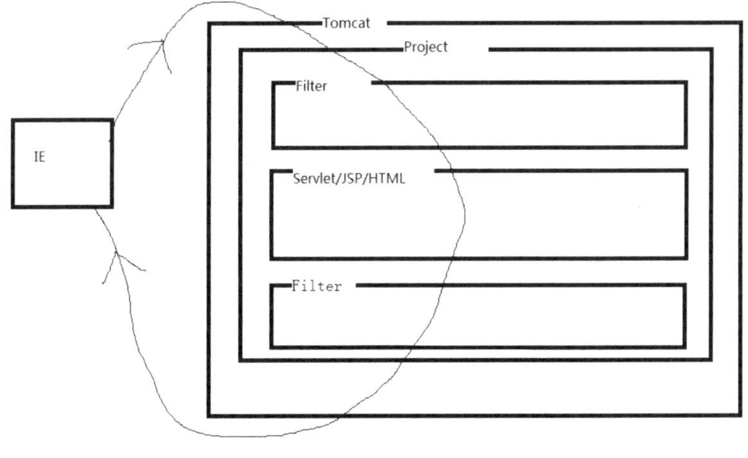


图 9-1

Filter 主要作用是减少项目中冗余的代码，将具有复用性的功能代码进行集中处理，比如实现权限验证、日志、统一编码处理等场景。

# 9.1 Filter的使用

Filter 在 API 中的表现形式为接口，也就是说如果要想使用过滤器，则程序员必须要实现 Filter 接口。Filter 接口的 API 列表如图 9-2 所示。

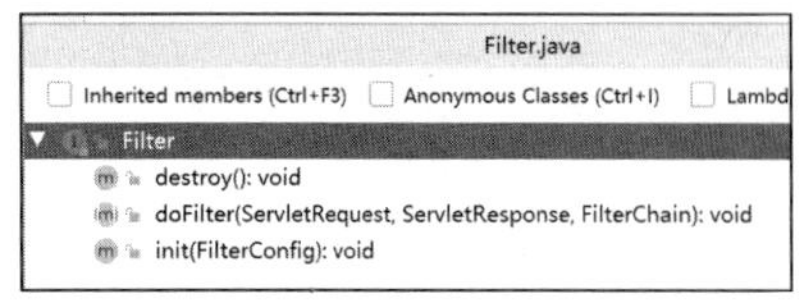


图 9-2

在这 3 个方法中，最常用的方法当属 public void doFilter(ServletRequest request, ServletResponse response, FilterChain chain)方法了，此方法是在过滤器正在过滤时执行的方法，业务代码要写在此方法中。下面来看一下 Filter 的基本使用示例。

创建 Web 项目，添加过滤器类 MyFilter.java 代码如下：

```
package myfilter;

import java.io.IOException;

import javax.servlet.Filter;
import javax.servlet.FilterChain;
import javax.servlet.FilterConfig;
import javax.servlet.ServletException;
import javax.servlet.ServletRequest;
import javax.servlet.ServletResponse;

public class MyFilter implements Filter {
    @Override
    public void init(FilterConfig filterConfig) throws ServletException {
    }
    @Override
    public void doFilter(ServletRequest request, ServletResponse response,
FilterChain chain)
            throws IOException, ServletException {
        System.out.println("执行了 doFilter()方法 " +
System.currentTimeMillis());
    }
    @Override
    public void destroy() {
    }
}
```

在目前的情况下，MyFilter.java 类只是一个普通的 Java 类，并没有参与到对 Web 请求或响应进行过滤的任务，故要将此类在 web.xml 文件中进行注册，注册代码如下：

```
<?xml version="1.0" encoding="UTF-8"?>
<web-app xmlns:xsi="http://www.w3.org/2001/XMLSchema-instance"
    xmlns="http://java.sun.com/xml/ns/javaee"
    xsi:schemaLocation="http://java.sun.com/xml/ns/javaee
http://java.sun.com/xml/ns/javaee/web-app_2_5.xsd"
    id="WebApp_ID" version="2.5">

    <filter>
        <filter-name>anyFilter</filter-name>
        <filter-class>myfilter.MyFilter</filter-class>
    </filter>

    <filter-mapping>
        <filter-name>anyFilter</filter-name>
        <url-pattern>/*</url-pattern>
    </filter-mapping>

</web-app>
```

配置代码<url-pattern>/*</url-pattern>代表要对所有的路径进行拦截过滤。配置代码设计完毕后在 Tomcat 中输入如下两个网址：

http://localhost:8081/test10/abc

http://localhost:8081/test10/no.html

这两个网址的访问资源 abc 和 no.html 并不存在，但通过控制台输出的信息来看，过滤器的确是拦劫了，如图 9-3 所示。说明前面从创建过滤器类到配置过滤器类的操作是正确的。那如果真的存在一个名称为 index.html 文件会出现什么样的情况呢？下面开始创建 index.html 文件，代码如下：

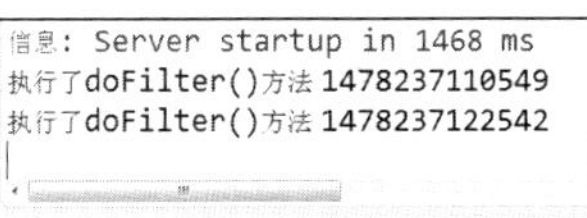


图 9-3

```
<!DOCTYPE html>
<html>
   <head>
      <meta charset="UTF-8">
   </head>
   <body>
      我是 index.html 文件
```

```
    </body>
</html>
```

项目结构如图 9-4 所示。这时，在浏览器上多次执行网址 http://localhost:8081/test10/index.html，控制台输出了过滤器执行后的信息，如图 9-5 所示。

图 9-4　　　　图 9-5

这时，浏览器中并没有出现应该出现的“我是 index.html 文件”文字，显示的是空白的界面。如何解决过滤器能被调用，而且还能正确访问目的资源的问题呢？更改过滤器代码如下：

```
package myfilter;

import java.io.IOException;

import javax.servlet.Filter;
import javax.servlet.FilterChain;
import javax.servlet.FilterConfig;
import javax.servlet.ServletException;
import javax.servlet.ServletRequest;
import javax.servlet.ServletResponse;

public class MyFilter implements Filter {

    @Override
    public void init(FilterConfig filterConfig) throws ServletException {
    }

    @Override
    public void doFilter(ServletRequest request, ServletResponse response,
FilterChain chain)
            throws IOException, ServletException {
        System.out.println("执行了doFilter()方法" +
System.currentTimeMillis());
        chain.doFilter(request, response);//放行
    }
```

```
    @Override
    public void destroy() {
    }

}
```

在过滤类中添加了代码：

```
chain.doFilter(request, response);
```

此代码的主要作用就是将请求继续交给其他过滤器，如果没有其他过滤器则执行目的资源，重启 Tomcat 再打开网址 http://localhost:8081/test10/index.html，控制台输出了过滤器执行后的信息，而且 index.html 文件中的文字也显示了出来，如图 9-6 所示。

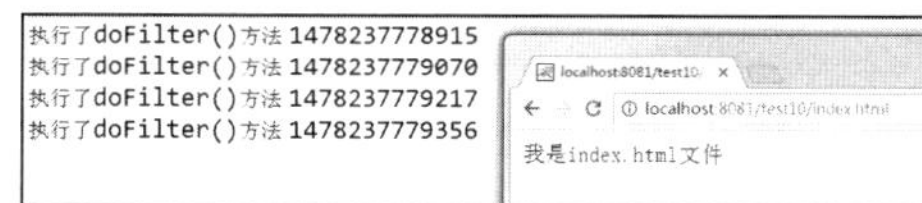


图 9-6

## 9.2 Filter的生命周期

Filter 也像 Servlet 一样，具有自己的生命周期。Filter 生命周期过程如下。

（1）实例化：创建 Filter 对象。

（2）初始化：调用 public void init(FilterConfig filterConfig)方法。

（3）过滤：调用 public void doFilter(ServletRequest request, ServletResponse response, FilterChain chain)方法。

（4）销毁：调用 public void destroy()方法。

更改 Filter 代码如下：

```
package myfilter;

import java.io.IOException;

import javax.servlet.Filter;
import javax.servlet.FilterChain;
import javax.servlet.FilterConfig;
import javax.servlet.ServletException;
import javax.servlet.ServletRequest;
import javax.servlet.ServletResponse;
```

```
public class MyFilter implements Filter {
    public MyFilter() {
        System.out.println("public MyFilter()");
    }
    @Override
    public void init(FilterConfig filterConfig) throws ServletException {
        System.out.println("public void init(FilterConfig filterConfig)");
    }
    @Override
    public void doFilter(ServletRequest request, ServletResponse response, 
FilterChain chain)
            throws IOException, ServletException {
        System.out.println("public void doFilter(ServletRequest request, 
ServletResponse response, FilterChain chain)");
        chain.doFilter(request, response);
    }
    @Override
    public void destroy() {
        System.out.println("public void destroy()");
    }

}
```

重启 Tomcat 后，在控制台输出了 Filter 的构造方法 public MyFilter()和初始化方法 public void init(FilterConfig filterConfig)，这两者是被 Tomcat 自动调用所输出的日志，如图 9-7 所示。

当多次打开网址 http://localhost:8081/test10/index.html 时，控制台输出了 doFilter()方法被多次调用的信息如图 9-8 所示。说明构造方法 public MyFilter()和初始化方法 public void init(FilterConfig filterConfig)只被调用一次，Filter 在当前项目中和 Servlet 是一样的，均为单例。

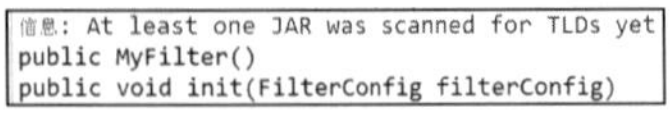

```
信息: At least one JAR was scanned for TLDs yet
public MyFilter()
public void init(FilterConfig filterConfig)
```

图 9-7

```
public void doFilter(ServletRequest request, ServletResponse response, FilterChain chain)
public void doFilter(ServletRequest request, ServletResponse response, FilterChain chain)
public void doFilter(ServletRequest request, ServletResponse response, FilterChain chain)
public void doFilter(ServletRequest request, ServletResponse response, FilterChain chain)
public void doFilter(ServletReque
public void doFilter(ServletReque
public void doFilter(ServletReque
public void doFilter(ServletReque
public void doFilter(ServletReque
```

localhost:8081/test10/index.html

我是index.html文件

图 9-8

那么，Filter 什么时候被销毁呢？关闭 Tomcat 的时候就销毁了，单击“Stop”菜单，如图 9-9 所示。控制台输出过滤器被销毁的信息，如图 9-10 所示。

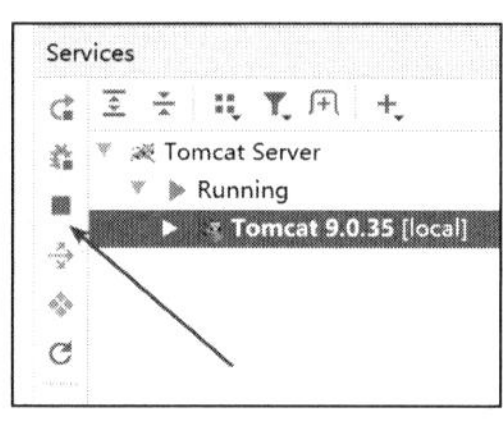


图 9-9

```
信息: Stopping service Catalina
public void destroy()
```

图 9-10

以上过程就是 Filter 的生命周期，与 Servlet 的生命周期基本是一样的。

## 9.3　获取私有/公共init初始化参数

可以在配置文件 web.xml 中向 Filter 配置一些参数，然后在 Filter 中进行读取。配置文件 web.xml 中的代码如下：

```
<filter>
    <filter-name>anyFilter</filter-name>
    <filter-class>myfilter.MyFilter</filter-class>
    <init-param>
        <param-name>filterInitParamName</param-name>
        <param-value>我是过滤器初始化参数值</param-value>
    </init-param>
</filter>

<filter-mapping>
    <filter-name>anyFilter</filter-name>
    <url-pattern>/*</url-pattern>
</filter-mapping>
```

Filter 的代码如下：

```
package myfilter;

import java.io.IOException;

import javax.servlet.Filter;
import javax.servlet.FilterChain;
import javax.servlet.FilterConfig;
import javax.servlet.ServletException;
import javax.servlet.ServletRequest;
```

```
import javax.servlet.ServletResponse;

public class MyFilter implements Filter {

    public MyFilter() {
        System.out.println("public MyFilter()");
    }

    @Override
    public void init(FilterConfig filterConfig) throws ServletException {
        System.out.println("public void init(FilterConfig filterConfig)");
        String                          initValue                          =
filterConfig.getInitParameter("filterInitParamName");
        System.out.println(initValue);
    }

    @Override
    public void doFilter(ServletRequest request, ServletResponse response,
FilterChain chain)
            throws IOException, ServletException {
        System.out.println("public  void  doFilter(ServletRequest  request,
ServletResponse response, FilterChain chain)");
        chain.doFilter(request, response);
    }

    @Override
    public void destroy() {
        System.out.println("public void destroy()");
    }

}
```

```
public MyFilter()
public void init(FilterConfig filterConfig)
我是过滤器初始化参数值
```

图 9-11

重启 Tomcat 后，在控制台成功获取 Filter 初始化参数的值，如图 9-11 所示。刚才获取的 Filter 参数值是 Filter 私有的，可以在 Filter 中获取 ServletContext 公共参数值，使用如下代码：

```
public class MyFilter implements Filter {
    public void init(FilterConfig config) throws ServletException {
        config.getServletContext().getInitParameter("xx");
    }
```

```
    public    void    doFilter(ServletRequest    req,    ServletResponse    resp,
FilterChain chain) throws ServletException, IOException {
        req.getServletContext().getInitParameter("xx");
        chain.doFilter(req, resp);
    }
    public void destroy() {
    }
}
```

配置代码如下：

```
<context-param>
    <param-name>xx</param-name>
    <param-value>xxValue</param-value>
</context-param>
```

Servlet 可以获得 Servlet 私有初始化参数及 ServletContext 公共初始化参数；Filter 可以获得 Filter 私有初始化参数及 ServletContext 公共初始化参数。

## 9.4　使用注解声明Filter

使用注解声明 Filter 示例代码如下：

```
package filter;

import javax.servlet.*;
import javax.servlet.annotation.WebFilter;
import javax.servlet.annotation.WebInitParam;
import java.io.IOException;

@WebFilter(filterName  =  "MyFilter",  urlPatterns  =  "/*",  initParams  =
{@WebInitParam(name = "a", value = "aa")})
public class MyFilter implements Filter {
    public void init(FilterConfig config) throws ServletException {
        System.out.println("init a=" + config.getInitParameter("a"));
    }

    public    void    doFilter(ServletRequest    req,    ServletResponse    resp,
FilterChain chain) throws ServletException, IOException {
        chain.doFilter(req, resp);
```

```
    }

    public void destroy() {
    }
}
```

Tomcat 启动后，在控制台输出内容为：init a=aa

## 9.5 过滤链的顺序——xml方式

多个过滤器可以组成过滤链，每个过滤器执行不同的任务，以便进行层层过滤处理，多个过滤器的执行顺序取决于在 web.xml 文件中的配置顺序，示例代码如下：

```
<!-- anyFilter3 最先被执行 -->
<filter>
    <filter-name>anyFilter3</filter-name>
    <filter-class>extfilter.AnyFilter3</filter-class>
    <init-param>
        <param-name>myParam</param-name>
        <param-value>myValue</param-value>
    </init-param>
</filter>

<filter-mapping>
    <filter-name>anyFilter3</filter-name>
    <url-pattern>/*</url-pattern>
</filter-mapping>

<!-- anyFilter1 其次被执行 -->
<filter>
    <filter-name>anyFilter1</filter-name>
    <filter-class>extfilter.AnyFilter1</filter-class>
</filter>

<filter-mapping>
    <filter-name>anyFilter1</filter-name>
    <url-pattern>/*</url-pattern>
</filter-mapping>

<!-- anyFilter2 最后被执行 -->
```

```
<filter>
    <filter-name>anyFilter2</filter-name>
    <filter-class>extfilter.AnyFilter2</filter-class>
</filter>

<filter-mapping>
    <filter-name>anyFilter2</filter-name>
    <url-pattern>/*</url-pattern>
</filter-mapping>
```

3 个 Filter 类代码分别如下：

```
public class AnyFilter1 implements Filter {
    @Override
    public void doFilter(ServletRequest request, ServletResponse response,
FilterChain chain)
            throws IOException, ServletException {
        System.out.println("AnyFilter1 begin");
        chain.doFilter(request, response);
        System.out.println("AnyFilter1   end");
    }
}

public class AnyFilter2 implements Filter {
    @Override
    public void doFilter(ServletRequest request, ServletResponse response,
FilterChain chain)
            throws IOException, ServletException {
        System.out.println("AnyFilter2 begin");
        chain.doFilter(request, response);
        System.out.println("AnyFilter2   end");
    }
}

public class AnyFilter3 implements Filter {
    @Override
    public void doFilter(ServletRequest request, ServletResponse response,
FilterChain chain)
            throws IOException, ServletException {
        System.out.println("AnyFilter3 begin");
        chain.doFilter(request, response);
```

```
        System.out.println("AnyFilter3   end");
    }
}
```

在浏览器上输入任意网址，在控制台输出的信息如图 9-12 所示，调用顺序如图 9-13 所示。

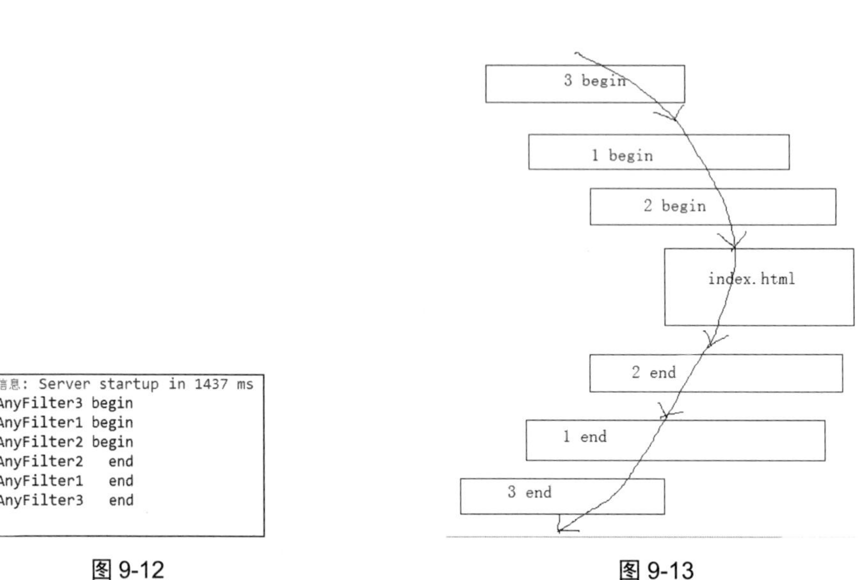


图 9-12

图 9-13

## 9.6 过滤链的顺序——annotation方式

在使用 annotation 方式实现过滤链时，执行顺序取决于过滤类的名称，按 ASCII 的顺序进行排序。

3 个过滤类代码如下：

```
@WebFilter("/*")
public class A implements Filter {
    public void doFilter(ServletRequest request, ServletResponse response,
FilterChain chain)
            throws IOException, ServletException {
        System.out.println("A begin");
        chain.doFilter(request, response);
        System.out.println("A   end");
    }
```

```
}

@WebFilter("/*")
public class Q implements Filter {
    public void doFilter(ServletRequest request, ServletResponse response,
FilterChain chain)
            throws IOException, ServletException {
        System.out.println("Q begin");
        chain.doFilter(request, response);
        System.out.println("Q   end");
    }
}

@WebFilter("/*")
public class Z implements Filter {
    public void doFilter(ServletRequest request, ServletResponse response,
FilterChain chain)
            throws IOException, ServletException {
        System.out.println("Z begin");
        chain.doFilter(request, response);
        System.out.println("Z   end");
    }
}
```

程序运行后，控制台输出结果如下：

```
A begin
Q begin
Z begin
Z   end
Q   end
A   end
```

## 9.7　使用Filter实现编码的处理

在 Servlet 中处理 post 提交中文乱码时，使用如下代码进行解决：

```
request.setCharacterEncoding("utf-8");
```

上面的代码需要在每一个 Servlet 中的第一行进行调用，代码出现严重的重复冗余，所以可以使用 Filter 来进行统一处理。创建字符编码过滤器，过滤器类代码如下：

```
package myfilter;

import java.io.IOException;

import javax.servlet.Filter;
import javax.servlet.FilterChain;
import javax.servlet.FilterConfig;
import javax.servlet.ServletException;
import javax.servlet.ServletRequest;
import javax.servlet.ServletResponse;

public class CharSetFilter implements Filter {
    private String encoding = null;

    public void init(FilterConfig fConfig) throws ServletException {
        encoding = fConfig.getInitParameter("encoding");
    }

    public void doFilter(ServletRequest request, ServletResponse response,
FilterChain chain)
            throws IOException, ServletException {
        System.out.println("CharSetFilter 设置的编码类型为：" + encoding);
        request.setCharacterEncoding(encoding);
        response.setCharacterEncoding(encoding);
        chain.doFilter(request, response);
    }

}
```

在 web.xml 文件中配置 Filter 代码如下：

```
<filter>
    <filter-name>CharSetFilter</filter-name>
    <filter-class>myfilter.CharSetFilter</filter-class>
    <init-param>
        <param-name>encoding</param-name>
        <param-value>utf-8</param-value>
    </init-param>
</filter>
<filter-mapping>
    <filter-name>CharSetFilter</filter-name>
```

```
    <url-pattern>/*</url-pattern>
</filter-mapping>
```

上面代码可以将编码进行统一的预处理，就不会出现中文乱码问题了。

## 9.8　Filter拦截请求和转发

如果 Filter 想正确拦截请求和转发，则需要进一步处理。编辑 Servlet 类 test1.java 代码如下：

```
public class test1 extends HttpServlet {
    protected void doGet(HttpServletRequest request, HttpServletResponse
response)
            throws ServletException, IOException {
        System.out.println("test1 run !");
        request.getRequestDispatcher("test2").forward(request, response);
    }
}
```

由 test1.java 转发到 test2.java。创建 Servlet 类 test2.java 代码如下：

```
public class test2 extends HttpServlet {
    protected void doGet(HttpServletRequest request, HttpServletResponse
response)
            throws ServletException, IOException {
        System.out.println("test2 run !");
    }
}
```

在 web.xml 文件中配置这两个 Servlet 代码如下：

```
<servlet>
    <servlet-name>test1</servlet-name>
    <servlet-class>controller.test1</servlet-class>
</servlet>
<servlet-mapping>
    <servlet-name>test1</servlet-name>
    <url-pattern>/test1</url-pattern>
</servlet-mapping>

<servlet>
```

```
    <servlet-name>test2</servlet-name>
    <servlet-class>controller.test2</servlet-class>
</servlet>
<servlet-mapping>
    <servlet-name>test2</servlet-name>
    <url-pattern>/test2</url-pattern>
</servlet-mapping>
```

部署项目并运行网址 http://localhost:8080/web13/test1，控制台输出结果如下：

```
test1 run !
test2 run !
```

这时，正确输出两个字符串，说明 Servlet 可以相互实现转发操作。当在项目中添加过滤器时，从 test1.java 以转发 forward 的方式访问 test2.java 时，在默认的情况下，过滤器并不对转发到 test2.java 的操作进行拦截。在项目中创建 Filter 代码如下：

```
public class ActionFilter implements Filter {
    @Override
    public void doFilter(ServletRequest request, ServletResponse response,
FilterChain chain)
            throws IOException, ServletException {
        String servletPath = ((HttpServletRequest) request).getServletPath();
        System.out.println("ActionFilter doFilter run ! Servlet Path =" +
servletPath);
        chain.doFilter(request, response);
    }
}
```

在 web.xml 文件中配置过滤器 Filter 代码如下：

```
<filter>
    <filter-name>actionFilter</filter-name>
    <filter-class>myfilter.ActionFilter</filter-class>
</filter>
<filter-mapping>
    <filter-name>actionFilter</filter-name>
    <url-pattern>/*</url-pattern>
</filter-mapping>
```

过滤器拦截 URL 的策略是对所有路径。重启 Tomcat 并运行网址 http://localhost:8080/

web13/test1，控制台输出结果如下：

```
ActionFilter doFilter run ! Servlet Path =/test1
test1 run !
test2 run !
```

从控制台输出的结果可以分析出，使用配置代码：

```
    <filter-mapping>
        <filter-name>actionFilter</filter-name>
        <url-pattern>/*</url-pattern>
    </filter-mapping>
```

当不添加其他配置时，Filter 默认拦截的是请求，并不包含转发 forward 的操作，故 Filter 并没有拦截转发到 test2.java 的操作，仅仅拦截了请求到 test1.java 的操作。

如果想实现在转发时 Filter 也要参与拦截，则需要在 web.xml 文件中添加<dispatcher>配置代码如下：

```
<filter>
    <filter-name>ActionFilter</filter-name>
    <filter-class>myfilter.ActionFilter</filter-class>
</filter>
<filter-mapping>
    <filter-name>ActionFilter</filter-name>
    <url-pattern>/*</url-pattern>
    <dispatcher>FORWARD</dispatcher>
</filter-mapping>
```

重启 Tomcat 并打开网址 http://localhost:8080/web13/test1，控制台输出结果如下：

```
test1 run !
ActionFilter doFilter run ! Servlet Path =/test2
test2 run !
```

过滤器成功对转发操作进行拦截，但从控制台输出的结果中可以分析出使用如下配置代码：

```
<filter-mapping>
    <filter-name>ActionFilter</filter-name>
    <url-pattern>/*</url-pattern>
    <dispatcher>FORWARD</dispatcher>
```

```
</filter-mapping>
```

Filter 不再参与请求拦截，原因是在 web.xml 文件中只配置了转发 forward 拦截，配置代码如下：

```
<dispatcher>FORWARD</dispatcher>
```

如果想实现在请求 test1 路径时，Filter 也要对请求的 test1.java 进行拦截，则需要在 web.xml 文件中添加拦截请求的配置，更改代码如下：

```
<filter>
    <filter-name>ActionFilter</filter-name>
    <filter-class>myfilter.ActionFilter</filter-class>
</filter>
<filter-mapping>
    <filter-name>ActionFilter</filter-name>
    <url-pattern>/*</url-pattern>
    <dispatcher>FORWARD</dispatcher>
    <dispatcher>REQUEST</dispatcher>
</filter-mapping>
```

重启 Tomcat 并打开网址 http://localhost:8080/web13/test1，控制台输出结果如下：

```
ActionFilter doFilter run ! Servlet Path =/test1
test1 run !
ActionFilter doFilter run ! Servlet Path =/test2
test2 run !
```

Filter 将请求和 forward 成功拦截。注解版本使用如下配置代码：

```
@WebFilter(filterName = "MyFilter", urlPatterns = "/*", initParams =
{@WebInitParam(name = "a", value = "aa")}, dispatcherTypes =
{DispatcherType.FORWARD, DispatcherType.REQUEST})
public class MyFilter implements Filter {
```

## 9.9 使用Filter实现权限验证

本章节将使用 Filter 来实现权限验证的功能。创建权限验证过滤器，代码如下：

```
package myfilter;
```

```
import java.io.IOException;
import java.util.ArrayList;
import java.util.List;

import javax.servlet.Filter;
import javax.servlet.FilterChain;
import javax.servlet.FilterConfig;
import javax.servlet.ServletException;
import javax.servlet.ServletRequest;
import javax.servlet.ServletResponse;
import javax.servlet.http.HttpServletRequest;
import javax.servlet.http.HttpServletResponse;

public class ValidateFilter implements Filter {

    private List validatePath = new ArrayList();

    @Override
    public void init(FilterConfig filterConfig) throws ServletException {
        validatePath.add("test1");
        validatePath.add("test2");
        // 以上代码说明 test1 和 test2 这两个路径必须经过登录验证
        // 只有登录了才可以访问这两个路径，其他路径可以未登录而直接访问
    }

    @Override
    public void doFilter(ServletRequest request, ServletResponse response,
FilterChain chain)
            throws IOException, ServletException {
        HttpServletRequest requestRef = (HttpServletRequest) request;
        HttpServletResponse responseRef = (HttpServletResponse) response;
        String servletPath = requestRef.getServletPath();
        servletPath = servletPath.substring(1);
        int findIndex = validatePath.indexOf(servletPath);
        if (findIndex >= 0) {
            String loginUser = (String) requestRef.getSession().
getAttribute("loginUser");
            if (loginUser == null) {
                responseRef.sendRedirect("login.jsp");
            } else {
                chain.doFilter(requestRef, responseRef);
```

```
            }
        } else {
            chain.doFilter(requestRef, responseRef);
        }
    }
}
```

创建登录界面 login.jsp，代码如下：

```
<body>
    show login ui!
    <form action="login" method="post">
        <input type="submit" value="submit" />
    </form>
</body>
```

创建名称为 test1，test2，test3 和 login 的 Servlet 代码如下：

```
public class test1 extends HttpServlet {
    protected void doGet(HttpServletRequest request, HttpServletResponse
response)
            throws ServletException, IOException {
        System.out.println("进入 test1");
    }
}

public class test2 extends HttpServlet {
    protected void doGet(HttpServletRequest request, HttpServletResponse
response)
            throws ServletException, IOException {
        System.out.println("进入 test2");
    }
}

public class test3 extends HttpServlet {
    protected void doGet(HttpServletRequest request, HttpServletResponse
response)
            throws ServletException, IOException {
        System.out.println("进入 test3");
    }
}
```

```
public class login extends HttpServlet {
    protected void doPost(HttpServletRequest request, HttpServletResponse
response)
            throws ServletException, IOException {
        request.getSession().setAttribute("loginUser", "XXX");
    }
}
```

第一步，执行名称为 test1 的 Servlet，请求先进入 Filter，在 Filter 中发现当执行 test1 这个路径时必须要登录过，但现在的情况是并未登录过，所以显示登录 login.jsp 界面。

第二步，执行名称为 test2 的 Servlet，请求先进入 Filter，在 Filter 中发现当执行 test2 这个路径时也必须要登录过，但现在的情况也是从未登录过，所以也显示登录 login.jsp 界面。

第三步，执行名称为 test3 的 Servlet，请求先进入 Filter，在 Filter 中发现 test3 是不需要登录即可访问的，所以执行了 test3 中的代码。

第四步，执行名称为 test1 的 Servlet，显示登录 login.jsp 界面，单击“submit 提交”按钮进行登录，再执行 test1 和 test2 这两个路径时就可以成功执行这两个 Servlet 中的代码。

## 9.10　综合使用Filter+ThreadLocal+Cookie实现解耦合

在前面章节创建的 CookieTools.java 类需要在构造方法中传递 HttpServletRequest 和 HttpServletResponse 这两个参数，造成 HttpServletRequest 和 HttpServletResponse 与 CookieTools.java 的构造方法产生紧耦合，后果就是代码不利于维护，具有对象之间的依赖性，破坏了 MVC 分层，如图 9-14 所示。

图 9-14

请求对象和响应对象是在 Servlet 类中产生的，为了让 CookieTools 能拥有这两个对象，需要 Service 以“接力”的方式将请求对象和响应对象传递给 CookieTools，造成 Service

与请求和响应产生紧耦合，Service 的代码严重“绑死”了 Web 环境，不便于 Service 的复用。

回顾一下，在前面章节中，对 Cookie 的操作封装了一个 Java 工具类，代码如下：

```
public class CookieToolsOld {
    private HttpServletRequest request;
    private HttpServletResponse response;
    public CookieToolsOld(HttpServletRequest request, HttpServletResponse
response) {
        super();
        this.request = request;
        this.response = response;
    }
    //其他方法的代码省略
}
```

上面的代码的确能实现对 Cookie 的操作，但该类中的 HttpServletRequest request 和 HttpServletResponse response 属性值是需要依赖于构造方法进行传入的，传入的顺序如下。

（1）Controller 层中的 doGet()方法或 doPost()方法拥有请求和响应两个对象。

（2）在 Controller 调用 Service 层时，需要将请求和响应传递给 Service 层。

（3）Service 层将传入的请求和响应再传入 CookieToolsOld.java 类中。

不规范的示例代码如图 9-15 所示。这造成了 Service 层和 CookieToolsOld.java 与请求和响应的紧耦合，不利于前期代码的设计与后期维护，基于这种情况，可以使用 Filter+ThreadLocal 进行解耦。

使用 Filter+ThreadLocal 可以解决上面出现的紧耦合问题，也就是在 Filter 中先拦截请求对象和响应对象，然后将这两个对象通过 ThreadLocal 类放入当前线程的 ThreadLocalMap 对象中，以后想获取自己的请求和响应时，只需要调用 ThreadLocal.get()方法即可。如果这么设计，则 CookieTools.java 的构造方法就不需要依赖 HttpServletRequest 和 HttpServletResponse 这两个对象了，紧耦合问题解决。

Login.java

```
package controller;

import java.io.IOException;

public class Login extends HttpServlet {
    protected void doPost(HttpServletRequest request, HttpServletResponse response)
            throws ServletException, IOException {
        LoginService loginService = new LoginService(request, response);
        boolean loginResult = loginService.login("a", "aa");
    }
}
```

LoginService.java

```
package service;

import javax.servlet.http.HttpServletRequest;

public class LoginService {

    private CookieToolsOld cookieToolsOld;

    public LoginService(HttpServletRequest request, HttpServletResponse response) {
        cookieToolsOld = new CookieToolsOld(request, response);
    }

    public boolean login(String username, String password) {
        if (1 == 1) {
            // 使用cookieToolsOld对象操作Cookie.
            // [illegible]
            return true;
        } else {
            return false;
        }
    }
```

CookieToolsOld.java

```
package cookietools;

import javax.servlet.http.HttpServletRequest;

public class CookieToolsOld {
    private HttpServletRequest request;
    private HttpServletResponse response;

    public CookieToolsOld(HttpServletRequest request, HttpServletResponse response) {
        super();
        this.request = request;
        this.response = response;
```

图 9-15

首先看一下使用 ThreadLocal 保存请求和响应工具类，代码如下：

```
package utils;

import javax.servlet.http.HttpServletRequest;
import javax.servlet.http.HttpServletResponse;

public class RequestResponseBox {

    private static ThreadLocal<HttpServletRequest> requestBox = new
ThreadLocal<>();
    private static ThreadLocal<HttpServletResponse> responseBox = new
ThreadLocal<>();

    public static void setRequest(HttpServletRequest request) {
        requestBox.set(request);
    }
```

```
    public static void setResponse(HttpServletResponse response) {
        responseBox.set(response);
    }

    public static HttpServletRequest getRequest() {
        return requestBox.get();
    }

    public static HttpServletResponse getResponse() {
        return responseBox.get();
    }

}
```

ThreadLocal 对象使用 static 进行声明，不会被 gc 垃圾回收器所回收，但 ThreadLocal.set(value)中的 value 值还是需要使用 ThreadLocal.remove()方法进行清除，释放内存资源。Filter 示例代码如下：

```
package extfilter;

import java.io.IOException;

import javax.servlet.Filter;
import javax.servlet.FilterChain;
import javax.servlet.FilterConfig;
import javax.servlet.ServletException;
import javax.servlet.ServletRequest;
import javax.servlet.ServletResponse;
import javax.servlet.http.HttpServletRequest;
import javax.servlet.http.HttpServletResponse;

import utils.RequestResponseBox;

public class Request_Response_Filter implements Filter {

    @Override
    public void init(FilterConfig filterConfig) throws ServletException {
    }

    @Override
    public void doFilter(ServletRequest request, ServletResponse response,
```

```
FilterChain chain)
            throws IOException, ServletException {
        RequestResponseBox.setRequest((HttpServletRequest) request);
        RequestResponseBox.setResponse((HttpServletResponse) response);
        chain.doFilter(request, response);
    }

    @Override
    public void destroy() {
    }

}
```

操作 Cookie 的工具类代码如下：

```
package cookietools;

import java.io.UnsupportedEncodingException;

import javax.servlet.http.Cookie;
import utils.RequestResponseBox;

public class CookieTools {

    public void save(String cookieName, String cookieValue, int maxAge) {
        try {
            cookieValue = java.net.URLEncoder.encode(cookieValue, "utf-8");
            Cookie cookie = new Cookie(cookieName, cookieValue);
            cookie.setMaxAge(maxAge);
            RequestResponseBox.getResponse().addCookie(cookie);
        } catch (UnsupportedEncodingException e) {
            e.printStackTrace();
        }
    }

    public String getValue(String cookieName) {
        String cookieValue = null;
        try {
            Cookie[]                          cookieArray                    =
RequestResponseBox.getRequest().getCookies();
            if (cookieArray != null) {
                for (int i = 0; i < cookieArray.length; i++) {
                    if (cookieArray[i].getName().equals(cookieName)) {
```

```
                        cookieValue = cookieArray[i].getValue();
                        cookieValue                                              =
java.net.URLDecoder.decode(cookieValue, "utf-8");
                        break;
                    }

                }
            }
        } catch (UnsupportedEncodingException e) {
            e.printStackTrace();
        }
        return cookieValue;
    }

    public void delete(String cookieName) {
        Cookie cookie = new Cookie(cookieName, "");
        cookie.setMaxAge(0);
        RequestResponseBox.getResponse().addCookie(cookie);
    }

}
```

在最新版本的 CookieTools.java 类中，构造方法并不依赖于请求和响应对象，就能实现本章的目标：在 CookieTools.java 类中的请求和响应不再使用构造方法进行传入，即可实现解耦。

## 9.11　使用Cookie实现购物车的核心逻辑

在使用 Cookie 实现购物车功能时，核心逻辑就是处理 String 值，因为购物车中的信息存储在 String 值中，String 值存储在 Cookie 中，所以解析 String 值就成了使用 Cookie 实现购物车功能的重点知识。解析 String 示例代码如下：

```
package web11;

public class Test1 {
    public static void main(String[] args) {
        String cartValue = "1_11-2_22-3_33";
        String[] bookinfoArray = cartValue.split("\\-");
        for (int i = 0; i < bookinfoArray.length; i++) {
            String eachBookString = bookinfoArray[i];
            String id = eachBookString.split("\\_")[0];
```

```
            String num = eachBookString.split("\\_")[1];
            System.out.println(id + "    " + num);
        }
    }
}
```

使用 split()方法必须要使用“\\”转义，防止发生特殊字符导致异常的情况，重现异常的代码如下：

```
package controller;

public class Test4 {
    public static void main(String[] args) {
        String myString = "1[2[3[4";
        String[] myArray = myString.split("[");
        for (int i = 0; i < myArray.length; i++) {
            System.out.println(myArray[i]);
        }
    }
}
```

程序运行后出现异常的信息如下：

```
Exception in thread "main" java.util.regex.PatternSyntaxException: Unclosed
character class near index 0
[
^
    at java.util.regex.Pattern.error(Pattern.java:1957)
    at java.util.regex.Pattern.clazz(Pattern.java:2550)
    at java.util.regex.Pattern.sequence(Pattern.java:2065)
    at java.util.regex.Pattern.expr(Pattern.java:1998)
    at java.util.regex.Pattern.compile(Pattern.java:1698)
    at java.util.regex.Pattern.<init>(Pattern.java:1351)
    at java.util.regex.Pattern.compile(Pattern.java:1028)
    at java.lang.String.split(String.java:2380)
    at java.lang.String.split(String.java:2422)
    at controller.Test4.main(Test4.java:6)
```

如果使用转义“\\”，则不会出现异常，代码如下：

```
package controller;
```

```
public class Test5 {
    public static void main(String[] args) {
        String myString = "1[2[3[4";
        String[] myArray = myString.split("\\[");
        for (int i = 0; i < myArray.length; i++) {
            System.out.println(myArray[i]);
        }
    }
}
```

程序运行结果如下：

```
1
2
3
4
```

在购物车中查询 bookId 是否存在，示例代码如下：

```
package web11;

public class Test2 {
    public static void main(String[] args) {
        String findBookId = "2";
        boolean isFindBookId = false;
        String cartValue = "1_11-2_22-3_33";
        String[] bookinfoArray = cartValue.split("\\-");
        for (int i = 0; i < bookinfoArray.length; i++) {
            String eachBookString = bookinfoArray[i];
            String id = eachBookString.split("\\_")[0];
            if (id.equals(findBookId)) {
                isFindBookId = true;
                break;
            }
        }
        System.out.println(isFindBookId);
    }
}
```

根据 bookId 从购物车中删除指定的书籍，示例代码如下：

```
package test;
```

```
public class Test4 {
    public static void main(String[] args) {
        String value = "1_11-2_22-3_33";
        String newValue = "";
        String[] bookinfoArray = value.split("\\-");
        String deleteBookId = "2";
        for (int i = 0; i < bookinfoArray.length; i++) {
            String eachBookinfo = bookinfoArray[i];
            String[] id_num_array = eachBookinfo.split("\\_");
            String bookId = id_num_array[0];
            String bookNum = id_num_array[1];
            if (bookId.equals(deleteBookId)) {
            } else {
                newValue = newValue + "-" + bookId + "_" + bookNum;
            }
        }

        if (!newValue.equals("") && newValue.charAt(0) == '-') {
            newValue = newValue.substring(1);
        }
        System.out.println(newValue);
    }

}
```

向购物车中存放书籍，如果 bookId 重复则数量不累加，如果 bookId 找不到则追加，示例代码如下：

```
package test;

public class Test4 {
    public static void main(String[] args) {
        String putBookId = "22";
        String cookieValue = "1_11-2_22-3_33";
        boolean isFindBookId = false;
        String[] bookinfoArray = cookieValue.split("\\-");
        for (int i = 0; i < bookinfoArray.length; i++) {
            String eachBookinfo = bookinfoArray[i];
            String bookId = eachBookinfo.split("\\_")[0];
            if (bookId.equals(putBookId)) {
```

```
                isFindBookId = true;
                break;
            }
        }
        if (isFindBookId == false) {
            cookieValue = cookieValue + "-" + putBookId + "_1";
        }
        if (cookieValue.charAt(0) == '-') {
            cookieValue = cookieValue.substring(1);
        }
        System.out.println(cookieValue);
    }
}
```

向购物车中存放书籍，如果 bookId 重复则数量加 1，如果 bookId 找不到则追加，示例代码如下：

```
package test;

public class Test5 {
    public static void main(String[] args) {
        String putBookId = "2";
        String cookieValue = "1_11-2_22-3_33";
        String newCookieValue = "";
        boolean isFindBookId = false;
        if (!"".equals(cookieValue)) {
            String[] bookinfoArray = cookieValue.split("\\-");
            for (int i = 0; i < bookinfoArray.length; i++) {
                String eachBookinfo = bookinfoArray[i];
                String bookId = eachBookinfo.split("\\_")[0];
                String bookNum = eachBookinfo.split("\\_")[1];
                if (bookId.equals(putBookId)) {
                    isFindBookId = true;
                    newCookieValue = newCookieValue + "-" + bookId + "_" +
((Integer.parseInt(bookNum)) + (1));
                } else {
                    newCookieValue = newCookieValue + "-" + bookId + "_" +
bookNum;
                }
            }
            if (isFindBookId == false) {
```

```
                newCookieValue = newCookieValue + "-" + putBookId + "_1";
            }
            if (newCookieValue.charAt(0) == '-') {
                newCookieValue = newCookieValue.substring(1);
            }
        } else {
            newCookieValue = putBookId + "_1";
        }
        System.out.println(newCookieValue);
    }
}
```

根据 bookId 修改购物车中书籍的数量，示例代码如下：

```
package web11;

public class Test6 {
    public static void main(String[] args) {
        String updateBookId = "1";
        String updateBookNum = "123";
        String cartValue = "1_11-2_22-3_33";
        String newCartValue = "";
        String[] bookinfoArray = cartValue.split("\\-");
        for (int i = 0; i < bookinfoArray.length; i++) {
            String eachBookString = bookinfoArray[i];
            String id = eachBookString.split("\\_")[0];
            String num = eachBookString.split("\\_")[1];
            if (id.equals(updateBookId)) {
                newCartValue = newCartValue + "-" + id + "_" + updateBookNum;
            } else {
                newCartValue = newCartValue + "-" + id + "_" + num;
            }
        }
        if (newCartValue.charAt(0) == '-') {
            newCartValue = newCartValue.substring(1);
        }
        System.out.println(newCartValue);
    }
}
```

# 第 10 章 Listener 接口

Listener 接口的作用是监听系统中部分组件的运行状态，接收发生事件的通知，可以在监听过程中对事件进行处理，比如实现日志记录、数据校验等。

## 10.1 HttpSessionActivationListener接口的使用

HttpSessionActivationListener接口的作用是可以在Web容器关闭时将session中的数据序列化，而Web容器启动时可以将数据进行反序列化，解决session中的数据在Web容器关闭后不丢失的问题。

若想要让 session 中的数据进行序列化与反序列化，则该数据对象必须要实现监听接口 HttpSessionActivationListener，还要同时实现 Serializable 接口。

实体类 Userinfo.java 代码如下：

```
package entity;

import javax.servlet.http.HttpSessionActivationListener;
import javax.servlet.http.HttpSessionEvent;
import java.io.Serializable;

//Userinfo 类必须实现序列化 Serializable 接口,不然 Userinfo 对象不能钝化到硬盘中的文件里
public class Userinfo implements Serializable,
HttpSessionActivationListener {
    private String username;

    public Userinfo(String username) {
        this.username = username;
```

```
    }

    public String getUsername() {
        return username;
    }

    public void setUsername(String username) {
        this.username = username;
    }

    @Override
    public void sessionDidActivate(HttpSessionEvent se) {
        System.out.println("激活 public void sessionDidActivate : " +
this.getUsername() + " " + this.hashCode());
    }

    @Override
    public void sessionWillPassivate(HttpSessionEvent se) {
        System.out.println("钝化 public void sessionWillPassivate : " +
this.getUsername() + " " + this.hashCode());
    }
}
```

解释两个术语：

（1）激活：将硬盘中的对象加载到内存中，相当于反序列化。

（2）钝化：将内存中的对象保存到硬盘中，相当于序列化。

创建向 session 中保存数据的 Servlet 代码如下：

```
package controller;

import entity.Userinfo;

import javax.servlet.ServletException;
import javax.servlet.annotation.WebServlet;
import javax.servlet.http.HttpServlet;
import javax.servlet.http.HttpServletRequest;
import javax.servlet.http.HttpServletResponse;
import javax.servlet.http.HttpSession;
import java.io.IOException;
```

```
@WebServlet(name = "test1", urlPatterns = "/test1")
public class test1 extends HttpServlet {
    protected void doGet(HttpServletRequest request, HttpServletResponse
response)
            throws ServletException, IOException {
        HttpSession session = request.getSession();
        for (int i = 0; i < 5; i++) {
            Userinfo userinfo = new Userinfo("username" + (i + 1));
            System.out.println("put  " + userinfo.getUsername() + " " +
userinfo.hashCode());
            session.setAttribute("userinfo" + (i + 1), userinfo);
        }
    }
}
```

创建从 session 中获取数据的 Servlet 代码如下：

```
package controller;

import entity.Userinfo;

import javax.servlet.ServletException;
import javax.servlet.annotation.WebServlet;
import javax.servlet.http.HttpServlet;
import javax.servlet.http.HttpServletRequest;
import javax.servlet.http.HttpServletResponse;
import javax.servlet.http.HttpSession;
import java.io.IOException;

@WebServlet(name = "test2", urlPatterns = "/test2")
public class test2 extends HttpServlet {
    protected void doGet(HttpServletRequest request, HttpServletResponse
response)
            throws ServletException, IOException {
        HttpSession session = request.getSession();
        Userinfo userinfo1 = (Userinfo) session.getAttribute("userinfo1");
        Userinfo userinfo2 = (Userinfo) session.getAttribute("userinfo2");
        Userinfo userinfo3 = (Userinfo) session.getAttribute("userinfo3");
        Userinfo userinfo4 = (Userinfo) session.getAttribute("userinfo4");
        Userinfo userinfo5 = (Userinfo) session.getAttribute("userinfo5");
        System.out.println("get  " + userinfo1.getUsername() + " " +
userinfo1.hashCode());
        System.out.println("get  " + userinfo2.getUsername() + " " +
```

```
userinfo2.hashCode());
        System.out.println("get  " + userinfo3.getUsername() + " " +
userinfo3.hashCode());
        System.out.println("get  " + userinfo4.getUsername() + " " +
userinfo4.hashCode());
        System.out.println("get  " + userinfo5.getUsername() + " " +
userinfo5.hashCode());
    }
}
```

在 Tomcat/conf 文件夹中的 context.xml 文件的<Context>节点下添加以下内容：

```
<Manager className="org.apache.catalina.session.PersistentManager"
saveOnRestart="true">
    <Store className="org.apache.catalina.session.FileStore" directory=
"C:\apache-tomcat-9.0.35\sessionData"/>
</Manager>
```

先启动 Tomcat，然后执行地址 test1，控制台输出信息如下：

```
put username1 1556610046
put username2 1152811305
put username3 667785802
put username4 694428867
put username5 341100017
```

控制台输出的信息说明创建了 5 个 Userinfo 类的对象，并分别将它们放入 session 中，再执行地址 test2，控制台输出信息如下：

```
get username1 1556610046
get username2 1152811305
get username3 667785802
get username4 694428867
get username5 341100017
```

通过查看 hashCode 值可以说明对 session 执行 set 和 get 操作是同一批 5 个 Userinfo 对象。下一步单击 Tomcat 服务的“Stop”键将 Tomcat 服务停止，如图 10-1 所示。

图 10-1

这时控制台输出信息如下：

```
钝化public void sessionWillPassivate : username5 341100017
钝化public void sessionWillPassivate : username3 667785802
```

```
钝化public void sessionWillPassivate : username4 694428867
钝化public void sessionWillPassivate : username1 1556610046
钝化public void sessionWillPassivate : username2 1152811305
```

Tomcat 服务停止后，将 session 中的 5 个对象钝化到硬盘中，序列化文件位置在如图 10-2 所示的路径中。当再次启动 Tomcat 时，硬盘.session 文件中的 session 数据重新被激活到内存中，如图 10-3 所示。

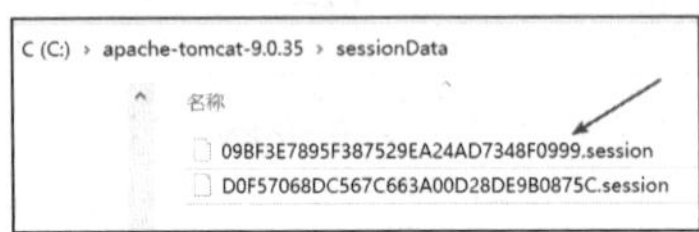


图 10-2

```
激活public void sessionDidActivate : username5 677546875
激活public void sessionDidActivate : username3 872171250
激活public void sessionDidActivate : username4 1927537736
激活public void sessionDidActivate : username1 1163393291
激活public void sessionDidActivate : username2 2035452809
```

图 10-3

从激活结果中可以发现，Tomcat 创建了新的 Userinfo 类的对象，但原有的 username 属性值却没有丢失，被放入新创建的 Userinfo 对象中。这时再执行地址 test2，还可以从 session 中获取钝化前的数据，username 值并未丢失，结果如下：

```
get username1 1163393291
get username2 2035452809
get username3 872171250
get username4 1927537736
get username5 677546875
```

## 10.2 HttpSessionAttributeListener接口的使用

HttpSessionAttributeListener 接口的作用是当向 session 中添加、删除或更新数据时，该监听接口可以接收到相应的通知。创建 HttpSessionAttributeListener 接口的实现类 MyHttpSessionAttributeListener.java，代码如下：

```
package __listenerTest2;

import javax.servlet.http.HttpSessionAttributeListener;
import javax.servlet.http.HttpSessionBindingEvent;

public class MyHttpSessionAttributeListener implements
HttpSessionAttributeListener {

    @Override
    public void attributeAdded(HttpSessionBindingEvent se) {
```

```
        System.out.println("attributeAdded  " + se.getName() + "  " +
se.getValue());
    }

    @Override
    public void attributeRemoved(HttpSessionBindingEvent se) {
        System.out.println("attributeRemoved  " + se.getName() + "  " +
se.getValue());
    }

    @Override
    public void attributeReplaced(HttpSessionBindingEvent se) {
        System.out.println("attributeReplaced  " + se.getName() + "  " +
se.getValue() + "  "
                + se.getSession().getAttribute(se.getName()));
    }

}
```

下一步要把这个监听类在 web.xml 文件中进行注册，配置代码如下：

```
<listener>
<listener-class>__listenerTest2.MyHttpSessionAttributeListener</listener
-class>
</listener>
```

创建向 session 中添加数据的 Servlet，代码如下：

```
public class test extends HttpServlet {
    protected void doGet(HttpServletRequest request, HttpServletResponse
response)
            throws ServletException, IOException {
        request.getSession().setAttribute("sessionKey", "sessionValue");
    }
}
```

创建向 session 中更新数据的 Servlet，代码如下：

```
public class test2 extends HttpServlet {
    protected void doGet(HttpServletRequest request, HttpServletResponse
response)
            throws ServletException, IOException {
        request.getSession().setAttribute("sessionKey",
```

```
"newSessionValue");
    }
}
```

创建从 session 中删除数据的 Servlet，代码如下：

```
public class test3 extends HttpServlet {
    protected void doGet(HttpServletRequest request, HttpServletResponse
response)
            throws ServletException, IOException {
        request.getSession().removeAttribute("sessionKey");
    }
}
```

打开网址：http://localhost:8081/__listenerTest2/test，控制台输出信息如图 10-4 所示。

```
attributeAdded sessionKey sessionValue
```

图 10-4

打开网址：http://localhost:8081/__listenerTest2/test2，控制台输出信息如图 10-5 所示。

```
attributeReplaced sessionKey sessionValue newSessionValue
```

图 10-5

打开网址：http://localhost:8081/__listenerTest2/test3，控制台输出信息如图 10-6 所示。

```
attributeRemoved sessionKey newSessionValue
```

图 10-6

使用注解方式实现代码如下：

```
@WebListener
public class MyListener implements HttpSessionAttributeListener {
```

## 10.3 HttpSessionBindingListener接口的使用

如果放入 session 中的对象实现了 HttpSessionBindingListener 接口，则将该对象放入 session 中时会执行 HttpSessionBindingListener 接口中的相应方法，达到记录某对象被放入 session 中的事件状态。创建监听接口 HttpSessionBindingListener 的实现类 MyHttpSessionBindingListener.java，代码如下：

```
package entity;

import javax.servlet.http.HttpSessionBindingEvent;
import javax.servlet.http.HttpSessionBindingListener;

public class Userinfo implements HttpSessionBindingListener {

    @Override
    public void valueBound(HttpSessionBindingEvent event) {
        System.out.println("valueBound 有绑定发生： " + event.getName() + "  "
+ event.getValue());

    }

    @Override
    public void valueUnbound(HttpSessionBindingEvent event) {
        System.out.println("valueUnbound 有反绑发生： " + event.getName() + "
" + event.getValue());
    }

}
```

创建将 Userinfo 对象放入 session 中的 Servlet 代码如下：

```
public class test extends HttpServlet {
    protected void doGet(HttpServletRequest request, HttpServletResponse
response)
            throws ServletException, IOException {
        HttpSession session = request.getSession();
        session.setAttribute("userinfo", new Userinfo());
    }
}
```

创建从 session 中删除 Userinfo 对象的 Servlet 代码如下：

```
public class test2 extends HttpServlet {
    protected void doGet(HttpServletRequest request, HttpServletResponse
response)
            throws ServletException, IOException {
        HttpSession session = request.getSession();
        session.removeAttribute("userinfo");
    }
}
```

打开网址：http://localhost:8081/__listenerTest3/test，控制台输出信息如图 10-7 所示。

```
valueBound有绑定发生： userinfo   entity.Userinfo@2f04beaf
```

图 10-7

打开网址：http://localhost:8081/__listenerTest3/test2，控制台输出信息如图 10-8 所示。

```
valueUnbound有反绑发生： userinfo   entity.Userinfo@2f04beaf
```

图 10-8

## 10.4 HttpSessionListener接口的使用

HttpSessionListener 接口的作用是当创建或销毁 session 时该监听接口可以接收到相应的通知。创建 HttpSessionListener 接口的实现类 MyHttpSessionListener.java，代码如下：

```
package __listenerTest4;

import javax.servlet.http.HttpSessionEvent;
import javax.servlet.http.HttpSessionListener;

public class MyHttpSessionListener implements HttpSessionListener {

    @Override
    public void sessionCreated(HttpSessionEvent se) {
        System.out.println("sessionCreated");
    }

    @Override
    public void sessionDestroyed(HttpSessionEvent se) {
        System.out.println("sessionDestroyed");
    }

}
```

在 web.xml 文件中配置该监听，代码如下：

```
<listener>
    <listener-class>__listenerTest4.MyHttpSessionListener</listener-class
>
```

```
</listener>
```

创建 Servlet，作用是创建出一个新的 session，示例代码如下：

```
public class test extends HttpServlet {
    protected void doGet(HttpServletRequest request, HttpServletResponse
response)
            throws ServletException, IOException {
        request.getSession();
    }
}
```

继续创建 Servlet，作用是将当前关联的 session 进行销毁，示例代码如下：

```
public class test2 extends HttpServlet {
    protected void doGet(HttpServletRequest request, HttpServletResponse
response)
            throws ServletException, IOException {
        request.getSession().invalidate();
    }
}
```

打开网址：http://localhost:8081/__listenerTest4/test，控制台输出信息如图 10-9 所示。打开网址：http://localhost:8081/__listenerTest4/test2，控制台输出信息如图 10-10 所示。

图 10-9

sessionDestroyed

图 10-10

注解版写法代码如下：

```
@WebListener
public class MyListener implements HttpSessionListener {
```

## 10.5 ServletContextAttributeListener接口的使用

ServletContextAttributeListener 接口的作用是当向 ServletContext 中添加、删除或更新数据时，该监听接口可以接收到相应的通知。创建 ServletContextAttributeListener 接口的实现类 MyServletContextAttributeListener.java，代码如下：

```
package __listenerTest5;

import javax.servlet.ServletContextAttributeEvent;
import javax.servlet.ServletContextAttributeListener;

public class MyServletContextAttributeListener implements
ServletContextAttributeListener {

    @Override
    public void attributeAdded(ServletContextAttributeEvent scae) {
        System.out.println("attributeAdded " + scae.getName() + " " +
scae.getValue());
    }

    @Override
    public void attributeRemoved(ServletContextAttributeEvent scae) {
        System.out.println("attributeRemoved " + scae.getName() + " " +
scae.getValue());

    }

    @Override
    public void attributeReplaced(ServletContextAttributeEvent scae) {
        System.out.println("attributeReplaced " + scae.getName() + " " +
scae.getValue() + " "
                + scae.getServletContext().getAttribute(scae.getName()));
    }

}
```

下一步要把这个监听类在 web.xml 文件中进行注册，配置代码如下：

```
<listener>
<listener-class>__listenerTest5.MyServletContextAttributeListener</liste
ner-class>
</listener>
```

创建向 SevletContext 中添加数据的 Servlet，代码如下：

```
public class test extends HttpServlet {
    protected void doGet(HttpServletRequest request, HttpServletResponse
response)
```

```
            throws ServletException, IOException {
        request.getServletContext().setAttribute("applicationKey",
"applicationValue");
    }
}
```

创建向 SevletContext 中更新数据的 Servlet，代码如下：

```
public class test2 extends HttpServlet {
    protected void doGet(HttpServletRequest request, HttpServletResponse
response)
            throws ServletException, IOException {
        request.getServletContext().setAttribute("applicationKey",
"newApplicationValue");
    }
}
```

创建从 SevletContext 中删除数据的 Servlet，代码如下：

```
public class test3 extends HttpServlet {
    protected void doGet(HttpServletRequest request, HttpServletResponse
response)
            throws ServletException, IOException {
        request.getServletContext().removeAttribute("applicationKey");
    }
}
```

打开网址：http://localhost:8081/__listenerTest5/test，控制台输出信息如图 10-11 所示。

```
attributeAdded applicationKey applicationValue
```

图 10-11

打开网址：http://localhost:8081/__listenerTest5/test2，控制台输出信息如图 10-12 所示。

```
attributeReplaced applicationKey applicationValue newApplicationValue
```

图 10-12

打开网址：http://localhost:8081/__listenerTest5/test3，控制台输出信息如图 10-13 所示。

```
attributeRemoved applicationKey newApplicationValue
```

图 10-13

注解版的写法如下：

```
@WebListener
public class MyListener implements ServletContextAttributeListener {
```

## 10.6 ServletContextListener接口的使用

ServletContextListener 接口的作用是当创建或销毁 ServletContext 时，该接口可以接收到相应的通知。创建 ServletContextListener 接口的实现类 MyServletContextListener.java，代码如下：

```
package __listenerTest6;

import javax.servlet.ServletContextEvent;
import javax.servlet.ServletContextListener;

public class MyServletContextListener implements ServletContextListener {
    @Override
    public void contextInitialized(ServletContextEvent sce) {
        System.out.println("contextInitialized");
    }

    @Override
    public void contextDestroyed(ServletContextEvent sce) {
        System.out.println("contextDestroyed");
    }
}
```

在 web.xml 文件中配置该监听接口，配置代码如下：

```
<listener>
    <listener-class>__listenerTest6.MyServletContextListener</listener-cl
ass>
</listener>
```

Tomcat 启动时创建 ServletContext 对象，控制台输出信息如图 10-14 所示。Tomcat

关闭时销毁 ServletContext 对象，控制台输出信息如图 10-15 所示。

```
信息: At least one JAR
contextInitialized
```

图 10-14

```
信息: ContextListener: contextDestroyed()
contextDestroyed
```

图 10-15

ServletContextListener 接口在实际的软件项目中十分有用，在 Tomcat 启动的时候做一些初始化的工作，web.xml 文件示例代码如下：

```
<listener>
    <listener-class>mylistener.MyListener</listener-class>
</listener>

<context-param>
    <param-name>initParamName</param-name>
    <param-value>c:\mydir\my.xml</param-value>
</context-param>
```

ServletContextListener 接口实现类 MyListener 的代码如下：

```
package mylistener;

import javax.servlet.ServletContextEvent;
import javax.servlet.ServletContextListener;

public class MyListener implements ServletContextListener {
    public void contextDestroyed(ServletContextEvent sce) {
        System.out.println("contextDestroyed");
    }

    public void contextInitialized(ServletContextEvent sce) {
        System.out.println(
                "contextInitialized              initParamName="               +
sce.getServletContext().getInitParameter("initParamName"));
    }
}
```

Tomcat 在启动时可以读取到<context-param>配置中的参数值，如图 10-16 所示。我们可以在<param-value></param-value>中配置 XML 文件所在路径，在 Tomcat 启动的时候就解析此 XML 文

```
信息: Starting Servlet engine: [Apache Tomcat/9.(
contextInitialized initParamName=c:\mydir\my.xml
九月 18, 2019 3:58:47 下午 org.apache.catalina.st
```

图 10-16

件，此种用法就是 SSM 或 SSH 整合时使用的方法。注解的写法代码如下：

```
@WebListener
public class MyListener implements ServletContextListener {
```

那么，如何用注解的写法实现下面 XML 配置代码中初始化参数的效果呢？代码如下：

```
<context-param>
    <param-name>name</param-name>
    <param-value>value</param-value>
</context-param>
```

Listener 类的代码如下：

```
package mylistener;

import javax.servlet.ServletContextEvent;
import javax.servlet.ServletContextListener;
import javax.servlet.annotation.WebListener;

@WebListener
public class MyServletContextListener implements ServletContextListener {
    public void contextDestroyed(ServletContextEvent sce) {
    }

    public void contextInitialized(ServletContextEvent sce) {
        sce.getServletContext().setInitParameter("publicKey",
"publicValue");

System.out.println(sce.getServletContext().getInitParameter("publicKey")
);
    }
}
```

以下两个 Servlet 类都可以获取公共的参数值，代码如下：

```
@WebServlet("/test1")
public class test1 extends HttpServlet {
    protected void doGet(HttpServletRequest request, HttpServletResponse
response)
```

```
            throws ServletException, IOException {
        System.out.println("test1                    "                    +
this.getServletContext().getInitParameter("publicKey"));
    }
}

@WebServlet("/test2")
public class test2 extends HttpServlet {
    protected void doGet(HttpServletRequest request, HttpServletResponse
response)
            throws ServletException, IOException {
        System.out.println("test2                    "                    +
this.getServletContext().getInitParameter("publicKey"));
    }
}
```

## 10.7　ServletRequestAttributeListener接口的使用

ServletRequestAttributeListener 接口的作用是当向请求中添加、删除或更新数据时，该监听接口可以接收到相应的通知。创建 ServletRequestAttributeListener 接口的实现类 MyServletRequestAttributeListener.java，代码如下：

```
package  listenerTest7;

import javax.servlet.ServletRequestAttributeEvent;
import javax.servlet.ServletRequestAttributeListener;

public      class      MyServletRequestAttributeListener      implements
ServletRequestAttributeListener {

    @Override
    public void attributeAdded(ServletRequestAttributeEvent srae) {
        System.out.println("attributeAdded " + srae.getName() + " " +
srae.getValue());
    }

    @Override
    public void attributeRemoved(ServletRequestAttributeEvent srae) {
        System.out.println("attributeRemoved " + srae.getName() + " " +
srae.getValue());
    }
```

```
    @Override
    public void attributeReplaced(ServletRequestAttributeEvent srae) {
        System.out.println("attributeReplaced " + srae.getName() + " " +
srae.getValue() + " "
                + srae.getServletRequest().getAttribute(srae.getName()));
    }

}
```

下一步要把这个监听类在 web.xml 文件中进行注册，配置代码如下：

```
    <listener>
    <listener-class>__listenerTest7.MyServletRequestAttributeListener</li
stener-class>
    </listener>
```

创建向请求中添加数据的 Servlet，代码如下：

```
public class test extends HttpServlet {
    protected void doGet(HttpServletRequest request, HttpServletResponse
response)
            throws ServletException, IOException {
        request.setAttribute("requestKey", "requestValue");
    }
}
```

创建向请求中更新数据的 Servlet，代码如下：

```
public class test2 extends HttpServlet {
    protected void doGet(HttpServletRequest request, HttpServletResponse
response)
            throws ServletException, IOException {
        request.setAttribute("requestKey", "requestValue");
        request.setAttribute("requestKey", "newRequestValue");
    }
}
```

创建从请求中删除数据的 Servlet，代码如下：

```
public class test3 extends HttpServlet {
    protected void doGet(HttpServletRequest request, HttpServletResponse
response)
```

```
            throws ServletException, IOException {
        request.setAttribute("requestKey", "requestValue");
        request.removeAttribute("requestKey");
    }
}
```

打开网址：http://localhost:8081/__listenerTest7/test，控制台输出信息如图 10-17 所示。

```
attributeReplaced org.apache.catalina.ASYNC_SUPPORTED true false
attributeAdded requestKey requestValue
```

图 10-17

打开网址：http://localhost:8081/__listenerTest7/test2，控制台输出信息如图 10-18 所示。

```
attributeReplaced org.apache.catalina.ASYNC_SUPPORTED true false
attributeAdded requestKey requestValue
attributeReplaced requestKey requestValue newRequestValue
```

图 10-18

打开网址：http://localhost:8081/__listenerTest7/test3，控制台输出信息如图 10-19 所示。

```
attributeReplaced org.apache.catalina.ASYNC_SUPPORTED true false
attributeAdded requestKey requestValue
attributeRemoved requestKey requestValue
```

图 10-19

注解的使用方式代码如下：

```
@WebListener
public class MyListener implements ServletRequestAttributeListener {
```

## 10.8 ServletRequestListener接口的使用

ServletRequestListener 接口的作用是当创建或销毁请求时该监听接口可以接收到相应的通知。创建 ServletRequestListener 接口的实现类 MyServletRequestListener.java，代码如下：

```
package __listenerTest8;

import javax.servlet.ServletRequestEvent;
import javax.servlet.ServletRequestListener;

public class MyServletRequestListener implements ServletRequestListener {
```

```
    @Override
    public void requestDestroyed(ServletRequestEvent sre) {
        System.out.println("requestDestroyed");
    }

    @Override
    public void requestInitialized(ServletRequestEvent sre) {
        System.out.println("requestInitialized");
    }
}
```

在 web.xml 文件中配置该监听，代码如下：

```
<listener>
    <listener-class>__listenerTest8.MyServletRequestListener</listener-cl
ass>
</listener>
```

```
Tomcat v7.0 Server at localhost [Apache To
requestInitialized
requestDestroyed
```

图 10-20

显示项目中的 abc.html 文件，打开网址：http://localhost:8081/__listenerTest8/abc.html，控制台输出信息如图 10-20 所示。注解的写法代码如下：

```
@WebListener
public class MyListener implements ServletRequestListener {
```

## 10.9 HttpSessionIdListener接口的使用

HttpSessionIdListener 接口的作用是当 HttpSession 的 ID 改变时该监听接口可以接收到相应的通知。创建 HttpSessionIdListener 接口的实现类 MyHttpSessionIdListener.java，代码如下：

```
package mylistener;

import javax.servlet.http.HttpSessionEvent;
import javax.servlet.http.HttpSessionIdListener;

public class MyHttpSessionIdListener implements HttpSessionIdListener {
    @Override
    public void sessionIdChanged(HttpSessionEvent se, String oldSessionId)
{
```

```
        System.out.println("sessionIdChanged        oldSessionId="        +
oldSessionId);
    }
}
```

在 web.xml 文件中配置该监听类，代码如下：

```
<listener>
    <listener-class>mylistener.MyHttpSessionIdListener</listener-class>
</listener>
```

创建名称为 test1 的 Servlet 代码如下：

```
package controller;

import java.io.IOException;

import javax.servlet.ServletException;
import javax.servlet.http.HttpServlet;
import javax.servlet.http.HttpServletRequest;
import javax.servlet.http.HttpServletResponse;
import javax.servlet.http.HttpSession;

public class test1 extends HttpServlet {
    protected void doGet(HttpServletRequest request, HttpServletResponse
response)
            throws ServletException, IOException {
        HttpSession session = request.getSession();
        session.setAttribute("a", "avalue");
        System.out.println("A " + session.getId());
        request.changeSessionId();
        System.out.println("B " + session.getId());
    }
}
```

创建名称为 test2 的 Servlet 代码如下：

```
package controller;

import java.io.IOException;

import javax.servlet.ServletException;
```

```
import javax.servlet.http.HttpServlet;
import javax.servlet.http.HttpServletRequest;
import javax.servlet.http.HttpServletResponse;

public class test2 extends HttpServlet {
    protected void doGet(HttpServletRequest request, HttpServletResponse
response)
            throws ServletException, IOException {
        System.out.println(request.getSession().getId() + " getValue=" +
request.getSession().getAttribute("a"));
    }
}
```

首先执行 test1，控制台输出信息如下：

```
A DB3DAB5AA9DB2BF876F7FFE180D8BA41
sessionIdChanged oldSessionId=DB3DAB5AA9DB2BF876F7FFE180D8BA41
B 2605E75D37FDB2CB4D44409AE6113B70
```

控制台输出的结果说明 session 中以 a 为 key 将对应的值 avalue 绑定到 sessionId 为 BA41 上，然后在 Servlet 中调用 request.changeSessionId()方法将 sessionId 由原来的 BA41 改成 3B70，在这个过程中仅仅是将 sessionId 值由原来的 BA41 改成 3B70。

再执行 test2，控制台完整的输出信息如下：

```
2605E75D37FDB2CB4D44409AE6113B70 getValue=avalue
```

从控制台输出的结果来看，是从 sessionId 为 3B70 上找到 key 为 a 对应的值 avalue 的。注解版示例代码如下：

```
@WebListener
public class MyListener implements HttpSessionIdListener {
```

## 10.10 使用HttpSessionListener接口实现在线人数统计

创建保存在线人数的工具类 OnlineList.java，代码如下：

```
package tools;

import java.util.ArrayList;
```

```
import java.util.List;

public class OnlineList {
    public static List list = new ArrayList();
}
```

创建 HttpSessionListener 接口的实现类 MyHttpSessionListener.java，代码如下：

```
package __listenerOnlineNum;

import javax.servlet.http.HttpSessionEvent;
import javax.servlet.http.HttpSessionListener;

import tools.OnlineList;

public class MyHttpSessionListener implements HttpSessionListener {

    @Override
    public void sessionCreated(HttpSessionEvent se) {
        OnlineList.list.add(se.getSession().getId());
    }

    @Override
    public void sessionDestroyed(HttpSessionEvent se) {
    }

}
```

需要在 web.xml 文件中配置该监听类，代码如下：

```
<listener>
    <listener-class>__listenerOnlineNum.MyHttpSessionListener</listener-c
lass>
</listener>
```

创建产生新的 session 的 Servlet 代码如下：

```
public class createSession extends HttpServlet {
    protected void doGet(HttpServletRequest request, HttpServletResponse
response)
            throws ServletException, IOException {
        request.getSession();
```

```
    }
}
```

创建显示在线人数的 Serlvet 代码如下：

```
public class test extends HttpServlet {
    protected void doGet(HttpServletRequest request, HttpServletResponse
response)
            throws ServletException, IOException {
        System.out.println("在线人数：" + OnlineList.list.size());
    }
}
```

在不同种类的浏览器中打开 3 次网址 http://localhost:8081/__listenerOnlineNum/createSession，模拟 3 个人在线的效果。然后再打开网址 http://localhost:8081/__listenerOnlineNum/test，控制台显示在线人数为 3，如图 10-21 所示。

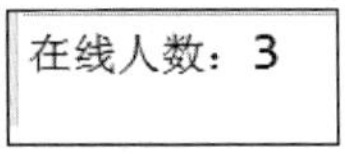

在线人数：3

图 10-21

由于一台电脑可以生成多个 HttpSession 对象，所以使用 HttpSessionListener 实现在线人数统计并不精确。

# 第 11 章 JSP-JSTL-EL 必备技术

JSP（Java Server Page）是一种服务端生成动态页面的技术，在 MVC 中属于 V 视图层。

## 11.1 JSP技术

### 11.1.1 使用 Servlet 生成网页

使用 PrintWriter 对象可以创建前台网页，示例代码如下：

```
package controller;

import java.io.IOException;
import java.io.PrintWriter;

import javax.servlet.ServletException;
import javax.servlet.http.HttpServlet;
import javax.servlet.http.HttpServletRequest;
import javax.servlet.http.HttpServletResponse;

public class test1 extends HttpServlet {
    protected void doGet(HttpServletRequest request, HttpServletResponse
response)
            throws ServletException, IOException {

        StringBuffer buffer = new StringBuffer();
        buffer.append("<table border='1'>");
```

```
        buffer.append("<tr>");
        buffer.append("<td>第一行第一列");
        buffer.append("</td>");
        buffer.append("<td>第一行第二列");
        buffer.append("</td>");
        buffer.append("</tr>");

        buffer.append("<tr>");
        buffer.append("<td>第二行第一列");
        buffer.append("</td>");
        buffer.append("<td>第二行第二列");
        buffer.append("</td>");
        buffer.append("</tr>");

        buffer.append("</table>");

        response.setContentType("text/html");
        response.setCharacterEncoding("utf-8");
        PrintWriter out = response.getWriter();
        out.print(buffer.toString());
        out.flush();
        out.close();
    }

}
```

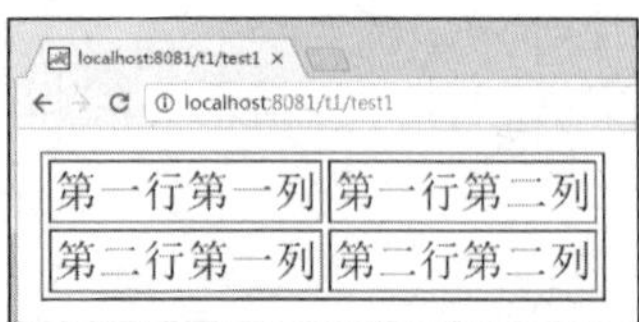


图 11-1

程序运行效果如图 11-1 所示。

虽然使用 Servlet 能生成前台网页，但过程还是比较烦琐的，需要在 Servlet 中拼接 HTML 代码再输出，十分不方便。JSP 简化了这个过程，下面来看看 JSP 是如何解决输出界面复杂问题的。

### 11.1.2 使用 JSP 生成网页

创建 JSP 文件，代码如下：

```
<%@ page language="java" contentType="text/html; charset=utf-8"
pageEncoding="utf-8" %>
 <!DOCTYPE HTML PUBLIC "-//W3C//DTD HTML 4.01 Transitional//EN"
"http://www.w3.org/TR/html4/loose.dtd">
```

```
<html>
    <head>
        <meta http-equiv="Content-Type" content="text/html; charset=utf-8">
        <title>Insert title here</title>
    </head>
    <body>
        jspPage:
        <br/>
        <table border='1'>
            <tr>
                <td>
                    第一行第一列
                </td>
                <td>
                    第一行第二列
                </td>
            </tr>
            <tr>
                <td>
                    第二行第一列
                </td>
                <td>
                    第二行第二列
                </td>
            </tr>
        </table>
    </body>
</html>
```

程序运行效果如图 11-2 所示。看来使用 JSP 也能实现前台用户界面的制作，而 JSP 和普通的 HTML 在最终显示的效果上并没有什么本质的区别。使用 HTML 文件也能显示 table 表格，示例代码如下：

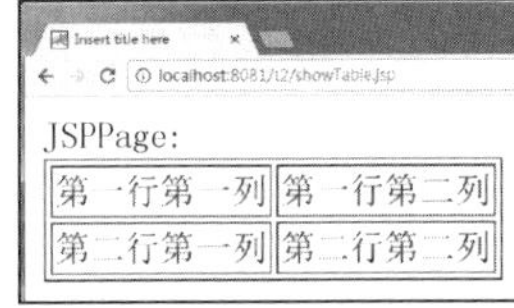


图 11-2

```
<!DOCTYPE HTML PUBLIC "-//W3C//DTD HTML 4.01 Transitional//EN"
"http://www.w3.org/TR/html4/loose.dtd">
<html>
    <head>
        <meta http-equiv="Content-Type" content="text/html; charset=utf-8">
        <title>Insert title here</title>
```

```
    </head>
    <body>
        htmlPage:
        <br/>
        <table border='1'>
            <tr>
                <td>
                    第一行第一列
                </td>
                <td>
                    第一行第二列
                </td>
            </tr>
            <tr>
                <td>
                    第二行第一列
                </td>
                <td>
                    第二行第二列
                </td>
            </tr>
        </table>
    </body>
</html>
```

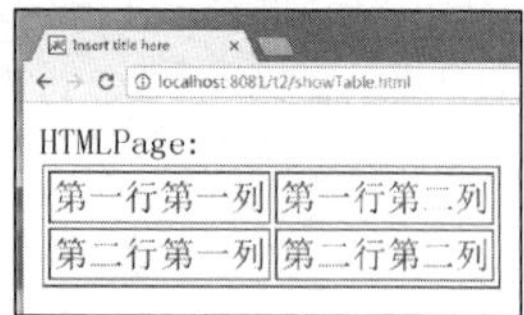


图 11-3

HTML 文件运行效果如图 11-3 所示。那么，HTML 文件和 JSP 文件到底有什么区别呢？其实区别很小，但影响非常大，那就是可以在 JSP 文件中写 Java 代码，就能让 JSP 文件显示的内容具有动态性，并可以通过 Java 代码对输出的内容进行控制，这是 HTML 文件做不到的。那么，如何在 JSP 文件中写 Java 程序呢？使用<%Java 程序代码%>代码段即可。

### 11.1.3 在 JSP 中执行 Java 程序

在 JSP 中运行 Java 程序的 JSP 文件源代码如下：

```
<%@ page language="java" contentType="text/html; charset=utf-8"
pageEncoding="utf-8" %>
<%@ page import="java.util.*" %>
 <!DOCTYPE html PUBLIC "-//W3C//DTD HTML 4.01 Transitional//EN"
"http://www.w3.org/TR/html4/loose.dtd">
```

```
<html>
    <head>
        <meta http-equiv="Content-Type" content="text/html; charset=utf-8">
        <title>Insert title here</title>
    </head>
    <body>
        jspPage:
        <br/>
        <table border='1'>
            <tr>
                <td>
                    第一行第一列
                </td>
                <td>
                    第一行第二列
                </td>
            </tr>
            <tr>
                <td>
                    第二行第一列
                </td>
                <td>
                    第二行第二列
                </td>
            </tr>
        </table>
        <br/>
        <br/>
        <%
        Date date = new Date();
        out.println(date+"-----能在 JSP 中写 Java 代码"); %>
    </body>
</html>
```

只要你在 JSP 文件的<%%>内部设计了 Java 代码，在打开 JSP 文件时就可以运行 Java 程序了，如图 11-4 所示。

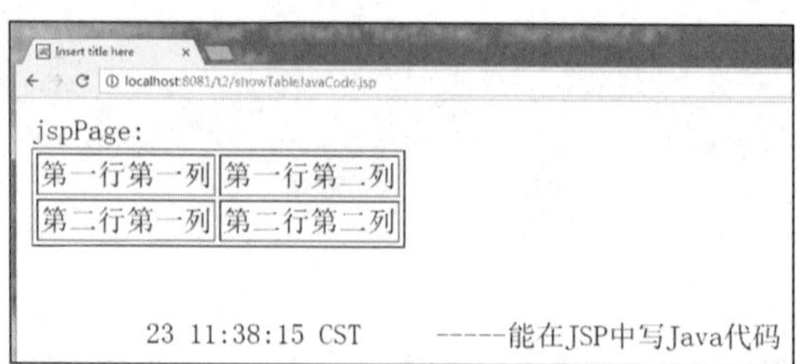


图 11-4

### 11.1.4 JSP 本质上是 Servlet

在最终运行时，JSP 会转换成一个 Servlet 类。JSP 文件示例代码如下：

```
<%@ page language="java" contentType="text/html; charset=utf-8"
    pageEncoding="utf-8"%>
<!DOCTYPE html PUBLIC "-//W3C//DTD HTML 4.01 Transitional//EN"
"http://www.w3.org/TR/html4/loose.dtd">
<html>
<head>
<meta http-equiv="Content-Type" content="text/html; charset=utf-8">
<title>Insert title here</title>
</head>
<body>
    我是静态的
    <br />
    <%
        out.println("我是动态的");
    %>
</body>
</html>
```

在执行 JSP 文件后，到指定文件夹中找到 Java 文件，可见在如图 11-5 所示的路径中生成了一个.java 文件。

› Administrator › AppData › Local › JetBrains › IntelliJIdea2020.1 › tomcat › _Default_(Template)_Project › work › Catalina › localhost › nomvc_war_exploded › org › apache › jsp

index_jsp.class
index_jsp.java

图 11-5

笔者的文件地址如下：

C:\Users\Administrator\AppData\Local\JetBrains\IntelliJIdea2020.1\tomcat_Default_(Template)_Project\work\Catalina\localhost\nomvc_war_exploded\org\apache\jsp

该地址可以参考 Tomcat 启动时控制台的输出信息：

```
Using CATALINA_BASE:
"C:\Users\Administrator\AppData\Local\JetBrains\IntelliJIdea2020.1\tomca
t\_Default_(Template)_Project"
```

index_jsp.java 文件中的所有源代码如下：

```
package controller;

import javax.servlet.*;
import javax.servlet.http.*;
import javax.servlet.jsp.*;

public final class index_jsp extends org.apache.jasper.runtime.HttpJspBase
    implements org.apache.jasper.runtime.JspSourceDependent,
                 org.apache.jasper.runtime.JspSourceImports {

  private static final JspFactory _jspxFactory =
          JspFactory.getDefaultFactory();

  private static java.util.Map<String, Long> _jspx_dependants;

  private static final java.util.Set<String> _jspx_imports_packages;

  private static final java.util.Set<String> _jspx_imports_classes;

  static {
    _jspx_imports_packages = new java.util.HashSet<>();
    _jspx_imports_packages.add("javax.servlet");
    _jspx_imports_packages.add("javax.servlet.http");
    _jspx_imports_packages.add("javax.servlet.jsp");
    _jspx_imports_classes = null;
  }

  private volatile javax.el.ExpressionFactory _el_expressionfactory;
  private volatile org.apache.tomcat.InstanceManager _jsp_instancemanager;

  public java.util.Map<String, Long> getDependants() {
```

```
    return _jspx_dependants;
  }

  public java.util.Set<String> getPackageImports() {
    return _jspx_imports_packages;
  }

  public java.util.Set<String> getClassImports() {
    return _jspx_imports_classes;
  }

  public javax.el.ExpressionFactory _jsp_getExpressionFactory() {
    if (_el_expressionfactory == null) {
      synchronized (this) {
        if (_el_expressionfactory == null) {
          _el_expressionfactory                                                      =
_jspxFactory.getJspApplicationContext(getServletConfig().getServletConte
xt()).getExpressionFactory();
        }
      }
    }
    return _el_expressionfactory;
  }

  public org.apache.tomcat.InstanceManager _jsp_getInstanceManager() {
    if (_jsp_instancemanager == null) {
      synchronized (this) {
        if (_jsp_instancemanager == null) {
          _jsp_instancemanager                                                       =
org.apache.jasper.runtime.InstanceManagerFactory.getInstanceManager(getS
ervletConfig());
        }
      }
    }
    return _jsp_instancemanager;
  }

  public void _jspInit() {
  }

  public void _jspDestroy() {
```

```
  }

  public void _jspService(final HttpServletRequest request, final
HttpServletResponse response)
      throws java.io.IOException, ServletException {

    if (!DispatcherType.ERROR.equals(request.getDispatcherType())) {
      final String _jspx_method = request.getMethod();
      if ("OPTIONS".equals(_jspx_method)) {
        response.setHeader("Allow","GET, HEAD, POST, OPTIONS");
        return;
      }
      if (!"GET".equals(_jspx_method) && !"POST".equals(_jspx_method)
&& !"HEAD".equals(_jspx_method)) {
        response.setHeader("Allow","GET, HEAD, POST, OPTIONS");
        response.sendError(HttpServletResponse.SC_METHOD_NOT_ALLOWED, "JSP
只允许 GET、POST 或 HEAD。Jasper 还允许 OPTIONS");
        return;
      }
    }

    final PageContext pageContext;
    HttpSession session = null;
    final ServletContext application;
    final ServletConfig config;
    JspWriter out = null;
    final Object page = this;
    JspWriter _jspx_out = null;
    PageContext _jspx_page_context = null;


    try {
      response.setContentType("text/html;charset=UTF-8");
      pageContext = _jspxFactory.getPageContext(this, request, response,
            null, true, 8192, true);
      _jspx_page_context = pageContext;
      application = pageContext.getServletContext();
      config = pageContext.getServletConfig();
      session = pageContext.getSession();
      out = pageContext.getOut();
      _jspx_out = out;
```

```
    out.write("\n");
    out.write("\n");
    out.write("<html>\n");
    out.write("    <head>\n");
    out.write("        <title>$Title$</title>\n");
    out.write("    </head>\n");
    out.write("    <body>\n");
    out.write("        我是静态的\n");
    out.write("        <br/>\n");
    out.write("        ");

          out.println("我是动态的");

    out.write("\n");
    out.write("    </body>\n");
    out.write("</html>\n");
   } catch (Throwable t) {
    if (!(t instanceof SkipPageException)){
      out = _jspx_out;
      if (out != null && out.getBufferSize() != 0)
        try {
          if (response.isCommitted()) {
            out.flush();
          } else {
            out.clearBuffer();
          }
        } catch (java.io.IOException e) {}
      if (_jspx_page_context != null) _jspx_page_
context.handlePageException(t);
      else throw new ServletException(t);
    }
   } finally {
    _jspxFactory.releasePageContext(_jspx_page_context);
   }
  }
}
```

从源代码中可以发现，JSP 页面上生成的所有内容最终都要使用 out 对象进行输出。index_jsp.java 类的父类是 org.apache.jasper.runtime.HttpJspBase，而 HttpJspBase（org.apache.jasper.runtime）的父类却是 HttpServlet，如图 11-6 所示。根据继承结构可知，JSP 文件最终要转化成一个 Servlet 类。

图 11-6

## 11.1.5　JSP 文件的内容

JSP 文件的内容主要有 HTML 标签、css 样式、JavaScript 脚本和 Java 代码等，示例代码如下：

```
<%@    page    language="java"    contentType="text/html;    charset=utf-8"
pageEncoding="utf-8" %>
<%@ page import="java.util.*" %>
 <!DOCTYPE   html   PUBLIC   "-//W3C//DTD   HTML   4.01   Transitional//EN"
"http://www.w3.org/TR/html4/loose.dtd">
<html>
   <head>
      <meta http-equiv="Content-Type" content="text/html; charset=utf-8">
      <title>Insert title here</title>
      <style type="text/css">
         .mySpanStyle {
            color: blue;
            font-size: 50px;
         }
      </style>
      <script type="text/javascript">
         function testMethod(){
            alert("testMethod run !");
         }
      </script>
   </head>
   <body>
      我是文本的！
      <br/>
      <a href="http://www.baidu.com">我是 HTML 中的 a 标签</a>
      <br/>
      <span style="color: red">我是 span1，可以加样式</span>
```

```
        <br/>
        <span class="mySpanStyle">我是 span1，可以加样式</span>
        <br/>
        <a href="javascript:testMethod()">JSP 中可以有 javascript 代码</a>
        <br/>
        <br/>
        <%
        for (int i = 0; i < 4; i++) {
        out.println((i + 1) + "__我是 Java 代码段，JSP 内置对象 out 负责输出数据<br/>");
        } %>
        <br/>
        <br/>
        <%="我是直接被打印的，是 Java 表达式" %>
        <br/>
        <%=1 + 100 %>
        <br/>
        <%=true && false %>
        <br/>
        <br/>
        <%
        out.println("getParameter=" + request.getParameter("username")); %>
    </body>
</html>
```

程序运行后的效果如图 11-7 所示。

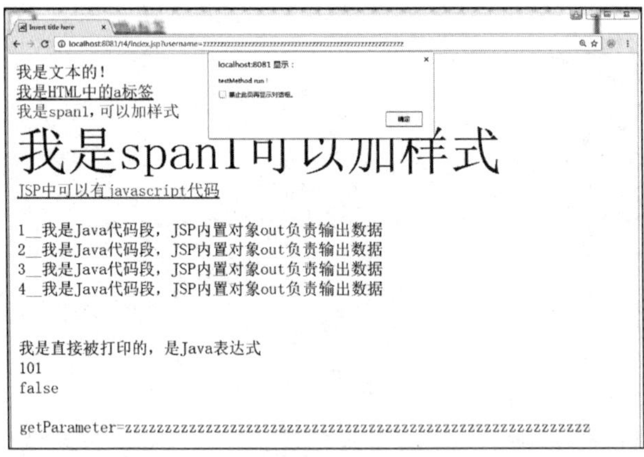


图 11-7

### 11.1.6 JSP 的指令

JSP 的指令可以定义 JSP 文件中的一些特性，比如设置编码、导入 JSTL 标签、导包、在 JSP 页面中导入其他资源等功能。JSP 的指令格式如下：

```
<%@ 指令名称 指令属性="指令属性值"%>
```

#### 1. page指令的import属性

该属性的主要作用是在 JSP 中导入指定类，以便在<%%>代码块中引用 Java 类并使用。示例指令代码如下：

```
<%@ page import="java.util.*"%>
```

如果导入多个包，则使用逗号“,”进行分隔，示例如下：

```
<%@ page import="entity.*,java.util.*"%>
```

#### 2. page指令的contentType属性

该属性的主要作用是定义服务器端返回 HTML 代码的编码类型。示例指令代码如下：

```
<%@ page language="java" contentType="text/html; charset=iso-8859-1"
   pageEncoding="utf-8"%>
```

JSP 运行后，在生成的 Servlet 源代码中创建了如图 11-8 所示的代码。

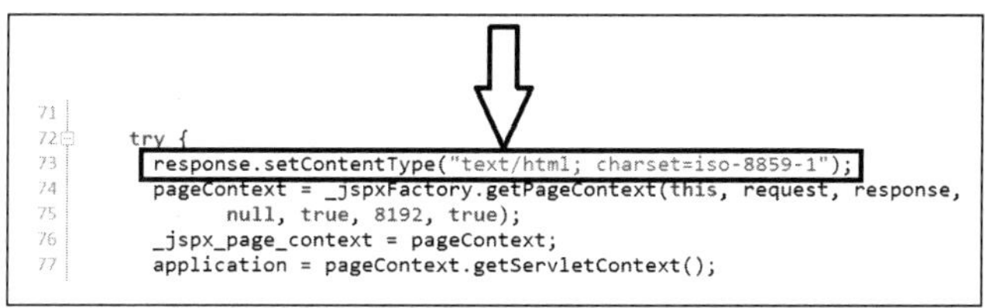


图 11-8

#### 3. page指令的pageEncoding属性

示例指令代码如下：

```
<%@ page language="java" pageEncoding="GBK"%>
```

该属性的主要作用是定义 Web 服务器读取 JSP 文件的编码类型，其设置关系到汉字是不是乱码的问题，更具体来讲，是先将 JSP 文件以字节输入流 InputStream 的形式读入内存，再使用 GBK 编码结合 InputStreamReader 将其转换成字符输入流 Reader，其中 InputStreamReader 类的父类是 Reader。这些知识点可以通过调试 Tomcat 源代码的方式来证明，即在源代码中搜索“pageEncoding”找到线索。通过调试源代码可以发现，编码是在 ParserController.java 类中进行处理的，如图 11-9 所示。

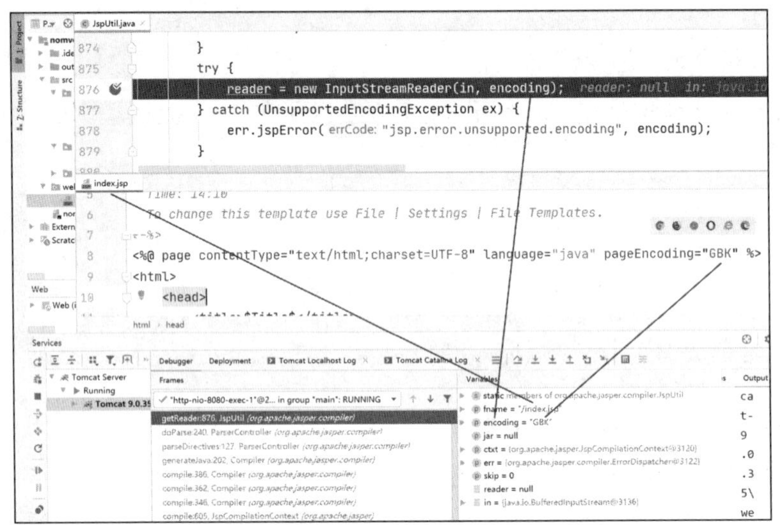

图 11-9

注意：在调试源代码时，getReader()方法被执行两次才可以看到 GBK 编码，第一次是以 iso-8859-1 编码读取 JSP 文件，形成字符输入流。之所以使用 iso-8859-1，是因为该编码最具有通用性，即使用 iso-8859-1 编码把全英文的配置信息 pageEncoding="GBK" 解析出来。第二次使用 GBK 编码将 iso-8859-1 编码的字节输入流转换成 GBK 编码的字符输入流。

### 11.1.7　几种指令的区别

如果使用 EditPlus 查看 HTML 文件的编码类型，则需要进行设置，取消“工具”菜单“首选项”中的“启动时加载上次正在编辑的文件”打钩。

（1）JSP 文件中的<%@ page contentType=""%>指令会在 Servlet 中生成 response.setContentType()代码，该指令在 JSP 文件中使用。

（2）JSP 文件中<%@ page pageEncoding=""%>指令的主要作用是定义 Web 服务器中 IO 流读取 JSP 文件的编码类型。该指令在 JSP 文件中使用。

（3）<meta charset="">定义 HTML 文件的编码类型，并不定义 JSP 文件的编码类型。在 IDEA 工具中，我们可以使用<meta charset="utf-8">对 HTML 文件进行配置，内容上一定要添加一些新的中文，并且执行保存时开发工具会将 HTML 文件的编码类型指定为 utf-8 编码（可以使用 EditPlus 软件查看 HTML 文件的编码类型）。反之，如果改成<meta charset="GBK">，则内容上一定要添加一些新的中文并且执行保存时会将 HTML 文件的编码类型指定为 GBK 编码。也就是说，IDEA 在保存 HTML 文件前，会先取出<meta charset="">中的编码类型，然后按照指定的编码类型对 HTML 文件进行保存（记事本不具有此功能）。由于一个标准的 JSP 文件要有<%@ page contentType="text/html;charset=XXX" language="java" %>指令，因此如果在一个标准的 JSP 文件中使用<meta charset="">，则该配置会被忽略，优先使用<%@ page contentType= "text/html;charset=XXX" language= "java" %>中的编码类型。如果单独使用<%@ page contentType="text/html; charset=XXX" language="java"%>指令或单独使用<%@ page language="java" pageEncoding="XXX" %>指令，则在保存 JSP 文件时会将 JSP 文件的编码类型设置为 XXX。当 contentType 和 pageEncoding 属性都存在时，pageEncoding 优先。需要注意的是，Tomcat 在内部不处理 HTML 文件中的<meta charset="">编码类型，只是将 HTML 文件的 byte[]传给浏览器，浏览器接收到 byte[]后再解析<meta charset="">中的编码，最后根据<meta charset="">编码类型进行处理。

### 11.1.8　验证 Servlet 使用 write()方法和 print()方法进行输出

JSP 在内部会转换成 Servlet，而在 Servlet 中会调用 write()方法和 print()方法输出相关的数据。

#### 1. 在Servlet内部执行write()方法或print()方法生成视图

在将 JSP 文件转换成 Servlet 后，内部会使用不同的方法进行输出。

（1）<%=%>和<%out.print();%>使用 print()方法输出。

（2）普通文字和 HTML 代码使用 write()方法输出。

示例代码如下：

```
<%@ page language="java" contentType="text/html; charset=utf-8"%>
<!DOCTYPE html PUBLIC "-//W3C//DTD HTML 4.01 Transitional//EN"
"http://www.w3.org/TR/html4/loose.dtd">
<html>
<body>
```

```
    <%="中国 1"%>
    <br />
    <%
        out.println("中国 2");
    %>
    <br /> 中国 3
</body>
</html>
```

程序运行后，Servlet 源代码如图 11-10 所示。

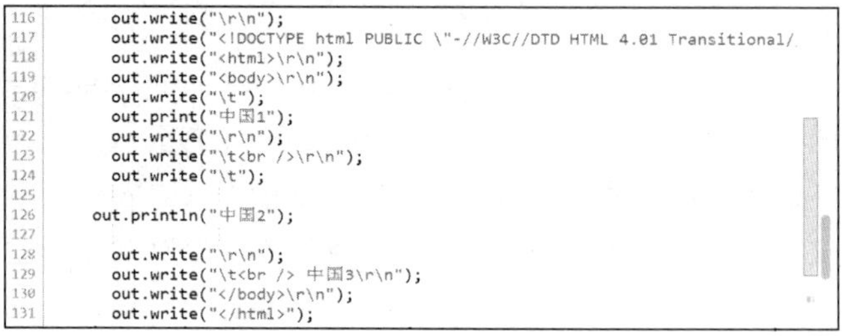

```
        out.write("\r\n");
        out.write("<!DOCTYPE html PUBLIC \"-//W3C//DTD HTML 4.01 Transitional/
        out.write("<html>\r\n");
        out.write("<body>\r\n");
        out.write("\t");
        out.print("中国1");
        out.write("\r\n");
        out.write("\t<br />\r\n");
        out.write("\t");

      out.println("中国2");

        out.write("\r\n");
        out.write("\t<br /> 中国3\r\n");
        out.write("</body>\r\n");
        out.write("</html>");
```

图 11-10

### 2. write()方法和print()方法的区别

在介绍 write()方法和 print()方法的区别之前，我们先来看看 write()方法和 print()方法的源代码，如下：

```
private java.io.Writer writer;
@Override
public void print(String s) throws IOException {
    if (s == null) s = "null";
    if (writer != null) {
        writer.write(s);
    } else {
        write(s);
    }
}

public void write(String str) throws IOException {
    write(str, 0, str.length());
}
```

（1）print()方法分析：在 print()方法中会对 s 参数进行 NULL 值的判断，因为<%=%>

和<%out.print();%>可以将 NULL 值传入 print()方法，所以就要使用 if 语句判断 s 参数值的情况，是 NULL 值还是非 NULL 值。

（2）write()方法：在使用 write()方法输出时，并没有使用 if 语句进行判断。如果 str 参数的值是 NULL，则会因为调用 str.length()方法而导致空指针异常。但在 JSP 页面中，因为普通文字和 HTML 代码没有 NULL 值，所以不会出现空指针异常的情况。

在被调用时，write()方法和 print()方法执行的区别要从两个角度进行总结。

（1）JSP 输出（隐式执行）：JSP 中的<%=%>和<%out.print();%>会调用 print()方法输出，而普通文字和 HTML 代码使用 write()方法输出，JSP 在内部调用这两种方法时都不会输出空指针异常。

（2）显式执行：当以显式的方式输出 NULL 值时，这两种方法是有区别的。write()方法会抛出异常，print()方法会对 NULL 值输出“NULL”字符串。

使用 write()方法的示例代码如下：

```
public class test1 extends HttpServlet {
    protected void doGet(HttpServletRequest request, HttpServletResponse
response)
            throws ServletException, IOException {
        String myname = null;
        PrintWriter out = response.getWriter();
        out.write(myname);
        out.flush();
        out.close();
    }
}
```

运行 Servlet，在控制台出现空指针异常，如下：

```
java.lang.NullPointerException
at org.apache.catalina.connector.CoyoteWriter.write(CoyoteWriter.java:180)
at controller.test1.doGet(test1.java:16)
```

使用 print()方法的示例代码如下：

```
public class test2 extends HttpServlet {
    protected void doGet(HttpServletRequest request, HttpServletResponse
response)
            throws ServletException, IOException {
```

```
        String myname = null;
        PrintWriter out = response.getWriter();
        out.print(myname);
        out.flush();
        out.close();
    }
}
```

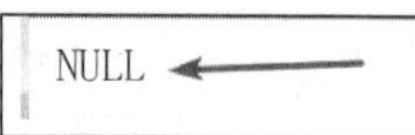

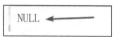

图 11-11

程序运行后，页面上输出“NULL”字符串，如图 11-11 所示。

### 11.1.9 从 Servlet 转发到 JSP 文件

在前面讲解转发时，都是从 Servlet 转发到 Servlet，下面将演示从 Servlet 转发到 JSP 文件，即真正实现 MVC 开发方式。

（1）M：代表数据模型或业务模型，这里 M 作为业务模型。

（2）V：将 JSP 文件或 HTML 文件作为 V 层，负责显示界面。

（3）C：将 Servlet 作为 C 层，实现流程控制。

Servlet 示例代码如下：

```
public class test1 extends HttpServlet {
    protected void doGet(HttpServletRequest request, HttpServletResponse
response)
            throws ServletException, IOException {
        List list = new ArrayList();
        list.add("中国 1");
        list.add("中国 2");
        list.add("中国 3");
        request.setAttribute("list", list);
        request.getRequestDispatcher("index4.jsp").forward(request,
response);
    }
}
```

index4.jsp 文件的代码如下：

```
<%@ page language="java" contentType="text/html; charset=utf-8"
    pageEncoding="utf-8"%>
<%@ page import="java.util.*"%>
<!DOCTYPE html PUBLIC "-//W3C//DTD HTML 4.01 Transitional//EN"
```

```
"http://www.w3.org/TR/html4/loose.dtd">
<html>
<body>
    <%
        List list = (List) request.getAttribute("list");
        for (int i = 0; i < list.size(); i++) {
            out.println(list.get(i) + "<br/>");
        }
    %>
</body>
</html>
```

打开网址 http://localhost:8081/t5/test1，控制层成功转发到 index4.jsp 文件并显示 List 中的数据，如图 11-12 所示。在掌握了 JSP 技术以后，我们不要在 Servlet 中使用 print()方法输出 HTML 代码来生成用户界面，因为那是在破坏 MVC 分层结构，会使代码后期的维护变得更难。

中国1
中国2
中国3

图 11-12

## 11.1.10 Java 代码块<%%>和<%!%>的区别

在 JSP 的<%%>和<%!%>代码块中都可以写 Java 代码，但两者有非常大的区别，示例代码如下：

```
<body>
    <%
        String username = "我是中国人";
    %>
</body>
```

JSP 文件生成的核心 Java 代码如下：

```
 public   void   _jspService(final   javax.servlet.http.HttpServletRequest
request, final javax.servlet.http.HttpServletResponse response)
{
        String username = "我是中国人";
}
```

可见，在源代码的_jspService()方法中创建了一个局部变量 username，这就是<%%>语句块的特点，即<%%>语句块中的代码要放在_jspService()方法中。根据这个特性，<%%>语句块中不允许声明方法，因为在 Java 语言中方法不允许嵌套方法，错误的代码如图 11-13 所示。

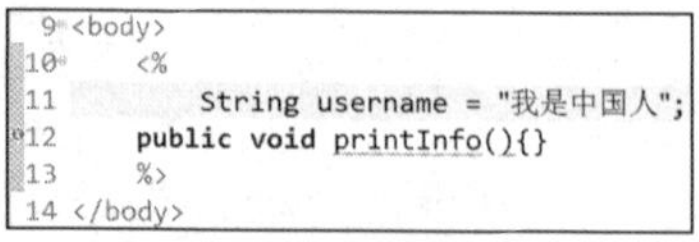

```
<body>
    <%
        String username = "我是中国人";
    public void printInfo(){}
    %>
</body>
```

图 11-13

如果使用<%!%>代码块，则程序在源代码中会声明实例变量和创建新的方法，示例代码如下：

```
<body>
    <%!
String username = "我是中国人实例变量";
    public void printInfo() {
    }%>
</body>
```

JSP 生成的核心源代码如下：

```
public final class test9_jsp extends org.apache.jasper.runtime.HttpJspBase
   implements org.apache.jasper.runtime.JspSourceDependent,
               org.apache.jasper.runtime.JspSourceImports {
String username = "我是中国人实例变量";
    public void printInfo() {
    }
    ......
```

可见，在 test9_jsp 类中创建了新的实例变量 username，以及添加了新的 printInfo()方法。根据 Java 的语法规则，我们可以分析出图 11-14 中的代码是错误的。那么，如何才能调用新添加的这个方法呢？使用如下代码即可：

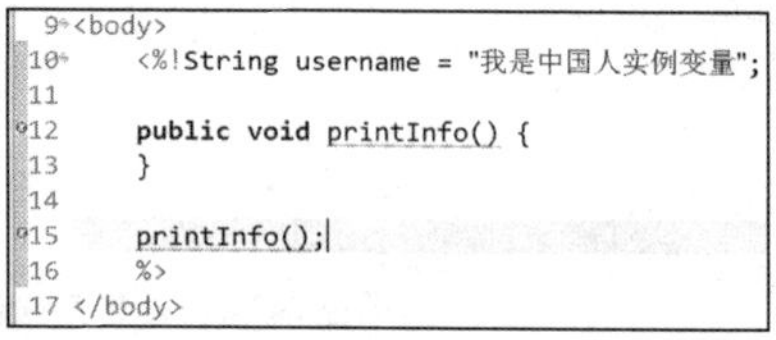

```
<body>
    <%!String username = "我是中国人实例变量";

    public void printInfo() {
    }

    printInfo();
    %>
</body>
```

图 11-14

```
<body>
    <%!String username = "我是中国人实例变量";

    public void printInfo() {
        System.out.println("执行了public void printInfo()");
    }%>

    <%
        printInfo();//此方法被_jspService()方法调用
    %>
</body>
```

执行程序后，成功在控制台上输出相关信息，如图 11-15 所示。

```
执行了public void printInfo()
执行了public void printInfo()
执行了public void printInfo()
执行了public void printInfo()
```

图 11-15

这时，生成的源代码的确是在_jspService()方法中调用了 printInfo()方法，源代码如下：

```
 public  void  _jspService(final  javax.servlet.http.HttpServletRequest
request, final javax.servlet.http.HttpServletResponse response)
     {
       printInfo();//此方法被_jspService()方法调用
      ......
}
```

### 11.1.11 内置对象 pageContext 的使用

在 JSP 文件中可以使用内置对象，内置对象就是 JSP 页面默认提供的对象，不需要实例化就可以直接使用，如 pageContext。pageContext 的作用是在当前的页面中存取数据，而数据不能在多个 JSP 页面之间互相传递，这点与请求不同。示例代码如下：

```
<%@ page language="java" contentType="text/html; charset=utf-8"
pageEncoding="utf-8" %>
 <!DOCTYPE  html  PUBLIC  "-//W3C//DTD  HTML  4.01  Transitional//EN"
"http://www.w3.org/TR/html4/loose.dtd">
<html>
   <body>
       <%
       request.setAttribute("requestKey", "requestValue");
       pageContext.setAttribute("pageContextKey", "pageContextValue");
       request.getRequestDispatcher("index4.jsp").forward(request,
response); %>
   </body>
</html>
```

index4.jsp 文件的代码如下：

```
<%@ page language="java" contentType="text/html; charset=utf-8"
pageEncoding="utf-8" %>
```

```
<!DOCTYPE html PUBLIC "-//W3C//DTD HTML 4.01 Transitional//EN"
"http://www.w3.org/TR/html4/loose.dtd">
<html>
   <body>
      <%
      out.println("index4.jsp requestValue=" +
request.getAttribute("requestKey") + "<br/>");
      out.println("index4.jsp pageContextValue=" +
pageContext.getAttribute("pageContextKey")); %>
   </body>
</html>
```

程序运行后的效果如图 11-16 所示。

```
index4.jsp requestValue=requestValue
index4.jsp pageContextValue=null
```

图 11-16

从运行结果可以得知，pageContext 中的数据的确不可以在多个 JSP 页面中共享，而请求却可以，故 pageContext 中的数据只可以在当前 JSP 页面中使用，示例代码如下：

```
<%@ page language="java" contentType="text/html; charset=utf-8"
pageEncoding="utf-8" %>
<!DOCTYPE html PUBLIC "-//W3C//DTD HTML 4.01 Transitional//EN"
"http://www.w3.org/TR/html4/loose.dtd">
<html>
   <head>
      <meta http-equiv="Content-Type" content="text/html; charset=utf-8">
      <title>Insert title here</title>
   </head>
   <body>
      <%
      pageContext.setAttribute("pageContextKey", "pageContextValue"); %>
      <%
      out.println(pageContext.getAttribute("pageContextKey")); %>
   </body>
</html>
```

```
pageContextValue
```

图 11-17

程序运行效果如图 11-17 所示。JSP 的内置对象其实就是生成的 Servlet 类_jspService()方法中声明的对象，在_jspService()方法中可以使用这些对象进行业务处理，验证代码如图 11-18 所示。

```
public void _jspService(final javax.servlet.http.HttpServletRequest request,
                        final javax.servlet.http.HttpServletResponse response
                        ){

  final javax.servlet.jsp.PageContext pageContext;
  javax.servlet.http.HttpSession session = null;
  final javax.servlet.ServletContext application;
  final javax.servlet.ServletConfig config;
  javax.servlet.jsp.JspWriter out = null;
  final java.lang.Object page = this;
  javax.servlet.jsp.JspWriter _jspx_out = null;
  javax.servlet.jsp.PageContext _jspx_page_context = null;
```

图 11-18

### 11.1.12 常用内置对象的使用

JSP 中常用的内置对象有 out，request，response，session，application，pageContext 和 config。这些内置对象的基本使用示例代码如下：

```
<%@ page language="java" contentType="text/html; charset=utf-8"
pageEncoding="utf-8" %>
<!DOCTYPE html PUBLIC "-//W3C//DTD HTML 4.01 Transitional//EN"
"http://www.w3.org/TR/html4/loose.dtd">
<html>
   <head>
      <meta http-equiv="Content-Type" content="text/html; charset=utf-8">
      <title>Insert title here</title>
   </head>
   <body>
      <%
      //out 主要负责输出
      String username = request.getParameter("username");
      out.println(username + "<br/>");
      //
      pageContext.setAttribute("pageContextKey", "pageContextValue");
      out.println(pageContext.getAttribute("pageContextKey") + "<br/>");
      //
      request.setAttribute("requestKey", "requestValue");
      out.println(request.getAttribute("requestKey") + "<br/>");
      //
      Cookie cookie = new Cookie("a", "aa");
      response.addCookie(cookie);
      //
      session.setAttribute("sessionKey", "sessionValue");
      out.println(session.getAttribute("sessionKey") + "<br/>");
      //application 相当于 ServletContext 对象
```

```
        application.setAttribute("applicationKey", "applicationValue");
        out.println(application.getAttribute("applicationKey") + "<br/>");
        //config 获得当前 Servlet 的配置信息
        out.println(config.getServletName() + "<br/>"); %>
        判断是不是真正的 MVC 模式，就看 JSP 中有没有<%%>或<%! %>代码，
        <br/>
        如果有，则彻底不符合 MVC 的开发方式，因为可以在 JSP 的<%%>或<%! %>代码块中调用业务代码，
        <br/>
        而 JSP 文件仅仅是做显示数据用的，并不是去调用业务，
        <br/>
        业务类是被 Servlet 调用的，再将处理后的结果显示到 JSP 中，
        <br/>
        这才真正的 MVC 开发方式。
    </body>
</html>
```

打开网址 http://localhost:8081/t1/index2.jsp?username=zzzzzzzzzzzzzzzzzzzzz，这时 JSP 页面显示效果如图 11-19 所示。

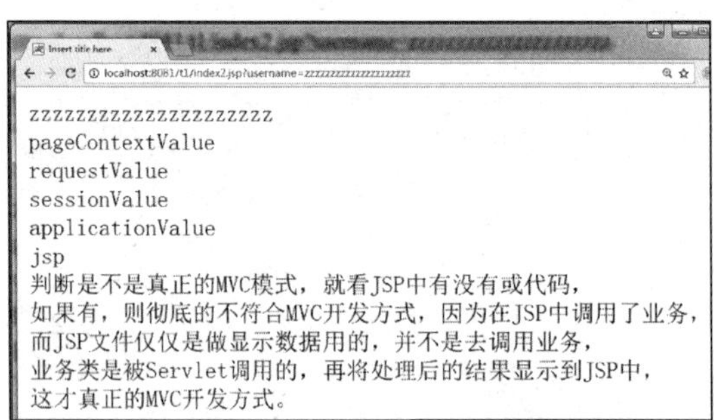


图 11-19

回顾一下 MVC 层之间的调用关系，如下。

（1）实现登录功能，如图 11-20 所示。

图 11-20

（2）显示数据列表，如图 11-21 所示。

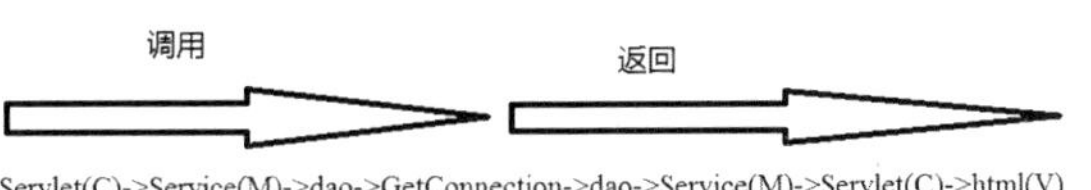


图 11-21

### 11.1.13　使用 pageContext 向不同作用域中存取值

使用 pageContext 可以向不同的作用域中存取值，示例代码如下：

```
<%@ page language="java" contentType="text/html; charset=utf-8"
pageEncoding="utf-8" %>
 <!DOCTYPE   html   PUBLIC   "-//W3C//DTD   HTML   4.01   Transitional//EN"
"http://www.w3.org/TR/html4/loose.dtd">
<html>
   <body>
      <%
      pageContext.setAttribute("pageContextKey1", "pageContextValue1");
      pageContext.setAttribute("pageContextKey2",    "pageContextValue2",
PageContext.PAGE_SCOPE);
      pageContext.setAttribute("requestKey",              "requesttValue",
PageContext.REQUEST_SCOPE);
      pageContext.setAttribute("sessionKey",               "sessionValue",
PageContext.SESSION_SCOPE);
      pageContext.setAttribute("applicationKey",     "applicationValue",
PageContext.APPLICATION_SCOPE);
      //
      out.println("pageContextKey1="                                      +
pageContext.getAttribute("pageContextKey1") + "<br/>");
      out.println("pageContextKey2="                                      +
pageContext.getAttribute("pageContextKey2") + "<br/>");
      out.println("requestKey="  +  request.getAttribute("requestKey")  +
"<br/>");
      out.println("sessionKey="  +  session.getAttribute("sessionKey")  +
"<br/>");
      out.println("applicationKey="                                       +
application.getAttribute("applicationKey") + "<br/>"); %>
   </body>
</html>
```

程序运行结果如图 11-22 所示。

```
pageContextKey1=pageContextValue1
pageContextKey2=pageContextValue2
requestKey=requesttValue
sessionKey=sessionValue
applicationKey=applicationValue
```

图 11-22

### 11.1.14 使用<%@ include file=""%>指令静态导入其他资源

index6.jsp 文件的示例代码如下：

```
<%@ page language="java" contentType="text/html; charset=utf-8"
pageEncoding="utf-8" %>
<!DOCTYPE html PUBLIC "-//W3C//DTD HTML 4.01 Transitional//EN"
"http://www.w3.org/TR/html4/loose.dtd">
<html>
   <head>
      <meta http-equiv="Content-Type" content="text/html; charset=utf-8">
      <title>Insert title here</title>
   </head>
   <body>
      index6.jsp
      <br/>
      <%@ include file="index7.jsp" %>
   </body>
</html>
```

index7.jsp 文件的示例代码如下：

```
<%@ page language="java" contentType="text/html; charset=utf-8"
pageEncoding="utf-8" %>
<!DOCTYPE html PUBLIC "-//W3C//DTD HTML 4.01 Transitional//EN"
"http://www.w3.org/TR/html4/loose.dtd">
<html>
   <head>
      <meta http-equiv="Content-Type" content="text/html; charset=utf-8">
      <title>Insert title here</title>
   </head>
   <body>
      index7.jsp
   </body>
</html>
```

打开 URL 网址 http://localhost:8081/t1/index6.jsp，显示界面如图 11-23 所示。利用此种方法导入资源，在 Java 源代码中属于静态导入，因为 HTML 代码是一同进行输出的，如图 11-24 所示。

```
index6.jsp
index7.jsp
```

图 11-23

```
out.write("\r\n");
out.write("<!DOCTYPE html PUBLIC \"-//W3C//DTD HTML 4.01 Transitiona
out.write("<html>\r\n");
out.write("<head>\r\n");
out.write("<meta http-equiv=\"Content-Type\" content=\"text/html; ch
out.write("<title>Insert title here</title>\r\n");
out.write("</head>\r\n");
out.write("<body>\r\n");
out.write("\tindex6.jsp\r\n");
out.write("\t<br />\r\n");
out.write("\t");
out.write("\r\n");
out.write("<!DOCTYPE html PUBLIC \"-//W3C//DTD HTML 4.01 Transitiona
out.write("<html>\r\n");
out.write("<head>\r\n");
out.write("<meta http-equiv=\"Content-Type\" content=\"text/html; ch
out.write("<title>Insert title here</title>\r\n");
out.write("</head>\r\n");
out.write("<body>index7.jsp\r\n");
out.write("</body>\r\n");
out.write("</html>\r\n");
```

图 11-24

### 11.1.15　使用<jsp:include page=" ">动态导入其他资源

index9.jsp 文件的示例代码如下：

```
<%@ page language="java" contentType="text/html; charset=utf-8"
pageEncoding="utf-8" %>
<!DOCTYPE html PUBLIC "-//W3C//DTD HTML 4.01 Transitional//EN"
"http://www.w3.org/TR/html4/loose.dtd">
<html>
   <head>
      <meta http-equiv="Content-Type" content="text/html; charset=utf-8">
      <title>Insert title here</title>
   </head>
   <body>
      index9.jsp
      <br/>
      <jsp:include page="index10.jsp">
         <jsp:param name="username" value="xyz" />
      </jsp:include>
   </body>
</html>
```

index10.jsp 文件的示例代码如下：

```
<%@ page language="java" contentType="text/html; charset=utf-8"
pageEncoding="utf-8" %>
<!DOCTYPE html PUBLIC "-//W3C//DTD HTML 4.01 Transitional//EN"
"http://www.w3.org/TR/html4/loose.dtd">
<html>
    <head>
        <meta http-equiv="Content-Type" content="text/html; charset=utf-8">
        <title>Insert title here</title>
    </head>
    <body>
        <%
        out.println("index10.jsp username=" +
request.getParameter("username")); %>
    </body>
</html>
```

index9. jsp
index10. jsp username=xyz

图 11-25

打开网址 http://localhost:8081/t1/index9.jsp，程序运行结果如图 11-25 所示。

之所以说使用<jsp:include page=" ">是动态导入，是因为在源代码中是使用 include()方法进行导入的，并不是 HTML 代码的 String 拼接输出，如图 11-26 所示。

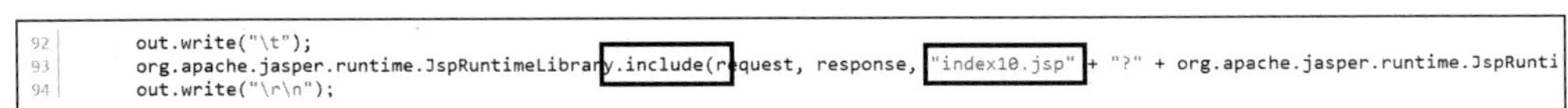

```
out.write("\t");
org.apache.jasper.runtime.JspRuntimeLibrary.include(request, response, "index10.jsp" + "?" + org.apache.jasper.runtime.JspRunti
out.write("\r\n");
```

图 11-26

### 11.1.16 JSP 的注释

在 JSP 中可以使用两种注释，示例代码如下：

```
<%@ page language="java" contentType="text/html; charset=utf-8"
pageEncoding="utf-8" %>
<!DOCTYPE html PUBLIC "-//W3C//DTD HTML 4.01 Transitional//EN"
"http://www.w3.org/TR/html4/loose.dtd">
<html>
    <head>
        <meta http-equiv="Content-Type" content="text/html; charset=utf-8">
        <title>Insert title here</title>
    </head>
    <body>
```

```
        <!-- 我是注释 1，我在 response 中 -->
        <br/>
        <br/>
        <%-- 我是注释 2，我不在 response 中 --%>
    </body>
</html>
```

注释<!--　-->是 HTML 注释，注释<%--　--%>是 JSP 注释，它们之间的区别从响应返回的 HTML 代码中可以得知，如图 11-27 所示。

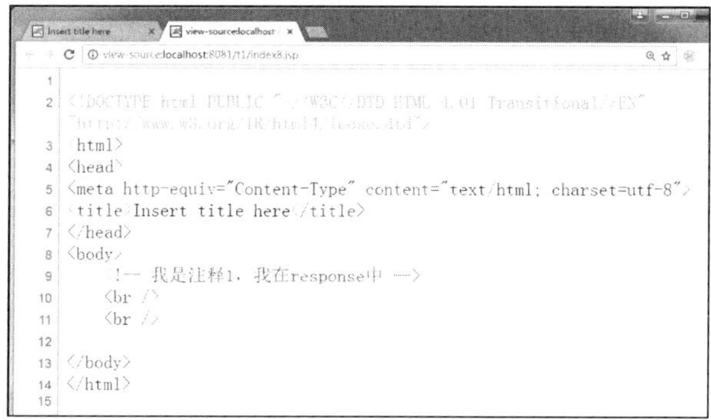

图 11-27

## 11.1.17　使用<jsp:useBean>，<jsp:setProperty>和<jsp:getProperty>访问类信息

示例代码如下：

```
<%@ page language="java" contentType="text/html; charset=utf-8"
pageEncoding="utf-8" %>
 <!DOCTYPE html PUBLIC "-//W3C//DTD HTML 4.01 Transitional//EN"
"http://www.w3.org/TR/html4/loose.dtd">
<html>
    <head>
        <meta http-equiv="Content-Type" content="text/html; charset=utf-8">
        <title>Insert title here</title>
    </head>
    <body>
        <%
      //使用 new 实例化可以直接实例化对象，也可以使用<jsp:useBean>
        entity.Userinfo userinfo = new entity.Userinfo();
```

```
        userinfo.setUsername("中国");
        out.println(userinfo.getUsername() + "<br/>"); %>
        <br/>
        <br/>
        <jsp:useBean       id="userinfoBean"       class="entity.Userinfo"
scope="request">
        </jsp:useBean>
        <jsp:setProperty name="userinfoBean" property="username" value="美
国" /><jsp:getProperty name="userinfoBean" property="username" />
        <br/>
        <br/>
        <br/>
        此案例中使用的技术在实际开发中并不常用，
        <br/>
        因为它们破坏了 MVC 模式，
        <br/>
        在 V 视图层中 new 实例化了对象，
        <br/>
        纯正的 MVC 模式不允许在 JSP 中 new 实例化任何对象！
    </body>
</html>
```

创建实体类 Userinfo.java 的代码如下：

```
package entity;

public class Userinfo {

    private String username;

    public String getUsername() {
        return username;
    }

    public void setUsername(String username) {
        this.username = username;
    }

}
```

运行结果如图 11-28 所示。

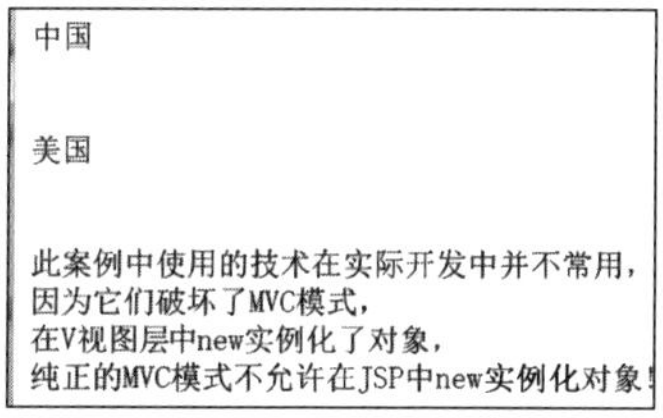

中国

美国

此案例中使用的技术在实际开发中并不常用，
因为它们破坏了MVC模式，
在V视图层中new实例化了对象，
纯正的MVC模式不允许在JSP中new实例化对象！

图 11-28

### 11.1.18　使用<jsp:forward page="">实现转发

在 index.jsp 文件中，设计如下代码：

```
<%@ page language="java" contentType="text/html; charset=utf-8"
pageEncoding="utf-8" %>
 <!DOCTYPE  html  PUBLIC  "-//W3C//DTD  HTML  4.01  Transitional//EN"
"http://www.w3.org/TR/html4/loose.dtd">
<html>
   <head>
      <meta http-equiv="Content-Type" content="text/html; charset=utf-8">
      <title>Insert title here</title>
   </head>
   <body>
      <jsp:forward page="listUserinfo">
      </jsp:forward>
   </body>
</html>
```

名称为 listUserinfo 的 Servlet 代码如下：

```
public class listUserinfo extends HttpServlet {
   protected void doGet(HttpServletRequest request, HttpServletResponse
response)
         throws ServletException, IOException {
       System.out.println("listUserinfo run !");
   }
}
```

打开网址 http://localhost:8081/t1/，网址上直接写出项目名，没有 JSP 或 Servlet 路径。根据 web.xml 文件中<welcome- file-list>的配置会先默认定位到 index.jsp 文件，再实现转发的操作。控制台显示从 Servlet 输出的信息，如图 11-29 所示。

```
Tomcat v7.0 Server at localhost [Apache Tom
listUserinfo run !
listUserinfo run !
listUserinfo run !
listUserinfo run !
listUserinfo run !
listUserinfo run !
```

图 11-29

# 11.2 JSTL和EL表达式

JSTL（Java Standard Taglib，Java 标准标签库），它的主要作用是可以用标签的形式来代表一些逻辑语句，比如 for，if 语句等。虽然在 JSP 中可以使用 JSTL 处理一些逻辑，但并不建议在前台处理逻辑非常复杂的业务，因为前台仅仅以显示数据为主，复杂的业务处理并不属于 V 视图层的职责。

在 JSTL 没有出现之前，Java 代码是直接写在 JSP 文件的<%%>，<%!%>，<%=%>代码块中的，这样会导致 JSP 文件中代码的可读性差，不利于后期的代码维护。JSTL 的出现就是为了解决 MVC 分层不清的问题，它不允许在 V 视图层中有业务逻辑代码，会先将原来处理简单业务逻辑相关的代码，比如 for，if 语句等功能封装到标签类当中，然后在页面中使用相应的标签来代替<%%>，<%!%>，<%=%>代码块。

JSTL 这样做避免了前台 V 视图层使用<%%>、<%!%>或<%=%>的写法，强制程序员不能在前台做复杂的业务，不能 new 实例化对象，这样做符合 MVC 的设计方法。代码分层了，人员分工明确了，结果就是程序可读性好，可复用，维护性也好了。

EL（Expression Language，表达式语言），主要用于在 JSP 页面上打印数据，代替了 out.println()及<%=%>的写法。它在 JSP 文件中是默认被支持的技术，而 JSTL 需要导入标签，JSP 默认不支持 JSTL。EL 表达式的格式非常简单，如下：

```
${}
```

使用这种格式可以获取作用域中的常用数据类型和 JavaBean 的属性值，可以访问 List 中的数据等。

## 11.2.1 使用 EL 表达式获取字符串

服务器端的 Servlet 核心代码如下：

```
String username = "中国";
request.setAttribute("username", username);
```

JSP 文件的核心代码如下：

```
<body>
    <%
    out.println("username=" + request.getAttribute("username"));
```

```
  %>
   <br/>
   <%=("username=" + request.getAttribute("username")) %>
   <br/>
   username=${username}
</body>
```

从代码的数量上来看，使用 EL 表达式能极大地减小 Java 代码量，提高开发效率。

### 11.2.2　使用 EL 表达式获取 JavaBean 中的数据

服务器端 Servlet 核心代码如下：

```
Userinfo userinfo = new Userinfo();
userinfo.setUsername("美国");
userinfo.setAddress("美国人");
request.setAttribute("userinfo", userinfo);
```

JSP 文件核心代码如下：

```
<body>
   <br/>
   username=${userinfo.username}
   <br/>
   address=${userinfo.address}
</body>
```

### 11.2.3　使用 EL 表达式查找数据

EL 表达式支持自动从 page->request->session->application 作用域中查找目标数据，最开始从 page 作用域中找，如果在 page 作用域中找不到，则到 request 作用域中找；如果在 request 作用域中还找不到，则到 session 作用域中找，以此类推。服务器端 Servlet 核心代码如下：

```
Userinfo userinfo = new Userinfo();
userinfo.setUsername("美国");
userinfo.setAddress("美国人");
request.getSession().setAttribute("sessionUserinfo", userinfo);
```

JSP 文件核心代码如下：

```
<body>
    <br/>
    查询的范围 page->request->session->application
    <br/>
    sessionUsername=${sessionUserinfo.username}
    <br/>
    sessionAddress=${sessionUserinfo.address}
</body>
```

### 11.2.4 key 优先获取作用域小的 scope 值

作用域从小到大分别是 page->request->session->application。服务器端 Servlet 核心代码如下：

```
request.setAttribute("xyz", "request 显示我");
request.getSession().setAttribute("xyz", "sessionValue 不显示");
```

JSP 文件核心代码如下：

```
<body>
    ${xyz}
</body>
```

### 11.2.5 使用 EL 表达式获取指定作用域中的值

前面讲述的是获取 request 作用域中的值，如果你想获取 session 作用域中的值，则需要在 JSP 文件中使用如下 EL 表达式：

```
${sessionScope.xyz}
```

获取 request 作用域中的值可以使用如下代码：

```
${requestScope.xyz}
```

获取 application 作用域中的值可以使用如下代码：

```
${applicationScope.xyz}
```

### 11.2.6 使用 EL 表达式打印 property 属性名

在使用 EL 表达式获取属性值时，支持属性名称是变量的形式，只支持方括号，不

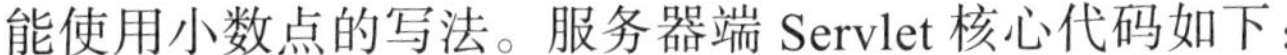

能使用小数点的写法。服务器端 Servlet 核心代码如下：

```
Userinfo userinfo = new Userinfo();
userinfo.setUsername("美国");
userinfo.setAddress("美国人");
request.setAttribute("userinfo", userinfo);
request.setAttribute("showKey1", "username");
request.setAttribute("showKey2", "address");
```

JSP 文件核心代码如下：

```
    <body>
    username1=${userinfo['username']}
    <br/>
    address1=${userinfo['address']}
    <br/>
    <br/>
    username3=${userinfo[showKey1]}
    <br/>
    address3=${userinfo[showKey2]}
</body>
```

### 11.2.7 使用 EL 表达式获取 List，array[]与 Map 中的数据

EL 表达式支持从数组及 List 中获取数据。向 List 和 array[]存储数据的 Servlet 核心代码如下：

```
List list1 = new ArrayList();
list1.add("中国 1");
list1.add("中国 2");
list1.add("中国 3");

List list2 = new ArrayList();
list2.add(new Userinfo("a1", "aa1"));
list2.add(new Userinfo("a2", "aa2"));
list2.add(new Userinfo("a3", "aa3"));

String[] stringArray = new String[] { "a", "b", "c" };
Userinfo[] userinfoArray = new Userinfo[] { new Userinfo("a1", "aa1"), new
Userinfo("a2", "aa2") };

request.setAttribute("list1", list1);
```

```
request.setAttribute("list2", list2);
request.setAttribute("stringArray", stringArray);
request.setAttribute("userinfoArray", userinfoArray);
```

JSP 文件核心代码如下：

```
<body>
    A__${list1[0]}__ ${list1[1]}
    <br/>
    <br/>
    <br/>
    B__${list2[0].username}__ ${list2[0].address}
    <br/>
    B__${list2[1].username}__ ${list2[1].address}
    <br/>
    B__${list2[2].username}__ ${list2[2].address}
    <br/>
    <br/>
    <br/>
    C__${stringArray[0]}_${stringArray[1]}_${stringArray[2]}
    <br/>
    <br/>
    <br/>
    D__${userinfoArray[0].username}__ ${userinfoArray[0].address}
    <br/>
    D__${userinfoArray[1].username}__ ${userinfoArray[1].address}
</body>
```

向 Map 中存储数据的 Servlet 核心代码如下：

```
public class test6 extends HttpServlet {
    protected void doGet(HttpServletRequest request, HttpServletResponse
response)
            throws ServletException, IOException {
        Map mapString = new HashMap();
        mapString.put("key1", "中国 1");
        mapString.put("key2", "中国 2");
        mapString.put("key3", "中国 3");

        Map<String, Userinfo> mapUserinfo = new HashMap<>();
        mapUserinfo.put("key1", new Userinfo("中国 1", "地址 1"));
        mapUserinfo.put("key2", new Userinfo("中国 2", "地址 2"));
        mapUserinfo.put("key3", new Userinfo("中国 3", "地址 3"));
```

```
        request.setAttribute("mapString", mapString);
        request.setAttribute("mapUserinfo", mapUserinfo);

        request.getRequestDispatcher("test17.jsp").forward(request,
response);
    }
}
```

JSP 文件核心代码如下：

```
<body>
    ${mapString.key1}_${mapString.key2}_${mapString.key3}
    <br/>
    <br/>
    ${mapString['key1']}_${mapString['key2']}_${mapString['key3']}
    <br/>
    <br/>
    ${mapUserinfo.key1.username}_${mapUserinfo.key1.address}
    <br/>
    ${mapUserinfo.key2.username}_${mapUserinfo.key2.address}
    <br/>
    ${mapUserinfo.key3.username}_${mapUserinfo.key3.address}
    <br/>
    <br/>
    ${mapUserinfo['key1']['username']}_${mapUserinfo['key1']['address']}
    <br/>
    ${mapUserinfo['key2']['username']}_${mapUserinfo['key2']['address']}
    <br/>
    ${mapUserinfo['key3']['username']}_${mapUserinfo['key3']['address']}
    <br/>
</body>
```

## 11.2.8　使用 EL 表达式输出 NULL 值

在处理 NULL 值时，EL 表达式输出的结果不是 NULL 值，而是空字符串。服务器端 Servlet 核心代码如下：

```
request.setAttribute("nullValue", null);
```

JSP 文件核心代码如下：

```
<body>
```

```
    |${nullValue}|
</body>
```

## 11.2.9 使用 EL 表达式打印嵌套中的值

创建 3 个 JavaBean，代码如图 11-30 所示。

```
package entity;

public class A {
    private B b = new B();

    public B getB() {
        return b;
    }

    public void setB(B b) {
        this.b = b;
    }

}
```

```
package entity;

public class B {
    private C c = new C();

    public C getC() {
        return c;
    }

    public void setC(C c) {
        this.c = c;
    }

}
```

```
package entity;

public class C {
    private String age = "100";

    public String getAge() {
        return age;
    }

    public void setAge(String age) {
        this.age = age;
    }

}
```

图 11-30

服务器端 Servlet 核心代码如下：

```
request.setAttribute("aKey", new A());
```

JSP 文件核心代码如下：

```
<body>
    ${aKey.b.c.age}
</body>
```

## 11.2.10 在 EL 表达式中使用 empty 进行空的判断

服务器端 Servlet 核心代码如下：

```
protected void doGet(HttpServletRequest request, HttpServletResponse response)
        throws ServletException, IOException {
    request.setAttribute("x1", "");// true
    request.setAttribute("x2", null);// true
    Userinfo userinfo = null;
    request.setAttribute("x3", userinfo);// true
    request.setAttribute("x4", new ArrayList());// true
    request.setAttribute("x5", new String[] {});// true

    List hasValue = new ArrayList();
    hasValue.add("anyValue");
```

```
    request.setAttribute("x6", hasValue);// false

    request.setAttribute("x7", new String[] { "abc" });// false

    List hasUserinfo = new ArrayList();
    hasUserinfo.add(new Userinfo());
    request.setAttribute("x8", hasUserinfo);// false

    List nullList = new ArrayList();
    nullList.add(null);
    request.setAttribute("x9", nullList);// false

    request.getRequestDispatcher("test18.jsp").forward(request,
response);
}
```

JSP 文件核心代码如下：

```
<body>
    ${empty x1}
    <br/>
    ${empty x2}
    <br/>
    ${empty x3}
    <br/>
    ${empty x4}
    <br/>
    ${empty x5}
    <br/>
    <br/>
    ${empty x6}
    <br/>
    ${empty x7}
    <br/>
    ${empty x8}
    <br/>
    ${empty x9}
</body>
```

程序运行结果如下：

```
true
true
```

```
true
true
true
false
false
false
false
```

在使用 empty 对 List 或数组 Array 进行判断时，运行的结果取决于有没有开辟存储数据的空间，如果开辟了存储数据的空间，则运行的结果就是 false，否则就是 true。

### 11.2.11　使用${param}获取 URL 中的参数值

JSP 核心代码如下：

```
<body>
    <%
    out.println("username1=" + request.getParameter("username")); %>
    <br/>
    username2=${param.username}
</body>
```

打开网址：http://localhost:8081/t2/index2.jsp?username=abcxyz。

### 11.2.12　使用 JSTL 表达式进行逻辑处理

在 JSTL 中，<c:if>标签的作用和 if 语句的一样，<c:choose>标签和 switch 语句的作用非常类似，<c:forEach>标签主要用来处理循环。

注意：在使用 JSTL 之前需要导入标签，在 JSP 中使用如下代码进行导入：

```
<%@ taglib uri="http://java.sun.com/jsp/jstl/core" prefix="c" %>
```

创建 Servlet 代码如下：

```
public class test7 extends HttpServlet {
    protected void doGet(HttpServletRequest request, HttpServletResponse
response)
            throws ServletException, IOException {
        List list1 = new ArrayList();
        list1.add("中国 1");
        list1.add("中国 2");
```

```
        list1.add("中国 3");
        list1.add("中国 4");
        request.setAttribute("list1", list1);

        List list2 = new ArrayList();
        list2.add(new Userinfo("账号 1", "地址 1"));
        list2.add(new Userinfo("账号 2", "地址 2"));
        request.setAttribute("list2", list2);

        request.setAttribute("username", "我使用 c:out 输出 3");

        request.setAttribute("myUserinfo", new Userinfo());

        request.getRequestDispatcher("test18.jsp").forward(request,
response);
    }
}
```

JSP 文件核心代码如下：

```
<body>
    test18.jsp
    <br/>
    <c:if test="${1==1}">
        1==1
    </c:if>
    <br/>
    <c:if test="${true==true}">
        true==true
    </c:if>
    <br/>
    c:choose 的作用和 switch+break 的作用一样。
    <c:choose>
        <c:when test="${1==100}">
            1==100
        </c:when>
        <c:when test="${1==1}">
            1==1
        </c:when>
        <c:when test="${1==200}">
            1==200
        </c:when>
        <c:otherwise>
```

```
        1==otherValue
    </c:otherwise>
</c:choose>
<br/>
<c:choose>
    <c:when test="${1==100}">
        1==100
    </c:when>
    <c:when test="${1==200}">
        1==200
    </c:when>
    <c:when test="${1==300}">
        1==300
    </c:when>
    <c:otherwise>
        1==otherValue
    </c:otherwise>
</c:choose>
<br/>
<c:forEach var="eachString" items="${list1}">
    ${eachString}
    <br/>
</c:forEach>
<br/>
<c:forEach var="eachUserinfo" items="${list2}">
    ${eachUserinfo.username}_${eachUserinfo.address}
    <br/>
</c:forEach>
<br/>
<c:forEach var="eachString" items="${list1}" varStatus="myStatus">
    ${myStatus.index+1}__${eachString}
    <br/>
</c:forEach>
<br/>
<c:forEach var="eachString" items="${list1}" varStatus="myStatus">
    <c:if test="${myStatus.index%2==0}">
        ${eachString}---
    </c:if>
    <c:if test="${myStatus.index%2!=0}">
        ${eachString}+++
    </c:if>
    <br/>
</c:forEach>
```

```
    <br/>
    <c:out value="我使用 c:out 输出 1">
    </c:out>
    <br/>
    <c:out value="${'我使用 c:out 输出 2'}">
    </c:out>
    <br/>
    <c:out value="${username}">
    </c:out>
    <br/>
    <c:set target="${myUserinfo}" property="address" value="zzzzzz">
    </c:set>
    ${myUserinfo.address}
</body>
```

### 11.2.13　对 Date 进行 String 格式化

对 Date 进行 String 格式化的 Servlet 代码如下：

```
Date nowDate = new Date();
request.setAttribute("nowDate", nowDate);
```

引入 fmt 声明：

```
<%@ taglib prefix="fmt" uri="http://java.sun.com/jsp/jstl/fmt" %>
```

在 JSP 中，使用 JSTL 对日期类型进行格式化，代码如下：

```
<fmt:formatDate value="${nowDate}" pattern="yyyy-MM-dd HH:mm:ss"/>
<br/>
<fmt:formatDate value="${nowDate}" pattern="yyyy 年 MM 月 dd 日 HH 时 mm 分 ss 秒
"/>
```

## 11.3　实现基于MVC的CURD增删改查

本案例使用 Servlet+Service+JSP+JDBC 实现基于 MVC 模式的 CURD 增删改查操作，各组件分工如下。

① M 业务层：Service.java，负责软件核心业务功能。

② V 视图层：.jsp 文件，负责显示数据。

③ C 控制层：Servlet.java，负责控制软件执行流程。

（1）创建 GetConnection.java，代码如下：

```
package dbtools;

import java.sql.Connection;
import java.sql.DriverManager;
import java.sql.SQLException;

public class GetConnection {
    public static Connection getConnection() throws ClassNotFoundException,
SQLException {
        String username = "y2";
        String password = "123";
        String url = "jdbc:oracle:thin:@localhost:1521:orcl";
        String driver = "oracle.jdbc.OracleDriver";
        Class.forName(driver);
        return DriverManager.getConnection(url, username, password);
    }
}
```

（2）创建 UserinfoDao.java，代码如下：

```
package dao;

import java.sql.Connection;
import java.sql.PreparedStatement;
import java.sql.ResultSet;
import java.sql.SQLException;
import java.sql.Timestamp;
import java.util.ArrayList;
import java.util.Date;
import java.util.List;

import dbtools.GetConnection;
import entity.Userinfo;

public class UserinfoDao {

    public   List<Userinfo>   selectAll()   throws   ClassNotFoundException,
SQLException {
        List list = new ArrayList();
```

```
        String sql = "select * from userinfo order by id asc";
        Connection conn = GetConnection.getConnection();
        PreparedStatement ps = conn.prepareStatement(sql);
        ResultSet rs = ps.executeQuery();
        while (rs.next()) {
            Userinfo userinfo = new Userinfo();

            int idDB = rs.getInt("id");
            String usernameDB = rs.getString("username");
            String passwordDB = rs.getString("password");
            int ageDB = rs.getInt("age");
            Date insertdateDB = rs.getDate("insertdate");

            userinfo.setId(idDB);
            userinfo.setUsername(usernameDB);
            userinfo.setPassword(passwordDB);
            userinfo.setAge(ageDB);
            userinfo.setInsertdate(insertdateDB);

            list.add(userinfo);

        }
        rs.close();
        ps.close();
        conn.close();

        return list;
    }

    public Userinfo selectById(int userId) throws ClassNotFoundException,
SQLException {
        Userinfo userinfo = null;

        String sql = "select * from userinfo where id=?";
        Connection conn = GetConnection.getConnection();
        PreparedStatement ps = conn.prepareStatement(sql);
        ps.setInt(1, userId);
        ResultSet rs = ps.executeQuery();
        while (rs.next()) {
            userinfo = new Userinfo();

            int idDB = rs.getInt("id");
```

```
            String usernameDB = rs.getString("username");
            String passwordDB = rs.getString("password");
            int ageDB = rs.getInt("age");
            Date insertdateDB = rs.getDate("insertdate");

            userinfo.setId(idDB);
            userinfo.setUsername(usernameDB);
            userinfo.setPassword(passwordDB);
            userinfo.setAge(ageDB);
            userinfo.setInsertdate(insertdateDB);

        }
        rs.close();
        ps.close();
        conn.close();

        return userinfo;
    }

    public void insertUserinfo(Userinfo userinfo) throws
ClassNotFoundException, SQLException {
        String sql = "insert into userinfo(id,username,password,age,
insertdate) values(idauto.nextval,?,?,?,?)";
        Connection conn = GetConnection.getConnection();
        PreparedStatement ps = conn.prepareStatement(sql);
        ps.setString(1, userinfo.getUsername());
        ps.setString(2, userinfo.getPassword());
        ps.setInt(3, userinfo.getAge());
        ps.setTimestamp(4, new Timestamp(userinfo.getInsertdate().getTime()));
        ps.executeUpdate();
        ps.close();
        conn.close();
    }

    public void updateById(Userinfo userinfo) throws ClassNotFoundException,
SQLException {
        String sql = "update userinfo set username=?,password=?,age=?,
insertdate=? where id=?";
        Connection conn = GetConnection.getConnection();
        PreparedStatement ps = conn.prepareStatement(sql);
        ps.setString(1, userinfo.getUsername());
        ps.setString(2, userinfo.getPassword());
        ps.setInt(3, userinfo.getAge());
```

```
        ps.setTimestamp(4, new Timestamp(userinfo.getInsertdate().getTime()));
        ps.setInt(5, userinfo.getId());
        ps.executeUpdate();
        ps.close();
        conn.close();
    }

    public void deleteById(int userId) throws ClassNotFoundException,
SQLException {
        String sql = "delete from userinfo where id=?";
        Connection conn = GetConnection.getConnection();
        PreparedStatement ps = conn.prepareStatement(sql);
        ps.setInt(1, userId);
        ps.executeUpdate();
        ps.close();
        conn.close();
    }

}
```

（3）创建 UserinfoService.java，代码如下：

```
package service;

import java.sql.SQLException;
import java.util.List;

import dao.UserinfoDao;
import entity.Userinfo;

public class UserinfoService {

    private UserinfoDao userinfoDao = new UserinfoDao();

    public List<Userinfo> selectAll() throws ClassNotFoundException,
SQLException {
        return userinfoDao.selectAll();
    }

    public Userinfo selectById(int userId) throws ClassNotFoundException,
SQLException {
        return userinfoDao.selectById(userId);
    }
```

```
    public void insertUserinfo(Userinfo userinfo) throws
ClassNotFoundException, SQLException {
        userinfoDao.insertUserinfo(userinfo);
    }

    public void updateById(Userinfo userinfo) throws ClassNotFoundException,
SQLException {
        userinfoDao.updateById(userinfo);
    }

    public void deleteById(int userId) throws ClassNotFoundException,
SQLException {
        userinfoDao.deleteById(userId);
    }

}
```

（4）创建 CharSetFilter.java，代码如下：

```
package extfilter;

import java.io.IOException;

import javax.servlet.Filter;
import javax.servlet.FilterChain;
import javax.servlet.FilterConfig;
import javax.servlet.ServletException;
import javax.servlet.ServletRequest;
import javax.servlet.ServletResponse;

public class CharSetFilter implements Filter {

    @Override
    public void init(FilterConfig filterConfig) throws ServletException {
    }

    @Override
    public void doFilter(ServletRequest request, ServletResponse response,
FilterChain chain)
            throws IOException, ServletException {
        request.setCharacterEncoding("utf-8");
        response.setCharacterEncoding("utf-8");
```

```
        chain.doFilter(request, response);
    }

    @Override
    public void destroy() {
    }

}
```

（5）创建 Userinfo.java，代码如下：

```
package entity;

import java.util.Date;

public class Userinfo {

    private int id;
    private String username;
    private String password;
    private int age;
    private Date insertdate;

    public Userinfo() {
    }

    public Userinfo(int id, String username, String password, int age, Date
insertdate) {
        super();
        this.id = id;
        this.username = username;
        this.password = password;
        this.age = age;
        this.insertdate = insertdate;
    }

   //省略 get()方法和 set()方法

}
```

（6）创建 index.jsp，代码如下：

```
<%@ page language="java" contentType="text/html; charset=utf-8"
```

```
pageEncoding="utf-8" %>
 <!DOCTYPE  html  PUBLIC  "-//W3C//DTD  HTML  4.01  Transitional//EN"
"http://www.w3.org/TR/html4/loose.dtd">
<html>
    <head>
        <meta http-equiv="Content-Type" content="text/html; charset=utf-8">
        <title>Insert title here</title>
    </head>
    <body>
        <jsp:forward page="listUserinfo"></jsp:forward>
    </body>
</html>
```

（7）创建 listUserinfo.jsp，代码如下：

```
<%@ page language="java" contentType="text/html; charset=utf-8"
pageEncoding="utf-8" %>
<%@ taglib uri="http://java.sun.com/jsp/jstl/core" prefix="c" %>
 <!DOCTYPE  html  PUBLIC  "-//W3C//DTD  HTML  4.01  Transitional//EN"
"http://www.w3.org/TR/html4/loose.dtd">
<html>
    <head>
        <meta http-equiv="Content-Type" content="text/html; charset=utf-8">
        <title>Insert title here</title>
        <script>
            function deleteById(){
                if (confirm('确认删除吗？')) {
                    return true;
                }
                else {
                    return false;
                }
            }
        </script>
    </head>
    <body>
        <form action="insertUserinfo" method="post">
            username:<input type="text" name="username" value="中国">
            <br/>
            password:<input type="text" name="password" value="中国人">
            <br/>
            age:<input type="text" name="age" value="100">
            <br/>
```

```
            insertdate:<input type="text" name="insertdate" value="2000-1-1">
            <br/>
            <input type="submit" value="submit">
        </form>
        <br/>
        <br/>
        <c:forEach var="eachUserinfo" items="${listUserinfo}">
            ${eachUserinfo.id}_
            ${eachUserinfo.username}_
            ${eachUserinfo.password}_
            ${eachUserinfo.age}_
            ${eachUserinfo.insertdate}_<a
href="deleteUserinfoById?id=${eachUserinfo.id}"
onclick="javascript:return                    deleteById();">delete</a>__<a
href="showUserinfoById?id=${eachUserinfo.id}">update</a>
            <br/>
        </c:forEach>
    </body>
</html>
```

（8）创建 showUserinfoById.jsp，代码如下：

```
<%@ page language="java" contentType="text/html; charset=utf-8"
pageEncoding="utf-8" %>
<%@ taglib uri="http://java.sun.com/jsp/jstl/core" prefix="c" %>
 <!DOCTYPE   html   PUBLIC   "-//W3C//DTD   HTML   4.01   Transitional//EN"
"http://www.w3.org/TR/html4/loose.dtd">
<html>
    <head>
        <meta http-equiv="Content-Type" content="text/html; charset=utf-8">
        <title>Insert title here</title>
    </head>
    <body>
        <form action="saveUpdateUserinfoById" method="post">
            <input    type="hidden"    name="id"    value="${userinfo.id}">
username:<input type="text" name="username" value="${userinfo.username}">
            <br/>
            password:<input type="text" name="password" value=
"${userinfo.password}">
            <br/>
            age:<input type="text" name="age" value="${userinfo.age}">
            <br/>
            insertdate:<input type="text" name="insertdate" value=
```

```
"${userinfo.insertdate}">
            <br/>
            <input type="submit" value="submit">
        </form>
    </body>
</html>
```

（9）创建 listUserinfo.java，代码如下：

```
package controller;

import java.io.IOException;
import java.sql.SQLException;
import java.util.List;

import javax.servlet.ServletException;
import javax.servlet.http.HttpServlet;
import javax.servlet.http.HttpServletRequest;
import javax.servlet.http.HttpServletResponse;

import service.UserinfoService;

public class listUserinfo extends HttpServlet {
    protected void doGet(HttpServletRequest request, HttpServletResponse
response)
            throws ServletException, IOException {
        try {
            UserinfoService userinfoService = new UserinfoService();
            List listUserinfo = userinfoService.selectAll();
            request.setAttribute("listUserinfo", listUserinfo);

    request.getRequestDispatcher("listUserinfo.jsp").forward(request,
response);
        } catch (ClassNotFoundException e) {
            e.printStackTrace();
        } catch (SQLException e) {
            e.printStackTrace();
        }
    }

}
```

（10）创建 insertUserinfo.java，代码如下：

```
package controller;

import java.io.IOException;
import java.sql.SQLException;
import java.text.ParseException;
import java.text.SimpleDateFormat;
import java.util.Date;

import javax.servlet.ServletException;
import javax.servlet.http.HttpServlet;
import javax.servlet.http.HttpServletRequest;
import javax.servlet.http.HttpServletResponse;

import entity.Userinfo;
import service.UserinfoService;

public class insertUserinfo extends HttpServlet {
    protected void doPost(HttpServletRequest request, HttpServletResponse
response)
            throws ServletException, IOException {
        try {
            String username = request.getParameter("username");
            String password = request.getParameter("password");
            String age = request.getParameter("age");
            String insertdate = request.getParameter("insertdate");

            SimpleDateFormat format = new SimpleDateFormat("yyyy-MM-dd");
            Date date = format.parse(insertdate);

            Userinfo userinfo = new Userinfo();
            userinfo.setUsername(username);
            userinfo.setPassword(password);
            userinfo.setAge(Integer.parseInt(age));
            userinfo.setInsertdate(date);

            UserinfoService userinfoService = new UserinfoService();
            userinfoService.insertUserinfo(userinfo);

            response.sendRedirect("listUserinfo");

        } catch (NumberFormatException e) {
            e.printStackTrace();
        } catch (ClassNotFoundException e) {
            e.printStackTrace();
```

```
        } catch (ParseException e) {
            e.printStackTrace();
        } catch (SQLException e) {
            e.printStackTrace();
        }

    }

}
```

（11）创建 deleteUserinfoById.java，代码如下：

```
package controller;

import java.io.IOException;
import java.sql.SQLException;

import javax.servlet.ServletException;
import javax.servlet.http.HttpServlet;
import javax.servlet.http.HttpServletRequest;
import javax.servlet.http.HttpServletResponse;

import service.UserinfoService;

public class deleteUserinfoById extends HttpServlet {
    protected void doGet(HttpServletRequest request, HttpServletResponse
response)
            throws ServletException, IOException {
        try {
            String id = request.getParameter("id");

            UserinfoService userinfoService = new UserinfoService();
            userinfoService.deleteById(Integer.parseInt(id));

            response.sendRedirect("listUserinfo");
        } catch (NumberFormatException e) {
            e.printStackTrace();
        } catch (ClassNotFoundException e) {
            e.printStackTrace();
        } catch (SQLException e) {
            e.printStackTrace();
        }
    }

}
```

（12）创建 showUserinfoById.java，代码如下：

```
package controller;

import java.io.IOException;
import java.sql.SQLException;

import javax.servlet.ServletException;
import javax.servlet.http.HttpServlet;
import javax.servlet.http.HttpServletRequest;
import javax.servlet.http.HttpServletResponse;

import entity.Userinfo;
import service.UserinfoService;

public class showUserinfoById extends HttpServlet {
    protected void doGet(HttpServletRequest request, HttpServletResponse
response)
            throws ServletException, IOException {
        try {
            String id = request.getParameter("id");

            UserinfoService userinfoService = new UserinfoService();
            Userinfo                      userinfo                      =
userinfoService.selectById(Integer.parseInt(id));
            request.setAttribute("userinfo", userinfo);


    request.getRequestDispatcher("showUserinfoById.jsp").forward(request,
response);
        } catch (NumberFormatException e) {
            e.printStackTrace();
        } catch (ClassNotFoundException e) {
            e.printStackTrace();
        } catch (SQLException e) {
            e.printStackTrace();
        }
    }

}
```

（13）创建 saveUpdateUserinfoById.java，代码如下：

```
package controller;
```

```
import java.io.IOException;
import java.sql.SQLException;
import java.text.ParseException;
import java.text.SimpleDateFormat;
import java.util.Date;

import javax.servlet.ServletException;
import javax.servlet.http.HttpServlet;
import javax.servlet.http.HttpServletRequest;
import javax.servlet.http.HttpServletResponse;

import entity.Userinfo;
import service.UserinfoService;

public class saveUpdateUserinfoById extends HttpServlet {
    protected void doPost(HttpServletRequest request, HttpServletResponse
response)
            throws ServletException, IOException {

        try {
            String id = request.getParameter("id");
            String username = request.getParameter("username");
            String password = request.getParameter("password");
            String age = request.getParameter("age");
            String insertdate = request.getParameter("insertdate");

            SimpleDateFormat format = new SimpleDateFormat("yyyy-MM-dd");
            Date date = format.parse(insertdate);

            Userinfo userinfo = new Userinfo();
            userinfo.setId(Integer.parseInt(id));
            userinfo.setUsername(username);
            userinfo.setPassword(password);
            userinfo.setAge(Integer.parseInt(age));
            userinfo.setInsertdate(date);

            UserinfoService userinfoService = new UserinfoService();
            userinfoService.updateById(userinfo);

            response.sendRedirect("listUserinfo");

        } catch (NumberFormatException e) {
            e.printStackTrace();
        } catch (ClassNotFoundException e) {
```

```
            e.printStackTrace();
        } catch (ParseException e) {
            e.printStackTrace();
        } catch (SQLException e) {
            e.printStackTrace();
        }

    }

}
```

（14）配置文件 web.xml，代码如下：

```
<?xml version="1.0" encoding="UTF-8"?>
<web-app xmlns:xsi="http://www.w3.org/2001/XMLSchema-instance"
    xmlns="http://java.sun.com/xml/ns/javaee"
    xsi:schemaLocation="http://java.sun.com/xml/ns/javaee
http://java.sun.com/xml/ns/javaee/web-app_2_5.xsd"
    id="WebApp_ID" version="2.5">
    <display-name>servletCURD__</display-name>
    <welcome-file-list>
        <welcome-file>index.html</welcome-file>
        <welcome-file>index.htm</welcome-file>
        <welcome-file>index.jsp</welcome-file>
        <welcome-file>default.html</welcome-file>
        <welcome-file>default.htm</welcome-file>
        <welcome-file>default.jsp</welcome-file>
    </welcome-file-list>
    <servlet>
        <servlet-name>insertUserinfo</servlet-name>
        <servlet-class>controller.insertUserinfo</servlet-class>
    </servlet>
    <servlet-mapping>
        <servlet-name>insertUserinfo</servlet-name>
        <url-pattern>/insertUserinfo</url-pattern>
    </servlet-mapping>

    <servlet>
        <servlet-name>listUserinfo</servlet-name>
        <servlet-class>controller.listUserinfo</servlet-class>
    </servlet>
    <servlet-mapping>
        <servlet-name>listUserinfo</servlet-name>
        <url-pattern>/listUserinfo</url-pattern>
    </servlet-mapping>
```

```
    <servlet>
        <servlet-name>deleteUserinfoById</servlet-name>
        <servlet-class>controller.deleteUserinfoById</servlet-class>
    </servlet>
    <servlet-mapping>
        <servlet-name>deleteUserinfoById</servlet-name>
        <url-pattern>/deleteUserinfoById</url-pattern>
    </servlet-mapping>

    <servlet>
        <servlet-name>showUserinfoById</servlet-name>
        <servlet-class>controller.showUserinfoById</servlet-class>
    </servlet>
    <servlet-mapping>
        <servlet-name>showUserinfoById</servlet-name>
        <url-pattern>/showUserinfoById</url-pattern>
    </servlet-mapping>

    <servlet>
        <servlet-name>saveUpdateUserinfoById</servlet-name>
        <servlet-class>controller.saveUpdateUserinfoById</servlet-class>
    </servlet>
    <servlet-mapping>
        <servlet-name>saveUpdateUserinfoById</servlet-name>
        <url-pattern>/saveUpdateUserinfoById</url-pattern>
    </servlet-mapping>

    <filter>
        <filter-name>charSetFilter</filter-name>
        <filter-class>extfilter.CharSetFilter</filter-class>
    </filter>
    <filter-mapping>
        <filter-name>charSetFilter</filter-name>
        <url-pattern>/*</url-pattern>
    </filter-mapping>

</web-app>
```

# 第 12 章
# 异步处理 AJAX 技术

异步通信的含义是客户端不需要等待服务器端的响应，而可以继续运行后面的程序代码，处理服务器端的响应大多数使用回调的方式。

AJAX（Asynchronous Javascript And XML，异步 JavaScript 和 XML）是一种创建交互式网页的网页开发技术，常用于创建快速和动态网页。快速的原因是前台与后台可以进行少量的数据交换，运行效率达到最大化；动态的原因是 AJAX 结合 JavaScript 可以实现对网页的部分更新，也被称为局部刷新。局部刷新是指可以在不重新加载整个网页的情况下，更新网页的部分内容，而传统的网页（不使用 AJAX）如果需要更新内容，则必须重新加载整个网页，造成传递的数据具有重复性，浪费大量网络资源。

AJAX API 的核心是 XMLHttpRequest 对象，该对象支持异步请求。所谓异步请求就是在使用 XMLHttpRequest 对象时向服务器发出请求并处理响应，而同时还可以执行后面的 JavaScript 程序代码，有些类似于多个线程同时执行的效果，优点是不阻塞用户的操作行为，大幅提升用户体验。虽然 AJAX 有这么大的优点，但直到 Web 2.0 时代，它的优势才被挖掘出来。Google 应用 AJAX 技术开发出了影响世界的 Google Map。

异步与同步的区别，如图 12-1 所示。同步是指发出数据后，必须等待响应结果，才可以发送下一条数据的通信方式；异步是指发出数据后，不需要等待响应结果，可继续发送下一条数据的通信方式。

在异步情况下，多个任务可以同时执行，而在同步情况下，多个任务之间是按顺序执行的。AJAX 是由浏览器端发起与服务器端交换少量数据并更新局部网页的异步通信技术。

图 12-1

## 12.1 实现无传参无返回值——get提交方式

服务器端代码如下：

```
public class ajax_test1 extends HttpServlet {
    protected void doGet(HttpServletRequest request, HttpServletResponse
response)
            throws ServletException, IOException {
        try {
            System.out.println("ajax_test1 begin " +
System.currentTimeMillis());
            Thread.sleep(3000);
            System.out.println("ajax_test1  end " +
System.currentTimeMillis());
        } catch (InterruptedException e) {
            e.printStackTrace();
        }
    }
}
```

客户端代码如下：

```
function sendAjax1(){
    var ajaxObject = new XMLHttpRequest();
    ajaxObject.open("get", "ajax_test1?t="+new Date().getTime(), true);
    ajaxObject.send();
}
```

使用 AJAXObject 对象向服务器发送请求需要执行 open()方法和 send()方法。

在 JSP 代码中，调用了 AJAXObject 对象的 open()方法，此方法具有 3 个参数，解释如下。

（1）第一个参数：提交类型。

（2）第二个参数：提交目标地址。为了防止浏览器从本地缓存中获取数据，需要添加一个参数 t，使 URL 的地址在每一次请求时一直在变化，因为时间一直在变化，这可以让浏览器认为该网址从未被访问过，以确保每次请求都到达服务器端，而不是从缓存中获取数据。

（3）第三个参数：是否以异步方式进行访问，其中传入 true 是异步，传入 false 是同步。

send()方法的作用是发起请求开始访问服务器端。我们可以使用如下 if 语句来处理浏览器的兼容性问题：

```
<script>
    // 创建 AJAX 对象
    var xhr = null;
    if (window.XMLHttpRequest) {
        //现代浏览器
        xhr = new XMLHttpRequest();
    } else {
        //低版本 IE
        xhr = new ActiveXObject("Microsoft.XMLHTTP");
    }
</script>
```

## 12.2　实现有传参无返回值——get提交方式

服务器端代码如下：

```
public class test2 extends HttpServlet {
    protected void doGet(HttpServletRequest request, HttpServletResponse
response)
            throws ServletException, IOException {
        String username = request.getParameter("username");
        String password = request.getParameter("password");
        System.out.println("username=" + username + " password=" + password);
    }
}
```

客户端代码如下：

```
function testMethod3(){
    var ajaxObject = new XMLHttpRequest();
    ajaxObject.open("get", "test3?t=" + new Date().getTime() + "&username=
中国&password=中国人", true);
    ajaxObject.send();
}
```

在 AJAX 中，当以 get 方式传递参数时，参数包含在 URL 网址中。

## 12.3 实现无传参无返回值——post提交方式

服务器端代码如下：

```
public class ajax_test3 extends HttpServlet {
    protected void doPost(HttpServletRequest request, HttpServletResponse
response)
            throws ServletException, IOException {
        System.out.println("ajax_test3 doPost");
    }
}
```

客户端代码如下：

```
function sendAjax3(){
    var ajaxObject = new XMLHttpRequest();
    ajaxObject.open("post", "ajax_test3?t="+new Date().getTime(), true);
    ajaxObject.setRequestHeader("Content-Type",
"application/x-www-form-urlencoded");
    ajaxObject.send();
}
```

在 AJAX 中，当使用 post 提交方式，并且请求中包含提交到服务器端的 param 参数时，必须设置请求头：

```
ajaxObject.setRequestHeader("Content-Type",
"application/x-www-form-urlencoded");
```

因为在 Tomcat 中对请求头中的内容是否为 application/x-www-form-urlencoded 进行

了判断，如果请求头有 key 与 value，则先去解析 request body 中的数据，并将数据填充到 parameters 参数中，然后使用 request.getParameter()方法进行获取。

在本示例中，因为客户端并未向服务器端传递参数，所以如下代码可以不写：

```
ajaxObject.setRequestHeader("Content-Type",
"application/x-www-form-urlencoded");
```

## 12.4　实现有传参无返回值——post提交方式

服务器端代码如下：

```
public class ajax_test4 extends HttpServlet {
    protected void doPost(HttpServletRequest request, HttpServletResponse
response)
            throws ServletException, IOException {
        request.setCharacterEncoding("utf-8");
        System.out.println(request.getParameter("username"));
        System.out.println(request.getParameter("age"));
    }
}
```

客户端代码如下：

```
function sendAjax4(){
   var ajaxObject = new XMLHttpRequest();
   ajaxObject.open("post", "ajax_test4?t="+new Date().getTime(), true);
   ajaxObject.setRequestHeader("Content-Type",
"application/x-www-form-urlencoded");//有传参，必须执行 setRequestHeader
   ajaxObject.send("username=中国人美国人&age=123");
}
```

将以 post 方式提交的数据传给 send()方法作为参数，参数格式为 key=value 键值对形式。以 post 方式提交的数据存放在 Form Data 中。

## 12.5　实现无传参有返回值String——get提交方式

本示例使用 get 提交方式处理返回值，其代码与 post 提交方式处理返回值的代码相同。服务器端代码如下：

```
public class test5 extends HttpServlet {
    protected void doGet(HttpServletRequest request, HttpServletResponse
response)
            throws ServletException, IOException {
        System.out.println("test5 run " + System.currentTimeMillis());
        response.setCharacterEncoding("utf-8");
        PrintWriter out = response.getWriter();
        out.print("我是返回值");
        out.flush();
        out.close();
    }
}
```

服务器返回字符串 String 数据。客户端代码如下：

```
var ajaxObject;
function testMethod(){
    ajaxObject = new XMLHttpRequest();
    ajaxObject.open("get", "test5?t=" + new Date().getTime(), true);
//onreadystatechange 是回调属性，属性名是固定的，值是一个回调 Function。
//onreadystatechange 的作用：监视 AJAXObject 对象的状态发生改变，以做不同的处理，使
用回调方法的目的是实现接收返回值
    ajaxObject.onreadystatechange = callbackFunction;
    ajaxObject.send();
}

function callbackFunction(){
    if (ajaxObject.readyState == 4) {//响应完成
        if (ajaxObject.status == 200) {//响应成功，无异常
            var getText = ajaxObject.responseText;
            alert(getText);
        }
    }
}
```

AJAXObject 对象的 readyState 和 status 属性的取值，如图 12-2 所示。

在 AJAXObject 对象的状态发生变化时，AJAXObject 对象的 onreadystatechange 事件执行关联的回调 Function 方法，在回调方法中进行处理业务。

| readyState | 存在 XMLHttpRequest 的状态。从 0 到 4 发生变化。<br>• 0: 请求未初始化<br>• 1: 服务器连接已建立<br>• 2: 请求已接收<br>• 3: 请求处理中<br>• 4: 请求已完成，且响应已就绪 |
|---|---|
| status | 200: "OK"<br>404: 未找到页面 |

图 12-2

## 12.6　实现无传参有返回值XML——get提交方式

服务器端代码如下：

```
public class test6 extends HttpServlet {
    protected void doGet(HttpServletRequest request, HttpServletResponse
response)
            throws ServletException, IOException {
        System.out.println("test6 run " + System.currentTimeMillis());
response.setContentType("text/xml;charset=utf-8");
        response.setCharacterEncoding("utf-8");
        PrintWriter out = response.getWriter();
        out.print("<userinfo><username>大中国</username></userinfo>");
        out.flush();
        out.close();
    }
}
```

服务器返回 XML 格式的字符串数据，并在前台使用 JQuery 框架对 XML 对象进行解析。

客户端代码如下：

```
var ajaxObject;
function testMethod(){
    ajaxObject = new XMLHttpRequest();
    ajaxObject.open("get", "test6?t=" + new Date().getTime(), true);
    ajaxObject.onreadystatechange = callbackFunction;
    ajaxObject.send();
}

function callbackFunction(){
    if (ajaxObject.readyState == 4) {//响应完成
        if (ajaxObject.status == 200) {//响应成功，无异常
```

```
            var xmlObject = ajaxObject.responseXML;
            var userinfoRef = $(xmlObject).find("userinfo");
            var getText = $(userinfoRef).find("username").text();
            alert(getText);
        }
    }
}
```

## 12.7 实现异步效果

异步就是每个请求走自己的业务，没有排队的效果。服务器端 test7.java 代码如下：

```
public class test7 extends HttpServlet {
    protected void doGet(HttpServletRequest request, HttpServletResponse
response)
            throws ServletException, IOException {
        try {
            System.out.println("test7 begin " +
System.currentTimeMillis());
            Thread.sleep(3000);
            System.out.println("test7   end " +
System.currentTimeMillis());
        } catch (InterruptedException e) {
            e.printStackTrace();
        }
    }
}
```

服务器端 test8.java 代码如下：

```
public class test8 extends HttpServlet {
    protected void doGet(HttpServletRequest request, HttpServletResponse
response)
            throws ServletException, IOException {
        try {
            System.out.println("test8 begin " +
System.currentTimeMillis());
            Thread.sleep(8000);
            System.out.println("test8   end " +
System.currentTimeMillis());
        } catch (InterruptedException e) {
```

```
            e.printStackTrace();
        }
    }
}
```

客户端代码如下：

```
var ajaxObject;
function testMethod(){
    sendAjax1();
    sendAjax2();
}

function sendAjax1(){
    ajaxObject = new XMLHttpRequest();
    ajaxObject.open("get", "test7?t=" + new Date().getTime(), true);
    ajaxObject.send();
}

function sendAjax2(){
    ajaxObject = new XMLHttpRequest();
    ajaxObject.open("get", "test8?t=" + new Date().getTime(), true);
    ajaxObject.send();
}
```

当 AJAXObject 对象的 open()方法的第三个参数传入结果为 true 时，以异步的方式访问服务器。

## 12.8　实现同步效果

同步就是以排队的方式，按顺序去访问服务器。

客户端代码如下：

```
var ajaxObject;
function testMethod(){
    sendAjax1();
    sendAjax2();
}

function sendAjax1(){
    ajaxObject = new XMLHttpRequest();
```

```
    //同步经常用在先删除后列表的情况下
    ajaxObject.open("get", "test7?t=" + new Date().getTime(), false);
    ajaxObject.send();
}

function sendAjax2(){
    ajaxObject = new XMLHttpRequest();
    ajaxObject.open("get", "test8?t=" + new Date().getTime(), true);
    ajaxObject.send();
}
```

当 AJAXObject 对象的 open()方法的第三个参数传入结果为 false 时，以同步的方式访问服务器。

使用异步的场景是两个 AJAX 请求没有关联，各自执行自己的业务，这提高了运行效率，提高了 CPU 的利用率。使用同步的场景是两个 AJAX 请求有关联，比如当删除成功之后才能列出最新版的数据时，要使用同步。

## 12.9 实现无刷新login登录案例

为了加深我们对 AJAX 技术的理解，可以先实现一个具有无刷新功能的 login 登录案例，体会一下 AJAX 的最简应用。服务器端代码如下：

```
public class test9 extends HttpServlet {
    protected void doPost(HttpServletRequest request, HttpServletResponse
response)
            throws ServletException, IOException {
        request.setCharacterEncoding("utf-8");
        String username = request.getParameter("username");
        String password = request.getParameter("password");
        boolean loginResult = false;
        if (username.equals("a") && password.equals("aa")) {
            loginResult = true;
        }
        PrintWriter out = response.getWriter();
        out.print(loginResult);// 注意：不要使用print()方法
        out.flush();
        out.close();
    }
}
```

客户端代码如下：

```
<!DOCTYPE html>
<html>
    <head>
        <meta charset="UTF-8">
        <title>Insert title here</title>
        <script src="jquery-3.1.0.js">
        </script>
        <script type="text/javascript">
            var ajaxObject;

            function login(){
                var usernameValue = $("#username").val();
                var passwordValue = $("#password").val();

                ajaxObject = new XMLHttpRequest();
                ajaxObject.open("post", "test9?t=" + new Date().getTime(),
true);
                ajaxObject.setRequestHeader("Content-Type",
"application/x-www-form-urlencoded");
                ajaxObject.onreadystatechange = callbackFunction;
                ajaxObject.send("username=" + usernameValue + "&password=" +
passwordValue);
            }

            function callbackFunction(){
                if (ajaxObject.readyState == 4) {//响应完成
                    if (ajaxObject.status == 200) {//响应成功，无异常
                        var loginResult = ajaxObject.responseText;
                        alert("|" + loginResult + "|");
                        if (loginResult == 'true') {
                            alert("登录成功");
                        }
                        else {
                            alert("登录失败！");
                        }
                    }
                }
            }
        </script>
    </head>
    <body>
```

```
        <div>
            username:<input type="text" id="username" />
            <br/>
            password:<input type="text" id="password" />
            <br/>
            <input type="button" onclick="javascript:login()" />
        </div>
    </body>
</html>
```

## 12.10　formdata和payload提交

本节知识点如下。

（1）当以 get 方式提交时，都是使用 URL 方式传输数据。

（2）当以 post 方式提交时，可以使用 formdata 或 payload 方式传输数据。

使用 URL 或 formdata 方式传输数据可以使用如下代码获取参数值：

```
String username = request.getParameter("username");
String password = request.getParameter("password");
```

使用 payload 方式传输数据，要使用如下代码：

```
BufferedReader reader = request.getReader();
```

创建 Web 项目 formdata_payload_test。创建接收 get 方式提交的 Servlet 代码如下：

```
package controller;

import java.io.IOException;

import javax.servlet.ServletException;
import javax.servlet.http.HttpServlet;
import javax.servlet.http.HttpServletRequest;
import javax.servlet.http.HttpServletResponse;

public class getTest1 extends HttpServlet {
    protected void doGet(HttpServletRequest request, HttpServletResponse
response)
            throws ServletException, IOException {
        String username = request.getParameter("username");
        String password = request.getParameter("password");
```

```
        System.out.println("getTest1 username=" + username + " password=" +
password);
    }
}
```

创建接收 post formdata 方式提交的 Servlet 代码如下：

```
package controller;

import java.io.IOException;

import javax.servlet.ServletException;
import javax.servlet.http.HttpServlet;
import javax.servlet.http.HttpServletRequest;
import javax.servlet.http.HttpServletResponse;

public class postTest1 extends HttpServlet {
    protected void doPost(HttpServletRequest request, HttpServletResponse
response)
            throws ServletException, IOException {
        String username = request.getParameter("username");
        String password = request.getParameter("password");
        System.out.println("postTest1 username=" + username + " password="
+ password);
    }
}
```

创建接收 post payload 方式提交的 Servlet 代码如下：

```
package controller;

import java.io.BufferedReader;
import java.io.IOException;

import javax.servlet.ServletException;
import javax.servlet.http.HttpServlet;
import javax.servlet.http.HttpServletRequest;
import javax.servlet.http.HttpServletResponse;

public class postTest2 extends HttpServlet {
    protected void doPost(HttpServletRequest request, HttpServletResponse
response)
            throws ServletException, IOException {
```

```
        StringBuffer buffer = new StringBuffer();
        BufferedReader reader = request.getReader();
        char[] charArray = new char[2];
        int readLength = reader.read(charArray);
        while (readLength != -1) {
            buffer.append(charArray);
            readLength = reader.read(charArray);
        }
        System.out.println(buffer.toString());
        reader.close();
    }
}
```

### 12.10.1 测试 get 方式传输数据需要依赖 URL

使用原生 AJAX API 发起 AJAX 的代码如下：

```
function testMethod1() {
    var ajaxObject = new XMLHttpRequest();
    var paramValue = "username=usernameValue1&password=passwordValue1";
    var url = "getTest1?t=" + new Date().getTime();
    url = url + "&" + paramValue;
    ajaxObject.open("get", url, true);
    ajaxObject.send();
}
```

代码执行效果如图 12-3 所示。

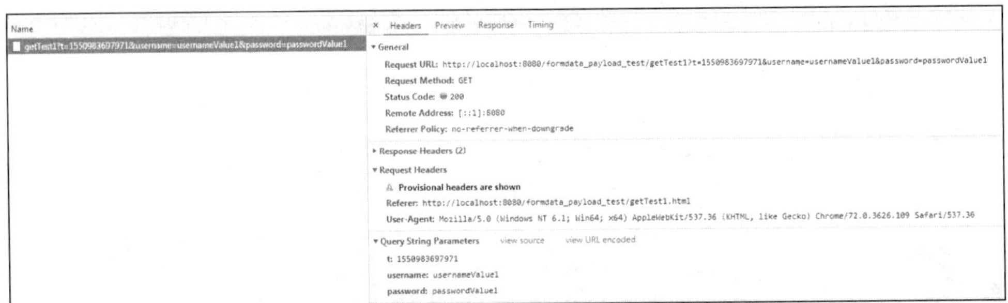

图 12-3

由此可见，服务器端可以获取参数值。

使用 jquery 的 AJAX 模块发起 AJAX 的代码如下：

```
function testMethod2() {
    var paramValue = "username=usernameValue1&password=passwordValue1";
    var url = "getTest1?t=" + new Date().getTime();
```

```
    url = url + "&" + paramValue;
    $.get(url);
}
```

代码执行效果如图 12-4 所示。

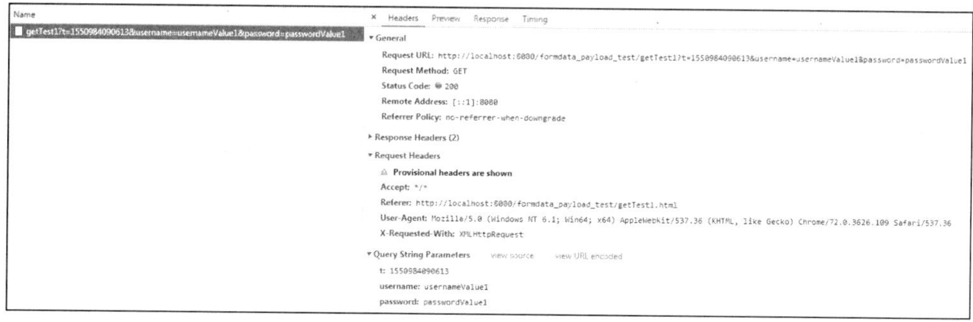


图 12-4

可见，服务器端可以获取参数值。

使用 jquery 的 AJAX 模块发起 AJAX 的代码如下：

```
function testMethod3() {
    var paramValue = "username=usernameValue1&password=passwordValue1";
    var url = "getTest1?t=" + new Date().getTime();
    url = url + "&" + paramValue;
    $.ajax({
        "url" : url,
        "type" : "get"
    });
}
```

代码执行效果如图 12-5 所示。

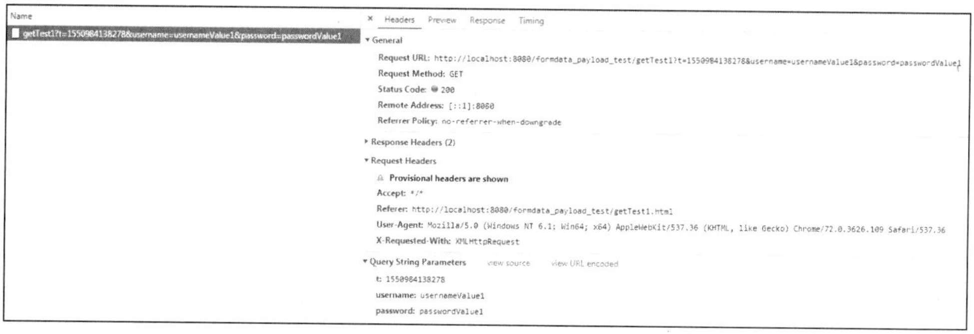

图 12-5

可见，服务器端可以获取参数值。

使用<form>实现 get 提交的代码如下：

```
<form action="getTest1" method="get">
    username:<input type="text" name="username" value="usernameValue2"><br
/>
    password:<input type="text" name="password" value="passwordValue2"><br
/>
    <input type="submit" value="get 提交_表单提交">
</form>
```

代码执行效果如图 12-6 所示。

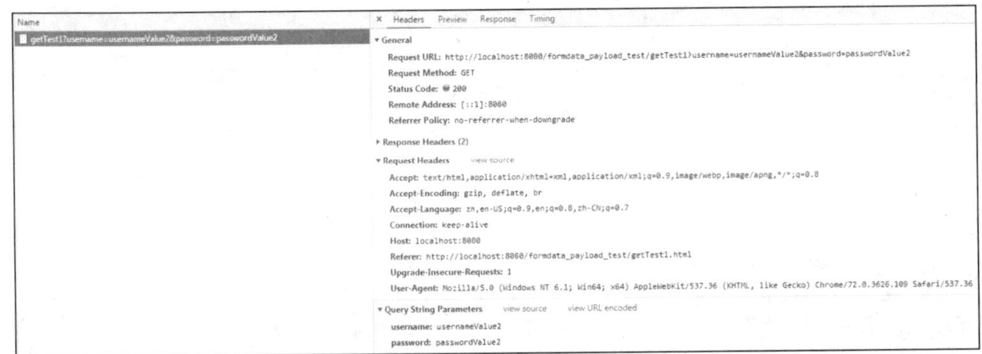

图 12-6

可见，服务器端可以获取参数值。

### 12.10.2　测试 post 提交使用 formdata 方式传输数据

使用原生 AJAX API 发起 AJAX 的代码如下：

```
function testMethod1() {
    var ajaxObject = new XMLHttpRequest();
    var paramValue = "username=usernameValue1&password=passwordValue1";
    var url = "postTest1?t=" + new Date().getTime();
    ajaxObject.open("post", url, true);
    ajaxObject.setRequestHeader("Content-Type",
            "application/x-www-form-urlencoded");
    ajaxObject.send(paramValue);
}
```

代码执行效果如图 12-7 所示。

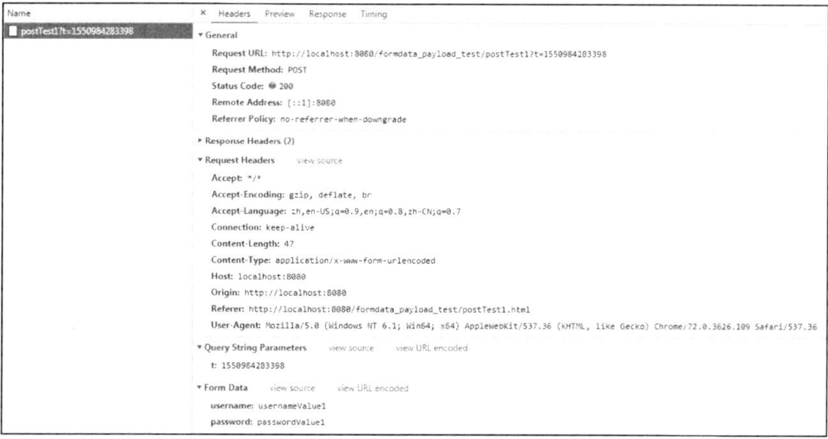


图 12-7

可见，服务器端可以获取参数值。

使用 jquery 的 AJAX 模块发起 AJAX 的代码如下：

```
function testMethod2() {
    var paramValue = "username=usernameValue1&password=passwordValue1";
    var url = "postTest1?t=" + new Date().getTime();
    $.post(url, paramValue);
}
```

代码执行效果如图 12-8 所示。

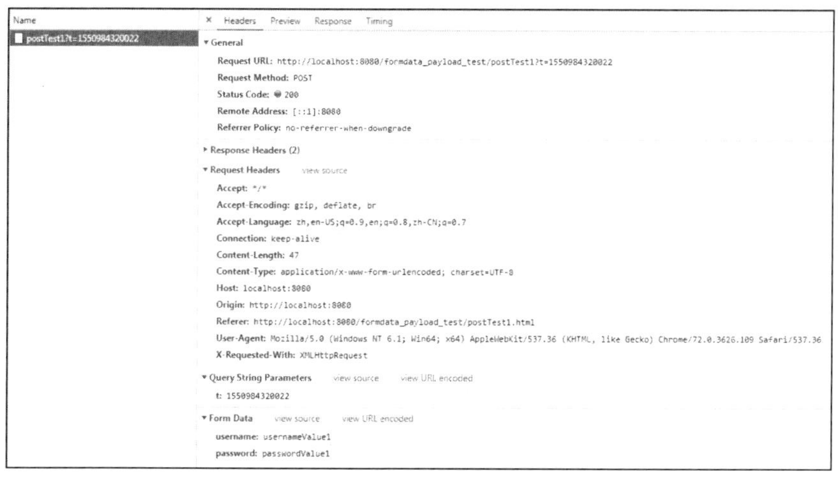


图 12-8

可见，服务器端可以获取参数值。

使用 jquery 的 AJAX 模块发起 AJAX 的代码如下：

```
function testMethod3() {
    var paramValue = "username=usernameValue1&password=passwordValue1";
    var url = "postTest1?t=" + new Date().getTime();
    $.ajax({
        "url" : url,
        "type" : "post",
        "contentType" : "application/x-www-form-urlencoded",
        "data" : {
            "username" : "usernameValue1",
            "password" : "passwordValue1"
        }
    });
}
```

代码执行效果如图 12-9 所示。

图 12-9

可见，服务器端可以获取参数值。使用<form>实现 post 提交的代码如下：

```
<form action="postTest1" method="post">
    username:<input type="text" name="username" value="usernameValue2"><br
/>
    password:<input type="text" name="password" value="passwordValue2"><br
/>
```

```
    <input type="submit" value="post 提交_表单提交">
</form>
```

代码执行效果如图 12-10 所示。

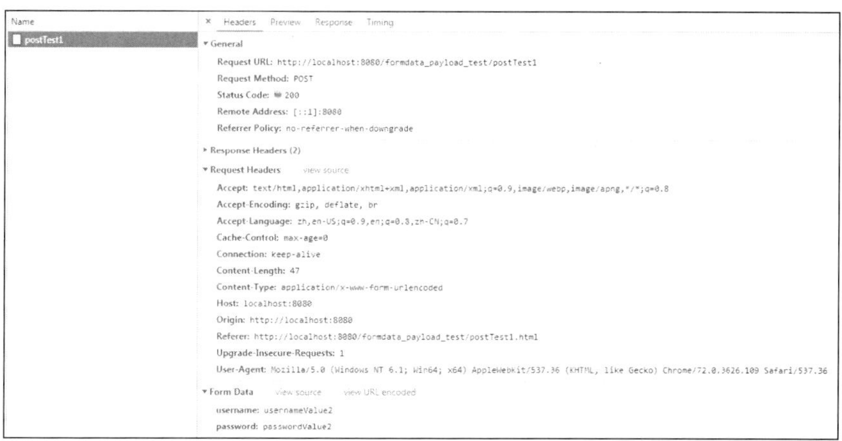


图 12-10

可见，服务器端可以获取参数值。上面所有测试都使用 formdata 方式传输数据到服务器端。

## 12.10.3　测试 post 提交使用 payload 方式传输数据

使用原生 AJAX API 发起 AJAX 的代码如下：

```
function Userinfo(username, password) {
    this.username = username;
    this.password = password;
}

function testMethod4() {
    var userinfo = new Userinfo("usernameValue", "passwordValue");
    var ajaxObject = new XMLHttpRequest();
    var url = "postTest1?t=" + new Date().getTime();
    //payload 只能用于 post 提交
    ajaxObject.open("post", url, true);
    //数据类型必须为 application/x-www-form-urlencoded 之外的类型
    ajaxObject.setRequestHeader("Content-Type",
            "application/json;charset=utf-8");
    ajaxObject.send(JSON.stringify(userinfo));
}
```

代码执行效果如图 12-11 所示。

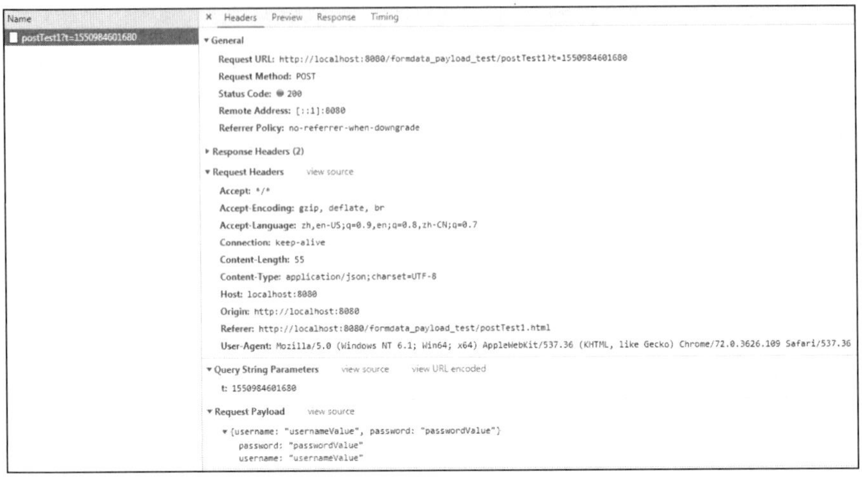

图 12-11

服务器端打印结果如下：

```
postTest1 username=null password=null
```

使用 jquery 的 AJAX 模块发起 AJAX 的代码如下：

```
function testMethod5() {
    var userinfo = new Userinfo("usernameValue", "passwordValue");
    var url = "postTest1?t=" + new Date().getTime();
    $.ajax({
        "url" : url,
        //payload 只能用于 post 提交
        "type" : "post",
        //数据类型必须为 application/x-www-form-urlencoded 之外的类型
        contentType : 'application/json;charset=utf-8',
        "data" : JSON.stringify(userinfo)
    });
}
```

代码执行效果如图 12-12 所示。

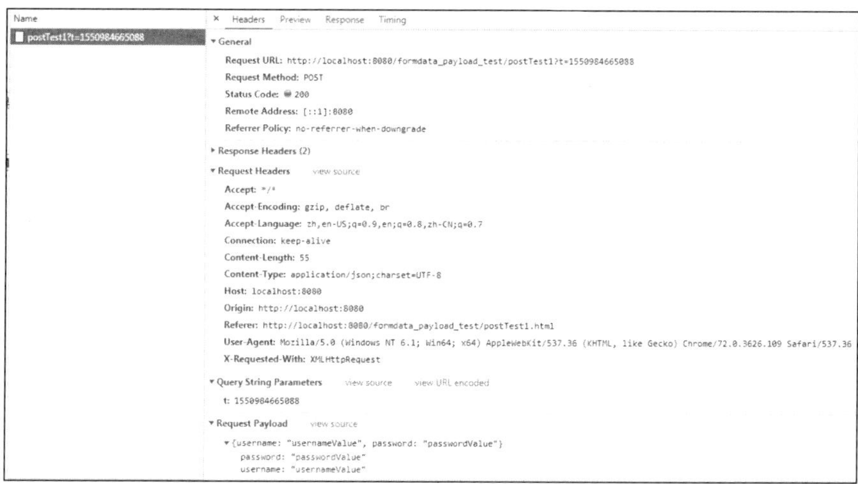


图 12-12

服务器端打印结果如下：

```
postTest1 username=null password=null
```

前面两段代码并未在服务器端将参数打印出来，那么如何获取使用 payload 方式传输的数据呢？将前面两段在服务器端没有打印结果的前台代码提交到类名为 postTest2.java 的 Servlet 即可获取。

总结以下 3 个方面。

（1）URL 传参：使用 getParameter()方法取参。

（2）form-data 传参：使用 getParameter()方法取参。

（3）payload 传参：使用 request.getReader()方法取参。

## 12.11　实现文件上传：<form>有刷新

前台 HTML 代码如下：

```
<form action="uploadTest" method="post" enctype="multipart/form-data">
    <input type="text" name="username" value="中国">
    <br>
    <input type="text" name="password" value="中国人">
    <br>
    <input type="file" name="uploadFile">
    <br>
```

```
    <input type="file" name="uploadFile">
    <br>
    <input type="file" name="uploadFile">
    <br>
    <input type="submit" value="submit">
</form>
```

服务器端上传文件示例代码如下：

```
@WebServlet(name = "uploadTest", urlPatterns = "/uploadTest")
@MultipartConfig
public class UploadTest extends HttpServlet {
    protected void doPost(HttpServletRequest request, HttpServletResponse
response) throws ServletException, IOException {
        //设置编码
        request.setCharacterEncoding("utf-8");
        String username = request.getParameter("username");
        String password = request.getParameter("password");
        System.out.println(username + " " + password);
        //获取上传的目标路径
        String uploadPath = this.getServletContext().getRealPath("/upload");
        System.out.println(uploadPath);
        //处理日期
        SimpleDateFormat format = new SimpleDateFormat
("yyyy_MM_dd_HH_mm_ss");
        Collection<Part> parts = request.getParts();
        Iterator<Part> iterator = parts.iterator();
        while (iterator.hasNext()) {
            Part part = iterator.next();
            InputStream fileContent = part.getInputStream();
            if (part.getSubmittedFileName() != null) {
                String uploadFileName = format.format(new Date());
                uploadFileName    =    uploadFileName    +    "_"    +
System.currentTimeMillis()   +   "_"   +   Math.random()   +   "_"   +
part.getSubmittedFileName();
                uploadFileName = uploadPath + "\\" + uploadFileName;
                System.out.println(uploadFileName);
                part.write(uploadFileName);
            }
          fileContent.close();
        }
    }
}
```

此段代码使用原生 Servlet3.0 API 进行实现。

## 12.12　实现文件上传：AJAX无刷新

前台 HTML 代码如下：

```
<!DOCTYPE html>
<html>
    <head>
        <meta charset="utf-8">
        <title>Insert title here</title>
        <script type="text/javascript" src="jquery-3.5.1.js"></script>
        <script type="text/javascript">
            function ajaxUpload() {
                var usernameValue = $("#username").val();
                var inputFile = $("#uploadFile")[0].files[0];

                var formData = new FormData();
                formData.append('username', usernameValue)
                formData.append('uploadFile', inputFile);

                $.ajax({
                    url: 'uploadTest?t='+new Date().getTime(),
                    type: 'POST',
                    data: formData,
                    processData: false,
                    contentType: false,
                    success: function() {
                        alert("上传成功！");
                    }
                });
            }
        </script>
    </head>
    <body>
        <div>
            username:<input type="text" id="username"><br>
            file:<input type="file" id="uploadFile"><br>
            <input  type="button"  value="button"  onclick=  "javascript:
ajaxUpload()"><br>
        </div>
    </body>
</html>
```

# 第 13 章 搭建 Maven Nexus 私服环境

搭建 Nexus 私服的主要原因如下：

（1）软件公司不同项目组写的公共 jar 包文件需要共享，但属于技术机密，又不想让它们通过互联网被访问。

（2）保证在局域网内上传和下载 jar 包文件的速度。

（3）自己的私服自己说了算。

Nexus 是一个强大的 Maven 仓库管理系统，具有极大地简化本地内部仓库的维护和外部仓库的访问、降低中央仓库的负荷、方便自己部署组件等优势。Nexus 是一套开箱即用的系统，不需要数据库，利用文件系统与 Lucene 框架结合的形式组织数据。

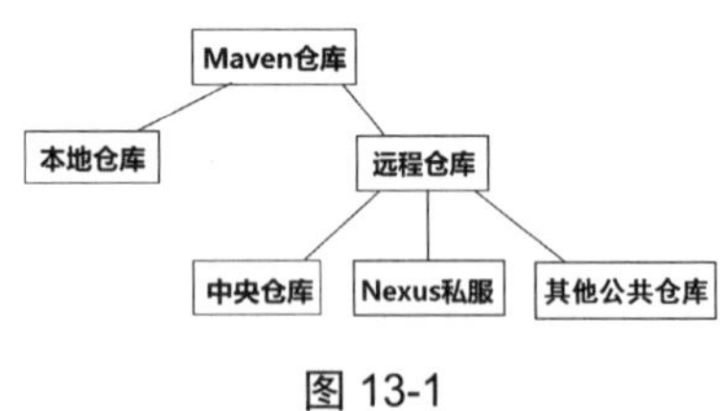


图 13-1

在 Maven 中，Repository 仓库存储项目生成组件，并根据组件的坐标到仓库中寻找这些组件，组件包含.jar 或.war 文件。Maven 中的 Repository 仓库主要分为两种，如图 13-1 所示。

当 Maven 根据组件坐标寻找组件时，首先从本地仓库寻找，如果本地仓库中存在该组件，则直接使用。如果本地仓库中不存在该组件，或者需要查看一下是否有更新的组件版本，则 Maven 就会去远程仓库中寻找，发现需要的组件之后，再下载到本地仓库使用。如果在本地仓库和远程仓库中都没有找到，则 Maven 会出现异常提示。

远程仓库主要分为三种：

（1）中央仓库是 Maven 自带的远程仓库，里面包含现在主流项目的 jar 包文件。

（2）Nexus 是一款管理私服的软件，是常用于公司内部共享公共组件的系统。

（3）其他公共仓库是指非 Maven 官方提供的仓库，如操作 Redis 的 Java Client API

客户端 Jedis 就提供了 Snapshot 版本的仓库，如图 13-2 所示。

**Snapshots**

```
<repositories>
  <repository>
    <id>snapshots-repo</id>
    <url>https://oss.sonatype.org/content/repositories/snapshots</url>
  </repository>
</repositories>
```

and

```
<dependencies>
  <dependency>
    <groupId>redis.clients</groupId>
    <artifactId>jedis</artifactId>
    <version>3.7.0-SNAPSHOT</version>
  </dependency>
</dependencies>
```

for upcoming major release

```
<dependencies>
  <dependency>
    <groupId>redis.clients</groupId>
    <artifactId>jedis</artifactId>
    <version>4.0.0-SNAPSHOT</version>
  </dependency>
</dependencies>
```

图 13-2

这是由 Jedis 提供的公共仓库，以上配置代码需要在 pom.xml 文件中添加。

## 13.1　下载Nexus OSS版本

Nexus 有两个版本，如图 13-3 所示。

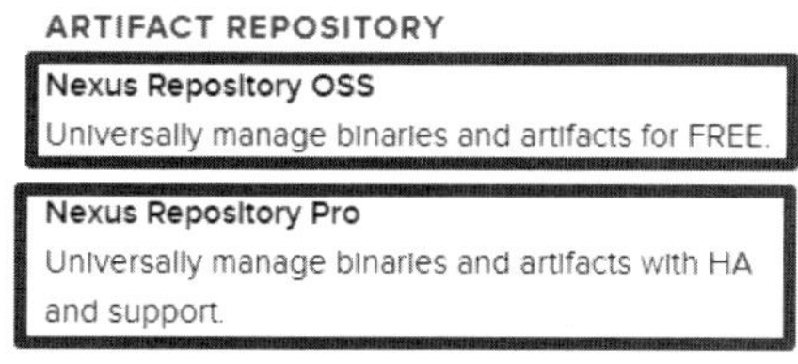


图 13-3

Professional Edition 版本是收费的，OSS Edition 是免费的,其中 OSS 是 Open Source Software 的缩写。

打开 Nexus 官网页面，单击“Nexus Reposltory OSS ”选项，如图 13-4 所示。

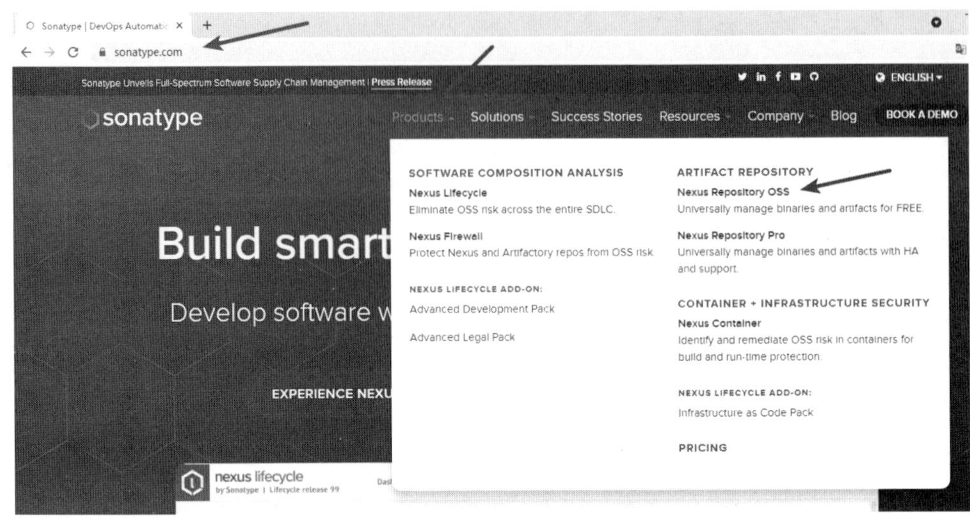


图 13-4

显示界面如图 13-5 所示，单击“GET REPOSITORY OSS”按钮。

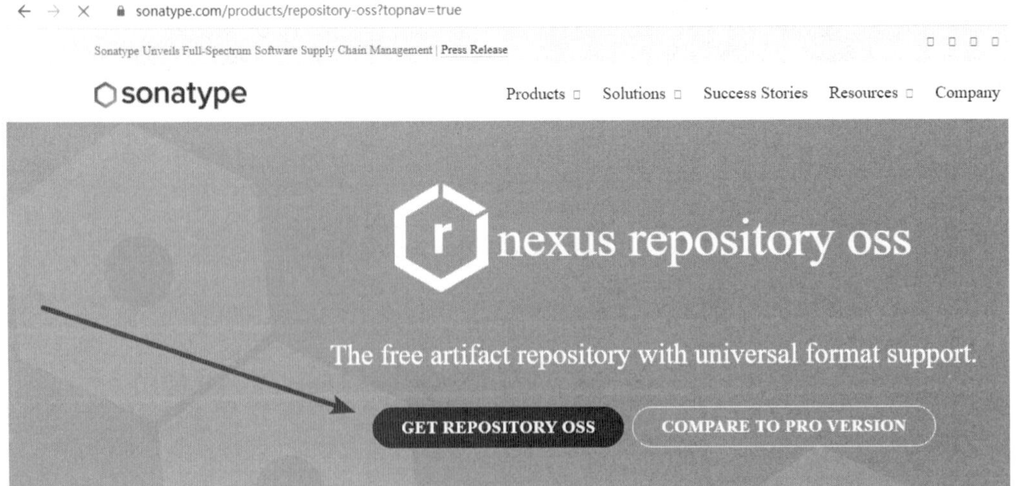


图 13-5

显示界面如图 13-6 所示，在输入邮箱地址后，会向邮箱发送使用手册的链接，单击“DOWNLOAD”按钮。

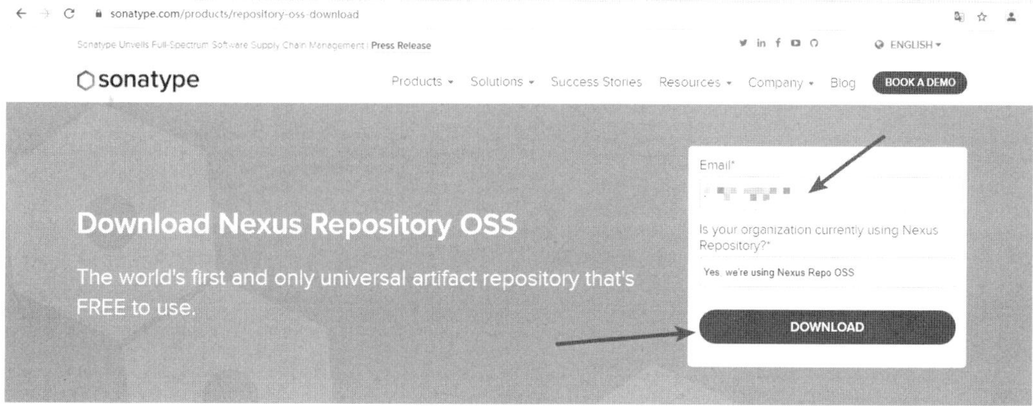


图 13-6

显示界面如图 13-7 所示，下载 Windows 版本的 Nexus。

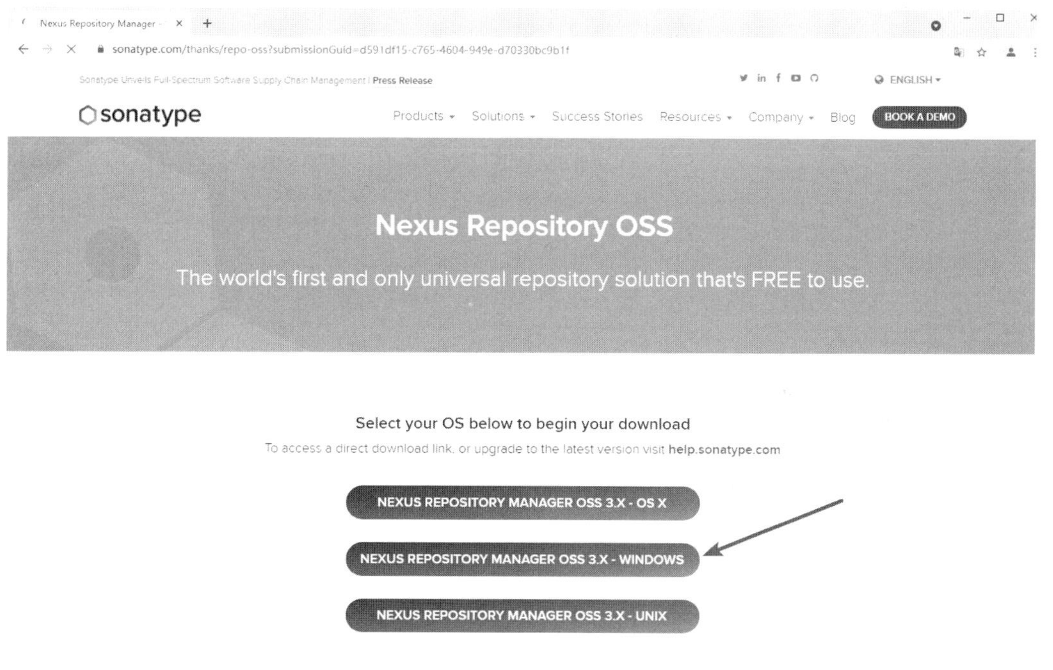


图 13-7

将下载的.zip 文件解压，如图 13-8 所示。

注意，解压的路径不要有中文，不然在启动 Nexus 服务时，会出现如图 13-9 所示的异常。

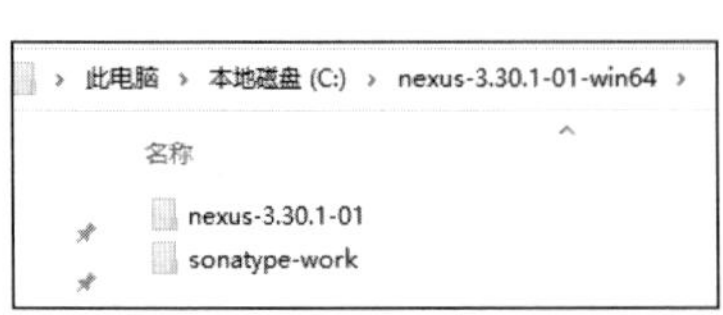

图 13-8

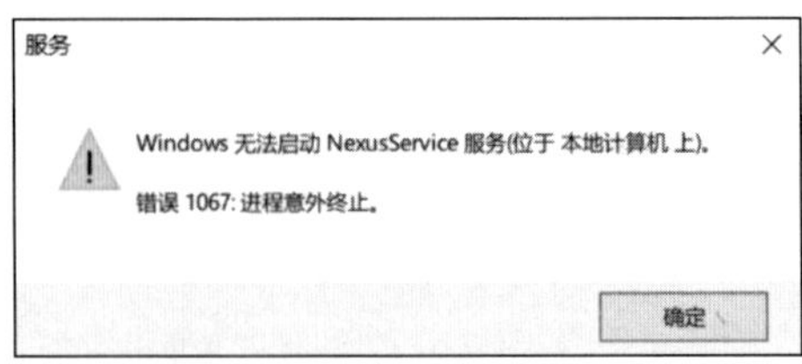


图 13-9

## 13.2 配置Nexus OSS环境变量

在路径中配置 Nexus 环境变量，如图 13-10 所示。

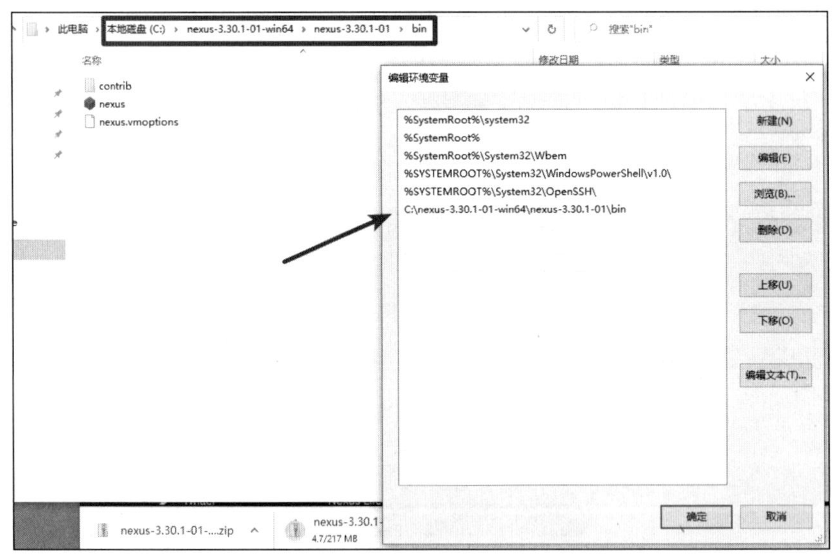


图 13-10

## 13.3 安装服务和启动服务

你一定要以管理员身份运行 CMD，如图 13-11 所示。不然，在执行命令时会出现异常：Could not open SCManager。

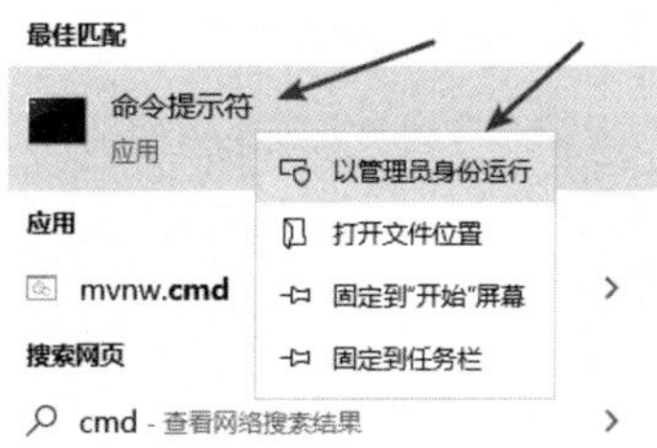


图 13-11

先在 CMD 中执行如下命令安装 Nexus 服务：

```
nexus /install NexusService
```

其中，参数值“NexusService”是自定义的服务名称。

提示：当在有中文的路径中执行 nexus /install 命令安装 Nexus 服务时，执行 nexus/start 命令会出现 1067 错误，这时，可以先使用 nexus /uninstall 命令卸载 Nexus 服务，然后将 Nexus 文件夹转移到全英文路径，再使用 nexus/install 命令重新安装 Nexus 服务。

再执行如下命令启动 NexusServer：

```
nexus /start NexusService
```

Windows 服务列表中的 Nexus 状态，如图 13-12 所示。

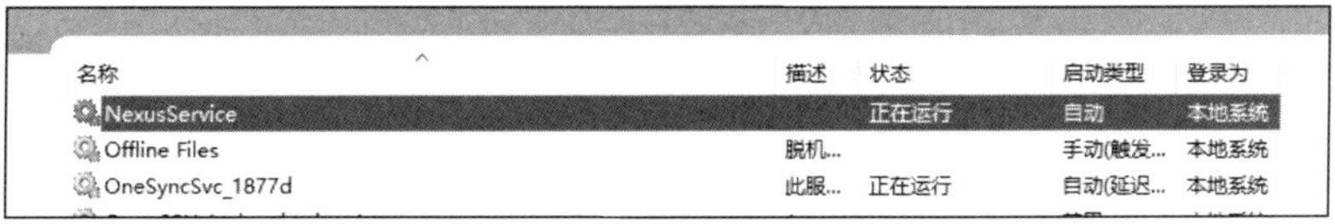


图 13-12　服务列表

## 13.4　登录Nexus

打开 http://localhost:8081，进入 Nexus 控制台界面，如图 13-13 所示。

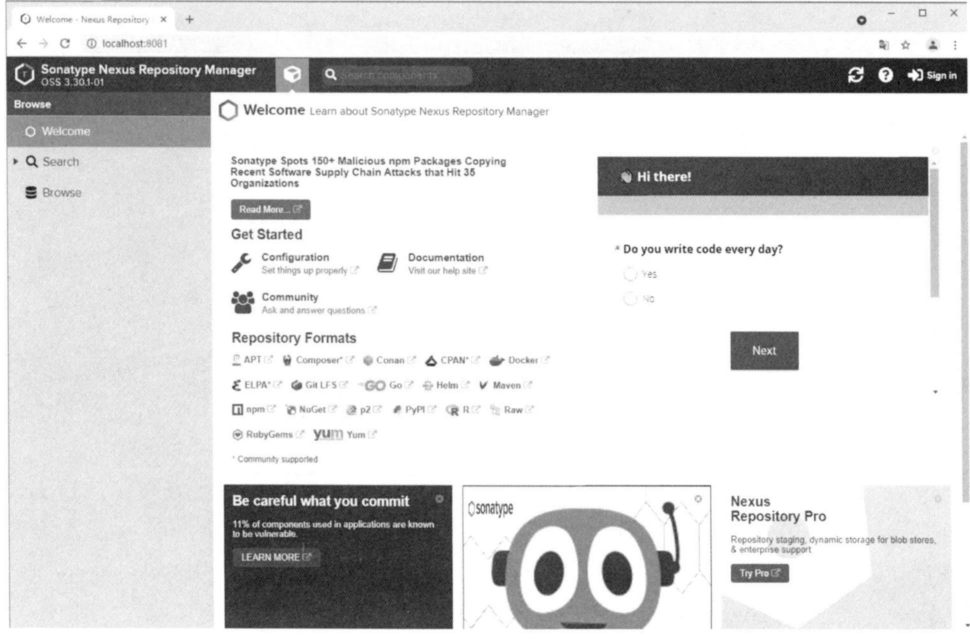


图 13-13

单击右上角的“Sign in”，输入账号和密码，如图 13-14 所示。

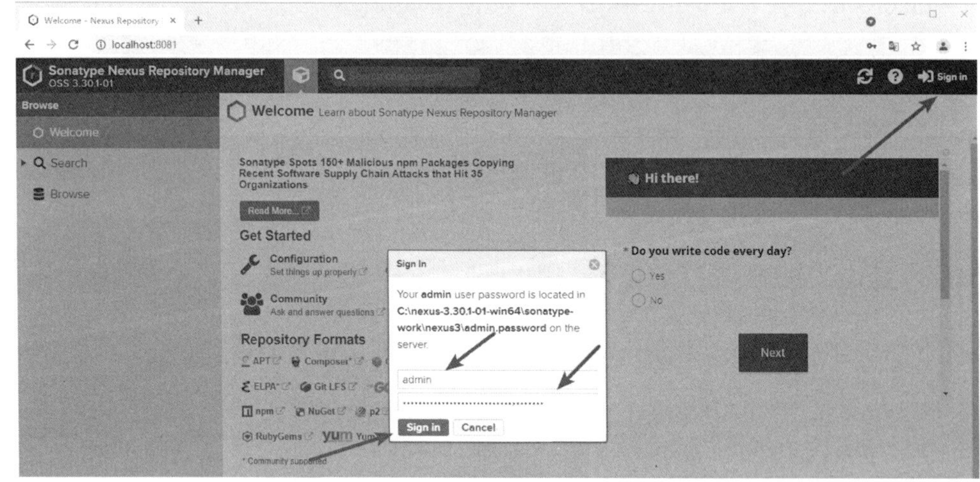


图 13-14

默认账号是 admin，默认密码保存在如图 13-15 所示的路径和文件中。

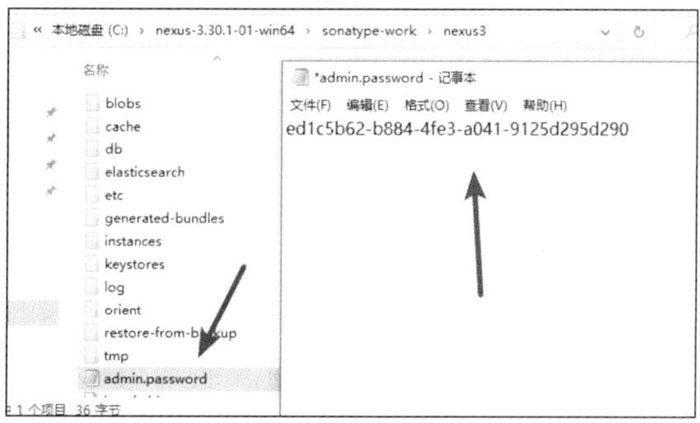


图 13-15

## 13.5　重置Nexus登录密码

输入正确的账号和密码后进行登录，显示配置界面如图 13-16 所示。

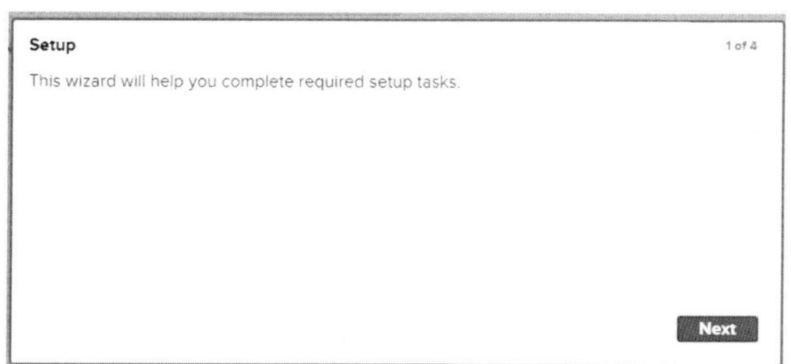


图 13-16

输入新的密码，如图 13-17 所示。

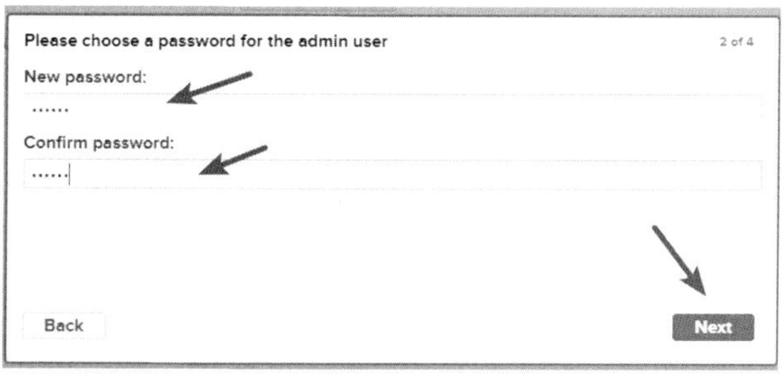


图 13-17

设置禁用匿名访问，如图 13-18 所示。

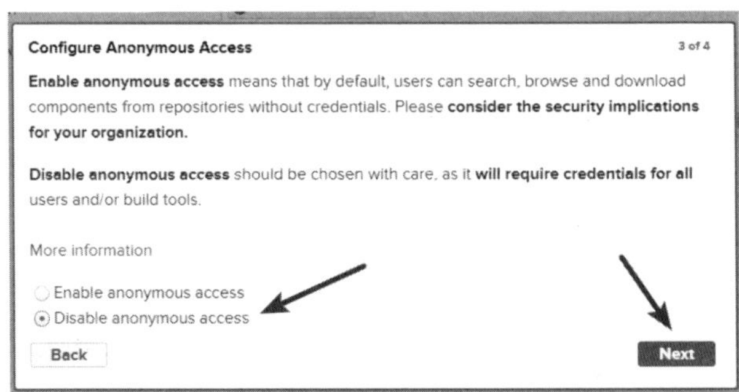


图 13-18

单击“Finish”按钮完成配置，如图 13-19 所示。

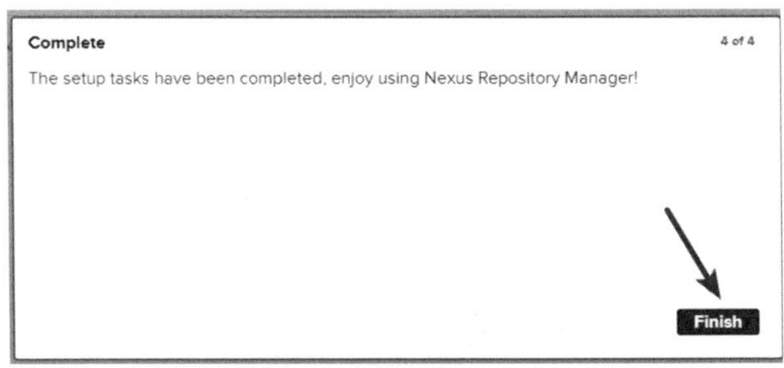


图 13-19

成功登录的效果，如图 13-20 所示。

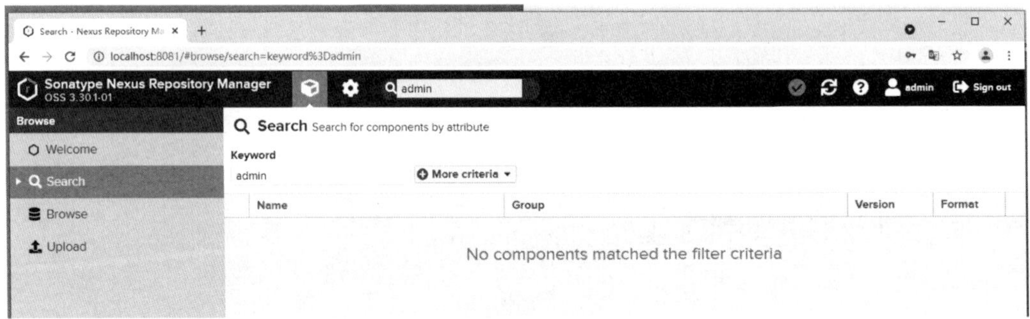


图 13-20

## 13.6 解决连接异常

查看日志，Nexus 的端口号为 8081，如图 13-21 所示。

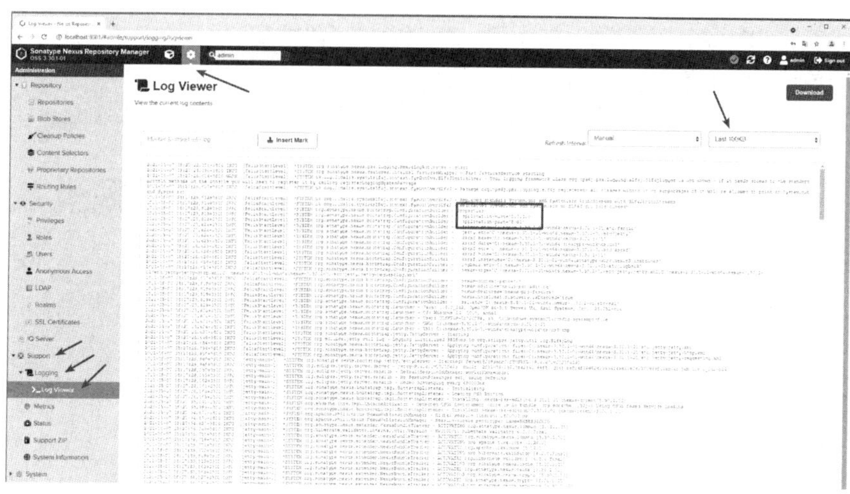


图 13-21

这时，在日志的下方出现连接远程服务器异常，如图 13-22 所示。

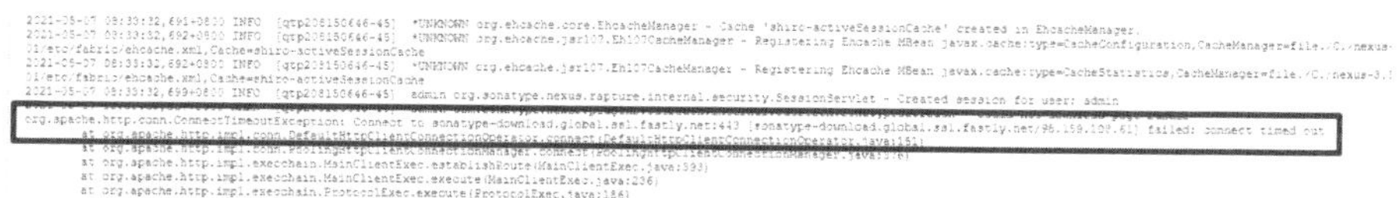


图 13-22

先单击“Outreach:Management”链接，如图 13-23 所示。

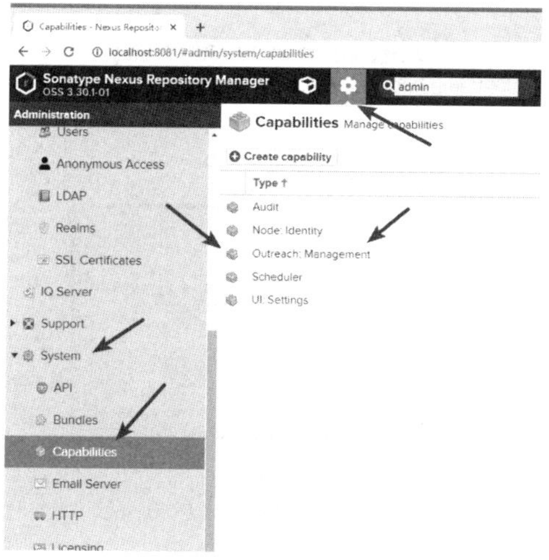


图 13-23

然后单击“Disable”按钮，如图 13-24 所示。这时图标变灰，成功进行 Disable 操作，禁止访问外部服务器，如图 13-25 所示。

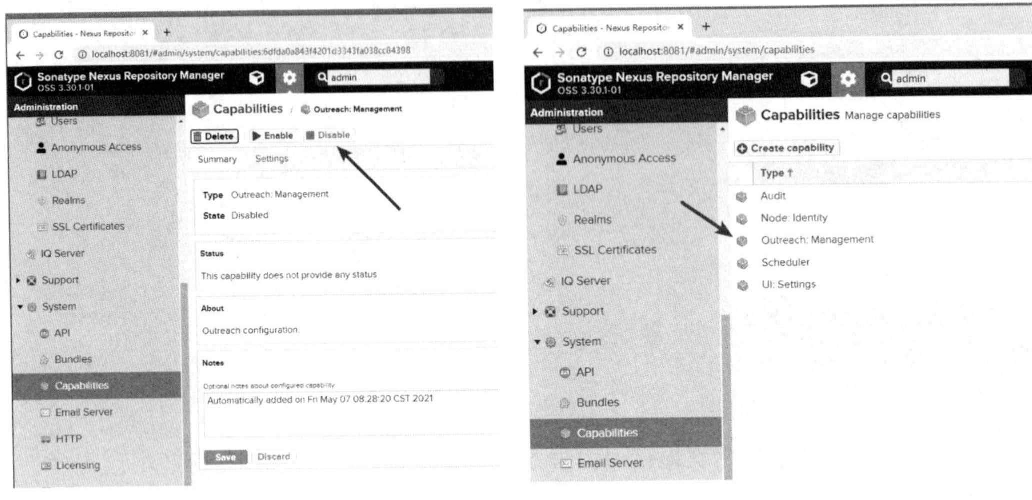

图 13-24　　　　图 13-25

## 13.7　仓库的类型

Nexus 自带了一些仓库，如图 13-26 所示。

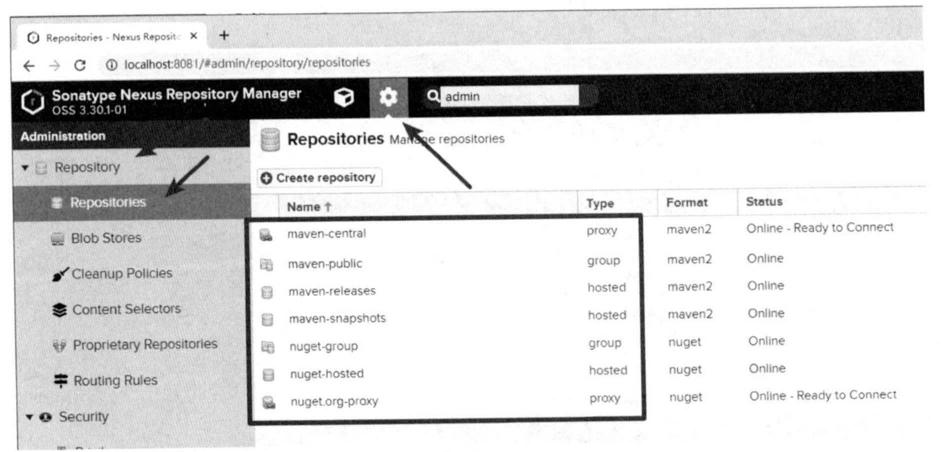

图 13-26

Nexus 主要提供三种不同类型的仓库：

（1）hosted（本地仓库）：用于部署公司或自己的 jar 包组件。

（2）proxy（代理仓库）：用于代理远程的公共仓库，如代理阿里云的 Maven 仓库。

（3）group（仓库组）：用于合并多个 hosted 和 proxy 仓库，如若项目同时在操作多个 repository 仓库，则不需要多次引用目标 repository 仓库，只需要引用一个 group 仓库组即可。

## 13.8　创建hosted类型的Maven仓库

hosted 类型的 Maven 仓库用于存储自己的 JAR 文件。单击“Create repository”按钮创建仓库，如图 13-27 所示。

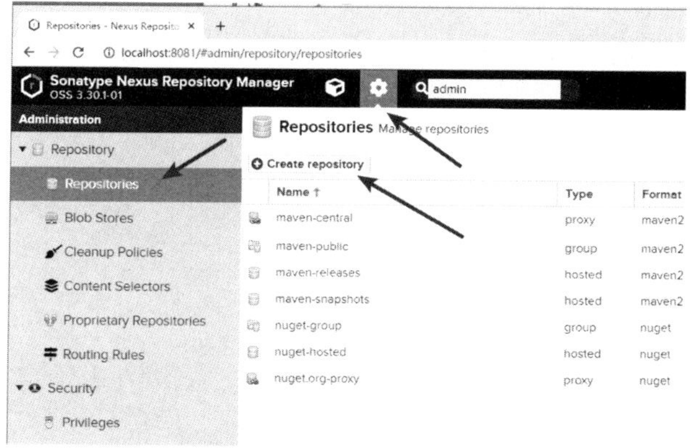


图 13-27

选择“maven2(hosted)”选项，如图 13-28 所示。

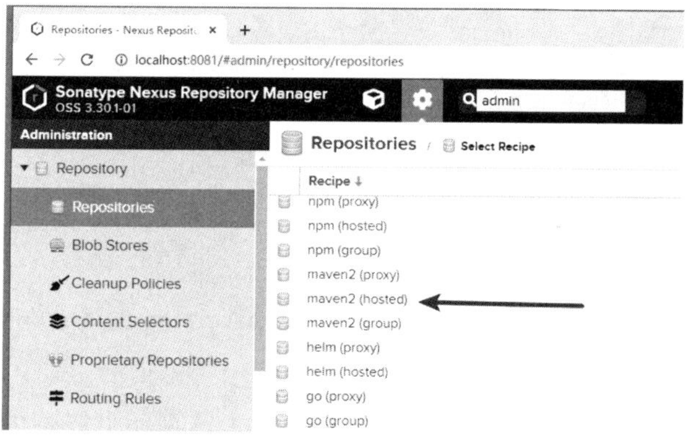


图 13-28

配置仓库信息，如图 13-29 所示。

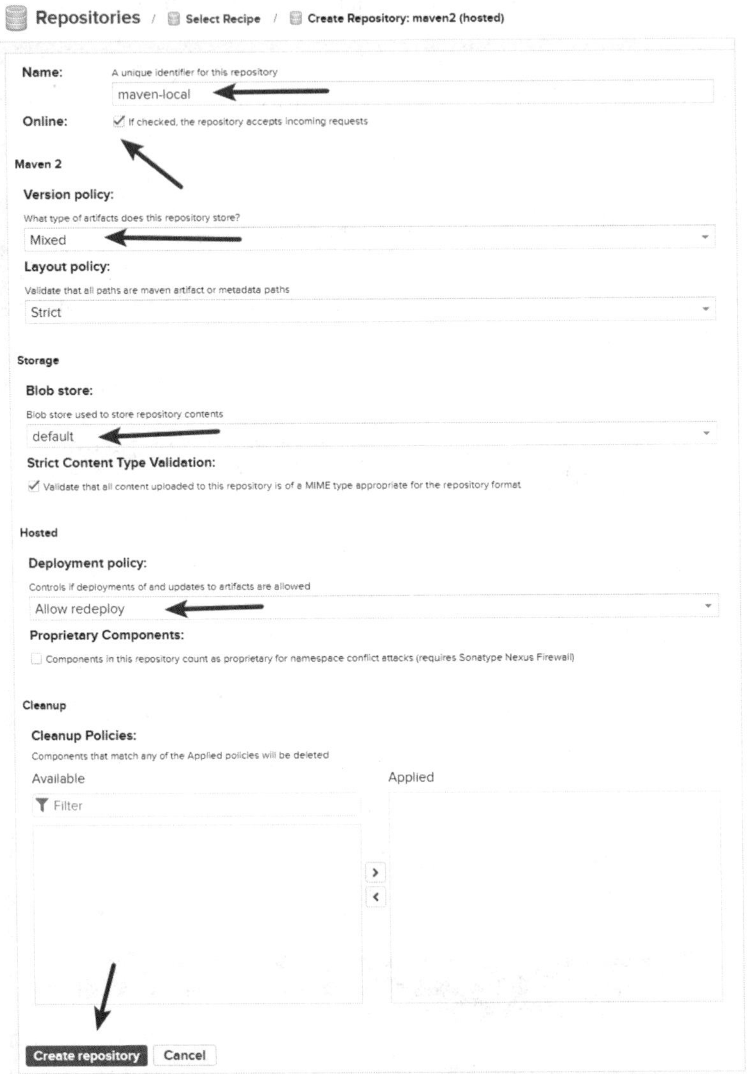

图 13-29

一个仓库可以存储两种版本的组件，而当使用 Version policy 版本策略进行配置时，主要有两个选项。

（1）Releases：负责存储项目中发布的 Releases 版本的组件，也就是稳定版本。比如，如果你完成了一个 jdbc 通用组件的项目，生成的组件名为 jdbc.jar，那么就可以把

此组件发布到 Releases 版本中。

（2）Snapshots：此仓库也比较常用，主要作用是存储非 Releases 版本的组件，也就是非稳定版本。比如，在 Releases 版本发布之前可能需要临时发布一个版本供同事使用，那么此时就可以将组件发布到 Snapshots 版本中。

在这里，我们选择混合版本，也就是当前创建的仓库可以保存 Releases 和 Snapshots 两个版本的组件。

单击“Create repository”按钮，开始创建 hosted 仓库。

## 13.9　创建proxy类型的Maven仓库

proxy 类型的 Maven 仓库的作用是将远程仓库中的 JAR 文件下载到本地的 Nexus 系统中，当需要 JAR 文件并且其也在 Nexus 系统中存在时，优先从 Nexus 系统中获取，而不远程获取，节省了带宽和下载 JAR 文件的时间。

选择“maven2(proxy)”选项，如图 13-30 所示。

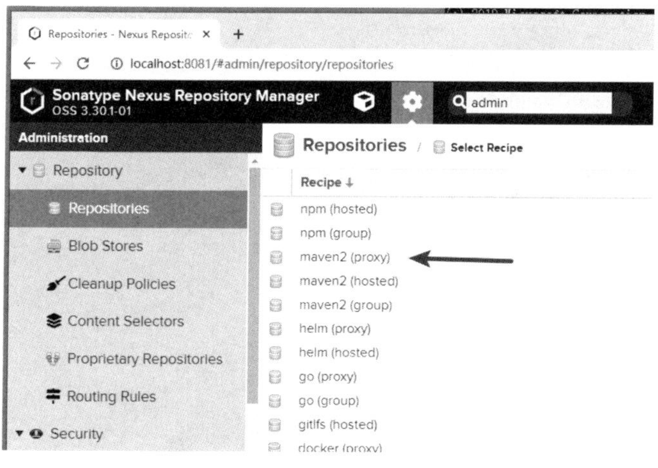


图 13-30

配置仓库信息，如图 13-31 所示。

此 proxy 仓库的作用是从阿里云 maven 镜像站获取 JAR 文件。

单击“Create repository”按钮，开始创建 proxy 仓库。

Repositories / Select Recipe / Create Repository: maven2 (proxy)

**Name:** A unique identifier for this repository

maven-proxy

**Online:** If checked, the repository accepts incoming requests

**Maven 2**

**Version policy:**

What type of artifacts does this repository store?

Mixed

**Layout policy:**

Validate that all paths are maven artifact or metadata paths

Strict

**Proxy**

**Remote storage:**

Location of the remote repository being proxied, e.g. https://repo1.maven.org/maven2/

http://maven.aliyun.com/nexus/content/groups/public

**Blocked:**

Block outbound connections on the repository

**Auto blocking enabled:**

Auto-block outbound connections on the repository if remote peer is detected as unreachable/unresponsive

**Maximum component age:**

How long (in minutes) to cache artifacts before rechecking the remote repository. Release repositories should use -1.

-1

**Maximum metadata age:**

How long (in minutes) to cache metadata before rechecking the remote repository.

1440

**Storage**

**Blob store:**

Blob store used to store repository contents

default

**Strict Content Type Validation:**

Validate that all content uploaded to this repository is of a MIME type appropriate for the repository format

**Routing Rule**

Choose a rule to restrict some requests from being served by this repository

None

**Negative Cache**

**Not found cache enabled:**

Cache responses for content not present in the proxied repository

**Not found cache TTL:**

How long to cache the fact that a file was not found in the repository (in minutes)

1440

**Cleanup**

**Cleanup Policies:**

Components that match any of the Applied policies will be deleted

Available　　Applied

Filter

图 13-31

## 13.10　创建group类型的Maven仓库

group 类型的 Maven 仓库是一个聚合类型的仓库，可以将前面创建的两个仓库聚合成一个 URL，对外提供服务，其具有前面两个仓库的功能，属于仓库功能的集合。

选择“maven2(group)”选项，如图 13-32 所示。

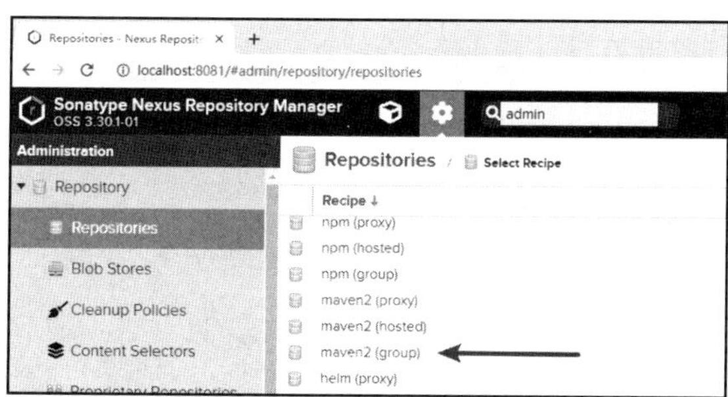


图 13-32

配置仓库信息，如图 13-33 所示。

图 13-33

在 Members 中关联 maven-local，maven-snapshots，maven-releases，maven-proxy 仓库。注意：仓库的排序一定要使用此顺序，目的是先在 maven-local 中查询.jar 文件，如果没有,则在 maven-snapshots，maven-releases 仓库中进行查询，如果还没有则通过 maven-proxy 进行远程下载。

单击“Create repository”按钮，开始创建 group 仓库。

至此，成功创建了 3 个仓库，如图 13-34 所示。

Repositories Manage repositories

Create repository

| Name ↑ | Type | Format | Status |
|---|---|---|---|
| maven-central | proxy | maven2 | Online - Ready to Connect |
| maven-group | group | maven2 | Online |
| maven-local | hosted | maven2 | Online |
| maven-proxy | proxy | maven2 | Online - Ready to Connect |
| maven-public | group | maven2 | Online |
| maven-releases | hosted | maven2 | Online |
| maven-snapshots | hosted | maven2 | Online |
| nuget-group | group | nuget | Online |
| nuget-hosted | hosted | nuget | Online |
| nuget.org-proxy | proxy | nuget | Online - Ready to Connect |

图 13-34

## 13.11 group-local-proxy仓库之间的关系

三种仓库的关系如图 13-35 所示。

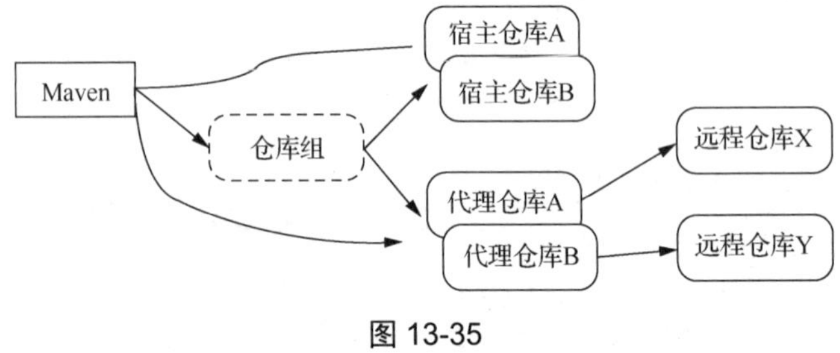


图 13-35

Maven 可以直接从本地仓库下载组件，也可以从代理仓库下载组件，而代理仓库会间接从远程仓库下载并缓存组件。为了方便，Maven 可以从仓库组下载组件，但仓库组没有实际内容，其需要转向本地仓库或者代理仓库获得组件内容。

## 13.12 配置Nexus私服URL

在 Maven 的 settings.xml 配置文件中注册私服 URL，如图 13-36 所示。

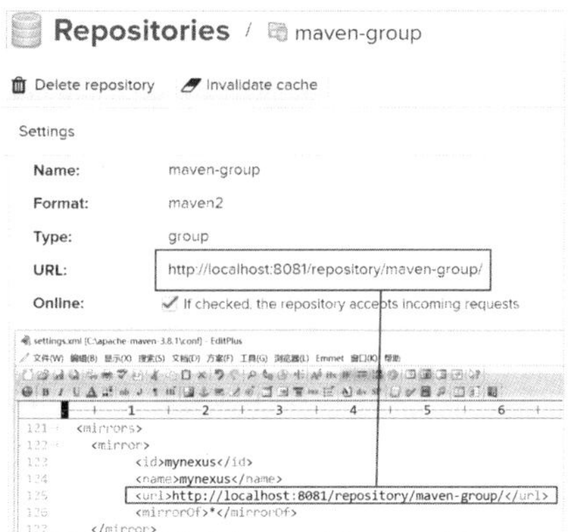


图 13-36

需要禁用 settings.xml 配置文件中的阿里云等第三方 Maven 镜像站，因为 Nexus 中的 proxy 仓库具有相同的作用。

配置代码如下：

```
<mirror>
    <id>mynexus</id>
    <name>mynexus</name>
    <url>http://localhost:8081/repository/maven-group/</url>
    <mirrorOf>*</mirrorOf>
</mirror>
```

其中，<mirrorOf>*</mirrorOf>中的*代表对远程仓库的所有请求都要转至该镜像。

## 13.13　配置登录Nexus的账号和密码

当 Maven 访问 Nexus 时，需要配置登录账号和密码，如图 13-37 所示。

```
<servers>
  <server>
    <id>mynexus</id>
    <username>admin</username>
    <password>xxxxxxxxxx</password>
  </server>
</servers>

<mirrors>
  <mirror>
      <id>mynexus</id>
      <name>mynexus</name>
      <url>http://localhost:8081/repository/maven-group/</url>
      <mirrorOf>*</mirrorOf>
  </mirror>
</mirrors>
```

图 13-37

配置代码如下：

```
<server>
 <id>mynexus</id>
 <username>admin</username>
 <password>××××××××</password>
</server>
```

密码“××××××××”要改成登录 Nexus 的登录密码，这里只是模拟的密码。

注意：<mirror>标签的<id>子标签中的内容和<server>标签的<id>子标签中的一样，原因是我们是根据 id 值把账号密码和私服进行对应的。

## 13.14　开启SNAPSHOT版本支持

因为 Nexus 官方建议使用 RELEASE 正式版本的依赖，所以在默认情况下，不支持下载 SNAPSHOT 版本的 JAR 文件，这时需要手动开启对 SNAPSHOT 版本下载的支持。

更改 Maven 的 settings.xml 配置文件，添加配置代码如下：

```
<profiles>
   <profile>
      <id>nexus</id>
      <repositories>
         <repository>
            <id>mynexus</id>
            <name>Nexus</name>
            <url>http://localhost:8081/repository/maven-group/</url>
            <releases>
               <enabled>true</enabled>
            </releases>
            <snapshots>
               <enabled>true</enabled>
            </snapshots>
         </repository>
      </repositories>
      <pluginRepositories>
         <pluginRepository>
            <id>mynexus</id>
            <name>Nexus</name>
            <url>http://localhost:8081/repository/maven-group/</url>
```

```
                <releases>
                    <enabled>true</enabled>
                </releases>
                <snapshots>
                    <enabled>true</enabled>
                </snapshots>
            </pluginRepository>
        </pluginRepositories>
    </profile>
</profiles>
<activeProfiles>
    <activeProfile>nexus</activeProfile>
</activeProfiles>
```

## 13.15　确认maven-group仓库内容为空

选择 maven-group 仓库，如图 13-38 所示。

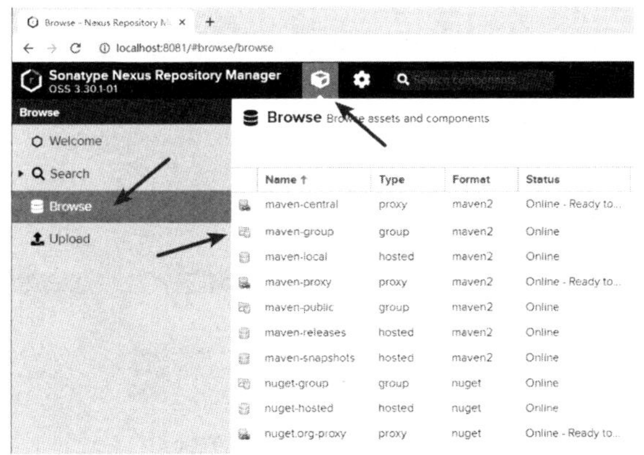


图 13-38

确认 maven-group 仓库内容为空，如图 13-39 所示。

图 13-39

## 13.16 在IDEA中创建测试用的项目

在 IDEA 中创建 maven quickstart 项目，并配置项目属性，如图 13-40 所示。

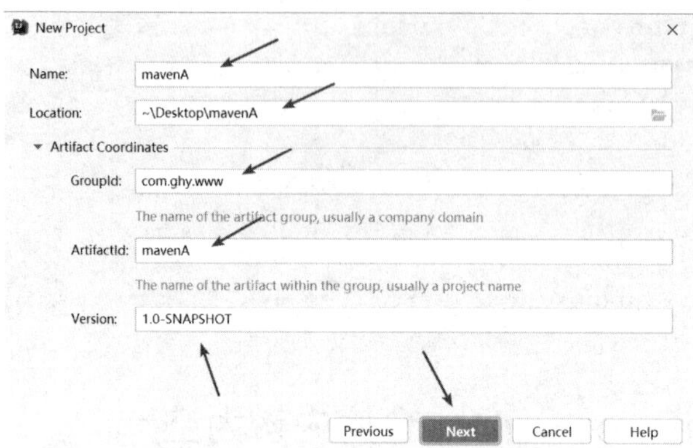


图 13-40

配置 Maven 属性，如图 13-41 所示。

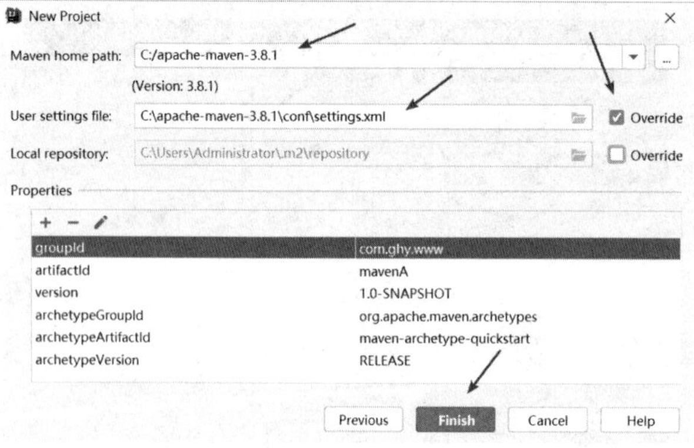


图 13-41

单击“Finish”按钮后，IDEA 开始通过 Nexus 私服中的 maven-proxy 仓库从阿里云 Maven 镜像中下载依赖的 JAR 文件，如图 13-42 所示。

```
Downloading from mynexus: http://172.20.10.2:8081/repository/maven-group/org/apache/maven/shared/maven-invoker/3.0.1/maven-invoker-3.0.1.pom
Downloaded from mynexus: http://172.20.10.2:8081/repository/maven-group/org/apache/maven/shared/maven-invoker/3.0.1/maven-invoker-3.0.1.pom (0 B at 0 B/s)
Downloading from mynexus: http://172.20.10.2:8081/repository/maven-group/org/apache/maven/shared/maven-shared-components/31/maven-shared-components-31.pom
Downloaded from mynexus: http://172.20.10.2:8081/repository/maven-group/org/apache/maven/shared/maven-shared-components/31/maven-shared-components-31.pom (0 B at 0 B/s)
Downloading from mynexus: http://172.20.10.2:8081/repository/maven-group/org/apache/maven/maven-parent/31/maven-parent-31.pom
Downloaded from mynexus: http://172.20.10.2:8081/repository/maven-group/org/apache/maven/maven-parent/31/maven-parent-31.pom (0 B at 0 B/s)
Downloading from mynexus: http://172.20.10.2:8081/repository/maven-group/org/apache/maven/shared/maven-shared-utils/3.2.1/maven-shared-utils-3.2.1.pom
Downloaded from mynexus: http://172.20.10.2:8081/repository/maven-group/org/apache/maven/shared/maven-shared-utils/3.2.1/maven-shared-utils-3.2.1.pom (0 B at 0 B/s)
```

图 13-42

项目创建成功，如图 13-43 所示。

```
[INFO] --------------------
[INFO] BUILD SUCCESS
[INFO] --------------------
```

图 13-43

这时，仓库 maven-group 中也有了 jar 包文件，如图 13-44 所示。

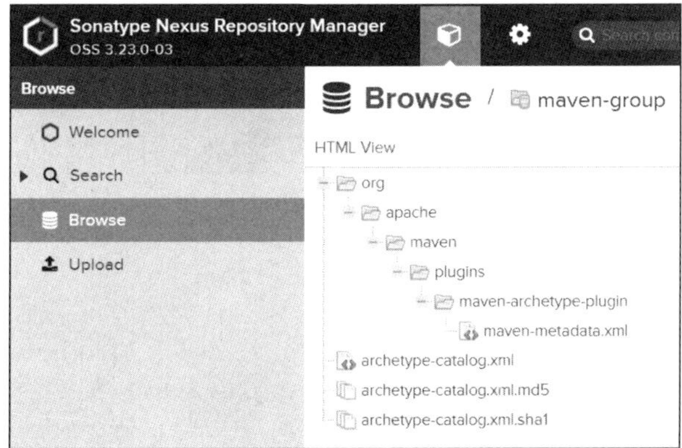


图 13-44

## 13.17　创建Java类和执行deploy操作

创建 Java 类，代码如图 13-45 所示。

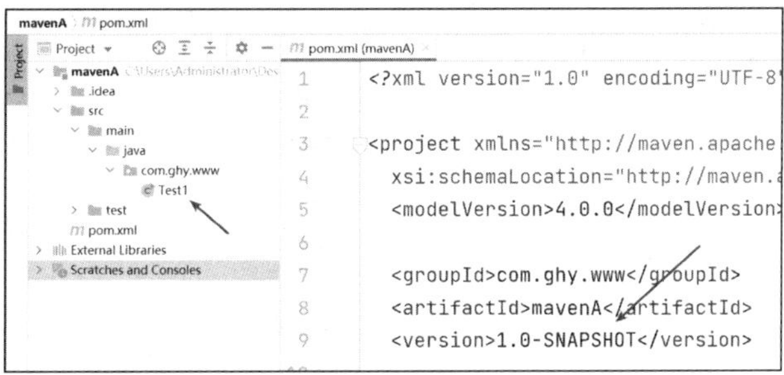


图 13-45

SNAPSHOT 版本中有一个 Java 类。

开始执行 deploy 部署操作，如图 13-46 所示。

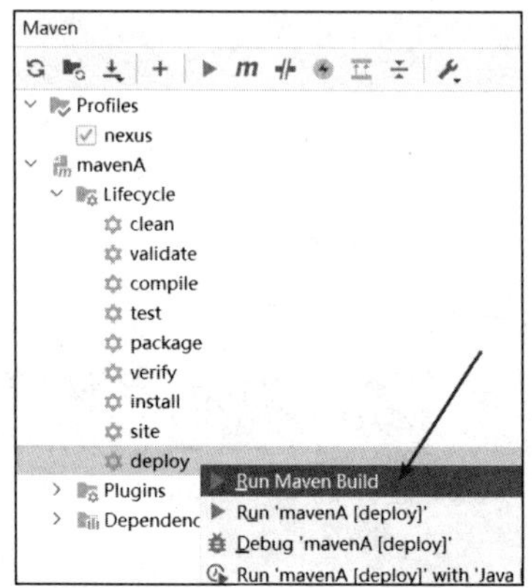


图 13-46

执行 deploy 操作后出现异常：

```
[ERROR] Failed to execute goal
org.apache.maven.plugins:maven-deploy-plugin:2.8.2:deploy (default-deploy)
on project nexus-test: Deployment failed: repository element was not specified
in the POM inside distributionManagement element or in
-DaltDeploymentRepository=id::layout::url parameter -> [Help 1]
```

如果想执行 deploy 操作，则需要在 pom.xml 文件中配置<distributionManagement>，添加如下配置：

```
<distributionManagement>
   <repository>
      <id>mynexus</id>
      <name>Nexus Release Repository</name>
      <url>http://localhost:8081/repository/maven-releases/</url>
   </repository>
   <snapshotRepository>
      <id>mynexus</id>
      <name>Nexus Release Repository</name>
      <url>http://localhost:8081/repository/maven-snapshots/</url>
   </snapshotRepository>
</distributionManagement>
```

这时，再次执行 deploy 操作，成功将项目打包成.jar 文件并上传到 maven-snapshots 仓库中，控制台显示日志如下：

```
Downloading from mynexus: http://localhost:8081/repository/
maven-snapshots/com/ ghy/www/mavenA/1.0-SNAPSHOT/maven-metadata.xml
```

```
Uploading to mynexus: http://localhost:8081/repository/maven-snapshots/
com/ghy/ www/mavenA/1.0-SNAPSHOT/mavenA-1.0-20210723.085945-1.jar
Uploaded to mynexus: http://localhost:8081/repository/maven-snapshots/
com/ghy/www/ mavenA/1.0-SNAPSHOT/mavenA-1.0-20210723.085945-1.jar (2.8 kB
at 44 kB/s)
Uploading to mynexus: http://localhost:8081/repository/maven-snapshots/
com/ghy/ www/mavenA/1.0-SNAPSHOT/mavenA-1.0-20210723.085945-1.pom
Uploaded to mynexus: http://localhost:8081/repository/maven-snapshots/
com/ghy/ www/mavenA/1.0-SNAPSHOT/mavenA-1.0-20210723.085945-1.pom (3.6 kB
at 64 kB/s)
Downloading from mynexus: http://localhost:8081/repository/maven-
snapshots/com/ ghy/www/mavenA/maven-metadata.xml
Uploading to mynexus: http://localhost:8081/repository/maven-snapshots/
com/ghy/ www/mavenA/1.0-SNAPSHOT/maven-metadata.xml
Uploaded to mynexus: http://localhost:8081/repository/maven-snapshots/
com/ghy/ www/mavenA/1.0-SNAPSHOT/maven-metadata.xml (761 B at 13 kB/s)
Uploading to mynexus: http://localhost:8081/repository/maven-snapshots/
com/ghy/ www/mavenA/maven-metadata.xml
Uploaded to mynexus: http://localhost:8081/repository/maven-snapshots/
com/ghy/www/ mavenA/maven-metadata.xml (275 B at 5.4 kB/s)
```

maven-snapshots 仓库中上传的.jar 文件，如图 13-47 所示。

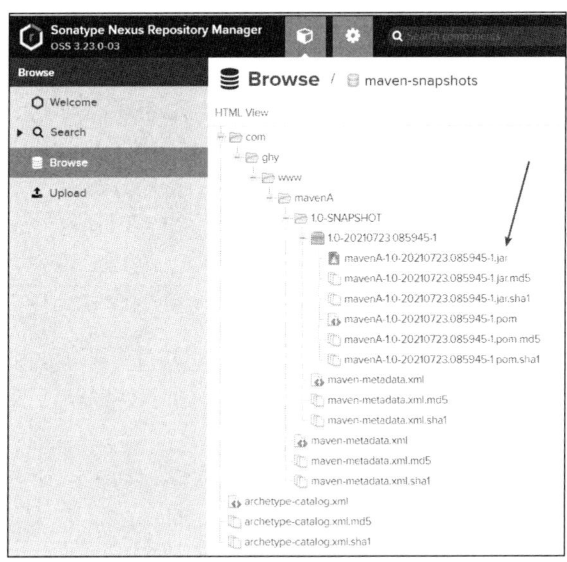


图 13-47

创建第二个 Java 类，将项目版本更改成 2.0-RELEASE，如图 13-48 所示。

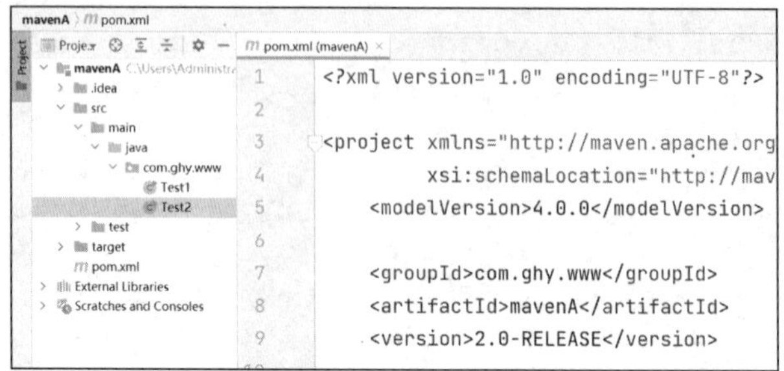


图 13-48

再次执行 deploy 操作， maven-releases 仓库中有.jar 文件了，如图 13-49 所示。

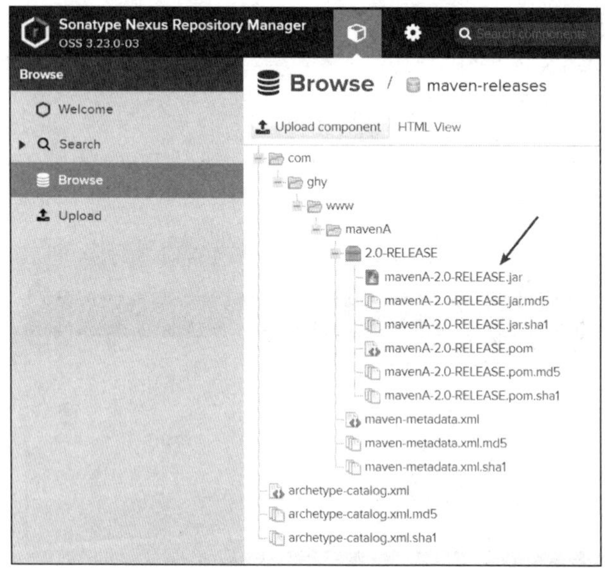


图 13-49

## 13.18 成功进行依赖

maven-snapshots 仓库中的依赖坐标，如图 13-50 所示。

maven-releases 仓库中的依赖坐标，如图 13-51 所示。

创建新的 maven quickstart 项目 mavenB，在 pom.xml 文件中引入 SNAPSHOT 版本的配置，效果如图 13-52 所示。

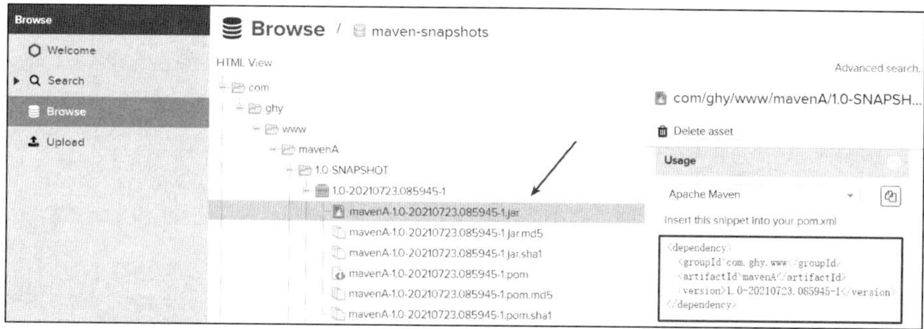


图 13-50

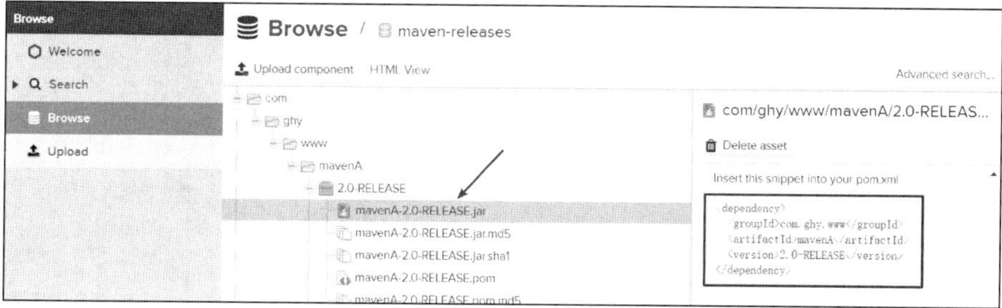


图 13-51

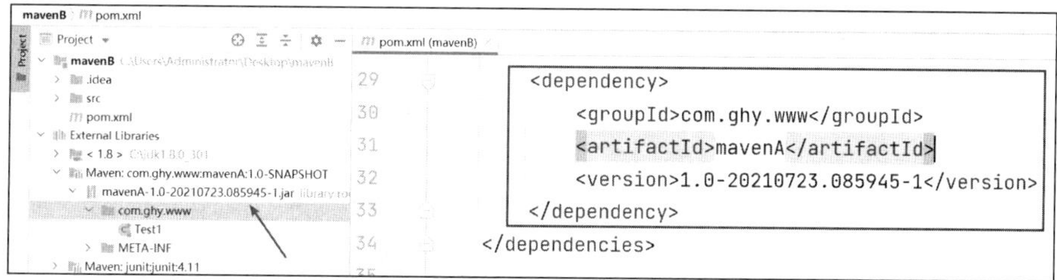


图 13-52

在 pom.xml 文件中引入 RELEASE 版本的配置，效果如图 13-53 所示。

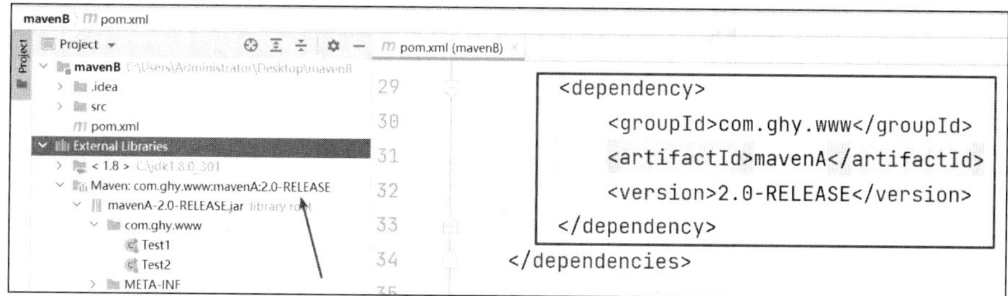


图 13-53

## 13.19 获取最新的RELEASE版本

首先编辑 pom.xml 文件和 Java 类，如图 13-54 所示，然后执行 deploy 操作。

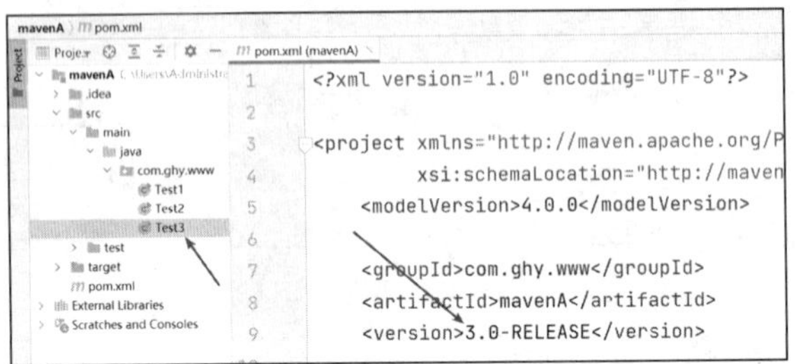

图 13-54

再次编辑 pom.xml 和 Java 类，如图 13-55 所示，然后执行 deploy 操作。

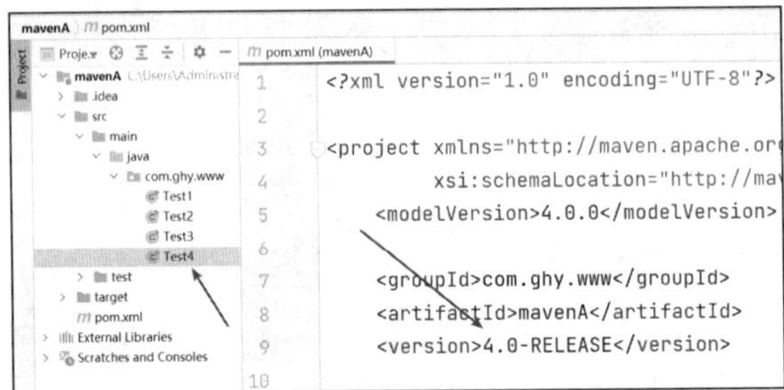

图 13-55

这时，maven-releases 仓库中有 3 个 RELEASE 版本，如图 13-56 所示。

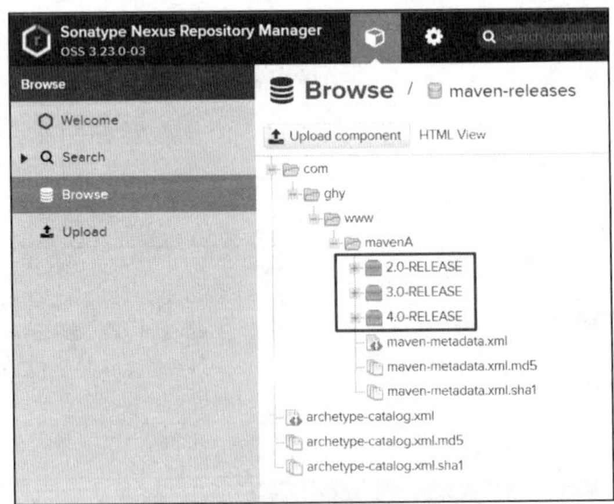

图 13-56

当实现依赖最新的 RELEASE 版本时，需要使用如下坐标配置：

```
<dependency>
   <groupId>com.ghy.www</groupId>
   <artifactId>mavenA</artifactId>
   <version>RELEASE</version>
</dependency>
```

程序运行结果如图 13-57 所示。

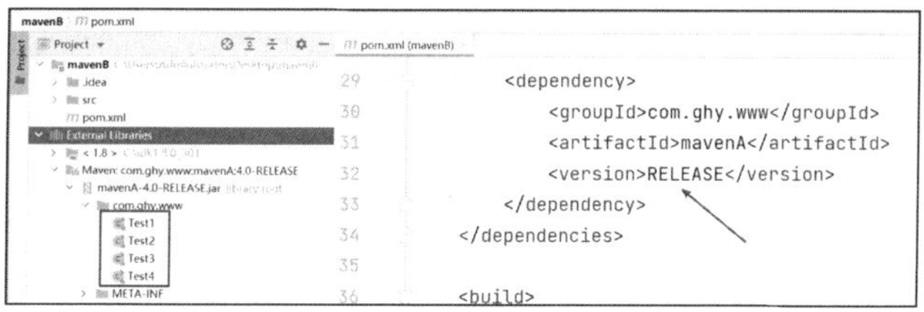


图 13-57

## 13.20　在Maven仓库中进行搜索

以阿里云的 Maven 镜像仓库作为示例，如图 13-58 所示。

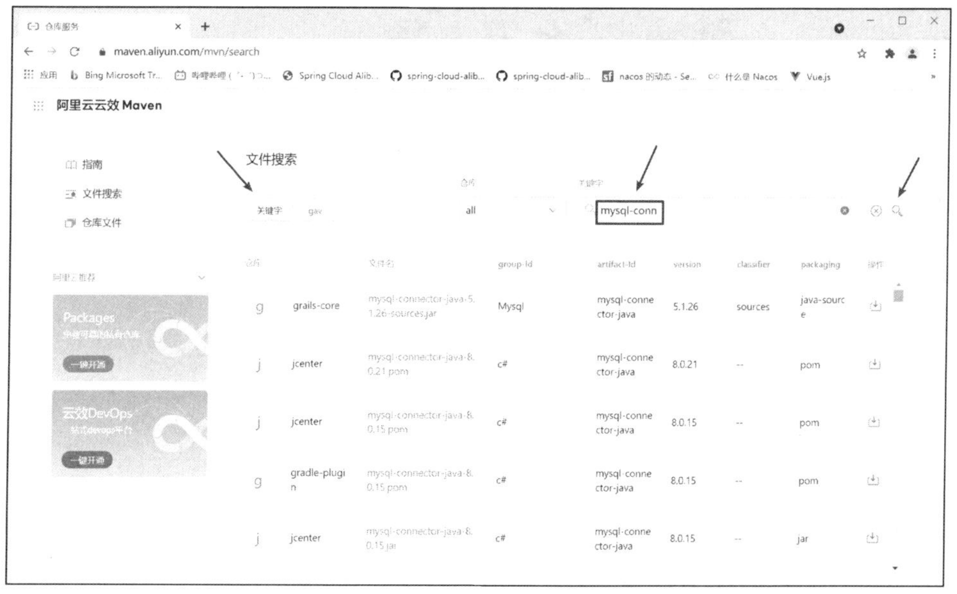


图 13-58

使用 GAV 进行搜索，G 代表 GroupId，A 代表 ArtifactId，V 代表 Version，如图 13-59 所示。

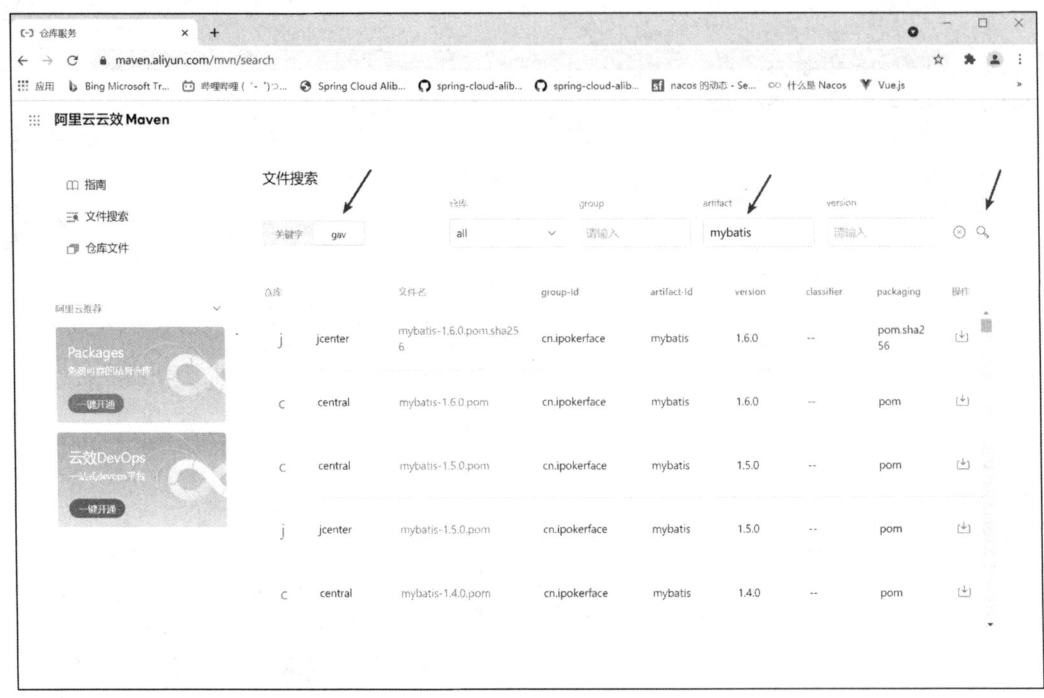


图 13-59

# 第 14 章 Maven 项目生命周期

## 14.1 生命周期

构建项目的生命周期是从敲项目代码的第一个字母开始到项目交付运行的完整过程，主要包含清理项目中无用的文件、编译、测试、打包和部署等环节。但由于每个公司或每个项目在管理软件生命周期时都有不同的处理方式，比如 A 公司使用手动编译，B 公司使用工具自动化编译。为了将这种过程进行标准化和统一化，Maven 抽象出了软件生命周期的过程，包含项目的清理、初始化、编译、测试、打包、集成测试、验证、部署等几乎所有的环节。绝大多数的软件项目，都能用 Maven 来管理生命周期。

由于 Maven 的生命周期是抽象的，因此具体执行是由插件进行的，即实现可插拔式的模块组装，方便生命周期中某个环节功能的扩展与更新，如清理功能由如下插件进行处理。

```
<plugin>
   <artifactId>maven-clean-plugin</artifactId>
   <version>3.1.0</version>
</plugin>
```

编译功能由如下插件进行处理。

```
<plugin>
   <artifactId>maven-compiler-plugin</artifactId>
   <version>3.8.0</version>
</plugin>
```

Maven提供了若干插件来实现生命周期中指定步骤的处理,一个插件实现一个功能。

我们可以把 Maven 的生命周期理解成一个整体，一个完整的从开始到结束的过程，但 Maven 又将这些步骤分为三大类。

（1）clean：负责清理项目。

（2）default：负责构建项目。

（3）site：负责建立项目站点，也就是以.html 文件的形式展示项目的简要报告。

在每一个大类中都包含若干阶段（phase），这些阶段按着规定的顺序执行，后面的阶段依赖前面的阶段，比如在某个大类中有 a，b，c 三个阶段，当执行 a 阶段时会执行 a 阶段；当执行 b 阶段时，会执行 a，b 阶段；当执行 c 阶段时，会执行 a，b，c 阶段。另外，每个大类都是独立存在的，当执行某一个大类中的某一个阶段时，不会同时执行其他大类中的某一个阶段，如当执行 default 生命周期时，不会自动执行 clean 生命周期。

### 14.1.1 clean 生命周期

clean 生命周期的作用就是清理项目中无用的文件，如 target 文件夹，包含 3 个阶段，如图 14-1 所示。

| phase | 介绍 |
|---|---|
| pre-clean | 在实际项目清理之前执行 |
| clean | 删除上一版本生成的所有文件 |
| post-clean | 项目清理结束后执行 |

图 14-1

运行效果如图 14-2 所示。

```
Microsoft Windows [版本 10.0.18363.1556]
(c) 2019 Microsoft Corporation。保留所有权利。
C:\Users\Administrator\Desktop\myproject\DAO>mvn post-clean
[INFO] Scanning for projects...
[INFO]
[INFO] ------------------------< com.ghy.www:DAO >----------------
[INFO] Building DAO 4.0-RELEASE
[INFO] --------------------------------[ jar ]--------------------
[INFO]
[INFO] --- maven-clean-plugin:3.1.0:clean (default-clean) @ DAO ---
[INFO] Deleting C:\Users\Administrator\Desktop\myproject\DAO\target
[INFO] ------------------------------------------------------------
[INFO] BUILD SUCCESS
[INFO] ------------------------------------------------------------
[INFO] Total time:  0.762 s
```

图 14-2

### 14.1.2 default 生命周期

default 生命周期定义了构建项目几乎所有的步骤，是 Maven 生命周期中最为核心的部分，其阶段的数量较多，如图 14-3 所示。

| phase | 描述 |
|---|---|
| validate | 验证项目是否正确，并且所有必要的信息均是有效的。 |
| initialize | 初始化构建状态，例如设置属性或创建目录。 |
| generate-sources | 生成源代码。 |
| process-sources | 处理源代码。 |
| generate-resources | 生成资源以包含在package包中。 |
| process-resources | 将资源复制并处理到目标目录中，以备打包。 |
| compile | 编译项目的源代码。 |
| process-classes | 对编译后生成的*.class文件进行二次处理，例如对Java类进行字节码增强。 |
| generate-test-sources | 生成测试源代码。 |
| process-test-sources | 处理测试源代码。 |
| generate-test-resources | 创建测试资源。 |
| process-test-resources | 将资源复制并处理到测试目标目录中。 |
| test-compile | 将测试源代码编译到测试目标目录中。 |
| process-test-classes | 对编译生成的.test文件进行后处理，例如对Java类进行字节码增强。 |
| test | 使用合适的单元测试框架运行测试，这些测试不应要求打包或部署代码。 |
| prepare-package | 在实际打包之前执行准备打包所需的任何操作，通常是已处理版本的未打包状态。 |
| package | 获取编译后的代码，并将其打包为可分发的格式，例如JAR。 |
| pre-integration-test | 在执行集成测试之前执行所需的操作，这可能涉及诸如设置所需环境的事情。 |
| integration-test | 处理该程序包并将其部署到可以运行集成测试的环境中（如有必要）。 |
| post-integration-test | 在执行集成测试后执行所需的操作，这可能包括清理环境。 |
| verify | 运行任何检查以验证package是否有效并符合质量标准。 |
| install | 将软件包安装到本地仓库中，以作为本地其他项目中的依赖项。 |
| deploy | 在集成或发布环境中完成后，最终将程序包复制到远程仓库，以便与其他开发人员和项目共享。 |

图 14-3

运行效果如图 14-4 所示。

```
C:\Users\Administrator\Desktop\mymaven1>mvn validate
[INFO] Scanning for projects...
[INFO]
[INFO] -----------------------< com.ghy.www:mymaven1 >-----------------------
[INFO] Building mymaven1 1.0-SNAPSHOT
[INFO] --------------------------------[ jar ]---------------------------------
[INFO] ------------------------------------------------------------------------
[INFO] BUILD SUCCESS
[INFO] ------------------------------------------------------------------------
[INFO] Total time:  0.106 s
[INFO] Finished at: 2021-07-23T13:52:11+08:00
[INFO] ------------------------------------------------------------------------

C:\Users\Administrator\Desktop\mymaven1>
```

图 14-4

### 14.1.3　site 生命周期

site 生命周期可以根据项目中的 pom.xml 配置文件生成.html 格式的项目介绍站点，如图 14-5 所示。

| phase | 描述 |
| --- | --- |
| pre-site | 在实际项目站点生成之前执行 |
| site | 生成项目的站点文档 |
| post-site | 站点生成结束后执行，以及准备部署站点 |
| site-deploy | 将生成的站点文档部署到指定的Web服务器 |

图 14-5

运行效果如图 14-6 所示。

```
C:\Users\Administrator\Desktop\myproject\DAO>mvn site
[INFO] Scanning for projects...
[INFO]
[INFO] --------------------------< com.ghy.www:DAO >--------------------------
[INFO] Building DAO 4.0-RELEASE
[INFO] --------------------------------[ jar ]---------------------------------
[INFO]
[INFO] --- maven-site-plugin:3.7.1:site (default-site) @ DAO ---
[INFO] configuring report plugin org.apache.maven.plugins:maven-project-info-reports-plugin:3.0.0
Downloading from mynexus: http://localhost:8081/repository/maven-group/org/apache/maven/plugins/maven-projec
roject-info-reports-plugin-3.0.0.pom
Downloaded from mynexus: http://localhost:8081/repository/maven-group/org/apache/maven/plugins/maven-project
oject-info-reports-plugin-3.0.0.pom (20 kB at 42 kB/s)
Downloading from mynexus: http://localhost:8081/repository/maven-group/org/apache/maven/plugins/maven-projec
roject-info-reports-plugin-3.0.0.jar
Downloaded from mynexus: http://localhost:8081/repository/maven-group/org/apache/maven/plugins/maven-project
oject-info-reports-plugin-3.0.0.jar (300 kB at 935 kB/s)
[INFO] Generating "Dependency Information" report --- maven-project-info-reports-plugin:3.0.0:dependency-info
[INFO] Generating "Dependency Management" report --- maven-project-info-reports-plugin:3.0.0:dependency-management
[INFO] Generating "Distribution Management" report --- maven-project-info-reports-plugin:3.0.0:distribution-management
[INFO] Generating "About" report         --- maven-project-info-reports-plugin:3.0.0:index
[INFO] Generating "Plugin Management" report --- maven-project-info-reports-plugin:3.0.0:plugin-management
Downloading from mynexus: http://localhost:8081/repository/maven-group/org/apache/maven/plugins/maven-release-plugin/2.5
-2.5.3.pom
Downloaded from mynexus: http://localhost:8081/repository/maven-group/org/apache/maven/plugins/maven-release-plugin/2.5.
2.5.3.pom (11 kB at 38 kB/s)
Downloading from mynexus: http://localhost:8081/repository/maven-group/org/apache/maven/release/maven-release/2.5.3/mave
Downloaded from mynexus: http://localhost:8081/repository/maven-group/org/apache/maven/release/maven-release/2.5.3/maven
 kB at 25 kB/s)
[INFO] Generating "Plugins" report       --- maven-project-info-reports-plugin:3.0.0:plugins
[INFO] Generating "Summary" report       --- maven-project-info-reports-plugin:3.0.0:summary
[INFO] ------------------------------------------------------------------------
[INFO] BUILD SUCCESS
[INFO] ------------------------------------------------------------------------
[INFO] Total time:  37.258 s
```

图 14-6

虽然 Maven 提供了很多阶段，但常见的阶段主要有 clean，test，install，deploy，site。切记，当执行某一个阶段时，前面所有的阶段都要执行一次，即前后具有依赖关系。

因为生命周期中的某个阶段依赖于插件，所以在 IDEA 中的 pom.xml 中声明了一些插件信息，配置代码如下：

```
<!—clean 生命周期-->
<!-- clean lifecycle, see https://maven.apache.org/ref/current/maven-core/
lifecycles.html#clean_Lifecycle -->
```

```
<plugin>
    <artifactId>maven-clean-plugin</artifactId>
    <version>3.1.0</version>
</plugin>
<!—default 生命周期-->
<!-- default lifecycle, jar packaging: see https://maven.apache.org/ref/
current/maven-core/default-bindings.html#Plugin_bindings_for_jar_packagi
ng -->
<plugin>
    <artifactId>maven-resources-plugin</artifactId>
    <version>3.0.2</version>
</plugin>
<plugin>
    <artifactId>maven-compiler-plugin</artifactId>
    <version>3.8.0</version>
</plugin>
<plugin>
    <artifactId>maven-surefire-plugin</artifactId>
    <version>2.22.1</version>
</plugin>
<plugin>
    <artifactId>maven-jar-plugin</artifactId>
    <version>3.0.2</version>
</plugin>
<plugin>
    <artifactId>maven-install-plugin</artifactId>
    <version>2.5.2</version>
</plugin>
<plugin>
    <artifactId>maven-deploy-plugin</artifactId>
    <version>2.8.2</version>
</plugin>
<!—site 生命周期-->
<!-- site lifecycle, see https://maven.apache.org/ref/current/maven-core/
lifecycles.html#site_Lifecycle -->
<plugin>
    <artifactId>maven-site-plugin</artifactId>
    <version>3.7.1</version>
</plugin>
<plugin>
    <artifactId>maven-project-info-reports-plugin</artifactId>
```

```
    <version>3.0.0</version>
</plugin>
```

后面会在 IDEA 中展示这些常用阶段的运行效果。

## 14.2 创建测试项目

创建 Maven quickstart 项目，如图 14-7 所示。

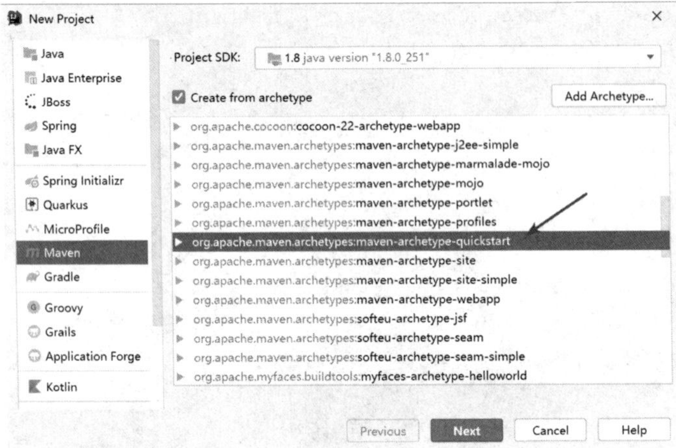


图 14-7

配置项目，如图 14-8 所示。

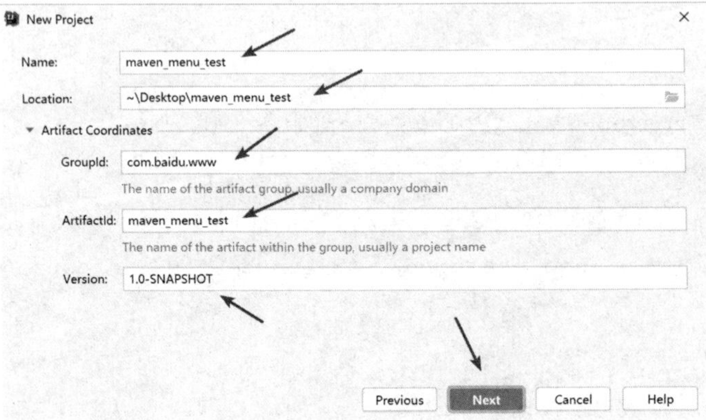


图 14-8

配置 Maven，如图 14-9 所示。

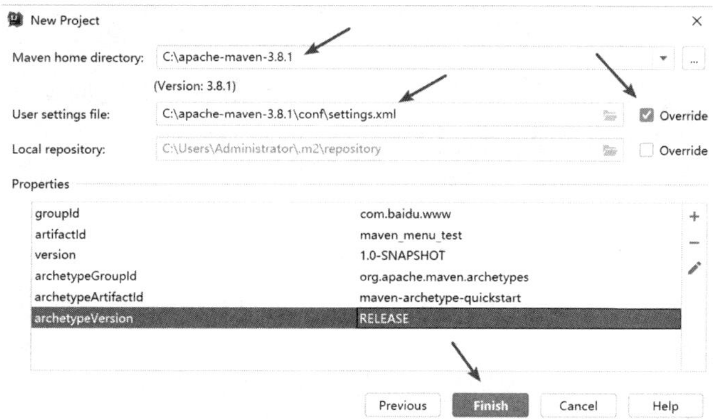


图 14-9

更改项目中的 pom.xml 配置文件，使用 1.8 版本，配置代码如下：

```
<properties>
    <project.build.sourceEncoding>UTF-8</project.build.sourceEncoding>
    <maven.compiler.source>1.8</maven.compiler.source>
    <maven.compiler.target>1.8</maven.compiler.target>
</properties>
```

创建的项目结构，如图 14-10 所示。

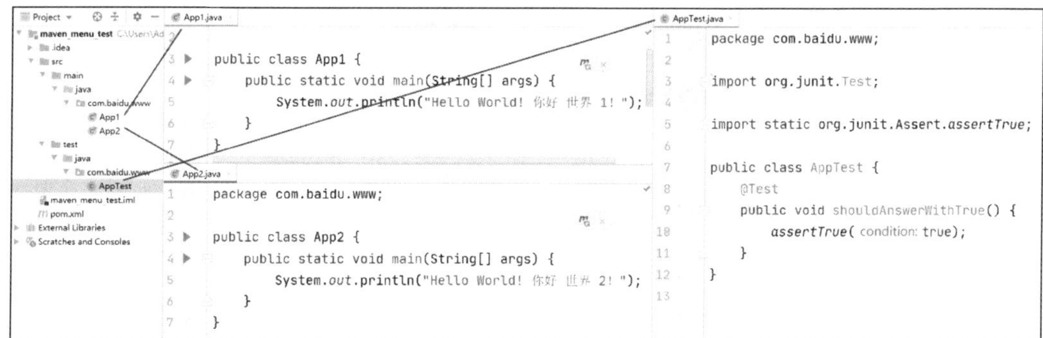

图 14-10

运行 App1.java 类，项目结构如图 14-11 所示。

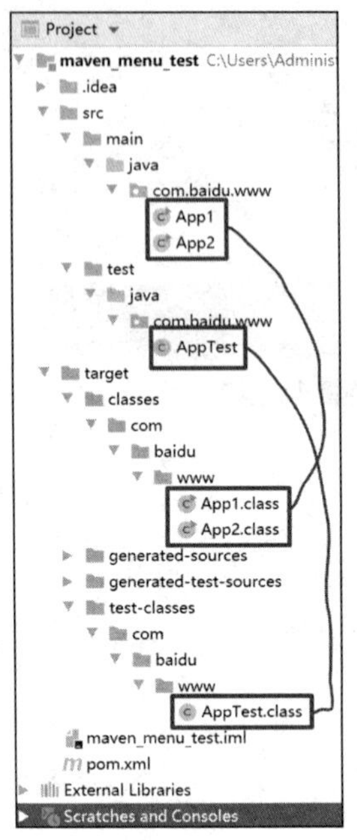


图 14-11

由此可见，项目增加了 target 文件夹，其包括所有.java 文件编译后的.class 文件。

在 IDEA 中的 Maven 面板中有若干菜单，如图 14-12 所示。

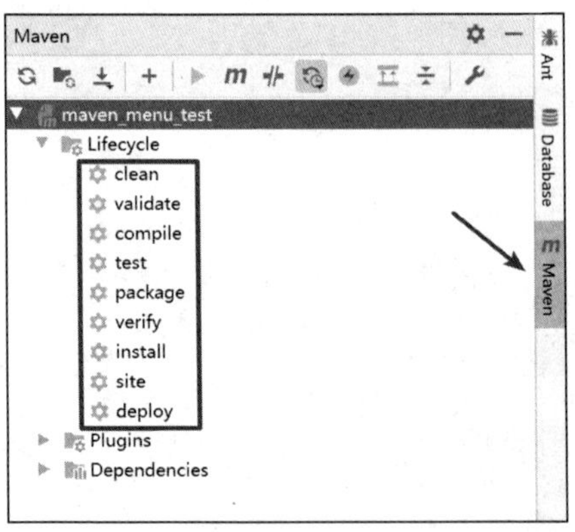


图 14-12

这些菜单都是开发软件项目过程中使用的，也是软件生命周期的体现，你可以按 Ctrl 键选择多个不同的阶段同时执行。

# 14.3　clean菜单的使用

clean 菜单的作用是删除项目中生成的 target 文件夹。当前项目中存在 target 文件夹，如图 14-13 所示。执行 clean 菜单，如图 14-14 所示。这时，target 文件夹被删除，如图 14-15 所示。

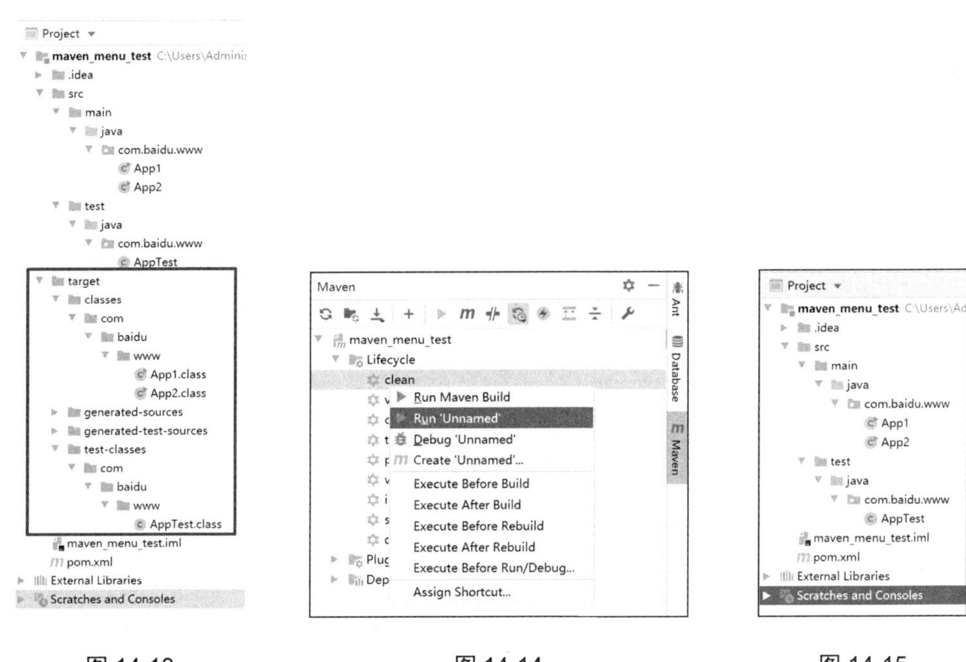


图 14-13　　图 14-14　　图 14-15

# 14.4　validate菜单的使用

validate 菜单的作用是验证项目环境是否正确，包含验证 pom.xml 配置文件。

首先我们将配置文件 pom.xml 改成错误的，如图 14-16 所示。执行 validate 菜单，如图 14-17 所示。

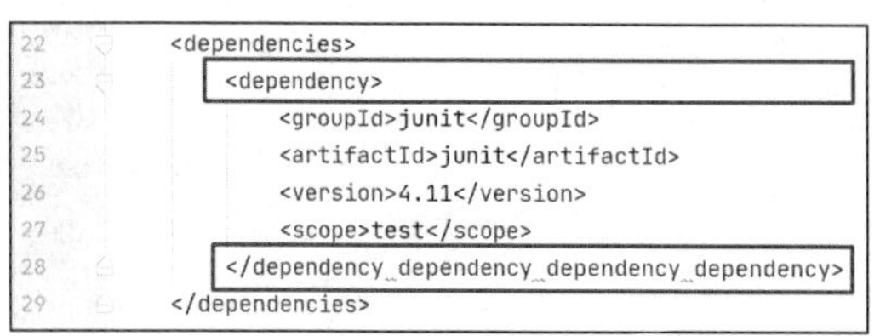

```
<dependencies>
    <dependency>
        <groupId>junit</groupId>
        <artifactId>junit</artifactId>
        <version>4.11</version>
        <scope>test</scope>
    </dependency_dependency_dependency_dependency>
</dependencies>
```

图 14-16

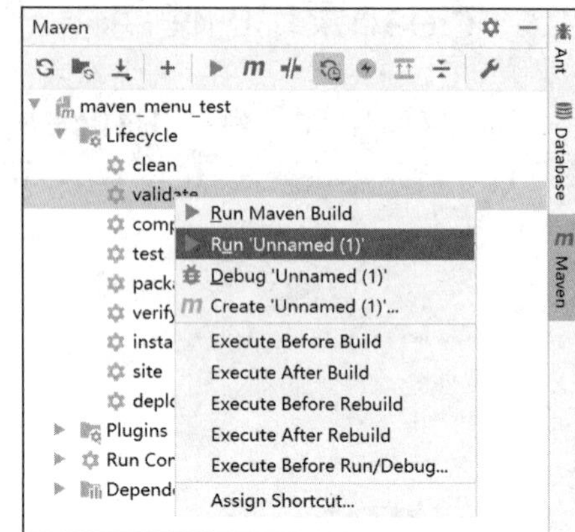


图 14-17

这时，validate 验证过程出现异常，如图 14-18 所示。

```
[INFO] Scanning for projects...
[ERROR] [ERROR] Some problems were encountered while processing the POMs:
[FATAL] Non-parseable POM C:\Users\Administrator\Desktop\maven_menu_test\pom.xml: expected > to finsh end tag not d from line 23 (position: TEXT seen
 ...</scope>\r\n        </dependency d... @28:23)  @ line 28, column 23
@
[ERROR] The build could not read 1 project -> [Help 1]
[ERROR]
[ERROR]   The project  (C:\Users\Administrator\Desktop\maven_menu_test\pom.xml) has 1 error
[ERROR]     Non-parseable POM C:\Users\Administrator\Desktop\maven_menu_test\pom.xml: expected > to finsh end tag not d from line 23 (position: TEXT seen
 ...</scope>\r\n        </dependency d... @28:23)  @ line 28, column 23 -> [Help 2]
[ERROR]
[ERROR] To see the full stack trace of the errors, re-run Maven with the -e switch.
[ERROR] Re-run Maven using the -X switch to enable full debug logging.
[ERROR]
[ERROR] For more information about the errors and possible solutions, please read the following articles:
[ERROR] [Help 1] http://cwiki.apache.org/confluence/display/MAVEN/ProjectBuildingException
[ERROR] [Help 2] http://cwiki.apache.org/confluence/display/MAVEN/ModelParseException
```

图 14-18

说明项目环境有问题，然后将 pom.xml 配置文件内容改成正确的，如图 14-19 所示。

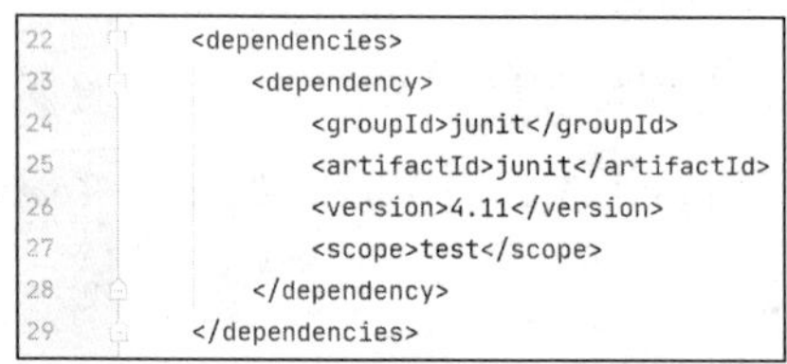

```
<dependencies>
    <dependency>
        <groupId>junit</groupId>
        <artifactId>junit</artifactId>
        <version>4.11</version>
        <scope>test</scope>
    </dependency>
</dependencies>
```

图 14-19

这次，执行 validate 菜单之后验证过程没有出现异常，如图 14-20 所示。

```
[INFO] Scanning for projects...
[INFO]
[INFO] ------------------< com.baidu.www:maven_menu_test >-------------------
[INFO] Building maven_menu_test 1.0-SNAPSHOT
[INFO] ------------------------------[ jar ]-------------------------------
[INFO] ------------------------------------------------------------------------
[INFO] BUILD SUCCESS
[INFO] ------------------------------------------------------------------------
[INFO] Total time:  0.104 s
```

图 14-20

## 14.5　compile菜单的使用

compile 菜单的作用是对 src/main 文件夹中的.java 文件进行编译，以生成.class 文件。

先执行 clean 菜单，恢复项目环境，当前项目的结构，如图 14-21 所示。可知项目结构中没有 target 文件夹，也就是没有.class 文件。

下面执行 compile 菜单，如图 14-22 所示。这时，生成了 target 文件夹，如图 14-23 所示。

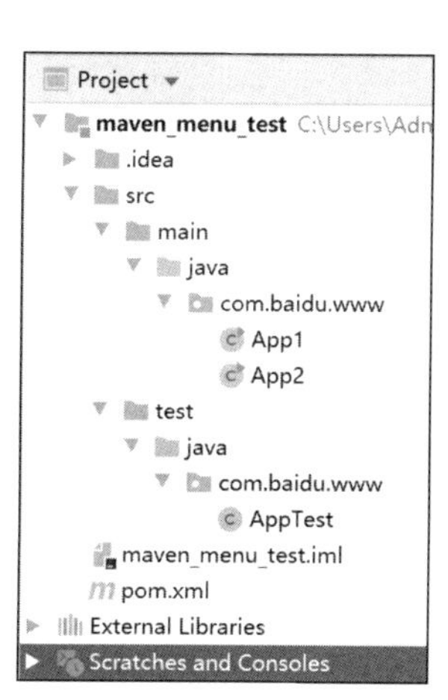


图 14-21

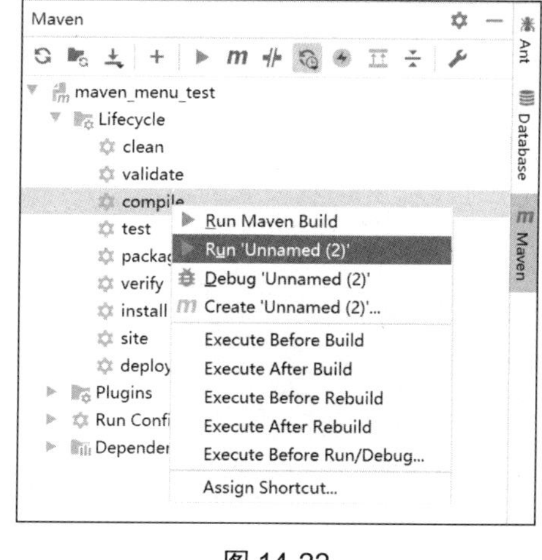


图 14-22

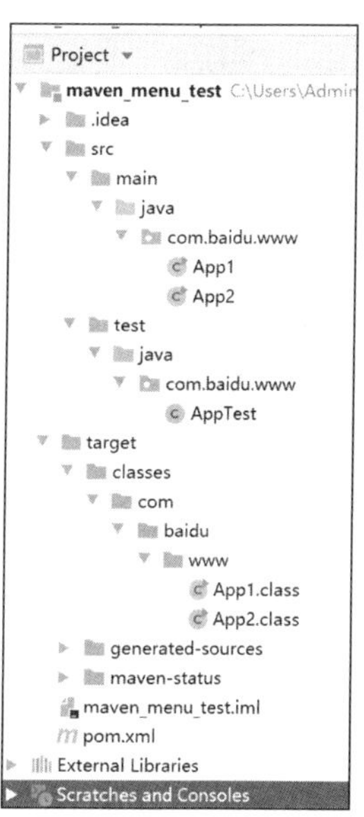


图 14-23

此时，target 文件夹有 src/main 文件夹中.java 文件对应的.class 文件，而没有 src/test 文件夹中.java 对应的.class 文件。也就是说，compile 菜单只编译 src/main 文件夹中的.java

文件，不对 src/test 文件夹中的.java 文件进行编译。

## 14.6 test菜单的使用

test 菜单的作用是执行单元测试。

首先执行 clean 菜单，恢复项目环境，然后执行 test 菜单，如图 14-24 所示。

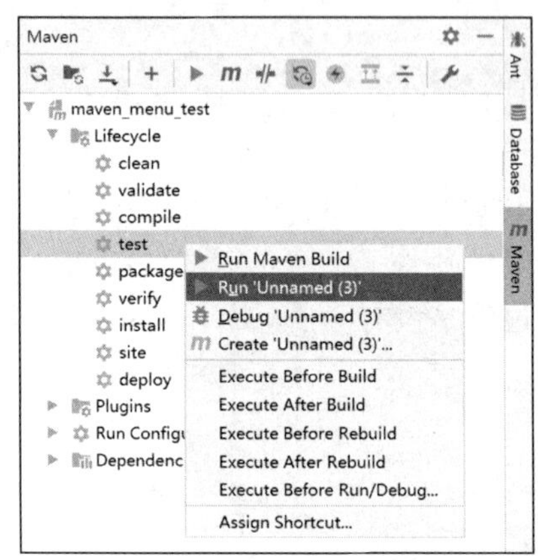


图 14-24

这时，项目通过测试，如图 14-25 所示。

```
[INFO]  T E S T S
[INFO] -------------------------------------------------------
[INFO] Running com.baidu.www.AppTest
[INFO] Tests run: 1, Failures: 0, Errors: 0, Skipped: 0, Time elapsed: 0.038 s - in com.baidu.www.AppTest
[INFO]
[INFO] Results:
[INFO]
[INFO] Tests run: 1, Failures: 0, Errors: 0, Skipped: 0
[INFO]
[INFO] ------------------------------------------------------------------------
[INFO] BUILD SUCCESS
[INFO] ------------------------------------------------------------------------
[INFO] Total time:  2.475 s
```

图 14-25

此时的项目结构如图 14-26 所示。

执行 test 菜单后，至少执行了 3 个步骤：

（1）对 src/main 中的.java 文件进行编译。

（2）对 src/test 中的.java 文件进行编译。

（3）执行 src/test 中的测试用例。

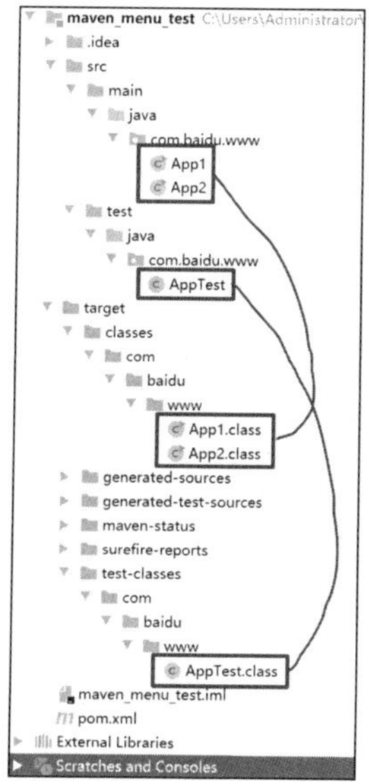

图 14-26　项目结构

## 14.7　package菜单的使用

package 菜单的作用是对项目进行打包。首先执行 clean 菜单，恢复项目环境，当前的项目结构，如图 14-27 所示。

然后执行 package 菜单，如图 14-28 所示。

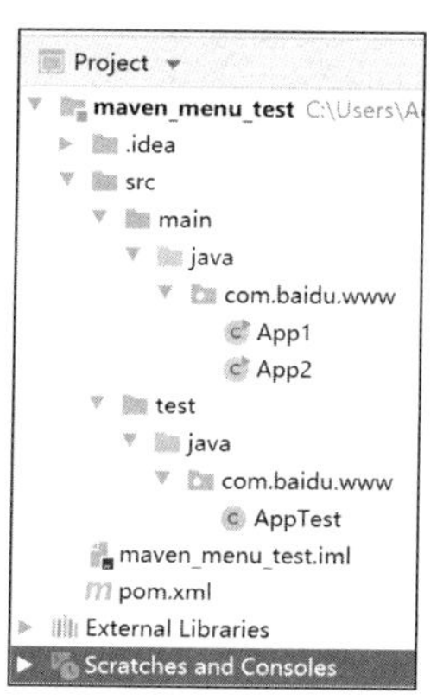

图 14-27

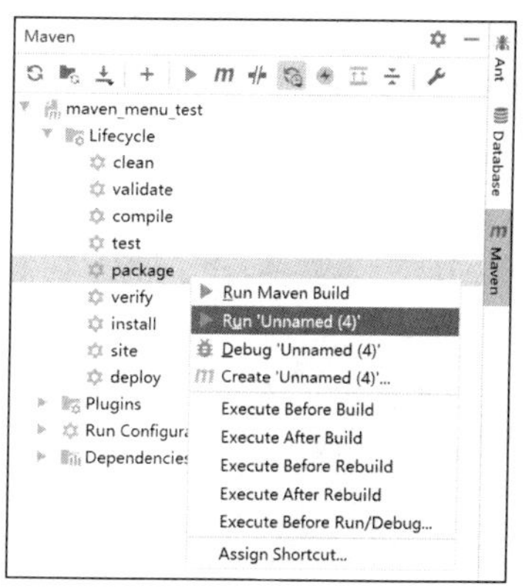

图 14-28

此时，项目被打包成 JAR 文件。解压 JAR 文件，并使用 tree /f 命令查看文件夹结构，如图 14-29 所示。

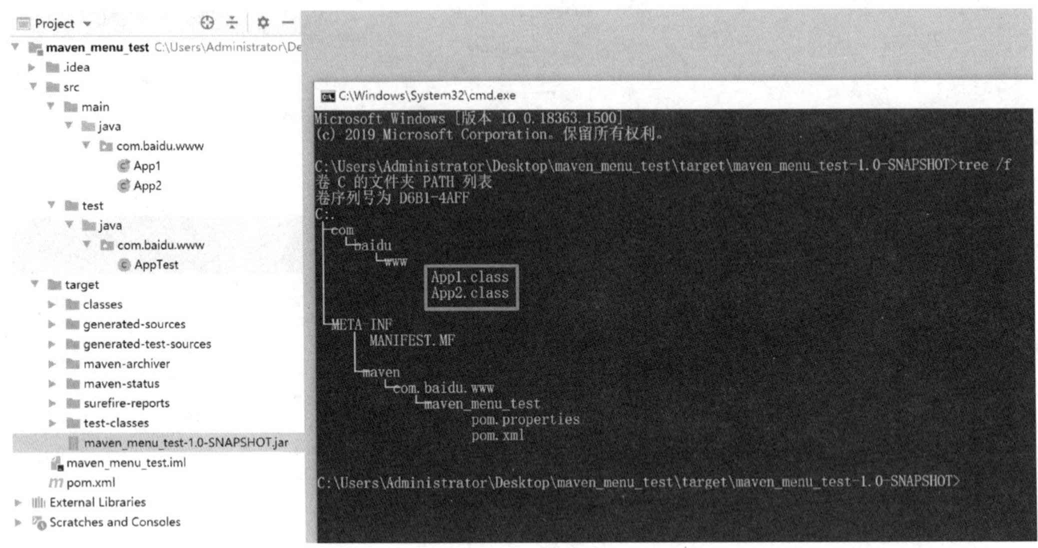


图 14-29

JAR 文件中只有 src/main 文件夹中.java 对应的.class 文件，没有 src/test 文件夹中.java 对应的.class 文件。

## 14.8 verify菜单的使用

verify 菜单具有 validate，compile，package 的功能。在大多数情况下，运行 verify 菜单与运行 package 菜单的效果相同。但是，如果有集成测试，则 verify 菜单也会执行这些测试。

## 14.9 install菜单的使用

install 菜单的作用是将 JAR 文件上传到本地仓库，以便让其他项目进行依赖引用。首先执行 clean 菜单，恢复项目环境，然后执行 install 菜单，如图 14-30 所示。

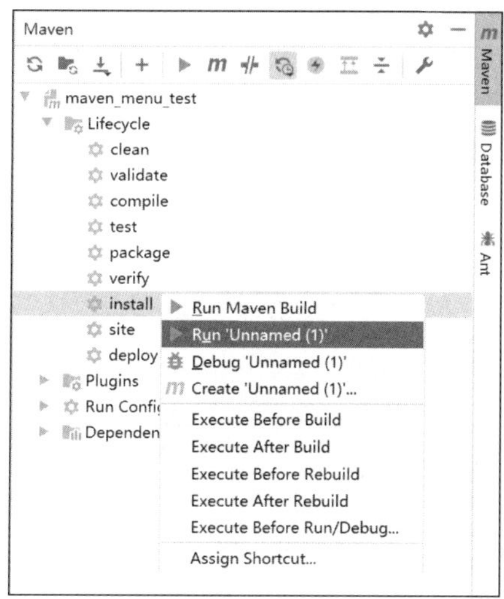


图 14-30

此时，JAR 文件被成功上传到本地仓库，如图 14-31 所示。

```
[INFO] Installing C:\Users\Administrator\Desktop\maven_menu_test\pom.xml to
 C:\mvn_repository\com\baidu\www\maven_menu_test\1.0-SNAPSHOT\maven_menu_test-1.0-SNAPSHOT.pom
```

图 14-31

仓库内容如图 14-32 所示。

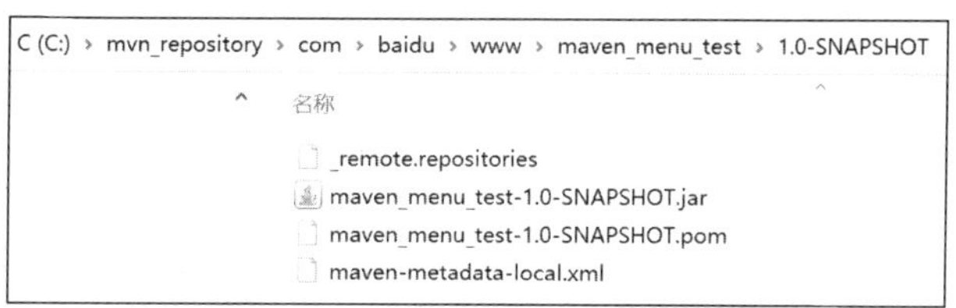


图 14-32

## 14.10　site菜单的使用

site 菜单的作用是生成帮助文档。

首先执行 clean 菜单，恢复项目环境，然后执行 site 菜单，如图 14-33 所示。

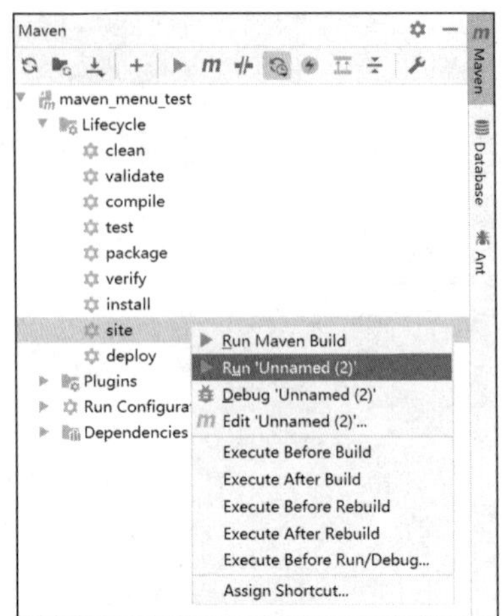


图 14-33

在项目中生成 site 文件夹，如图 14-34 所示，site 文件夹中都是项目相关的统计信息与其他信息的介绍。

图 14-34　生成 site 文件夹

## 14.11　deploy菜单的使用

deploy 菜单的作用是将 JAR 文件上传到远程仓库，以便让其他项目进行依赖引用。该菜单在前面章节介绍过，这里不再介绍。

# 第 15 章
# Maven 依赖的应用

## 15.1 依赖的范围

依赖具有范围性，依赖的范围就是控制 jar 包文件应用的作用域，如图 15-1 所示。

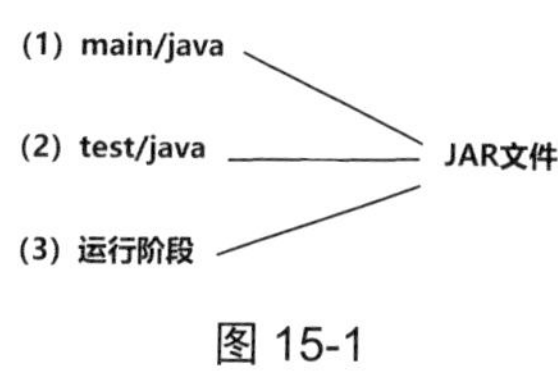

图 15-1

有三处需要依赖 jar 包文件：编译 main/java 文件夹中的.java 文件时，编译 test/java 文件夹中的.java 文件时，.java 文件运行时。

### 15.1.1 依赖范围：compile

依赖范围 compile 包含 main/java，test/java 和运行阶段三个范围。

编辑 pom.xml 配置代码如下：

```
<dependencies>
   <dependency>
      <groupId>mysql</groupId>
      <artifactId>mysql-connector-java</artifactId>
      <version>8.0.25</version>
      <scope>compile</scope>
   </dependency>
</dependencies>
```

这时，依赖范围为 compile，说明 JDBC 驱动 jar 包文件在 main/java，test/java 和运行阶段中都能被访问到。

添加如下代码，生成的 jar 包文件中会包含依赖。

```
<build>
    <plugins><!-- 打包可运行 jar 包文件 -->
        <plugin>
            <groupId>org.apache.maven.plugins</groupId>
            <artifactId>maven-assembly-plugin</artifactId>
            <version>3.3.0</version>
            <configuration>
                <archive>
                    <manifest>
                        <mainClass>com.ghy.www.App</mainClass>
                    </manifest>
                </archive>
                <descriptorRefs>
                    <descriptorRef>jar-with-dependencies</descriptorRef>
                </descriptorRefs>
            </configuration>
            <executions>
                <execution>
                    <id>make-assembly</id>
                    <phase>package</phase>
                    <goals>
                        <goal>single</goal>
                    </goals>
                </execution>
            </executions>
        </plugin>
    </plugins>
</build>
```

路径 main/java 中的.java 类代码，如图 15-2 所示。

```
App.java
package com.ghy.www;

import java.sql.SQLException;

public class App {
    public static void main(String[] args) throws SQLException {
        System.out.println("这是运行：" + new com.mysql.cj.jdbc.Driver());
    }
}
```

图 15-2

路径 test/java 中的.java 类代码，如图 15-3 所示。

```java
package com.ghy.www;

import org.junit.Test;

import java.sql.SQLException;

import static org.junit.Assert.assertTrue;

public class AppTest {
    @Test
    public void shouldAnswerWithTrue() throws SQLException {
        System.out.println("这是测试: " + new com.mysql.cj.jdbc.Driver());
        assertTrue( condition: true);
    }
}
```

图 15-3

可见，两个 Java 类均没有报错，说明都是正确的代码，都使用了依赖 jar 包中的 com.mysql.cj.jdbc.Driver 类。

执行 verify 菜单，生成两个.jar 文件，如图 15-4 所示。

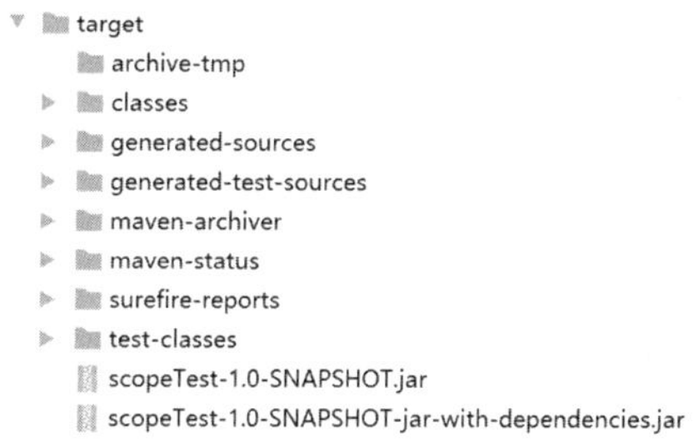


图 15-4

单击“Open in Terminal”菜单，如图 15-5 所示。

surefire-reports
test-classes
scopeTest-1.0-SNAPSHOT.jar
scopeTest-1.0-SNAPSHOT-jar-with-depen
pom.xml
External Libraries
Scratches and Consoles
New
Cut　Ctrl+X
Copy
Paste　Ctrl+V
Find Usages　Ctrl+G
Analyze
Refactor
Add to Favorites
Delete...　Delete
Run 'scopeTest-1.0-SNAPSH...'　Ctrl+Shift+F10
Debug 'scopeTest-1.0-SNAPSH...'
Run 'scopeTest-1.0-SNAPSH...' with 'Java Flight Recorder'
Create 'scopeTest-1.0-SNAPSH...'
Show in Explorer
File Path　Ctrl+Alt+F12
Open in Terminal
Local History
Reload from Disk
Terminal: Local　Local (2)　+
Microsoft Windows [版本 10.0.18
(c) 2019 Microsoft Corporation。
C:\Users\Administrator\Desktop\

图 15-5

输入 java -jar 命令，成功执行 App.java 类中的 main()方法，如图 15-6 所示。

```
C:\Users\Administrator\Desktop\scopeTest\target>java -jar scopeTest-1.0-SNAPSHOT-jar-with-dependencies.jar
这是运行：com.mysql.cj.jdbc.Driver@27d6c5e0
```

图 15-6

### 15.1.2 依赖范围：test

依赖范围 test 只在 test/java 范围中有效。

编辑 pom.xml 配置代码如下：

```
<dependencies>
    <dependency>
        <groupId>mysql</groupId>
        <artifactId>mysql-connector-java</artifactId>
        <version>8.0.23</version>
        <scope>test</scope>
    </dependency>
</dependencies>
```

这时，依赖范围为 test。路径 main/java 中的.java 类代码，如图 15-7 所示。

```
package com.ghy.www;

import java.sql.SQLException;

public class App {
    public static void main(String[] args) throws SQLException {
        System.out.println("这是运行：" + new com.mysql.cj.jdbc.Driver());
    }
}
```

Cannot resolve symbol 'Driver'

Add library 'Maven: mysql:mysql-connector-java:8.0.23' to classpath Alt+Shift+Enter

图 15-7

路径 test/java 中的.java 类代码，如图 15-8 所示。

```
package com.ghy.www;

import org.junit.Test;

import java.sql.SQLException;

import static org.junit.Assert.assertTrue;

public class AppTest {
    @Test
    public void shouldAnswerWithTrue() throws SQLException {
        System.out.println("这是测试：" + new com.mysql.cj.jdbc.Driver());
        assertTrue( condition: true);
    }
}
```

图 15-8 程序代码正确

执行 verify 菜单，出现异常，如图 15-9 所示。

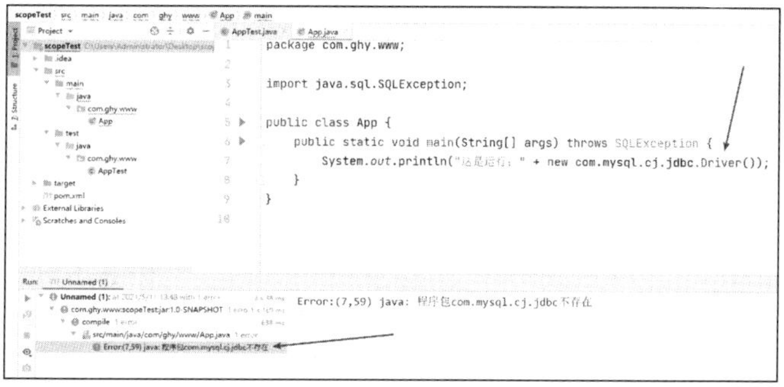

```
package com.ghy.www;

import java.sql.SQLException;

public class App {
    public static void main(String[] args) throws SQLException {
        System.out.println("这是运行：" + new com.mysql.cj.jdbc.Driver());
    }
}
```

图 15-9　出现异常

由于 main/java/com/ghy/www/App.java 类出现编译异常，因此依赖范围 test 在运行阶段无效。

### 15.1.3　依赖范围：provided

依赖范围 provided 只在 main/java 和 test/java 两个范围中有效。

编辑 pom.xml 配置代码如下：

```
<dependencies>
    <dependency>
        <groupId>mysql</groupId>
        <artifactId>mysql-connector-java</artifactId>
        <version>8.0.23</version>
        <scope>provided</scope>
    </dependency>
</dependencies>
```

这时，依赖范围为 provided。

路径 main/java 中的.java 类代码，如图 15-10 所示。

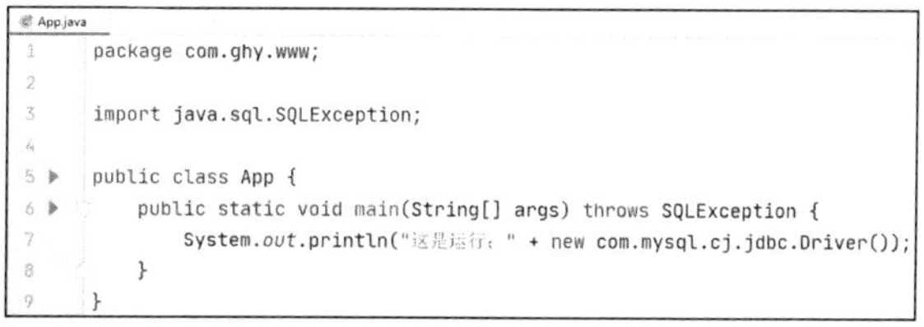

```
package com.ghy.www;

import java.sql.SQLException;

public class App {
    public static void main(String[] args) throws SQLException {
        System.out.println("这是运行：" + new com.mysql.cj.jdbc.Driver());
    }
}
```

图 15-10

路径 test/java 中的.java 类代码，如图 15-11 所示。

```
AppTest.java
package com.ghy.www;

import org.junit.Test;

import java.sql.SQLException;

import static org.junit.Assert.assertTrue;

public class AppTest {
    @Test
    public void shouldAnswerWithTrue() throws SQLException {
        System.out.println("这是测试：" + new com.mysql.cj.jdbc.Driver());
        assertTrue( condition: true);
    }
}
```

图 15-11　程序代码正确

执行 verify 菜单，创建两个.jar 文件并执行 java -jar 命令，输出如图 15-12 所示的结果。

```
C:\Users\Administrator\Desktop\scopeTest\target>java -jar scopeTest-1.0-SNAPSHOT-jar-with-dependencies.jar
Exception in thread "main" java.lang.NoClassDefFoundError: com/mysql/cj/jdbc/Driver
        at com.ghy.www.App.main(App.java:7)
Caused by: java.lang.ClassNotFoundException: com.mysql.cj.jdbc.Driver
        at java.net.URLClassLoader.findClass(URLClassLoader.java:382)
        at java.lang.ClassLoader.loadClass(ClassLoader.java:418)
        at sun.misc.Launcher$AppClassLoader.loadClass(Launcher.java:355)
        at java.lang.ClassLoader.loadClass(ClassLoader.java:351)
        ... 1 more

C:\Users\Administrator\Desktop\scopeTest\target>
```

图 15-12

基于依赖范围 provided 的特性，经常被用于 Servlet 相关的依赖中，因为大多数.war 文件是放在 Tomcat 中运行的，而 Tomcat 中有 Servlet 相关的 jar 包文件，不需要项目自带 Servlet 类的 jar 包文件。

### 15.1.4　依赖范围：runtime

依赖范围 runtime 在 test/java 和运行阶段两个范围中有效。

编辑 pom.xml 配置代码如下：

```
<dependencies>
    <dependency>
        <groupId>mysql</groupId>
        <artifactId>mysql-connector-java</artifactId>
        <version>8.0.23</version>
        <scope>runtime</scope>
    </dependency>
```

```
</dependencies>
```

这时，依赖范围为 runtime。

路径 main/java 中的.java 类代码，如图 15-13 所示。

```
App.java
package com.ghy.www;

import java.sql.Connection;
import java.sql.SQLException;

public class App {
    public static void main(String[] args) throws SQLException {
        Connection conn;
    }
}
```

图 15-13

不使用 jar 包文件中的类的主要原因是在使用 JDBC 技术时，使用的基本都是 JDK 中的 JDBC 接口，基本不会使用 mysql-jdbc-driver.jar 中的类，而使用 runtime 后就可以排除不需要导入.java 类的 jar 包文件。

路径 test/java 中的.java 类代码,如图 15-14 所示。

```
AppTest.java
package com.ghy.www;

import org.junit.Test;

import java.sql.SQLException;

import static org.junit.Assert.assertTrue;

public class AppTest {
    @Test
    public void shouldAnswerWithTrue() throws SQLException {
        System.out.println("这是测试：" + new com.mysql.cj.jdbc.Driver());
        assertTrue( condition: true);
    }
}
```

图 15-14

执行 verify 菜单，创建两个.jar 文件，文件 scopeTest-1.0-SNAPSHOT-jar-with-dependencies.jar 中的内容，如图 15-15 所示。

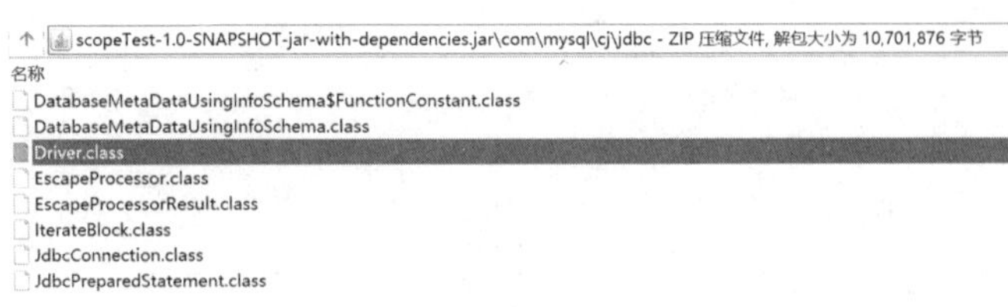


图 15-15

可见，在打包时已将 mysql 驱动中的类加入.jar 文件中了。

### 15.1.5 四种依赖范围总结

四种依赖范围的总结，如图 15-16 所示。

| 依赖范围 | main/java | test/java | 运行阶段 | 示例 |
|---|---|---|---|---|
| compile | Y | Y | Y | mybatis-jar |
| test | - | Y | - | Junit |
| provided | Y | Y | - | Servlet-API |
| runtime | - | Y | Y | JDBC驱动 |

图 15-16

（1）因为 mybatis-jar 文件中的类在 main/java，test/java 和运行阶段都会使用到，所以使用 compile 依赖范围。

（2）因为 Junit 文件中的类只需要在 test/java 中使用，所以使用 test 依赖范围。

（3）Servlet-API 文件中的类需要在 main/java 和 test/java 中使用，但不需要在运行阶段中使用，因为 Tomcat 中已经自带了 Servlet-API 相关的.jar 文件，没有必要重复引用。

（4）在大多数情况下，JDBC 驱动中的类在 main/java 中不会使用到，因为都是使用 JDK 中的 JDBC 接口进行开发的，所以使用 runtime 依赖范围。

## 15.2 传递性依赖和依赖范围

什么是传递性依赖？举个例子，比如有 A，B，C 三个项目，项目 B 依赖于项目 C，如果项目 A 依赖于项目 B，则项目 A 会自动依赖于项目 C，这就是传递性依赖。

有了传递性依赖，会逐项目自动依赖需要的 jar 包文件，而无须烦琐的手动指定，也不用担心会引用多余的文件，因为完全是按需自动引用的。但是多个依赖范围在混合使用时会互相影响，如图 15-17 所示。

| | compile | provided | runtime | test |
|---|---|---|---|---|
| compile | compile | - | runtime | - |
| provided | provided | - | provided | - |
| runtime | runtime | - | runtime | - |
| test | test | - | test | - |

图 15-17

可知，第一列和第一行的交叉点就是传递性依赖最终的依赖范围。

第一列代表第一阶段依赖，比如项目 A 使用 compile 依赖范围依赖于项目 B。

第一行代表第二阶段依赖，比如项目 B 使用 runtime 依赖范围依赖于项目 C。

因为第一列的 compile 和第一行的 runtime 交叉点就是 runtime，所以运算结果就是项目 A 以 runtime 依赖范围依赖于项目 C。

下面在 IDEA 中以创建项目示例的方式来验证前面计算的过程。

创建项目 B，创建 B 类。

创建项目 C，创建 C 类。

项目 B 和 C 的版本为<version>1.0-RELEASE</version>。

对项目 B 和 C 添加如下配置：

```
<distributionManagement>
    <repository>
        <id>mynexus</id>
        <name>nexus-releases</name>
        <url>http://localhost:8081/repository/maven-releases/</url>
    </repository>
    <snapshotRepository>
        <id>mynexus</id>
        <name>nexus-snapshot</name>
        <url>http://localhost:8081/repository/maven-snapshots/</url>
    </snapshotRepository>
</distributionManagement>
```

首先对项目 B 和 C 执行 deploy 菜单，并将.jar 文件上传到远程仓库中，如图 15-18 所示。

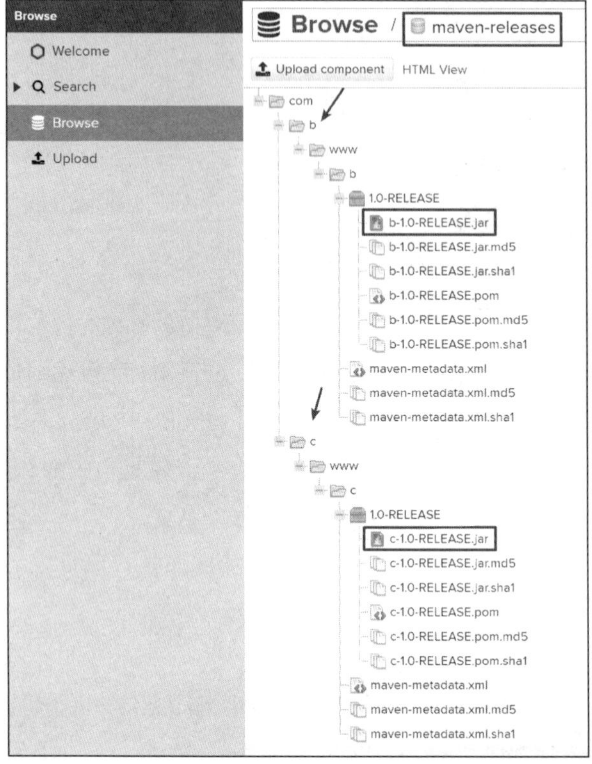


图 15-18

然后在项目 B 中以 runtime 依赖范围依赖于项目 C，配置代码如下：

```
<dependency>
    <groupId>com.c.www</groupId>
    <artifactId>c</artifactId>
    <version>1.0-RELEASE</version>
    <scope>runtime</scope>
</dependency>
```

对项目 B 重新执行 deploy 菜单，进行部署操作。

创建项目 A，创建 A 类，并且以 compile 依赖范围依赖于项目 B，配置代码如下：

```
<dependency>
    <groupId>com.b.www</groupId>
    <artifactId>b</artifactId>
    <version>1.0-RELEASE</version>
```

```
    <scope>compile</scope>
</dependency>
```

在项目 A 中添加自动运行 jar 包文件的插件，代码如下：

```
<build>
    <plugins><!-- 打包可运行 jar 包文件 -->
        <plugin>
            <groupId>org.apache.maven.plugins</groupId>
            <artifactId>maven-assembly-plugin</artifactId>
            <version>3.3.0</version>
            <configuration>
                <archive>
                    <manifest>
                        <mainClass>com.ghy.www.App</mainClass>
                    </manifest>
                </archive>
                <descriptorRefs>
                    <descriptorRef>jar-with-dependencies</descriptorRef>
                </descriptorRefs>
            </configuration>
            <executions>
                <execution>
                    <id>make-assembly</id>
                    <phase>package</phase>
                    <goals>
                        <goal>single</goal>
                    </goals>
                </execution>
            </executions>
        </plugin>
    </plugins>
</build>
```

最终，运算的结果是 runtime 依赖范围。因此，在项目 A 中，使用 C 类只在 test/java 和运行阶段两个范围中有效。

（1）测试 main/java 阶段：项目 A main/java 文件夹中的 A 类出现异常，如图 15-19 所示。

A.java

```
package com.a.www;

import com.c.www.C;

public class A {
    public static void main(String[] args) {
        C c;
    }
}

```

Cannot resolve symbol 'C'

Create type parameter 'C' Alt+Shift+Enter More actions... Alt+E

图 15-19

（2）测试 test/java 阶段：项目 A test/java 文件夹中的测试类引用 C 类是正确的，如图 15-20 所示。

```
package com.a.www;

import com.c.www.C;
import org.junit.Test;

import static org.junit.Assert.assertTrue;

public class AppTest {
    @Test
    public void shouldAnswerWithTrue() {
        C c;
        assertTrue( condition: true);
    }
}

```

图 15-20

（3）测试运行阶段：先更改项目 A main/java 中的 A 类代码，如图 15-21 所示。

A.java

```
package com.a.www;

public class A {
    public static void main(String[] args) throws ClassNotFoundException {
        System.out.println(Class.forName("com.c.www.C"));
    }
}
```

图 15-21

然后对项目 A 执行 verify 菜单，生成两个.jar 文件，文件中包含 C 类，如图 15-22 所示。

图 15-22

运行效果如图 15-23 所示。

```
Terminal:   Local    Local (2)   +
Microsoft Windows [版本 10.0.18363.1500]
(c) 2019 Microsoft Corporation。保留所有权利。

C:\Users\Administrator\Desktop\a\target>java -jar a-1.0-SNAPSHOT-jar-with-dependencies.jar
class com.c.www.C

C:\Users\Administrator\Desktop\a\target>
```

图 15-23

## 15.3　依赖调解

依赖调解是在依赖过程中出现多个不同版本的依赖处理策略。

### 15.3.1　最短路径

当间接的依赖出现不同版本时，Maven 采用最短路径的方式来决定最终使用的版本，如图 15-24 所示。

```
A
├── B
│   └── C
│       └── D 2.0
└── E
    └── D 1.0
```

图 15-24

因为 A 到 D 1.0 的路径距离比 A 到 D 2.0 的路径距离短，所以 A 项目最终依赖的是 D 1.0 版本。

本实验需要创建 a，b，c，d，e 五个项目：

（1）项目 a 依赖于项目 b，项目 b 依赖于项目 c，项目 c 依赖于项目 d 的 2.0-RELEASE 版本。

（2）项目 a 依赖于项目 e，项目 e 依赖于项目 d 的 1.0-RELEASE 版本。

项目 d 有两个 RELEASE 版本，其中 1.0-RELEASE 版本如图 15-25 所示，对项目 d 执行 deploy 菜单。2.0-RELEASE 版本如图 15-26 所示，对项目 d 执行 deploy 菜单。

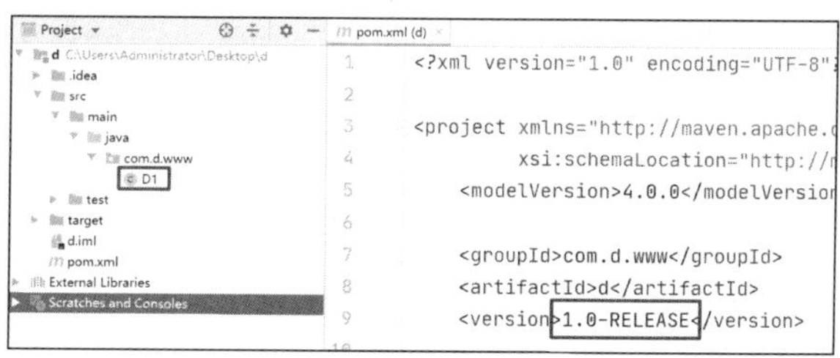


图 15-25

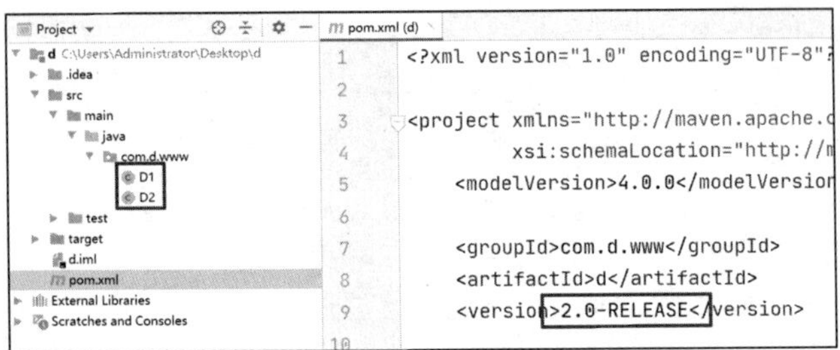

图 15-26

这时，仓库中d项目的版本如图15-27所示。因为项目c依赖于项目d的2.0-RELEASE版本，项目 a 依赖于项目 e，项目 e 又依赖于项目 d 的 1.0-RELEASE 版本，所以项目 a 使用的是项目 d 的 1.0-RELEASE 版本，如图 15-28 所示。

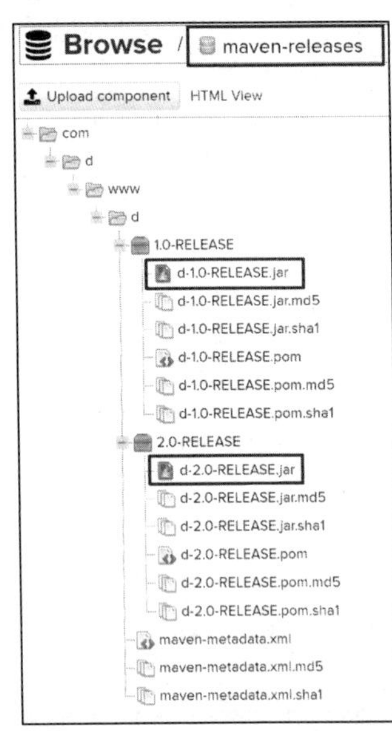

图 15-27

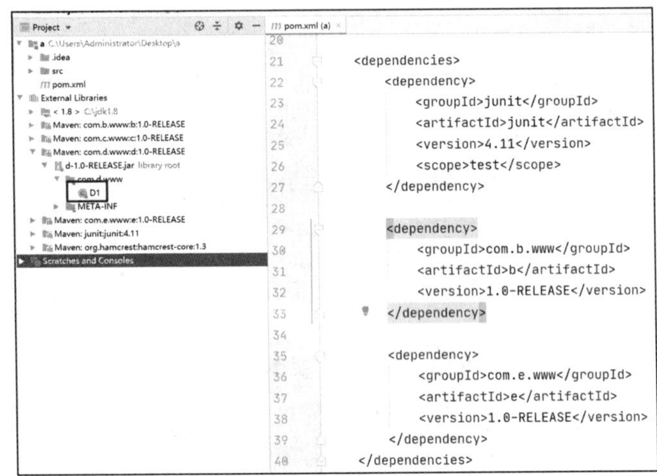

图 15-28

如果项目 a 想使用项目 d 的 2.0-RELEASE 版本，则可以在项目 a 中显式依赖项目 d 的 2.0-RELEASE 版本，如图 15-29 所示。

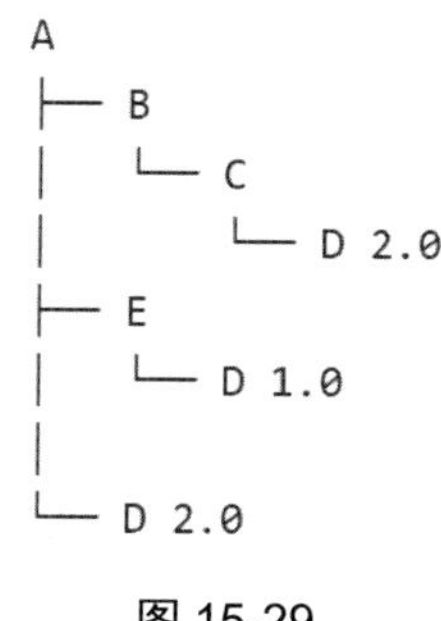


图 15-29

在项目 a 中，添加如下依赖配置：

```
<dependency>
    <groupId>com.d.www</groupId>
    <artifactId>d</artifactId>
    <version>2.0-RELEASE</version>
</dependency>
```

这时，项目 a 使用了项目 d 的 2.0-RELEASE 版本，如图 15-30 所示。

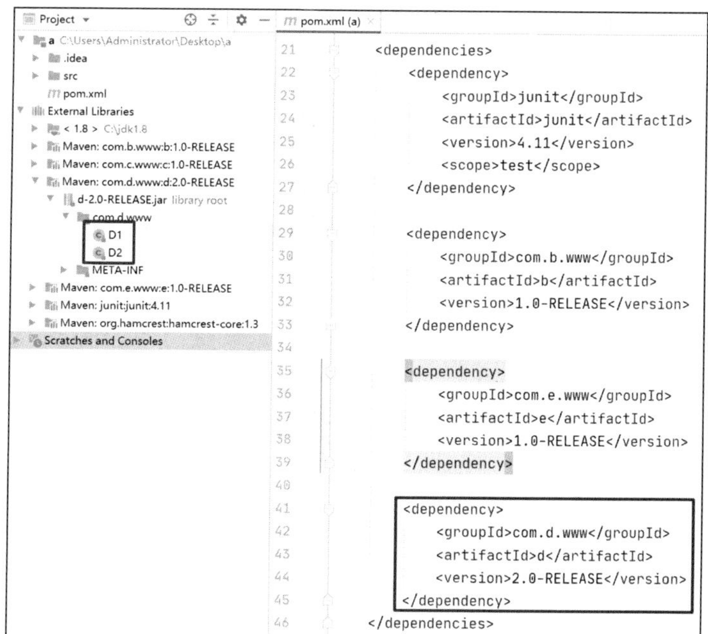


图 15-30

### 15.3.2 路径相同

如果出现路径长度相同的情况，如图 15-31 所示，则在 pom.xml 配置文件中按顺序决定使用的版本，即谁先声明，就优先使用谁。

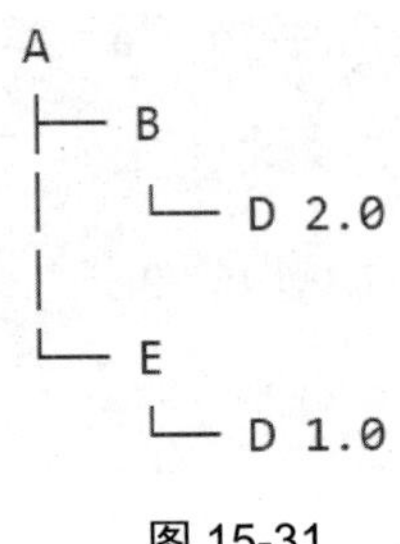


图 15-31

可知，项目 a 使用项目 d 的 2.0-RELEASE 版本，如图 15-32 所示。

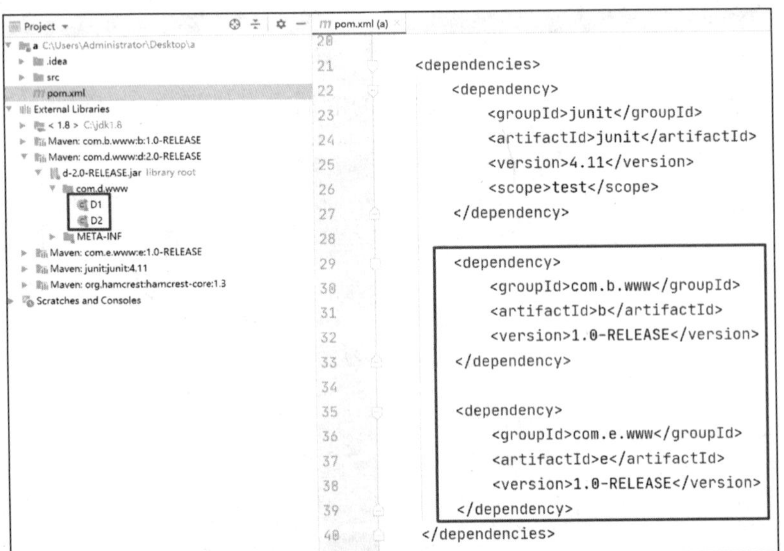


图 15-32

如果更改依赖顺序，则项目 a 使用项目 d 的 1.0-RELEASE 版本，如图 15-33 所示。

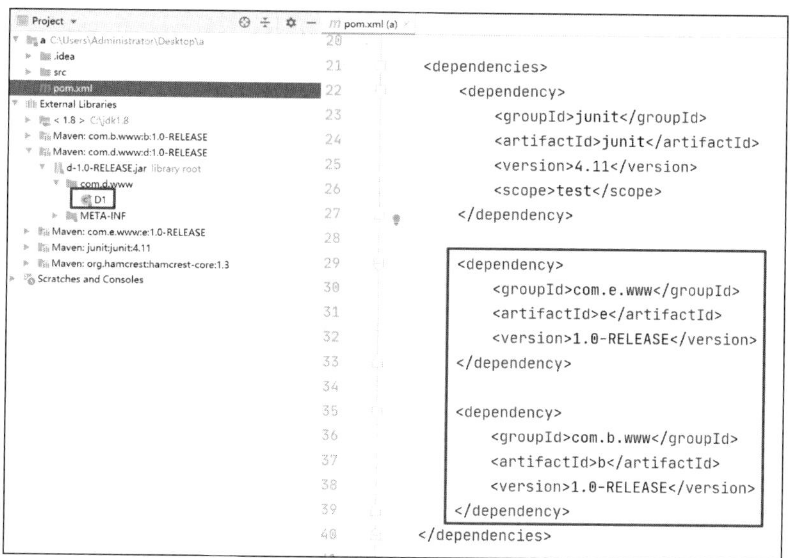

图 15-33

## 15.4　可选依赖

如图 15-34 所示，其中实线为 compile 依赖范围，虚线为可选依赖。

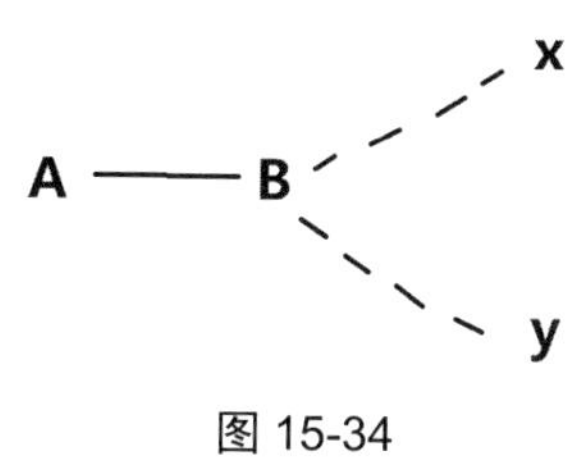


图 15-34

由图 15-34 可知，项目 A 依赖于项目 B，项目 B 依赖于项目 x 和项目 y，但项目 x 和项目 y 是项目 B 的两种解决方案的实现。比如，项目 x 和项目 y 都是针对项目 B 保存业务接口进行实现的，而项目 x 是将数据保存到数据库，项目 y 是将数据保存到.xml，那么在开发项目 B 时，因为使用的是保存业务的接口，其实现是在 x 和 y 这两个项目中，所以在项目 B 中对 x 和 y 两种实现都要进行依赖测试。

现在，假设将全部的虚线都变成实线，根据传递性依赖的特性，项目 A 会一起依赖于项目 x 和项目 y，但项目 A 不需要一起依赖于项目 x 和项目 y，因为这两种功能使用一种即可，这时就可以使用可选依赖，两条虚线出现了！在使用可选依赖后，项目 A 自已决定要依赖的项目。

先来看看无虚线的效果，如图 15-35 所示。

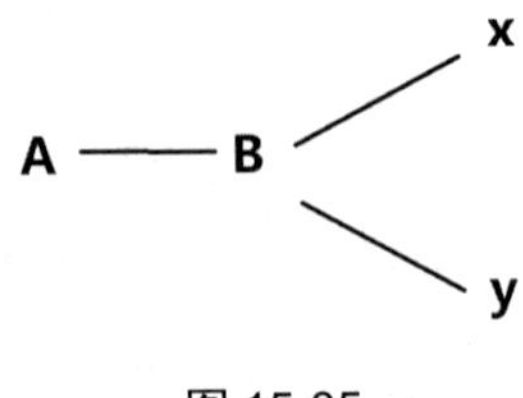


图 15-35

项目 B 的依赖信息，如图 15-36 所示。

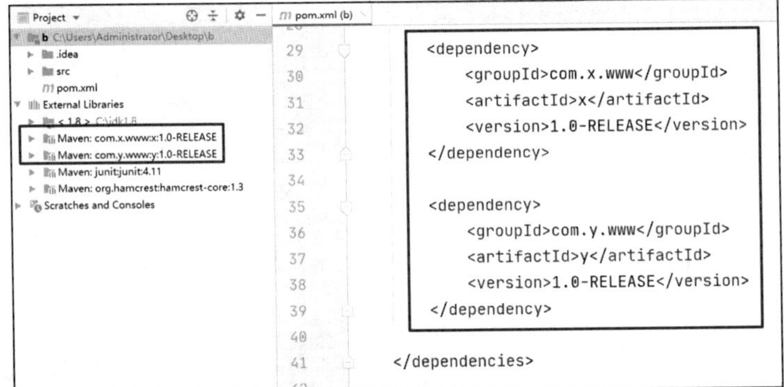


图 15-36

项目 A 的依赖信息，如图 15-37 所示。

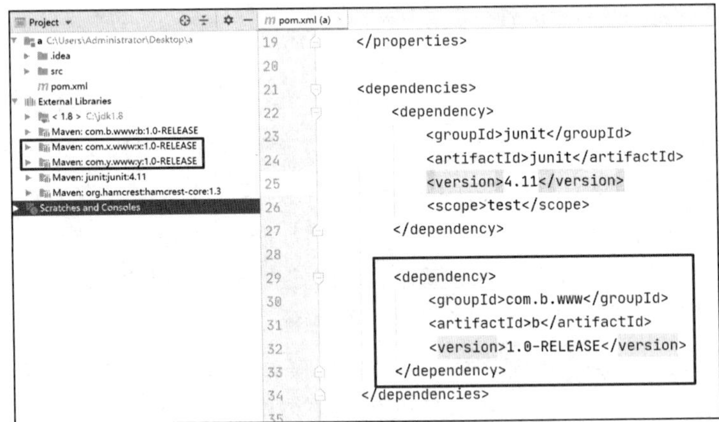


图 15-37

这时，项目 A 同时依赖了项目 x 和项目 y。我们添加可选依赖功能，也就是增加虚线，如图 15-38 所示。

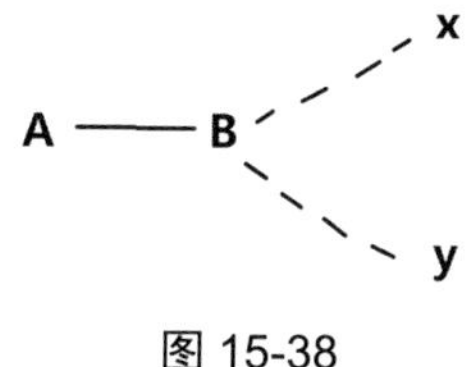


图 15-38

首先更改项目 B 中的 pom.xml 配置代码，如图 15-39 所示。

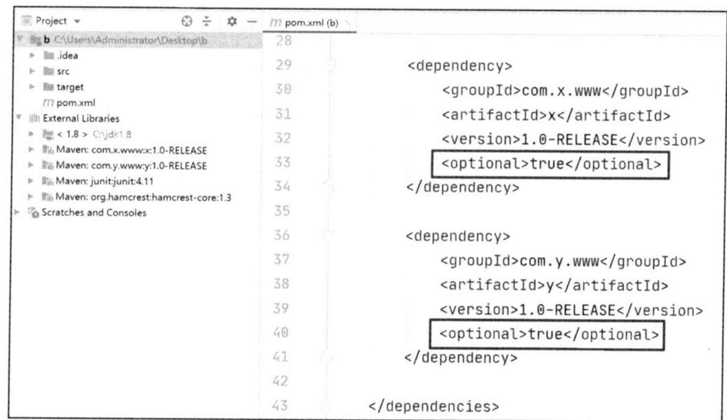


图 15-39

然后使用配置代码<optional>true</optional>实现可选依赖，这时项目 A 的结构如图 15-40 所示。

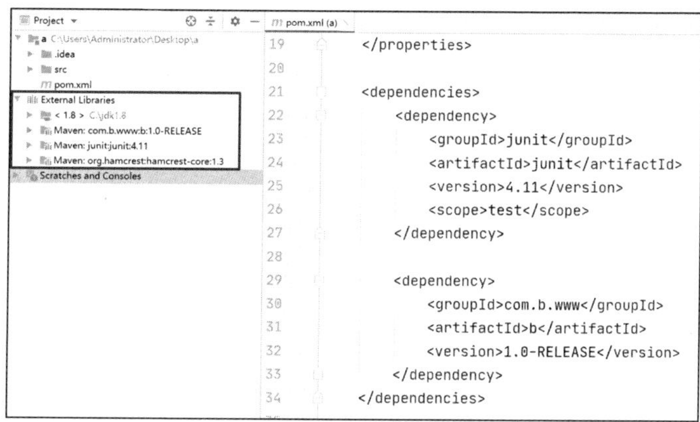


图 15-40

由项目 A 决定到底使用项目 x 还是项目 y，如图 15-41 所示，其中配置代码<version></version>不能缺少。

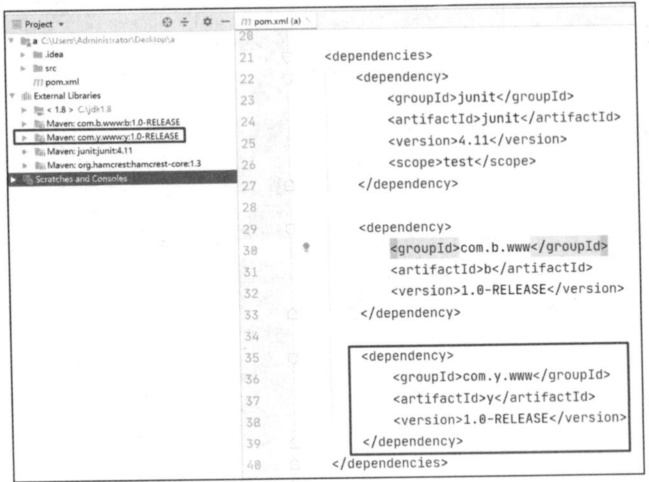


图 15-41

虽然上面的示例实现了可选依赖，但并不是最好的设计，因为项目 B 有两个功能，违反了职责单一原则。更好的做法是先创建 project-b-savedb 和 project-b-savexml 两个项目，然后项目 A 根据保存数据的需求来依赖不同功能的项目 B，在这种情况下并没有使用可选依赖，依赖关系简单，项目关系易理解且可读性强。

## 15.5 排除依赖

传递性依赖的原理，如图 15-42 所示。

**A — B — C 1.0**

图 15-42

如果项目 A 依赖于项目 B，项目 B 依赖于项目 C 的 1.0-RELEASE 版本，则项目 A 也依赖于项目 C 的 1.0-RELEASE 版本。

如果项目 B 的结构，如图 15-43 所示，则项目 A 的结构，如图 15-44 所示，可见项目 A 中包含项目 C 的 1.0-RELEASE 版本。

如果项目 A 想使用项目 C 的 2.0-RELEASE 版本，则需要在项目 A 的 pom.xml 文件中添加如图 15-45 所示的配置代码。

图 15-43

```
Project
a C:\Users\Administrator\Desktop\a
  .idea
  src
  pom.xml
External Libraries
  < 1.8 > C:\jdk1.8
  Maven: com.b.www:b:1.0-RELEASE
  Maven: com.c.www:c:1.0-RELEASE
    c-1.0-RELEASE.jar library root
      com.c.www
        C1
      META-INF
  Maven: junit:junit:4.11
  Maven: org.hamcrest:hamcrest-core:1.3
Scratches and Consoles
```

```xml
<dependencies>
    <dependency>
        <groupId>junit</groupId>
        <artifactId>junit</artifactId>
        <version>4.11</version>
        <scope>test</scope>
    </dependency>

    <dependency>
        <groupId>com.b.www</groupId>
        <artifactId>b</artifactId>
        <version>1.0-RELEASE</version>
    </dependency>
</dependencies>
```

图 15-44

图 15-45

这时，项目 A 中没有依赖项目 C 的 1.0-RELEASE 版本，再指定使用项目 C 的 2.0-RELEASE 版本，如图 15-46 所示。

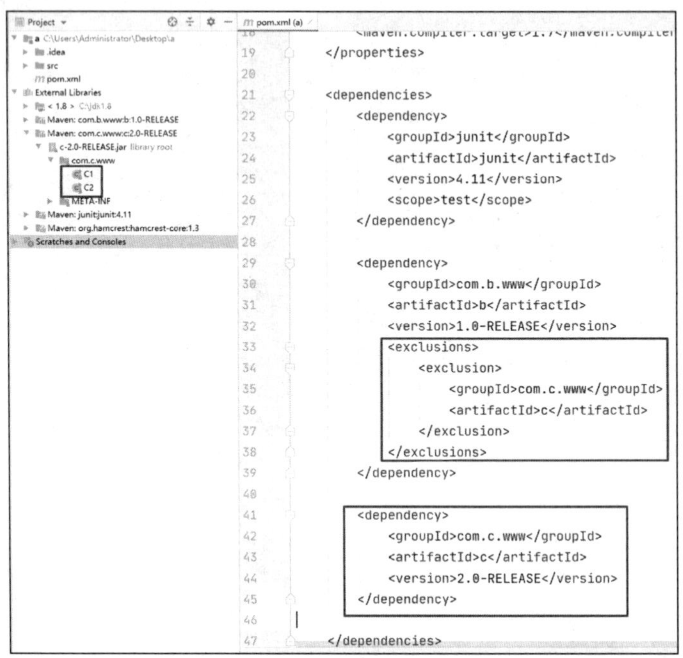


图 15-46

现在的依赖性传递效果如图 15-47 所示。

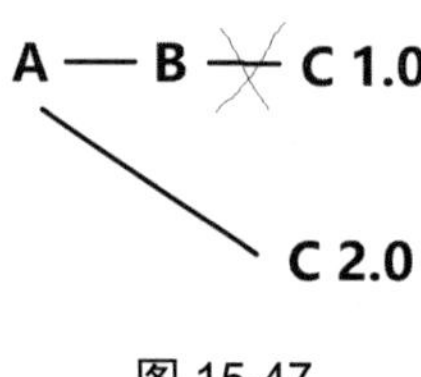


图 15-47

另外一种写法也可以实现，如图 15-48 所示，即不写<exclusions></exclusions>标签也可以实现同样的效果。

## 15.6 集中处理版本

由于项目中依赖的版本号使用的是字符串常量，如图 15-49 所示，因此我们可以集中对<version></version>版本号进行处理，形成一个类似于“字符串常量”的写法，更

改后的 pom.xml 配置代码如图 15-50 所示。

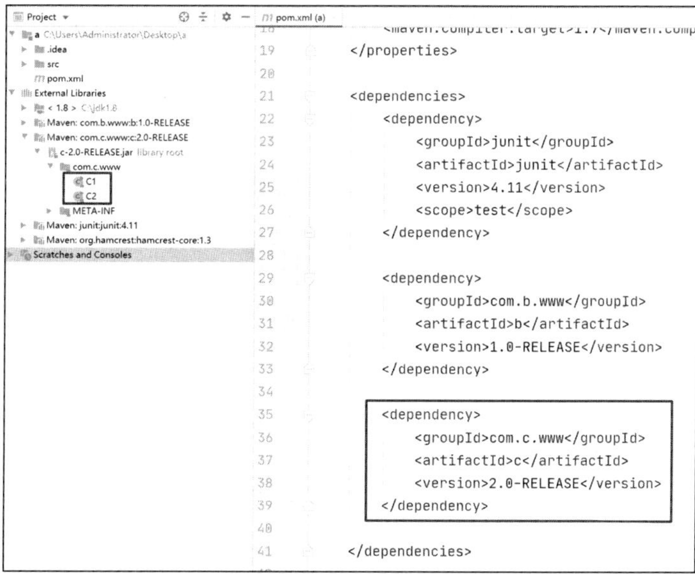


图 15-48

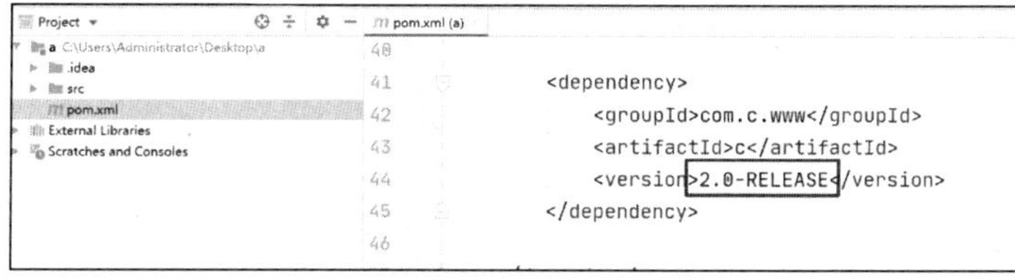


图 15-49

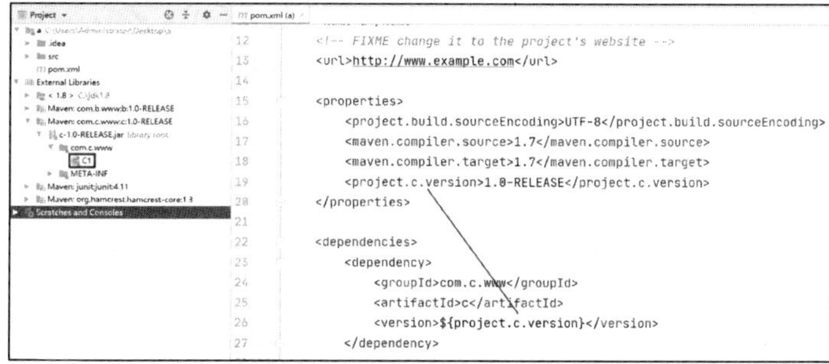


图 15-50

# 15.7 显示依赖结构

IDEA 支持以图形的方式显示依赖结构。

## 15.7.1 在 IDEA 中显示依赖结构

在 IDEA 中显示依赖结构，使用如下依赖配置代码：

```
<dependency>
    <groupId>org.springframework</groupId>
    <artifactId>spring-context</artifactId>
    <version>5.2.14.RELEASE</version>
</dependency>

<dependency>
    <groupId>org.springframework</groupId>
    <artifactId>spring-aop</artifactId>
    <version>5.2.14.RELEASE</version>
</dependency>

<dependency>
    <groupId>org.springframework</groupId>
    <artifactId>spring-beans</artifactId>
    <version>5.2.14.RELEASE</version>
</dependency>

<dependency>
    <groupId>org.springframework</groupId>
    <artifactId>spring-web</artifactId>
    <version>5.2.14.RELEASE</version>
</dependency>

<dependency>
    <groupId>org.springframework</groupId>
    <artifactId>spring-webmvc</artifactId>
    <version>5.2.14.RELEASE</version>
</dependency>
```

执行如图 15-51 所示的菜单。

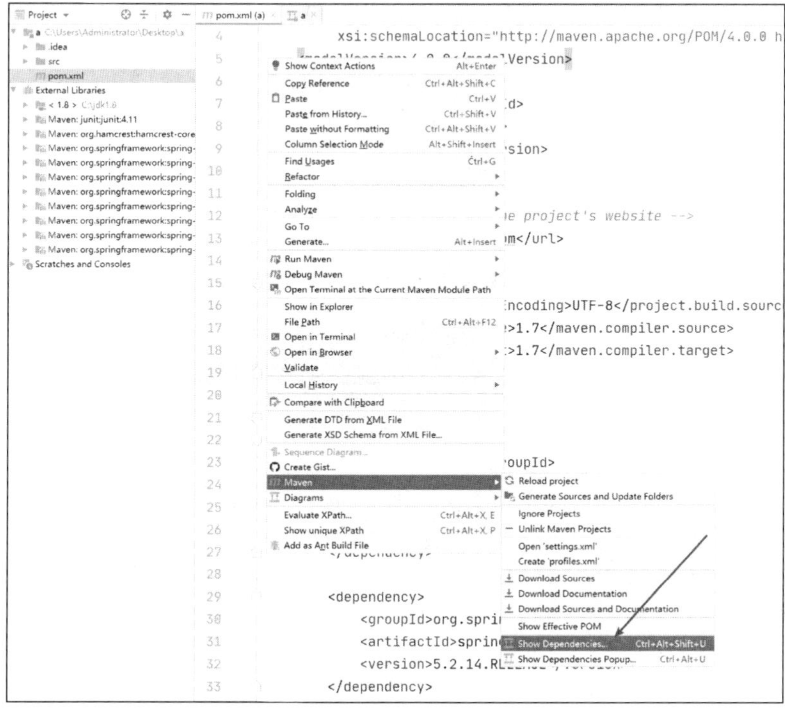


图 15-51

显示界面如图 15-52 所示。

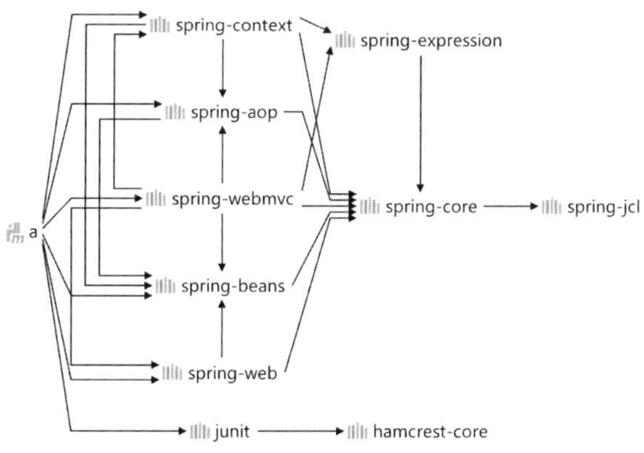


图 15-52

可以看到，显示的结构比较混乱，推荐使用如图 15-53 所示的布局。这时，显示内容如图 15-54 所示。

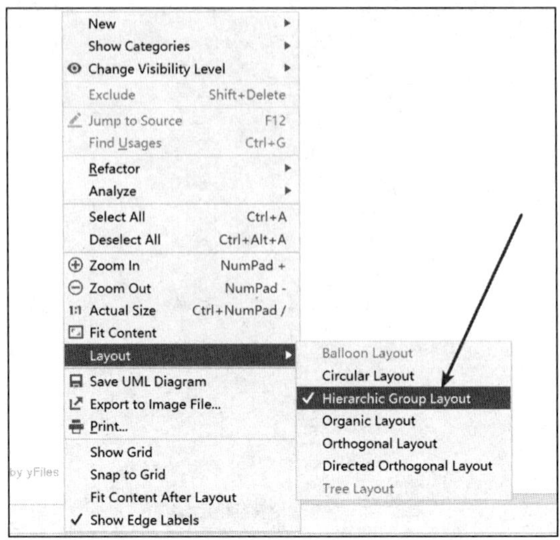


图 15-53

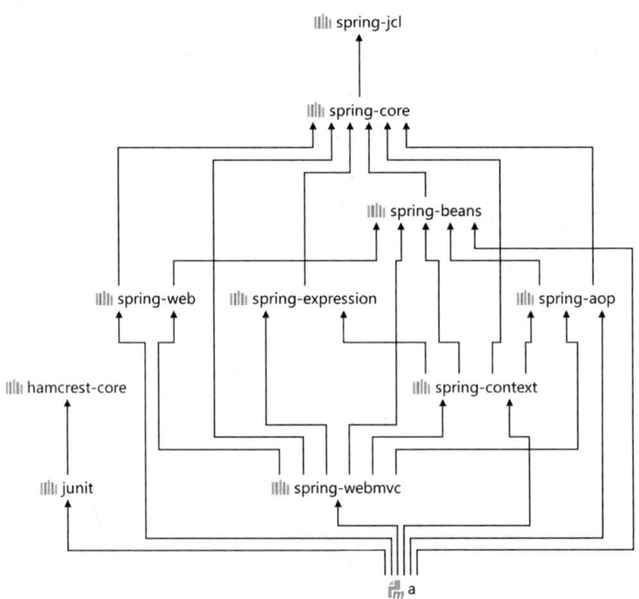


图 15-54

### 15.7.2 使用命令 mvn dependency:list 显示依赖列表

执行效果如图 15-55 所示。

```
C:\Users\Administrator\Desktop\a>mvn dependency:list
[INFO] Scanning for projects...
INFO]
[INFO] ---------------------------< com.a.www:a >------------------------
[INFO] Building a 1.0-SNAPSHOT
[INFO] --------------------------------[ jar ]----------------------------
[INFO]
[INFO] --- maven-dependency-plugin:2.8:list (default-cli) @ a ---
[INFO]
[INFO] The following files have been resolved:
[INFO]    junit:junit:jar:4.11:test
[INFO]    org.springframework:spring-expression:jar:5.2.14.RELEASE:compile
[INFO]    org.springframework:spring-context:jar:5.2.14.RELEASE:compile
[INFO]    org.springframework:spring-core:jar:5.2.14.RELEASE:compile
[INFO]    org.springframework:spring-webmvc:jar:5.2.14.RELEASE:compile
[INFO]    org.hamcrest:hamcrest-core:jar:1.3:test
[INFO]    org.springframework:spring-aop:jar:5.2.14.RELEASE:compile
[INFO]    org.springframework:spring-beans:jar:5.2.14.RELEASE:compile
[INFO]    org.springframework:spring-web:jar:5.2.14.RELEASE:compile
[INFO]    org.springframework:spring-jcl:jar:5.2.14.RELEASE:compile
[INFO]
[INFO] ------------------------------------------------------------------
[INFO] BUILD SUCCESS
[INFO] ------------------------------------------------------------------
[INFO] Total time:  1.895 s
```

图 15-55

### 15.7.3　使用命令 mvn dependency:tree 显示依赖树

执行效果如图 15-56 所示。

```
C:\Users\Administrator\Desktop\a>mvn dependency:tree
[INFO] Scanning for projects...
[INFO]
[INFO] ---------------------------< com.a.www:a >---------------------------
[INFO] Building a 1.0-SNAPSHOT
[INFO] --------------------------------[ jar ]-------------------------------
[INFO]
[INFO] --- maven-dependency-plugin:2.8:tree (default-cli) @ a ---
[INFO] com.a.www:a:jar:1.0-SNAPSHOT
[INFO] +- junit:junit:jar:4.11:test
[INFO] |  \- org.hamcrest:hamcrest-core:jar:1.3:test
[INFO] +- org.springframework:spring-context:jar:5.2.14.RELEASE:compile
[INFO] |  +- org.springframework:spring-core:jar:5.2.14.RELEASE:compile
[INFO] |  |  \- org.springframework:spring-jcl:jar:5.2.14.RELEASE:compile
[INFO] |  \- org.springframework:spring-expression:jar:5.2.14.RELEASE:compile
[INFO] +- org.springframework:spring-aop:jar:5.2.14.RELEASE:compile
[INFO] +- org.springframework:spring-beans:jar:5.2.14.RELEASE:compile
[INFO] +- org.springframework:spring-web:jar:5.2.14.RELEASE:compile
[INFO] \- org.springframework:spring-webmvc:jar:5.2.14.RELEASE:compile
[INFO] ---------------------------------------------------------------------
[INFO] BUILD SUCCESS
[INFO] ---------------------------------------------------------------------
[INFO] Total time:  1.622 s
```

图 15-56

（1）+-：代表依赖列表中除去最后一项依赖的其他依赖。

（2）\-：代表依赖列表中最后一项依赖。

（3）|：子依赖列表。

## 15.7.4 使用命令 mvn dependency:analyze 分析依赖

### 1. Unused declared dependencies found

Unused declared dependencies found 代表直接依赖的组件并没有使用到，执行效果如图 15-57 所示。

```
[INFO]
[INFO] --- maven-dependency-plugin:2.8:analyze (default-cli) @ a ---
[WARNING] Unused declared dependencies found:
[WARNING]    org.springframework:spring-context:jar:5.2.14.RELEASE:compile
[WARNING]    org.springframework:spring-aop:jar:5.2.14.RELEASE:compile
[WARNING]    org.springframework:spring-beans:jar:5.2.14.RELEASE:compile
[WARNING]    org.springframework:spring-web:jar:5.2.14.RELEASE:compile
[WARNING]    org.springframework:spring-webmvc:jar:5.2.14.RELEASE:compile
[INFO] ------------------------------------------------------------------------
[INFO] BUILD SUCCESS
[INFO] ------------------------------------------------------------------------
[INFO] Total time:  3.400 s
```

图 15-57

现在，项目中有 5 个 Unused declared dependencies found 未使用的依赖。

创建 Test1.java 类，并使用依赖中的 Class，代码如下：

```
package com.a.www;

import org.springframework.context.ApplicationContext;
import org.springframework.context.support.ClassPathXmlApplicationContext;

public class Test1 {
    public static void main(String[] args) {
        ApplicationContext ac = new ClassPathXmlApplicationContext();
        System.out.println(ac);
    }
}
```

再次执行 mvn dependency:analyze 命令，输出结果如图 15-58 所示。

```
[INFO] --- maven-dependency-plugin:2.8:analyze (default-cli) @ a ---
[WARNING] Unused declared dependencies found:
[WARNING]    org.springframework:spring-aop:jar:5.2.14.RELEASE:compile
[WARNING]    org.springframework:spring-beans:jar:5.2.14.RELEASE:compile
[WARNING]    org.springframework:spring-web:jar:5.2.14.RELEASE:compile
[WARNING]    org.springframework:spring-webmvc:jar:5.2.14.RELEASE:compile
```

图 15-58

经过对比可以发现，组件 org.springframework:spring-context:jar:5.2.14.RELEASE 已经被使用了，没有在 Unused declared dependencies found 未使用的依赖列表中显示。

那么，其他 4 个未使用的依赖就应该删除吗？不是的，因为 Maven 只会分析 main/java 和 test/java 中直接引用的依赖，不会发现间接的依赖，所以不能直接删除，要具体情况具体分析。不过，从 mvn dependency:analyze 命令中还是能大概分析出哪些依赖是当前项目没有使用的，有选择性地进行删除，这也是对依赖结构和关系的优化。

## 2. Used undeclared dependencies found

Used undeclared dependencies found 代表使用非直接依赖的组件，创建 Test2.java 类的代码如下：

```
package com.a.www;

import org.apache.commons.logging.impl.NoOpLog;

public class Test2 {
    public static void main(String[] args) {
        NoOpLog adapter = new NoOpLog();
    }
}
```

在 pom.xml 配置文件中，并没有直接依赖 org.apache.commons.logging，而是有间接依赖，但是在 Test2.java 类中却使用了间接依赖的类，这些依赖项会在 Used undeclared dependencies found 列表中显示，如图 15-59 所示。

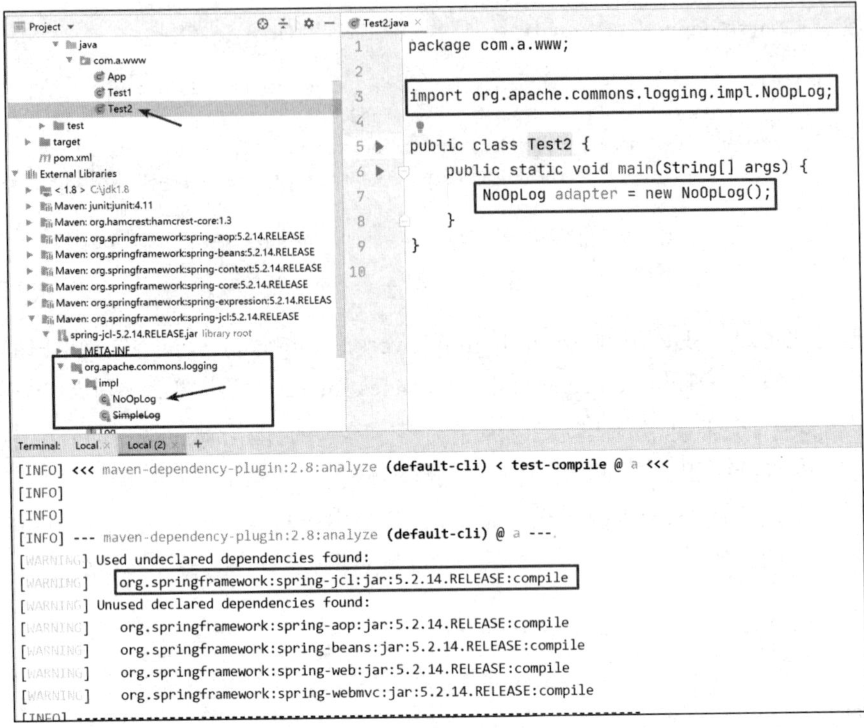


图 15-59

针对在 Used undeclared dependencies found 列表中出现的依赖需要提高一些警惕，因为间接依赖的组件会随直接依赖组件的升级而升级，导致在 main()方法中使用的写法是旧版本的间接依赖，在升级直接依赖版本时会出现编译错误。

## 15.8 依赖snapshot版本的自动更新特性

如果项目 A 依赖于项目 B，而项目 B 还在 snapshot 阶段，则项目 A 可以自动获取项目 B 的最新版组件。

项目 B 有两个 snapshot 版本，其中 1.0-SNAPSHOT 版本如图 15-60 所示，2.0-SNAPSHOT 版本如图 15-61 所示。

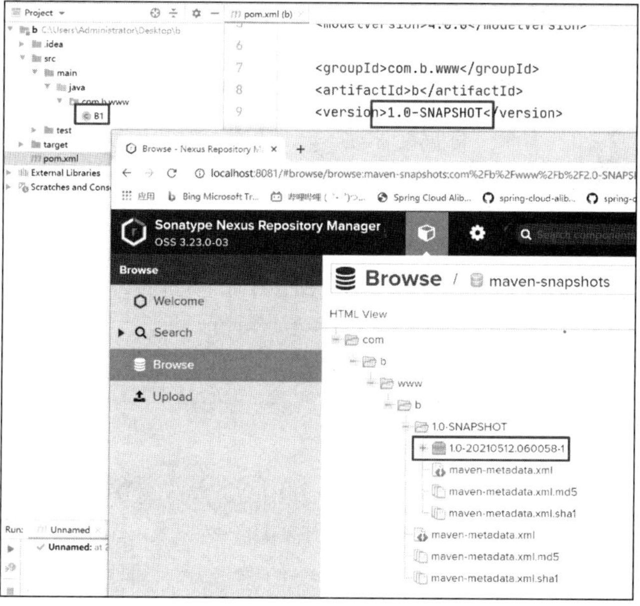

Project
pom.xml (b)
b C:\Users\Administrator\Desktop\b
.idea
src
main
java
com.b.www
B1
test
target
pom.xml
External Libraries
Scratches and Consoles
<groupId>com.b.www</groupId>
<artifactId>b</artifactId>
<version>1.0-SNAPSHOT</version>
Browse - Nexus Repository M
localhost:8081/#browse/browse:maven-snapshots:com%2Fb%2Fwww%2Fb%2F2.0-SNAPS
Sonatype Nexus Repository Manager
OSS 3.23.0-03
Browse
Welcome
Search
Browse
Upload
Browse / maven-snapshots
HTML View
com
b
www
b
1.0-SNAPSHOT
1.0-20210512.060058-1
maven-metadata.xml
maven-metadata.xml.md5
maven-metadata.xml.sha1
maven-metadata.xml
maven-metadata.xml.md5
maven-metadata.xml.sha1
Run: Unnamed


图 15-60

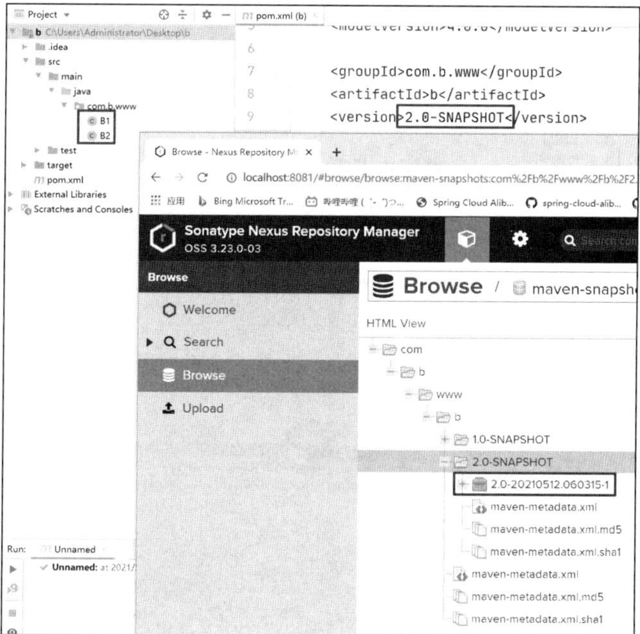

Project
pom.xml (b)
b C:\Users\Administrator\Desktop\b
.idea
src
main
java
com.b.www
B1
B2
test
target
pom.xml
External Libraries
Scratches and Consoles
<groupId>com.b.www</groupId>
<artifactId>b</artifactId>
<version>2.0-SNAPSHOT</version>
Browse - Nexus Repository M
localhost:8081/#browse/browse:maven-snapshots:com%2Fb%2Fwww%2Fb%2F2
Sonatype Nexus Repository Manager
OSS 3.23.0-03
Browse
Welcome
Search
Browse
Upload
Browse / maven-snapsh
HTML View
com
b
www
b
1.0-SNAPSHOT
2.0-SNAPSHOT
2.0-20210512.060315-1
maven-metadata.xml
maven-metadata.xml.md5
maven-metadata.xml.sha1
maven-metadata.xml
maven-metadata.xml.md5
maven-metadata.xml.sha1
Run: Unnamed


图 15-61

项目 A 使用如下依赖配置代码：

```
<dependency>
    <groupId>com.b.www</groupId>
    <artifactId>b</artifactId>
    <version>2.0-SNAPSHOT</version>
</dependency>
```

下面我们先对项目 B 中的第 2 个 SNAPSHOT 版本进行小幅更新，如图 15-62 所示。

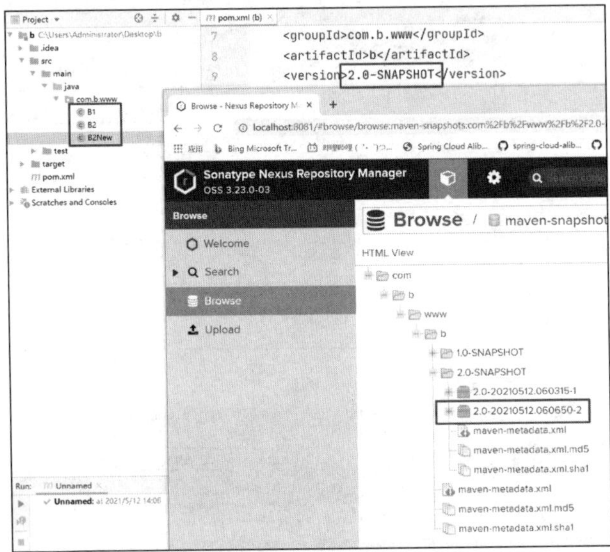

图 15-62

然后对项目 A 执行 Reload project 菜单，如图 15-63 所示。

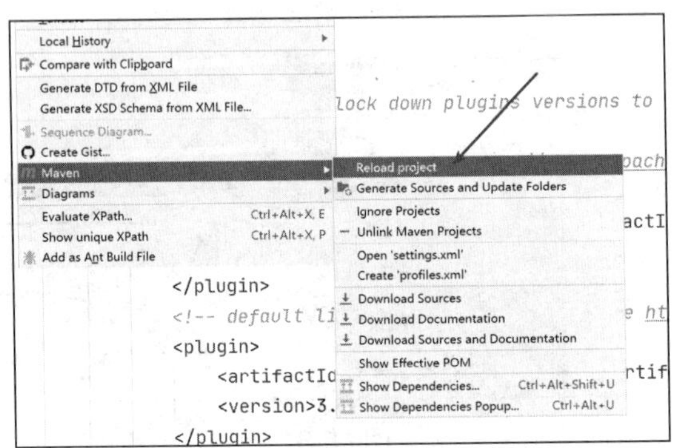

图 15-63

这时，项目 A 使用了 SNAPSHOT 最新的子版本组件，如图 15-64 所示。

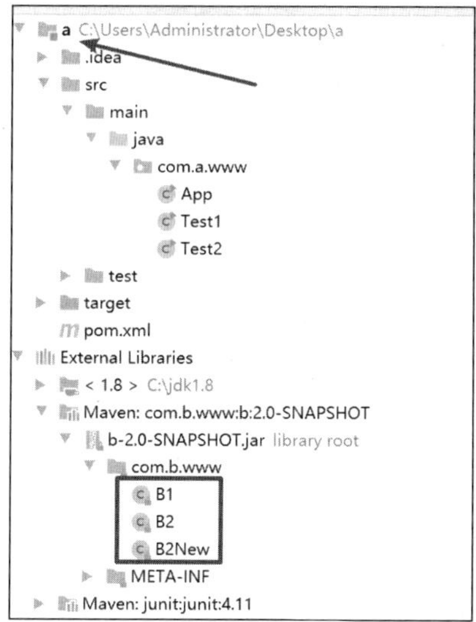


图 15-64

## 15.9　将 source 源代码打包并发布

在 pom.xml 配置文件中，添加如下配置：

```
<build>
    <plugins>
        <plugin>
            <groupId>org.apache.maven.plugins</groupId>
            <artifactId>maven-source-plugin</artifactId>
            <version>2.2.1</version>
            <executions>
                <execution>
                    <id>attach-sources</id>
                    <goals>
                        <goal>jar</goal>
                    </goals>
                </execution>
            </executions>
        </plugin>
    </plugins>
</build>
```

执行 deploy 菜单，仓库内容如图 15-65 所示。

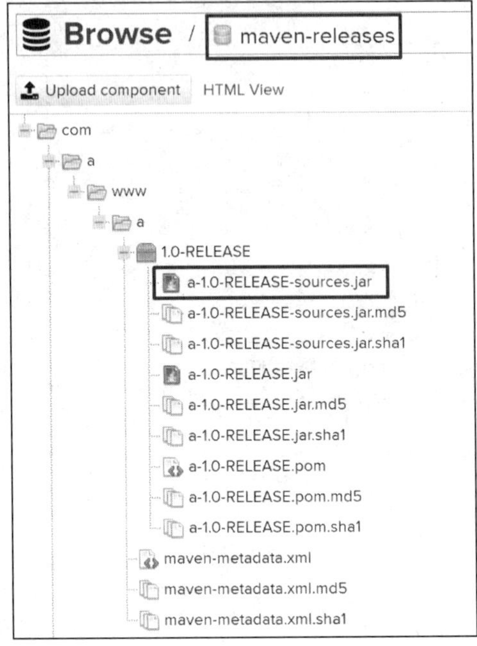


图 15-65

其中，a-1.0-RELEASE-sources.jar 文件的内容如图 15-66 所示。

a-1.0-RELEASE-sources.jar\com\a\www - ZIP 压缩文件, 解包大小为 499 字节
名称
..
A1.java
A2.java

图 15-66

## 15.10 跳过测试

在打包时，默认是执行了测试，执行 package 菜单如图 15-67 所示。这样比较耗时，我们可以先在 IDEA 中单击如图 15-68 所示的按钮，然后再次执行 package 菜单，这时测试被跳过，如图 15-69 所示。

```
[INFO] --- maven-compiler-plugin:3.8.0:testCompile (default-testCompile) @ a ---
[INFO] Nothing to compile - all classes are up to date
[INFO]
[INFO] --- maven-surefire-plugin:2.22.1:test (default-test) @ a ---
[INFO]
[INFO] -------------------------------------------------------
[INFO]  T E S T S
[INFO] -------------------------------------------------------
[INFO] Running com.a.www.AppTest
[INFO] Tests run: 1, Failures: 0, Errors: 0, Skipped: 0, Time elapsed: 0.024 s - in com.a.www.AppTest
[INFO]
[INFO] Results:
[INFO]
[INFO] Tests run: 1, Failures: 0, Errors: 0, Skipped: 0
[INFO]
[INFO]
[INFO] --- maven-jar-plugin:3.0.2:jar (default-jar) @ a ---
[INFO]
[INFO] >>> maven-source-plugin:2.2.1:jar (attach-sources) > generate-sources @ a >>>
[INFO]
[INFO] <<< maven-source-plugin:2.2.1:jar (attach-sources) < generate-sources @ a <<<
[INFO]
```

图 15-67

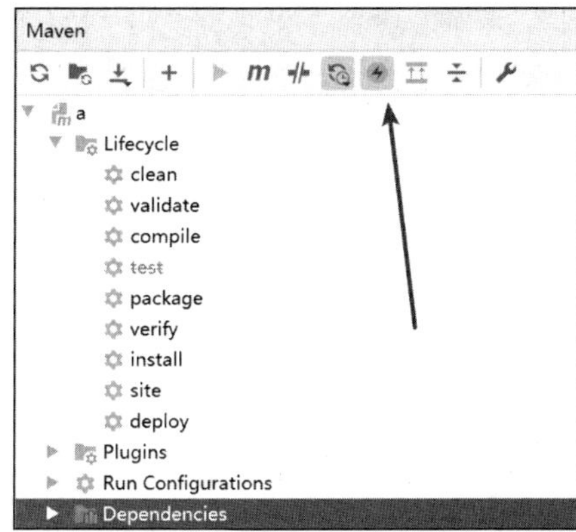


图 15-68

```
[INFO] skip non existing resourceDirectory C:\Users\Administrator\Desktop
[INFO]
[INFO] --- maven-compiler-plugin:3.8.0:testCompile (default-testCompile)
[INFO] Nothing to compile - all classes are up to date
[INFO]
[INFO] --- maven-surefire-plugin:2.22.1:test (default-test) @ a ---
[INFO] Tests are skipped.
[INFO]
[INFO] --- maven-jar-plugin:3.0.2:jar (default-jar) @ a ---
[INFO]
[INFO] >>> maven-source-plugin:2.2.1:jar (attach-sources) > generate-sou
[INFO]
[INFO] <<< maven-source-plugin:2.2.1:jar (attach-sources) < generate-sou
[INFO]
```

图 15-69

# 第 16 章 Maven 的聚合与继承

为了便于代码后期的维护，大型项目往往会被拆分成多个子模块，而在 Maven 中将多个子模块整合到一起的过程被称为聚合。聚合是为了对多个项目进行统一构建。

父项目集中处理公共配置，子项目可以复用，这在 Maven 中被称为继承。

## 16.1 项目的聚合

下面搭建项目聚合环境。

### 16.1.1 创建父项目

以 Empty Project 的方式创建父项目，如图 16-1 所示，单击“Next”按钮。

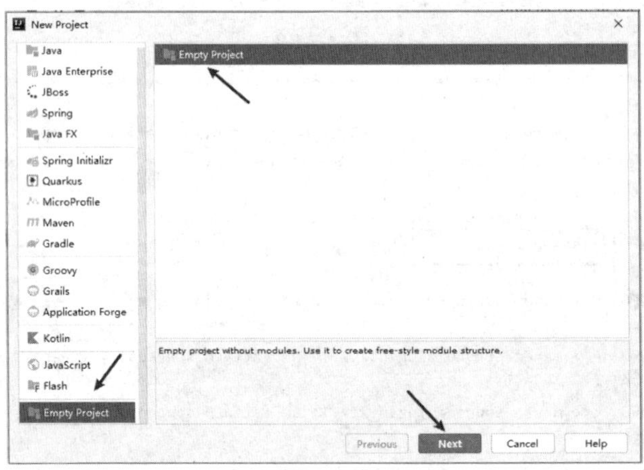


图 16-1

设置项目名称，如图 16-2 所示，单击“Finish”按钮完成父项目的创建。。

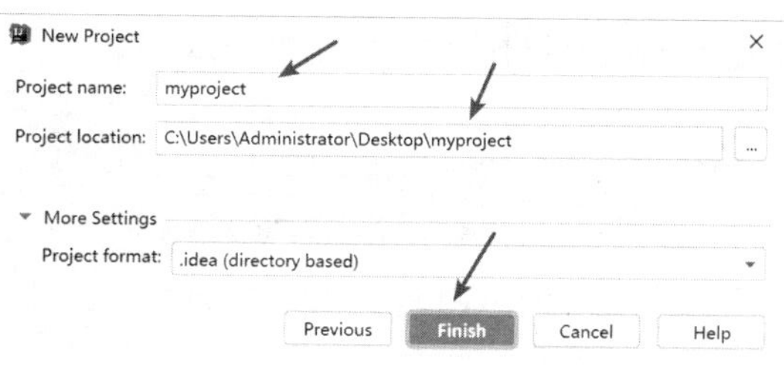


图 16-2

## 16.1.2　创建 DAO 子模块

如图 16-3 所示，新建一个 DAO 子模块。

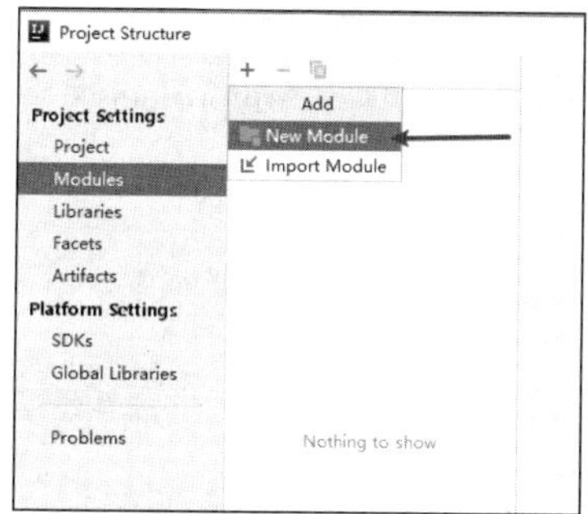


图 16-3

配置子模块，单击“Next”按钮，如图 16-4 所示。

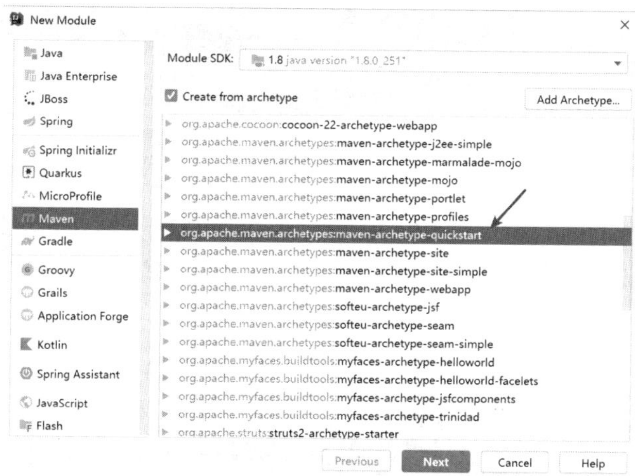


图 16-4

配置项目属性，单击“Next”按钮，如图 16-5 所示。

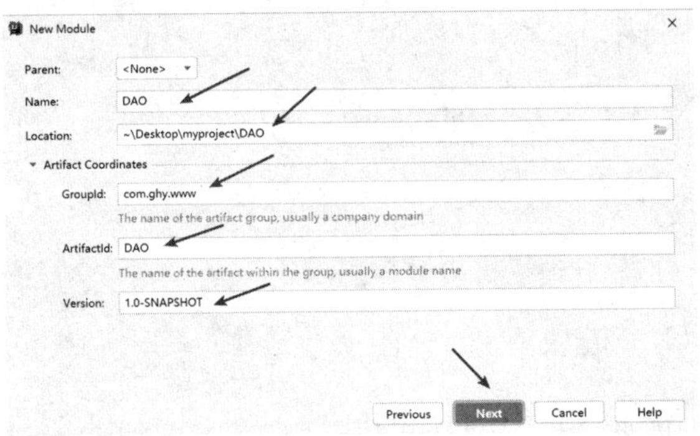


图 16-5

注意：子模块的保存路径是在父项目的 myproject 文件夹中。

继续配置，单击“Finish”按钮完成 DAO 子模块的创建，如图 16-6 所示。

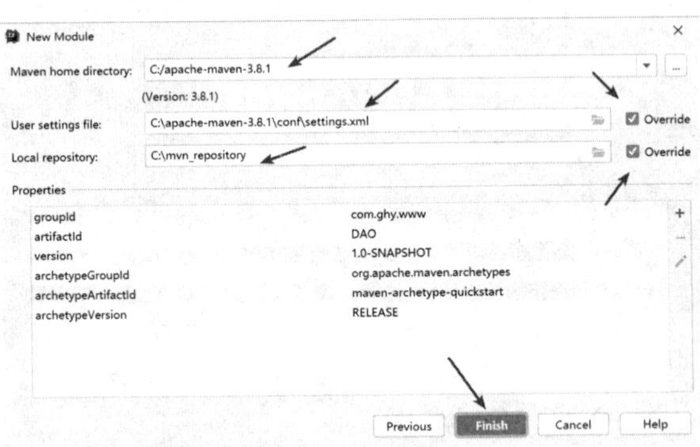


图 16-6

### 16.1.3 创建 Service 子模块和 Web 子模块

继续创建基于 Maven quickstart 的 Service 子模块和基于 Maven webapp 的 Web 子模块。这时，子模块的列表结构如图 16-7 所示。

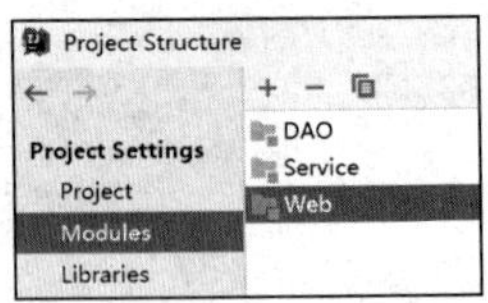


图 16-7

完成后的项目结构如图 16-8 所示，文件夹结构如图 16-9 所示。

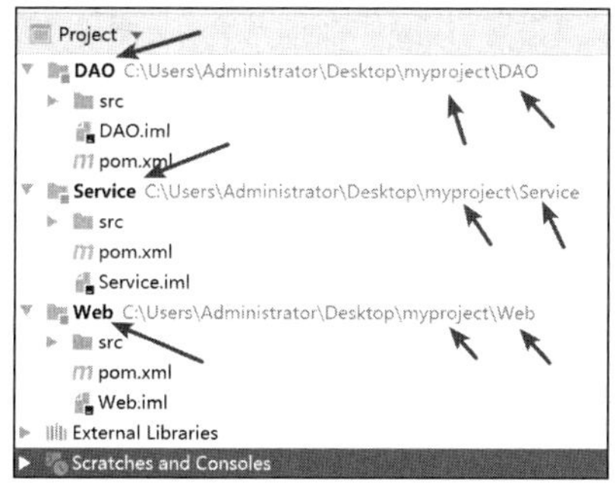


图 16-8

图 16-9

在 IDEA 中，父项目引用子模块的证据如图 16-10 所示。

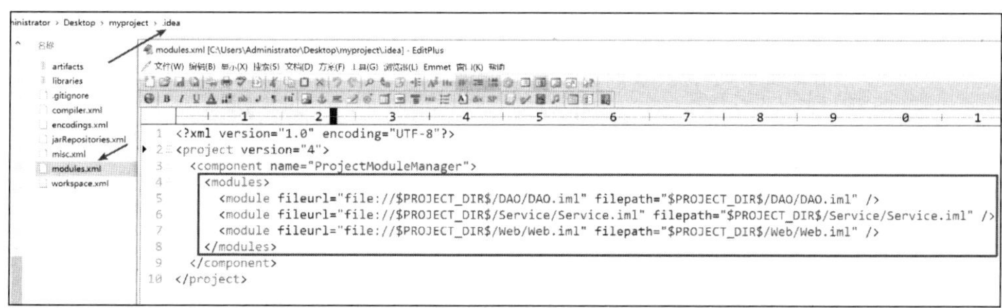


图 16-10

如果不用 IDEA，则在父项目的 pom.xml 配置文件中使用<modules><module></module></modules>标签来引用其他子模块，示例配置如下：

```
<modules>
   <module>anyproject-emailservice</module>
   <module>anyproject-userinfoservice</module>
</modules>
```

并且父项目 pom.xml 配置文件中的<package>值必须为 pom，示例配置如下：

```
<packaging>pom</packaging>
```

如果使用 IDEA，则父子模块一切的关联配置全部由 IDEA 进行处理，无须手动配置。另外，在 IDEA 中也可以采用手动方式使用<modules>标签引用子模块，但这在实际的软件开发中使用较少。

### 16.1.4 编辑 DAO 子模块中的 pom.xml 配置文件

pom.xml 配置文件的核心代码，如图 16-11 所示。

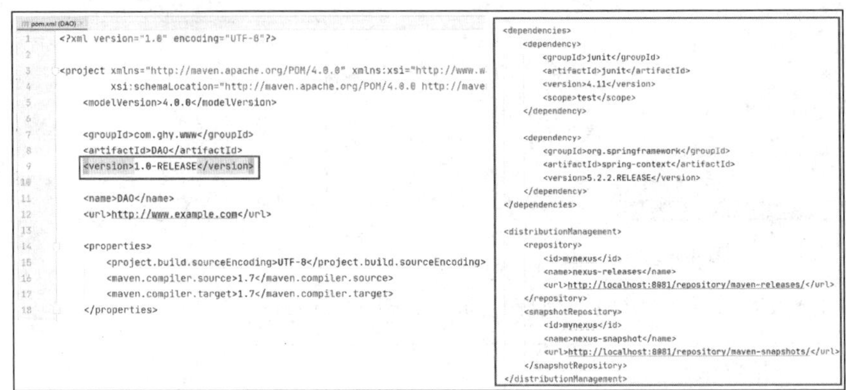


图 16-11

### 16.1.5 创建 DAO 类并发布

创建 DAO 类的代码如下：

```
package com.ghy.www.dao;

import org.springframework.context.support.ClassPathXmlApplicationContext;

public class UserinfoDAO {
    public String getUsernameByUserId(long userId) {
        return "返回的 username 值=" + new ClassPathXmlApplicationContext();
    }
}
```

对 DAO 子模块执行发布操作，如图 16-12 所示。

图 16-12

DAO 子模块成功发布，如图 16-13 所示。

图 16-13

## 16.1.6　创建 Service 类并引用 DAO 子模块

创建 Service 类的代码如下：

```
package com.ghy.www.service;

import com.ghy.www.dao.UserinfoDAO;

public class UserinfoService {
    private UserinfoDAO userinfoDAO = new UserinfoDAO();

    public String getUsernameByUserIdService() {
        return userinfoDAO.getUsernameByUserId(123);
    }
}
```

单击“Add dependency on module'DAO'”链接，实现 Service 子模块引用 DAO 子模块，如图 16-14 所示。

```
UserinfoService.java
package com.ghy.www.service;

import com.ghy.www.dao.UserinfoDAO;

public class UserinfoService {
    private UserinfoDAO userinfoDAO = new UserinfoDAO(
                        Cannot resolve symbol 'UserinfoDAO'
    public Strin        Add dependency on module 'DAO'  Alt+Shift+Enter   More actions...
        return userinfoDAO.getUsernameByUserId(123);
    }
}
```

图 16-14

Service 子模块引用 DAO 子模块是在 Service 子模块的 pom.xml 配置文件中使用如下配置实现的。

```
<dependencies>
    <dependency>
        <groupId>com.ghy.www</groupId>
        <artifactId>DAO</artifactId>
        <version>1.0-RELEASE</version>
        <scope>compile</scope>
    </dependency>
</dependencies>
```

## 16.1.7 编辑 Service 子模块中的 pom.xml 配置文件并发布

pom.xml 配置文件的核心代码，如图 16-15 所示。

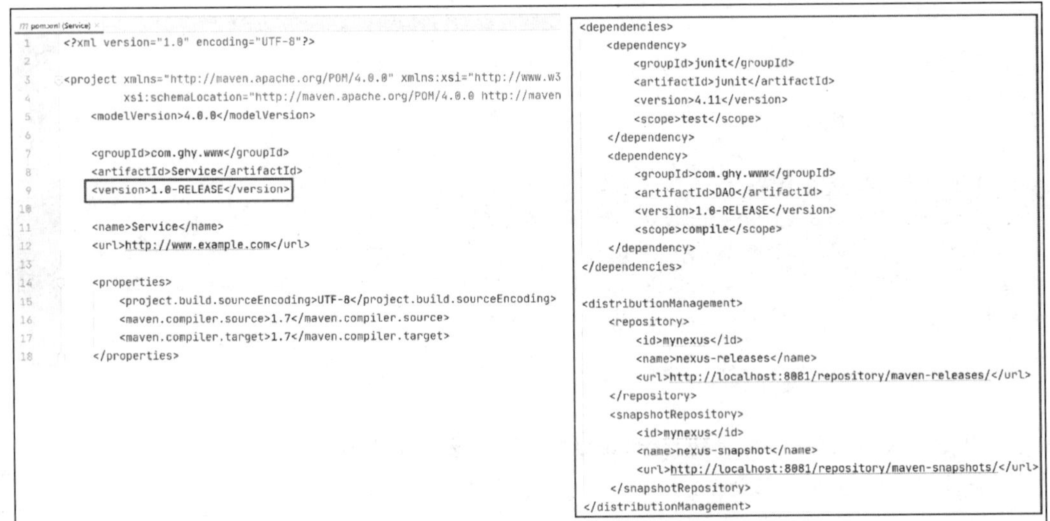

```
<?xml version="1.0" encoding="UTF-8"?>

<project xmlns="http://maven.apache.org/POM/4.0.0" xmlns:xsi="http://www.w3
         xsi:schemaLocation="http://maven.apache.org/POM/4.0.0 http://maven
    <modelVersion>4.0.0</modelVersion>

    <groupId>com.ghy.www</groupId>
    <artifactId>Service</artifactId>
    <version>1.0-RELEASE</version>

    <name>Service</name>
    <url>http://www.example.com</url>

    <properties>
        <project.build.sourceEncoding>UTF-8</project.build.sourceEncoding>
        <maven.compiler.source>1.7</maven.compiler.source>
        <maven.compiler.target>1.7</maven.compiler.target>
    </properties>
```

```
<dependencies>
    <dependency>
        <groupId>junit</groupId>
        <artifactId>junit</artifactId>
        <version>4.11</version>
        <scope>test</scope>
    </dependency>
    <dependency>
        <groupId>com.ghy.www</groupId>
        <artifactId>DAO</artifactId>
        <version>1.0-RELEASE</version>
        <scope>compile</scope>
    </dependency>
</dependencies>

<distributionManagement>
    <repository>
        <id>mynexus</id>
        <name>nexus-releases</name>
        <url>http://localhost:8081/repository/maven-releases/</url>
    </repository>
    <snapshotRepository>
        <id>mynexus</id>
        <name>nexus-snapshot</name>
        <url>http://localhost:8081/repository/maven-snapshots/</url>
    </snapshotRepository>
</distributionManagement>
```

图 16-15

对 Service 子模块执行发布操作，成功发布的结果如图 16-16 所示。

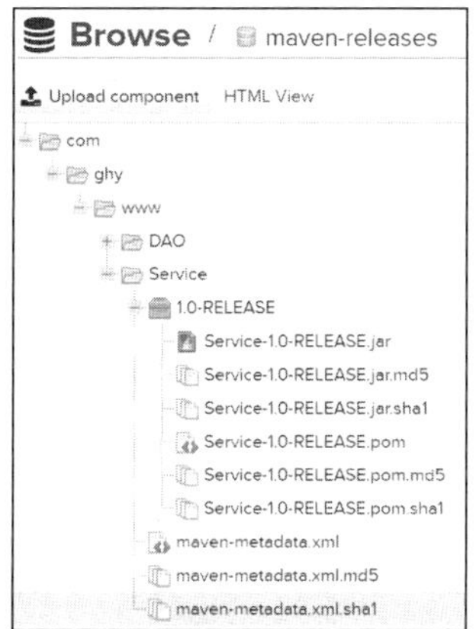


图 16-16

### 16.1.8　创建 java 和 resources 文件夹

在 Web 子模块中创建 java 和 resources 文件夹，如图 16-17 所示。

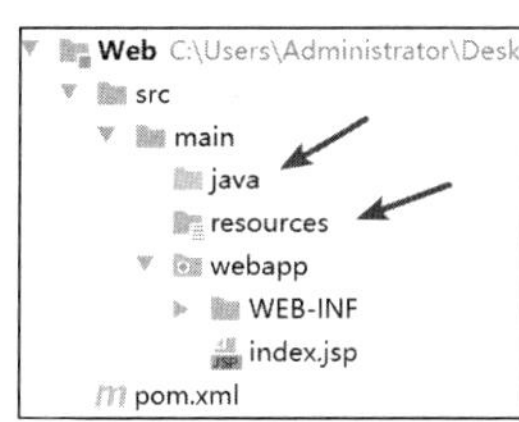


图 16-17

### 16.1.9　添加 Servlet 和 JSTL 依赖

在 Web 子模块的 pom.xml 配置文件中添加 Servlet 和 JSTL 依赖的配置代码如下：

```
<dependencies>
    <dependency>
        <groupId>javax.servlet</groupId>
        <artifactId>javax.servlet-api</artifactId>
        <version>4.0.1</version>
    </dependency>
    <dependency>
        <groupId>org.glassfish.web</groupId>
        <artifactId>jstl-impl</artifactId>
        <version>1.2</version>
    </dependency>
</dependencies>
```

### 16.1.10　创建 Servlet 类并引用 Service 子模块

创建 Servlet 类的代码如下：

```
@WebServlet(name = "Test1", urlPatterns = "/Test1")
public class Test1 extends HttpServlet {
    protected void doGet(HttpServletRequest request, HttpServletResponse response) throws ServletException, IOException {
        UserinfoService userinfoService = new UserinfoService();
        System.out.println("Test doGet 方法执行了！ = " + userinfoService.getUsernameByUserIdService());
    }
}
```

Web 子模块对 Service 子模块的引用是在 Web 子模块的 pom.xml 配置文件中使用如下配置实现的。

```
<dependencies>
    <dependency>
        <groupId>com.ghy.www</groupId>
        <artifactId>Service</artifactId>
        <version>1.0-RELEASE</version>
        <scope>compile</scope>
    </dependency>
</dependencies>
```

### 16.1.11 编辑 Web 子模块中的 pom.xml 配置文件并发布

pom.xml 配置文件的核心代码，如图 16-18 所示。

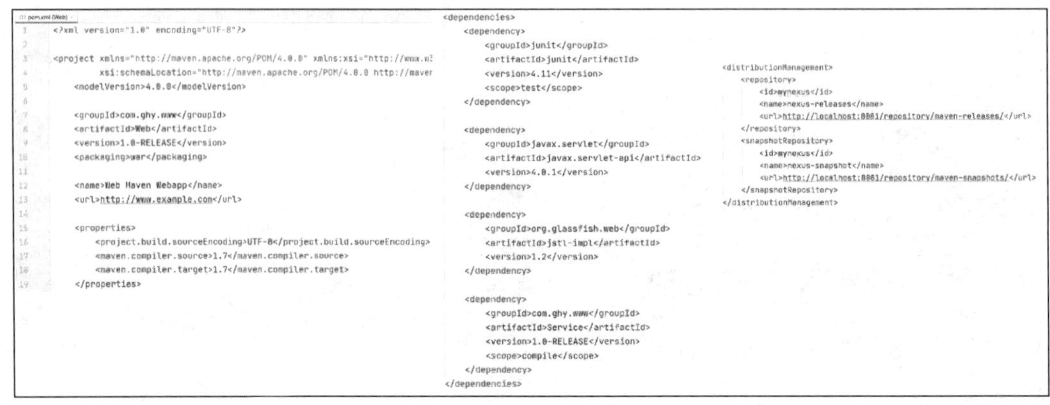

```
<?xml version="1.0" encoding="UTF-8"?>

<project xmlns="http://maven.apache.org/POM/4.0.0" xmlns:xsi="http://www.w3
         xsi:schemaLocation="http://maven.apache.org/POM/4.0.0 http://maven
    <modelVersion>4.0.0</modelVersion>

    <groupId>com.ghy.www</groupId>
    <artifactId>Web</artifactId>
    <version>1.0-RELEASE</version>
    <packaging>war</packaging>

    <name>Web Maven Webapp</name>
    <url>http://www.example.com</url>

    <properties>
        <project.build.sourceEncoding>UTF-8</project.build.sourceEncoding>
        <maven.compiler.source>1.7</maven.compiler.source>
        <maven.compiler.target>1.7</maven.compiler.target>
    </properties>

<dependencies>
    <dependency>
        <groupId>junit</groupId>
        <artifactId>junit</artifactId>
        <version>4.11</version>
        <scope>test</scope>
    </dependency>

    <dependency>
        <groupId>javax.servlet</groupId>
        <artifactId>javax.servlet-api</artifactId>
        <version>4.0.1</version>
    </dependency>

    <dependency>
        <groupId>org.glassfish.web</groupId>
        <artifactId>jstl-impl</artifactId>
        <version>1.2</version>
    </dependency>

    <dependency>
        <groupId>com.ghy.www</groupId>
        <artifactId>Service</artifactId>
        <version>1.0-RELEASE</version>
        <scope>compile</scope>
    </dependency>
</dependencies>

<distributionManagement>
    <repository>
        <id>mynexus</id>
        <name>nexus-releases</name>
        <url>http://localhost:8081/repository/maven-releases/</url>
    </repository>
    <snapshotRepository>
        <id>mynexus</id>
        <name>nexus-snapshot</name>
        <url>http://localhost:8081/repository/maven-snapshots/</url>
    </snapshotRepository>
</distributionManagement>
```

图 16-18

对 Web 子模块执行发布操作，成功发布的结果如图 16-19 所示。

图 16-19

## 16.1.12　运行项目

启动 Tomcat 后执行路径为 Test1 的 Servlet，程序成功运行，效果如图 16-20 所示。

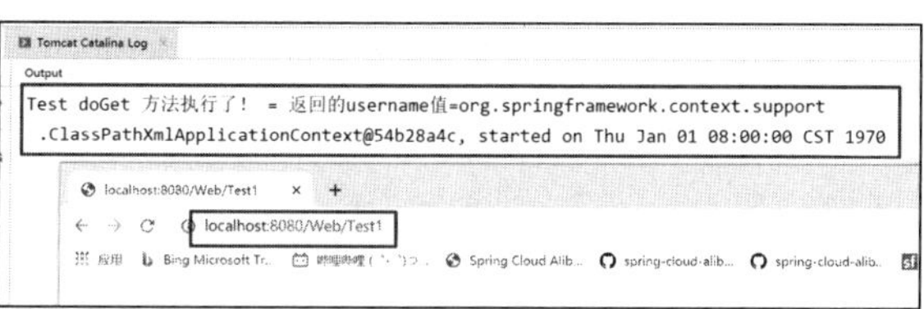


图 16-20

## 16.1.13　自动导出 war 包文件

启动 Tomcat 后 IDEA 会自动导出项目的 war 包文件，如图 16-21 所示。

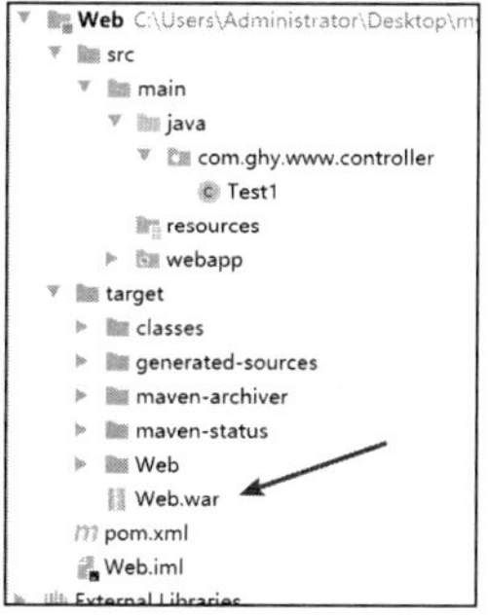


图 16-21

war 包中包含第三方的依赖.jar 文件，如 Servlet，JSLT，Spring，DAO，Service 等，如图 16-22 所示。

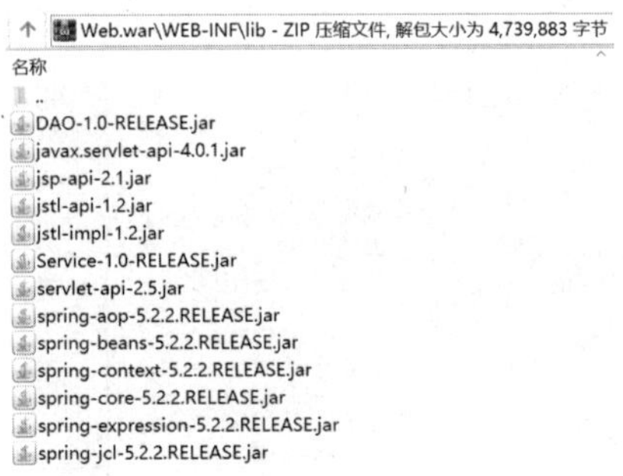


图 16-22

### 16.1.14　远程仓库中的内容

这时，远程仓库中的内容如图 16-23 所示。

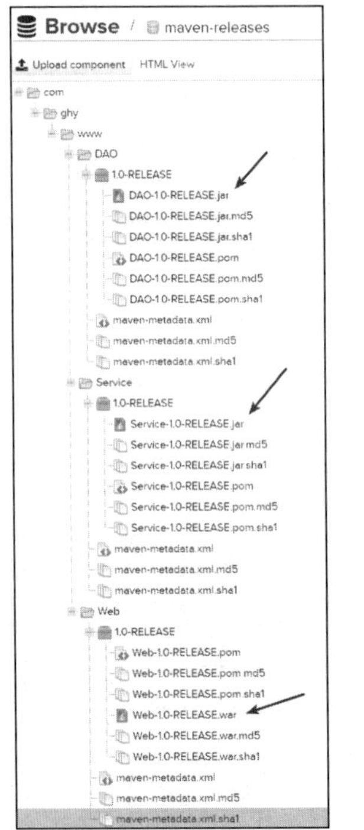


图 16-23

## 16.2　实现项目的继承

在 OOP 面向对象编程中，继承的目的是实现代码复用，而在 Maven 中，继承的目的是实现配置复用。如果 DAO，Service，Web 三个子模块都依赖相同的组件，如 junit，配置代码如下：

```
<dependency>
    <groupId>junit</groupId>
    <artifactId>junit</artifactId>
    <version>4.11</version>
    <scope>test</scope>
</dependency>
```

那么，这三个子模块的 pom.xml 配置文件中的<dependency></dependency>配置代码就出现了重复。这时，我们可以创建一个 Parent 子模块，作为 DAO，Service，Web 三个子模块公共配置的来源。

pom.xml 配置文件中能被继承的标签，如图 16-24 所示。

- groupId
- version
- description
- url
- inceptionYear
- organization
- licenses
- developers
- contributors
- mailingLists
- scm
- issueManagement
- ciManagement
- properties
- dependencyManagement
- dependencies
- repositories
- pluginRepositories
- build
    - plugin executions with matching ids
    - plugin configuration
    - etc.
- reporting
- profiles

图 16-24

不能被继承的标签，如图 16-25 所示。

- artifactId
- name
- prerequisites

图 16-25

### 16.2.1 搭建继承环境

本实验还是基于前面的 myproject 项目。

#### 1. 创建子模块Parent1

在 myproject 项目的基础上，继续创建 quickstart 类型的 Maven 子模块，名称为 Parent1，配置信息如图 16-26 所示。

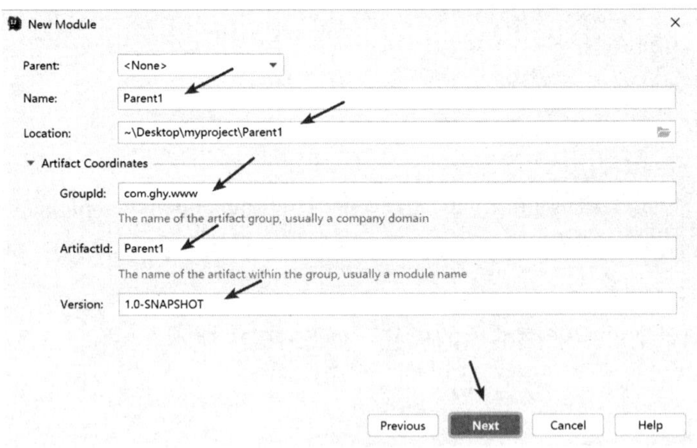


图 16-26

#### 2. 子模块列表

这时，子模块的列表如图 16-27 所示。

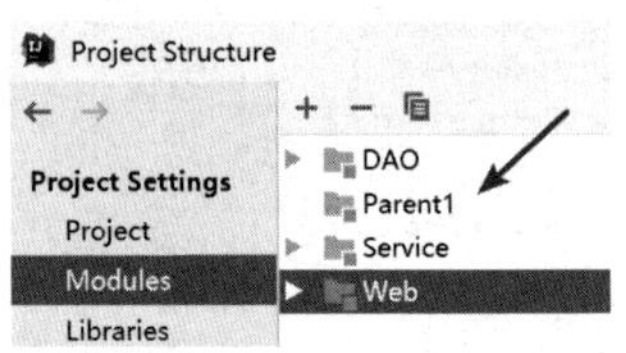


图 16-27

#### 3. Parent1子模块归属于myproject项目

Parent1 子模块归属于 myproject 项目，即项目关联了 Parent1 子模块，如图 16-28 所示。

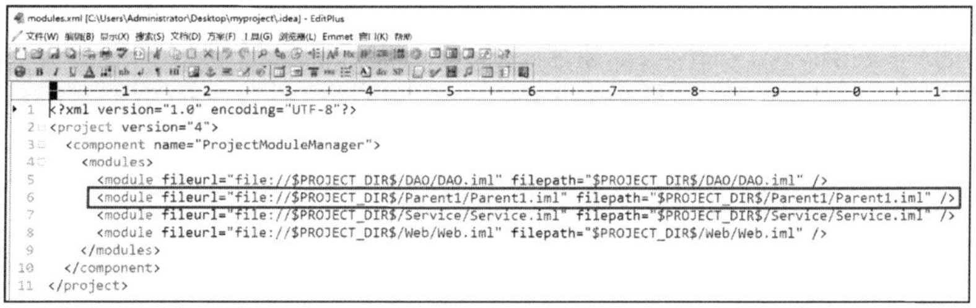

```
<?xml version="1.0" encoding="UTF-8"?>
<project version="4">
  <component name="ProjectModuleManager">
    <modules>
      <module fileurl="file://$PROJECT_DIR$/DAO/DAO.iml" filepath="$PROJECT_DIR$/DAO/DAO.iml" />
      <module fileurl="file://$PROJECT_DIR$/Parent1/Parent1.iml" filepath="$PROJECT_DIR$/Parent1/Parent1.iml" />
      <module fileurl="file://$PROJECT_DIR$/Service/Service.iml" filepath="$PROJECT_DIR$/Service/Service.iml" />
      <module fileurl="file://$PROJECT_DIR$/Web/Web.iml" filepath="$PROJECT_DIR$/Web/Web.iml" />
    </modules>
  </component>
</project>
```

图 16-28

#### 4. 删除子模块Parent1中的src节点

因为在 Parent1 子模块中不需要写.java 文件的代码，所以可以将 src 文件夹删除，如图 16-29 所示。

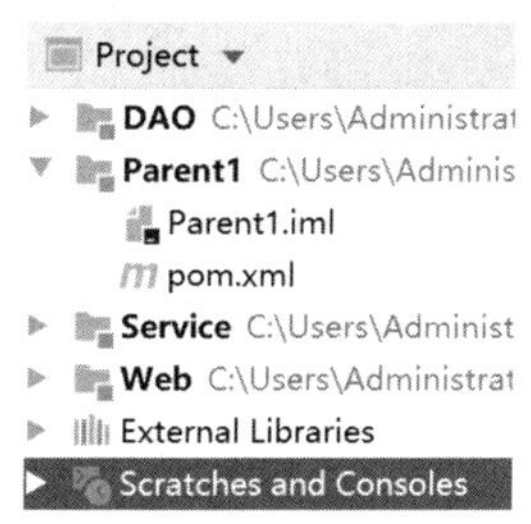


图 16-29

#### 5. 更改Parent1子模块<packaging>和<version>的配置

更改 Parent1 子模块的版本号为<version>2.0-RELEASE</version>，打包方式为<packaging>pom</packaging>，配置代码如图 16-30 所示。

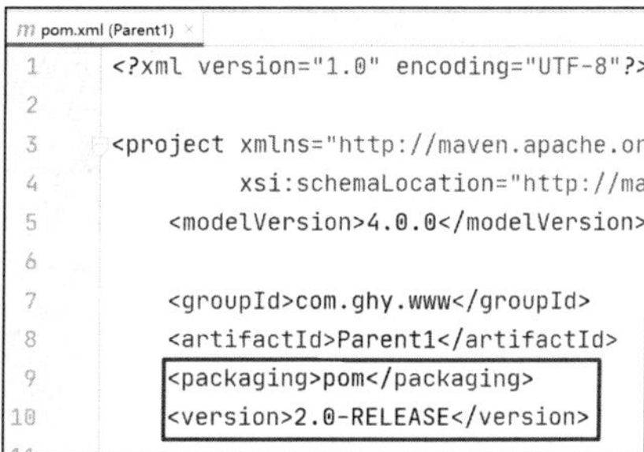


图 16-30

其中，配置代码<packaging>pom</packaging>代表项目类型是父项目。

### 6. 将DAO子模块中的依赖配置移至Parent1子模块中

将原来 DAO 子模块中的依赖配置代码：

```
<dependencies>
    <dependency>
        <groupId>junit</groupId>
        <artifactId>junit</artifactId>
        <version>4.11</version>
        <scope>test</scope>
    </dependency>
    <dependency>
        <groupId>org.springframework</groupId>
        <artifactId>spring-context</artifactId>
        <version>5.2.2.RELEASE</version>
    </dependency>
</dependencies>
```

移至 Parent1 子模块中，这时， Parent1 子模块的核心配置代码如图 16-31 所示。

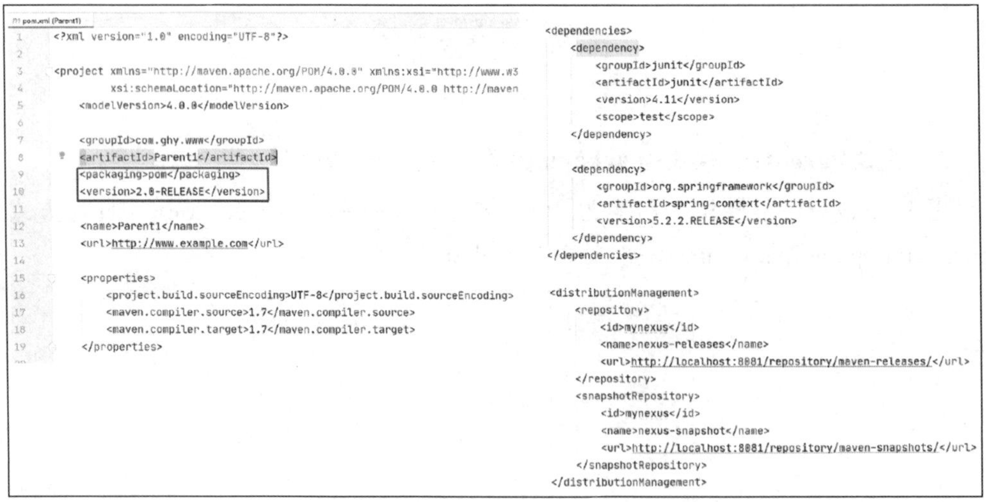

```
pom.xml (Parent1)
<?xml version="1.0" encoding="UTF-8"?>

<project xmlns="http://maven.apache.org/POM/4.0.0" xmlns:xsi="http://www.w3
        xsi:schemaLocation="http://maven.apache.org/POM/4.0.0 http://maven
    <modelVersion>4.0.0</modelVersion>

    <groupId>com.ghy.www</groupId>
    <artifactId>Parent1</artifactId>
    <packaging>pom</packaging>
    <version>2.0-RELEASE</version>

    <name>Parent1</name>
    <url>http://www.example.com</url>

    <properties>
        <project.build.sourceEncoding>UTF-8</project.build.sourceEncoding>
        <maven.compiler.source>1.7</maven.compiler.source>
        <maven.compiler.target>1.7</maven.compiler.target>
    </properties>

<dependencies>
    <dependency>
        <groupId>junit</groupId>
        <artifactId>junit</artifactId>
        <version>4.11</version>
        <scope>test</scope>
    </dependency>

    <dependency>
        <groupId>org.springframework</groupId>
        <artifactId>spring-context</artifactId>
        <version>5.2.2.RELEASE</version>
    </dependency>
</dependencies>

<distributionManagement>
    <repository>
        <id>mynexus</id>
        <name>nexus-releases</name>
        <url>http://localhost:8081/repository/maven-releases/</url>
    </repository>
    <snapshotRepository>
        <id>mynexus</id>
        <name>nexus-snapshot</name>
        <url>http://localhost:8081/repository/maven-snapshots/</url>
    </snapshotRepository>
</distributionManagement>
```

图 16-31

### 7. 发布Parent1子模块

发布 Parent1 子模块，如图 16-32 所示。

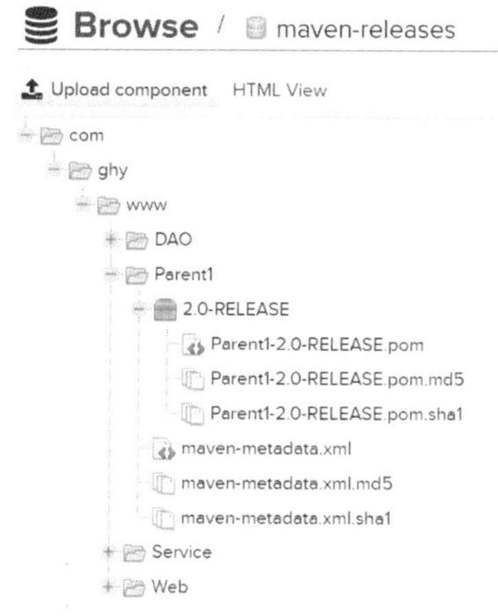


图 16-32

### 8. 编辑DAO子模块的pom.xml配置文件

DAO 子模块的配置代码<dependencies>为空，如图 16-33 所示，DAO 子模块版本更新为<version>2.0-RELEASE</version>。

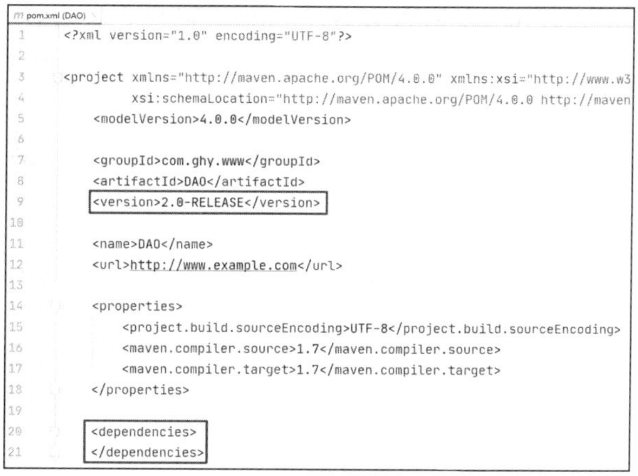

```
pom.xml (DAO)
<?xml version="1.0" encoding="UTF-8"?>

<project xmlns="http://maven.apache.org/POM/4.0.0" xmlns:xsi="http://www.w3
         xsi:schemaLocation="http://maven.apache.org/POM/4.0.0 http://maven
    <modelVersion>4.0.0</modelVersion>

    <groupId>com.ghy.www</groupId>
    <artifactId>DAO</artifactId>
    <version>2.0-RELEASE</version>

    <name>DAO</name>
    <url>http://www.example.com</url>

    <properties>
        <project.build.sourceEncoding>UTF-8</project.build.sourceEncoding>
        <maven.compiler.source>1.7</maven.compiler.source>
        <maven.compiler.target>1.7</maven.compiler.target>
    </properties>

    <dependencies>
    </dependencies>
```

图 16-33

### 9. 在DAO子模块中添加<parent>标签实现继承

首先对 DAO 子模块执行 Maven->Reload project 菜单，这时 DAO 子模块中的代码呈现错误状态，如图 16-34 所示，这是因为找不到.jar 文件，所以出现了编译异常。

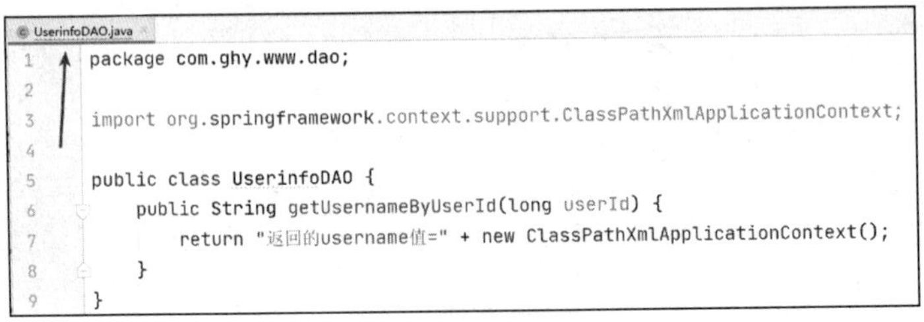

```java
package com.ghy.www.dao;

import org.springframework.context.support.ClassPathXmlApplicationContext;

public class UserinfoDAO {
    public String getUsernameByUserId(long userId) {
        return "返回的username值=" + new ClassPathXmlApplicationContext();
    }
}
```

图 16-34

然后编辑 DAO 子模块的 pom.xml 配置文件，添加配置代码如下：

```xml
<parent>
    <groupId>com.ghy.www</groupId>
    <artifactId>Parent1</artifactId>
    <version>2.0-RELEASE</version>
    <relativePath>../Parent1/pom.xml</relativePath>
</parent>
```

再次执行 Maven->Reload project 菜单，这时 DAO 子模块中的.java 代码编译错误消失，如图 16-35 所示，说明 DAO 子模块从 Parent1 子模块中成功继承了依赖配置。

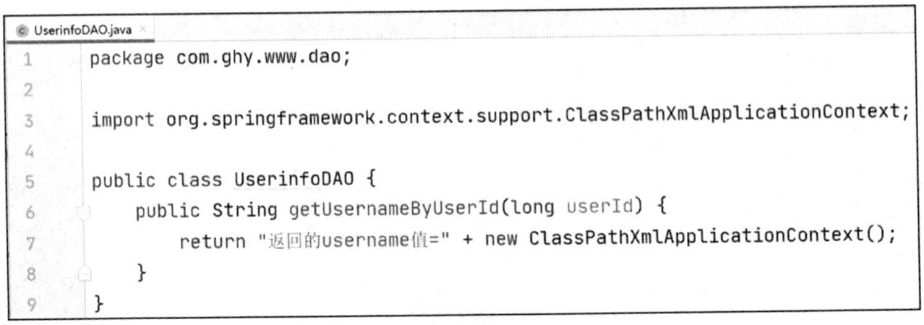

```java
package com.ghy.www.dao;

import org.springframework.context.support.ClassPathXmlApplicationContext;

public class UserinfoDAO {
    public String getUsernameByUserId(long userId) {
        return "返回的username值=" + new ClassPathXmlApplicationContext();
    }
}
```

图 16-35

### 10. 发布DAO子模块

成功发布 DAO 子模块，如图 16-36 所示。

### 11. 更改Service子模块的pom.xml配置

更改 Service 子模块的配置代码，如图 16-37 所示。

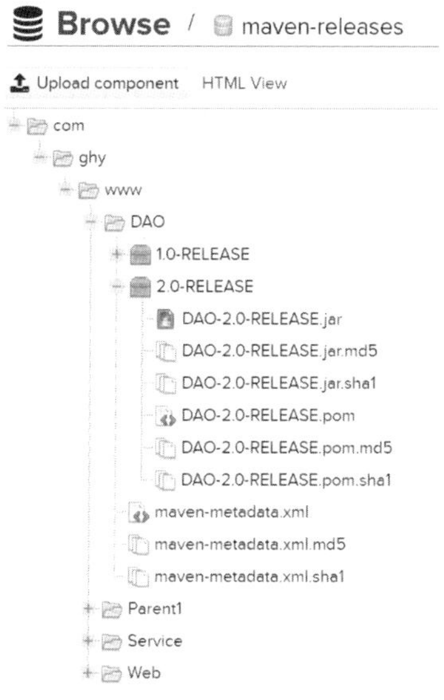


图 16-36

```
pom.xml (Service)
<?xml version="1.0" encoding="UTF-8"?>

<project xmlns="http://maven.apache.org/POM/4.0.0" xmlns:
         xsi:schemaLocation="http://maven.apache.org/POM/
    <parent>
        <groupId>com.ghy.www</groupId>
        <artifactId>Parent1</artifactId>
        <version>2.0-RELEASE</version>
        <relativePath>../Parent1/pom.xml</relativePath>
    </parent>

    <modelVersion>4.0.0</modelVersion>

    <groupId>com.ghy.www</groupId>
    <artifactId>Service</artifactId>
    <version>2.0-RELEASE</version>

    <name>Service</name>
    <url>http://www.example.com</url>
```

```
<properties>
    <project.build.sourceEncoding>UTF-8</project.build.sourceEncoding>
    <maven.compiler.source>1.7</maven.compiler.source>
    <maven.compiler.target>1.7</maven.compiler.target>
</properties>

<dependencies>
    <dependency>
        <groupId>com.ghy.www</groupId>
        <artifactId>DAO</artifactId>
        <version>2.0-RELEASE</version>
        <scope>compile</scope>
    </dependency>
</dependencies>
```

图 16-37

## 12. 发布Service子模块

成功发布 Service 子模块，如图 16-38 所示。

图 16-38

### 13. 编辑Servlet代码

编辑 Servlet 代码如下：

```
@WebServlet(name = "Test1", urlPatterns = "/Test1")
public class Test1 extends HttpServlet {
   protected void doGet(HttpServletRequest request, HttpServletResponse
response) throws ServletException, IOException {
       UserinfoService userinfoService = new UserinfoService();
       System.out.println("Test doGet 方法执行了！ = " +
userinfoService.getUsernameByUserIdService());
       System.out.println("版本为：2.0-RELEASE");
   }
}
```

### 14. 将Web子模块中的依赖配置移至Parent1子模块中

把如下 Web 子模块中的依赖配置代码移至 Parent1 子模块中：

```
<dependency>
   <groupId>javax.servlet</groupId>
   <artifactId>javax.servlet-api</artifactId>
   <version>4.0.1</version>
</dependency>
```

```
<dependency>
    <groupId>org.glassfish.web</groupId>
    <artifactId>jstl-impl</artifactId>
    <version>1.2</version>
</dependency>
```

最终 Web 子模块的配置代码，如图 16-39 所示。

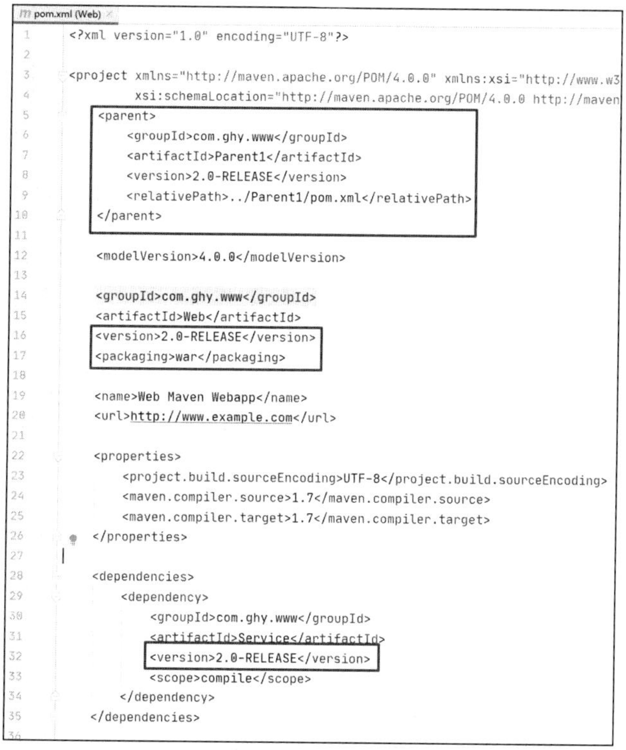

```
pom.xml (Web)
<?xml version="1.0" encoding="UTF-8"?>

<project xmlns="http://maven.apache.org/POM/4.0.0" xmlns:xsi="http://www.w3
         xsi:schemaLocation="http://maven.apache.org/POM/4.0.0 http://maven
  <parent>
      <groupId>com.ghy.www</groupId>
      <artifactId>Parent1</artifactId>
      <version>2.0-RELEASE</version>
      <relativePath>../Parent1/pom.xml</relativePath>
  </parent>

  <modelVersion>4.0.0</modelVersion>

  <groupId>com.ghy.www</groupId>
  <artifactId>Web</artifactId>
  <version>2.0-RELEASE</version>
  <packaging>war</packaging>

  <name>Web Maven Webapp</name>
  <url>http://www.example.com</url>

  <properties>
      <project.build.sourceEncoding>UTF-8</project.build.sourceEncoding>
      <maven.compiler.source>1.7</maven.compiler.source>
      <maven.compiler.target>1.7</maven.compiler.target>
  </properties>

  <dependencies>
      <dependency>
          <groupId>com.ghy.www</groupId>
          <artifactId>Service</artifactId>
          <version>2.0-RELEASE</version>
          <scope>compile</scope>
      </dependency>
  </dependencies>
```

图 16-39

子模块 Parent1 的依赖配置，如图 16-40 所示。

### 15. 发布Web子模块

成功发布 Web 子模块，如图 16-41 所示，至此，成功对 4 个子模块进行发布。。

```
<dependencies>
    <dependency>
        <groupId>junit</groupId>
        <artifactId>junit</artifactId>
        <version>4.11</version>
        <scope>test</scope>
    </dependency>

    <dependency>
        <groupId>org.springframework</groupId>
        <artifactId>spring-context</artifactId>
        <version>5.2.2.RELEASE</version>
    </dependency>

    <dependency>
        <groupId>javax.servlet</groupId>
        <artifactId>javax.servlet-api</artifactId>
        <version>4.0.1</version>
    </dependency>

    <dependency>
        <groupId>org.glassfish.web</groupId>
        <artifactId>jstl-impl</artifactId>
        <version>1.2</version>
    </dependency>
</dependencies>
```

图 16-40

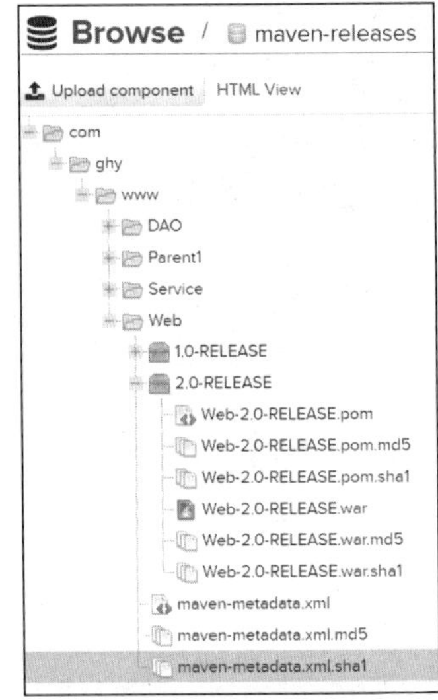


图 16-41

**16. 运行项目**

启动 Tomcat 后执行路径为 Test1 的 Servlet，程序成功运行，效果如图 16-42 所示。

```
Output
Test doGet 方法执行了！ = 返回的username值=org.springframework.context.support
 .ClassPathXmlApplicationContext@11c0c34, started on Thu Jan 01 08:00:00 CST 1970
版本为：2.0-RELEASE
```

图 16-42

## 16.2.2 配置<dependencyManagement></dependencyManagement>

现在，所有的子模块都能使用 Parent1 子模块中的依赖配置：

```
<dependencies>
    <dependency>
        <groupId>junit</groupId>
        <artifactId>junit</artifactId>
        <version>4.11</version>
        <scope>test</scope>
    </dependency>

    <dependency>
```

```
            <groupId>org.springframework</groupId>
            <artifactId>spring-context</artifactId>
            <version>5.2.2.RELEASE</version>
        </dependency>

        <dependency>
            <groupId>javax.servlet</groupId>
            <artifactId>javax.servlet-api</artifactId>
            <version>4.0.1</version>
        </dependency>

        <dependency>
            <groupId>org.glassfish.web</groupId>
            <artifactId>jstl-impl</artifactId>
            <version>1.2</version>
        </dependency>
    </dependencies>
```

因为除了 Parent1 子模块，所有其他子模块都使用如下代码继承 Parent1 子模块中的配置。

```
<parent>
    <groupId>com.ghy.www</groupId>
    <artifactId>Parent1</artifactId>
    <version>2.0-RELEASE</version>
    <relativePath>../Parent1/pom.xml</relativePath>
</parent>
```

这种写法的优点是自动继承依赖配置，缺点是“一股脑”式的继承依赖，不管子模块用不用，统统继承。如果你想实现子模块能自主选择需要的依赖，则可以在 Parent1 子模块中使用标签<dependencyManagement></dependencyManagement>。

<dependencyManagement></dependencyManagement>的作用是在 Parent1 子模块中对其他子模块提供一些“可使用的依赖列表”，这样其他子模块在继承 Parent1 子模块中的配置时，可以自行决定使用那些依赖。

### 1. 编辑Parent1子模块的pom.xml配置代码

首先在 Parent1 子模块的 pom.xml 文件中，添加 dependencyManagement 配置代码，如图 16-43 所示。

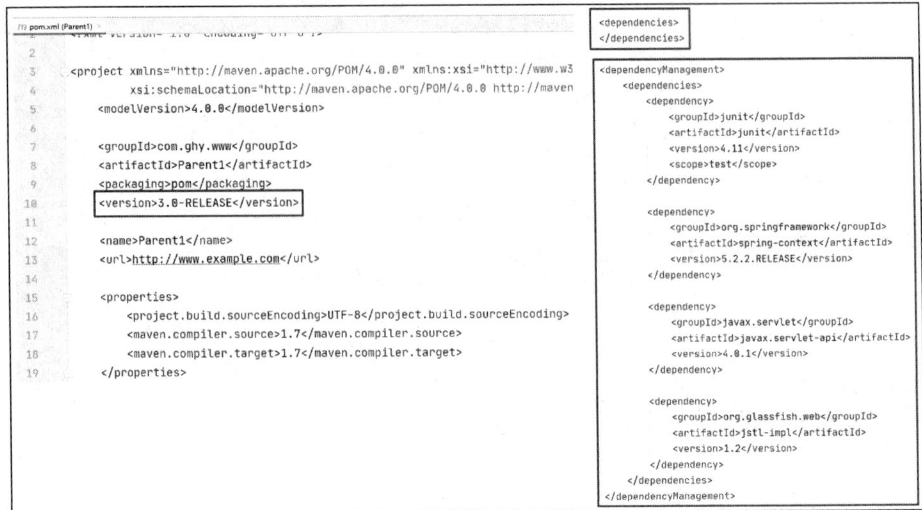

图 16-43

这时，Parent1 子模块的版本更改为<version>3.0-RELEASE</version>，并且还出现了<dependencies></dependencies>空标签。这样，之前所有的依赖配置都被放在标签<dependencyManagement></dependencyManagement>中了。

然后对 Parent1 子模块执行发布。

### 2. 更改DAO子模块的配置信息

更改 DAO 子模块的配置信息，如图 16-44 所示。

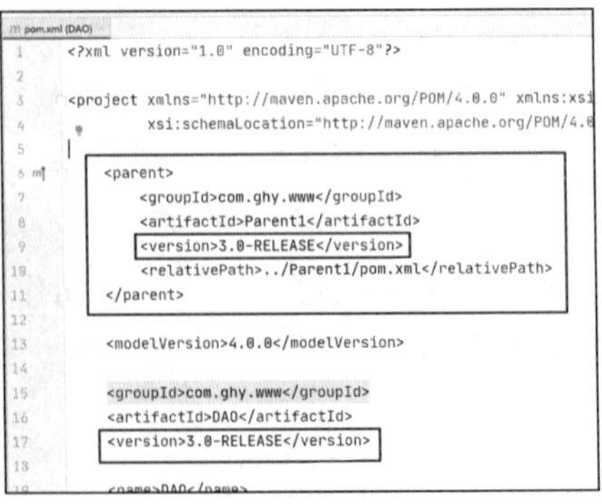

图 16-44

### 3. 解决DAO子模块代码编译错误问题

在执行 Maven->Reload project 菜单之后，DAO 子模块中的代码出现编译错误，如图 16-45 所示，这是因为 DAO 子模块没有显式添加依赖，不会自动使用<dependencyManagement>中的依赖配置。。

```
UserinfoDAO.java
package com.ghy.www.dao;

import org.springframework.context.support.ClassPathXmlApplicationContext;

public class UserinfoDAO {
    public String getUsernameByUserId(long userId) {
        return "返回的username值=" + new ClassPathXmlApplicationContext();
    }
}
```

图 16-45

下面编辑 DAO 子模块中的 pom.xml 配置代码，如图 16-46 所示， DAO 子模块显式指定依赖 spring-context。

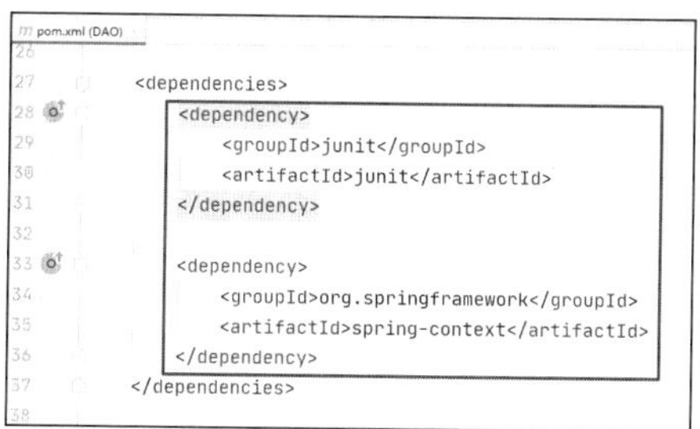

```
pom.xml (DAO)
<dependencies>
    <dependency>
        <groupId>junit</groupId>
        <artifactId>junit</artifactId>
    </dependency>

    <dependency>
        <groupId>org.springframework</groupId>
        <artifactId>spring-context</artifactId>
    </dependency>
</dependencies>
```

图 16-46

在配置代码<dependency>中并没有<version></version>版本相关的配置，默认使用 Parent1 子模块中的版本，这样可以减少 pom.xml 配置文件中的代码量。

再次执行 Maven->Reload project 菜单，这时 DAO 子模块中的 Java 代码是正确的，如图 16-47 所示。

```
UserinfoDAO.java
package com.ghy.www.dao;

import org.springframework.context.support.ClassPathXmlApplicationContext;

public class UserinfoDAO {
    public String getUsernameByUserId(long userId) {
        return "返回的username值=" + new ClassPathXmlApplicationContext();
    }
}
```

图 16-47

### 4. 自由添加其他依赖

DAO 子模块除了可以继承依赖，也可以自行添加依赖，这增加了子模块使用依赖的灵活性，如图 16-48 所示。

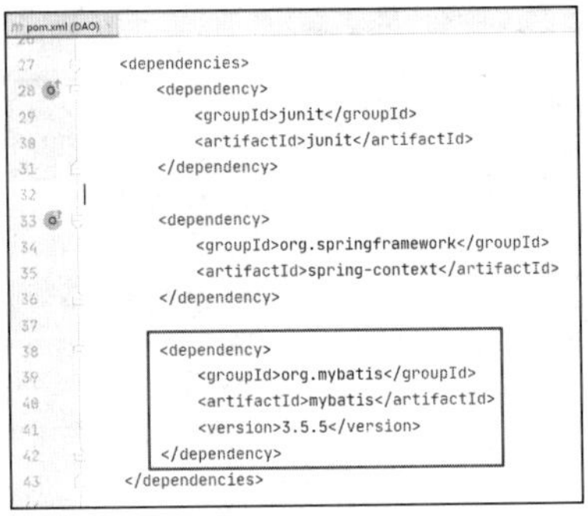

```
<dependencies>
    <dependency>
        <groupId>junit</groupId>
        <artifactId>junit</artifactId>
    </dependency>

    <dependency>
        <groupId>org.springframework</groupId>
        <artifactId>spring-context</artifactId>
    </dependency>

    <dependency>
        <groupId>org.mybatis</groupId>
        <artifactId>mybatis</artifactId>
        <version>3.5.5</version>
    </dependency>
</dependencies>
```

图 16-48

在实际的项目中，<dependencyManagement></dependencyManagement>配置会被优先推荐使用来管理继承的依赖，理由是子模块没有盲目地继承全部的依赖。另外，在子模块中以显式的方式声明使用的依赖，明确了子模块依赖的内容，具有依赖操作的自主性，同时也减少了子模块依赖的代码量，至少减少了<version>配置，版本管理更加集中化。

最后对 DAO 子模块进行发布操作。

### 5. 更改Sevlet代码

更改 Servlet 代码如下：

```
@WebServlet(name = "Test1", urlPatterns = "/Test1")
public class Test1 extends HttpServlet {
    protected void doGet(HttpServletRequest request, HttpServletResponse
response) throws ServletException, IOException {
        UserinfoService userinfoService = new UserinfoService();
        System.out.println("Test  doGet  方 法 执 行 了 !    =  "  +
userinfoService.getUsernameByUserIdService());
        System.out.println("版本为：3.0-RELEASE");
    }
}
```

### 6. 升级Service和Web模块的版本并发布

Service 子模块的核心配置代码，如图 16-49 所示，并发布 Service 子模块。

Web 子模块的核心配置代码，如图 16-50 所示，并发布 Web 子模块。。

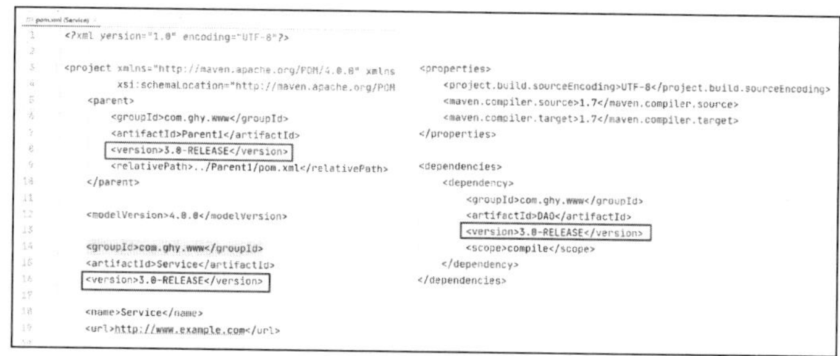


图 16-49

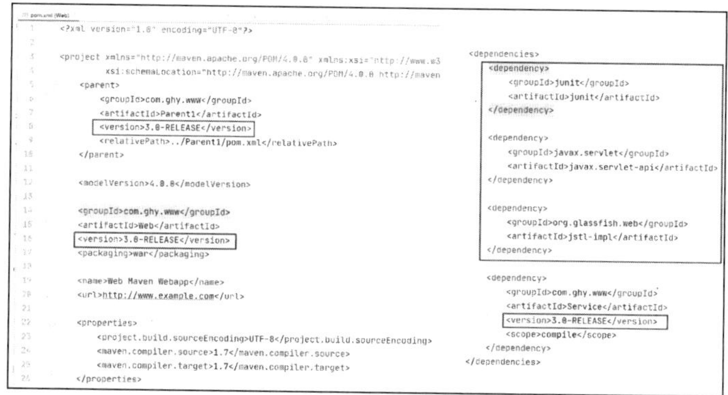


图 16-50

远程仓库的内容如图 16-51 所示，4 个模块都成功发布，版本均为 3.0-RELEASE。

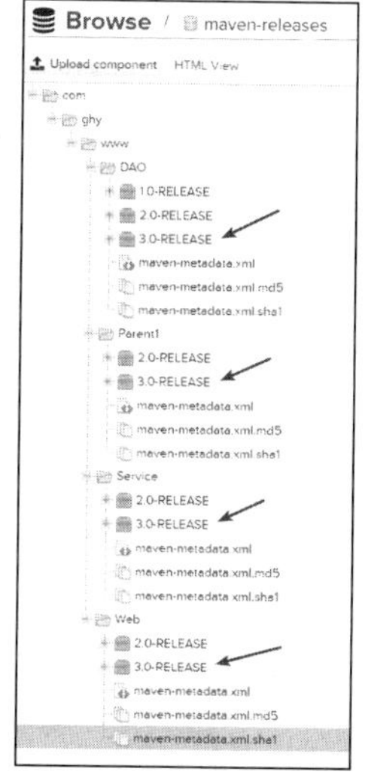


图 16-51

#### 7. 运行项目

启动 Tomcat 后执行路径为 Test1 的 Servlet，程序成功运行，效果如图 16-52 所示。

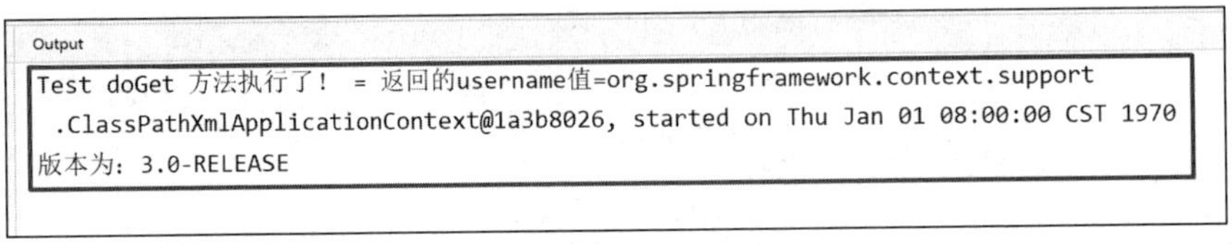

```
Output
Test doGet 方法执行了！ = 返回的username值=org.springframework.context.support
 .ClassPathXmlApplicationContext@1a3b8026, started on Thu Jan 01 08:00:00 CST 1970
版本为：3.0-RELEASE
```

图 16-52

### 16.2.3 配置<scope>import</scope>依赖范围

依赖范围<scope>import</scope>的作用是在两个不同的子模块间继承<dependencyManagement></dependencyManagement>中的依赖。

#### 1. 升级Parent1子模块的版本

更改 Parent1 子模块的版本如图 16-53 所示，并对其进行发布。

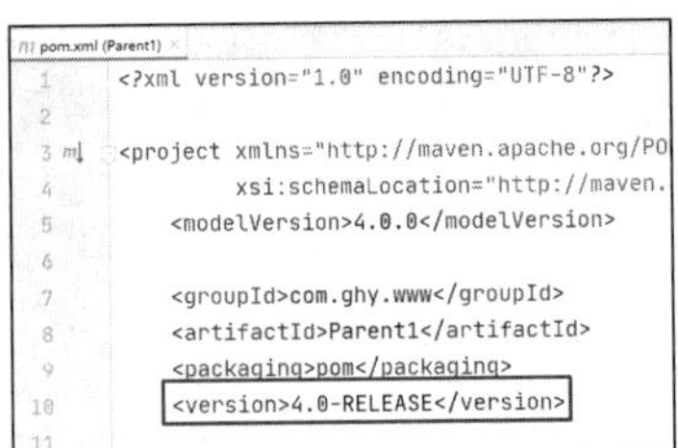


图 16-53

#### 2. 创建新的子模块Parent2

创建新的子模块 Parent2，如图 16-54 所示。

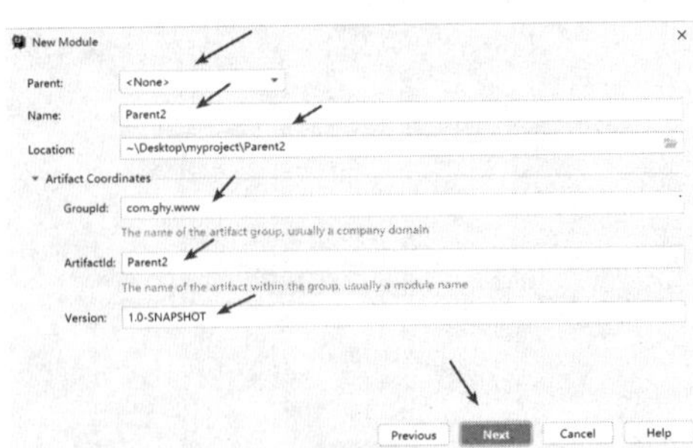


图 16-54

### 3. 将Parent1子模块的<dependencyManagement>导入Parent2子模块

将 Parent1 子模块的<dependencyManagement>导入 Parent2 子模块，如图 16-55 所示。

```
pom.xml (Parent2)
<?xml version="1.0" encoding="UTF-8"?>

<project xmlns="http://maven.apache.org/POM/4.0.0" xmlns:xsi="http://www.w3.
        xsi:schemaLocation="http://maven.apache.org/POM/4.0.0 http://maven.

    <modelVersion>4.0.0</modelVersion>

    <groupId>com.ghy.www</groupId>
    <artifactId>Parent2</artifactId>
    <packaging>pom</packaging>
    <version>4.0-RELEASE</version>

    <name>Parent2</name>
    <!-- FIXME change it to the project's website -->
    <url>http://www.example.com</url>

    <properties>
        <project.build.sourceEncoding>UTF-8</project.build.sourceEncoding>
        <maven.compiler.source>1.7</maven.compiler.source>
        <maven.compiler.target>1.7</maven.compiler.target>
    </properties>

<dependencies>
    <dependency>
        <groupId>mysql</groupId>
        <artifactId>mysql-connector-java</artifactId>
        <version>8.0.23</version>
    </dependency>
</dependencies>

<dependencyManagement>
    <dependencies>
        <dependency>
            <groupId>com.ghy.www</groupId>
            <artifactId>Parent1</artifactId>
            <version>4.0-RELEASE</version>
            <type>pom</type>
            <scope>import</scope>
        </dependency>
    </dependencies>
</dependencyManagement>
```

图 16-55

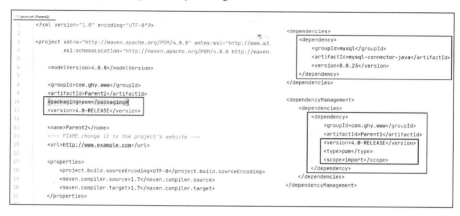

另外，Parent2 子模块还声明了一些<dependencies>配置。

```
<dependencies>
    <dependency>
        <groupId>mysql</groupId>
        <artifactId>mysql-connector-java</artifactId>
        <version>8.0.23</version>
    </dependency>
</dependencies>
```

继续在 Parent2 模块中添加如下配置：

```
<distributionManagement>
    <repository>
        <id>mynexus</id>
        <name>nexus-releases</name>
        <url>http://localhost:8081/repository/maven-releases/</url>
    </repository>
    <snapshotRepository>
        <id>mynexus</id>
        <name>nexus-snapshot</name>
        <url>http://localhost:8081/repository/maven-snapshots/</url>
    </snapshotRepository>
</distributionManagement>
```

### 4. 发布Parent2子模块

对 Parent2 子模块执行发布操作，完成后的仓库内容如图 16-56 所示。

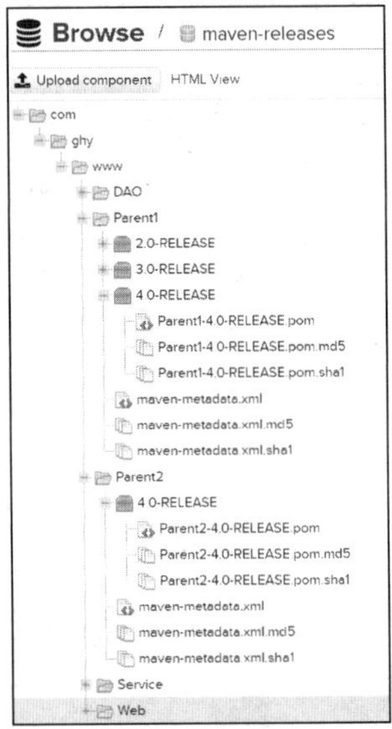


图 16-56

### 5. 编辑DAO子模块的配置代码

编辑 DAO 子模块的配置代码，如图 16-57 所示。

```
pom.xml (DAO)
<?xml version="1.0" encoding="UTF-8"?>

<project xmlns="http://maven.apache.org/POM/4.0.0" xmlns:x
        xsi:schemaLocation="http://maven.apache.org/POM/4
    <parent>
        <groupId>com.ghy.www</groupId>
        <artifactId>Parent2</artifactId>
        <version>4.0-RELEASE</version>
        <relativePath>../Parent2/pom.xml</relativePath>
    </parent>

    <modelVersion>4.0.0</modelVersion>

    <groupId>com.ghy.www</groupId>
    <artifactId>DAO</artifactId>
    <version>4.0-RELEASE</version>

    <name>DAO</name>
    <url>http://www.example.com</url>
```

```
<properties>
    <project.build.sourceEncoding>UTF-8</project.build.sourceEncoding>
    <maven.compiler.source>1.7</maven.compiler.source>
    <maven.compiler.target>1.7</maven.compiler.target>
</properties>

<dependencies>
    <dependency>
        <groupId>junit</groupId>
        <artifactId>junit</artifactId>
    </dependency>

    <dependency>
        <groupId>org.springframework</groupId>
        <artifactId>spring-context</artifactId>
    </dependency>

    <dependency>
        <groupId>org.mybatis</groupId>
        <artifactId>mybatis</artifactId>
        <version>3.5.5</version>
    </dependency>
</dependencies>
```

图 16-57

### 6. 编辑DAO子模块的类代码

在 DAO 子模块中，修改类的代码如下：

```
package com.ghy.www.dao;

import org.apache.ibatis.session.SqlSessionFactoryBuilder;
import org.springframework.context.support.ClassPathXmlApplicationContext;

public class UserinfoDAO {
    public String getUsernameByUserId(long userId) {
        return "返回的 username 值=" + new ClassPathXmlApplicationContext() +
" " + new SqlSessionFactoryBuilder();
    }
}
```

### 7. 发布DAO子模块

发布 DAO 子模块，如图 16-58 所示。

图 16-58

### 8. 编辑Service子模块的配置代码

编辑 Service 子模块的配置代码，如图 16-59 所示。

```
pom.xml (Service)
<?xml version="1.0" encoding="UTF-8"?>

<project xmlns="http://maven.apache.org/POM/4.0.0" xmlns:xsi="http://www.w3
         xsi:schemaLocation="http://maven.apache.org/POM/4.0.0 http://maven
    <parent>
        <groupId>com.ghy.www</groupId>
        <artifactId>Parent2</artifactId>
        <version>4.0-RELEASE</version>
        <relativePath>../Parent2/pom.xml</relativePath>
    </parent>

    <modelVersion>4.0.0</modelVersion>

    <groupId>com.ghy.www</groupId>
    <artifactId>Service</artifactId>
    <version>4.0-RELEASE</version>

    <name>Service</name>
    <url>http://www.example.com</url>

    <properties>
        <project.build.sourceEncoding>UTF-8</project.build.sourceEncoding>
        <maven.compiler.source>1.7</maven.compiler.source>
        <maven.compiler.target>1.7</maven.compiler.target>
    </properties>
```

```
<dependencies>
    <dependency>
        <groupId>junit</groupId>
        <artifactId>junit</artifactId>
    </dependency>

    <dependency>
        <groupId>com.ghy.www</groupId>
        <artifactId>DAO</artifactId>
        <version>4.0-RELEASE</version>
        <scope>compile</scope>
    </dependency>
</dependencies>
```

图 16-59

### 9. 发布Service子模块

发布 Service 子模块，如图 16-60 所示。

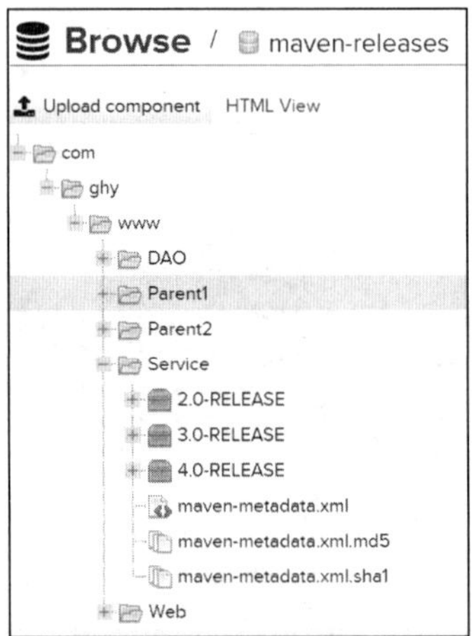


图 16-60

### 10. 编辑Web子模块的配置代码

编辑 Web 子模块的配置代码，如图 16-61 所示。

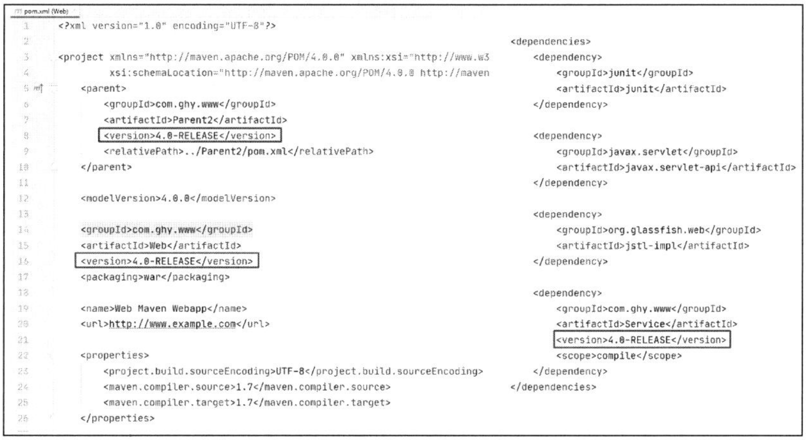

```
<?xml version="1.0" encoding="UTF-8"?>

<project xmlns="http://maven.apache.org/POM/4.0.0" xmlns:xsi="http://www.w3
         xsi:schemaLocation="http://maven.apache.org/POM/4.0.0 http://maven
    <parent>
        <groupId>com.ghy.www</groupId>
        <artifactId>Parent2</artifactId>
        <version>4.0-RELEASE</version>
        <relativePath>../Parent2/pom.xml</relativePath>
    </parent>

    <modelVersion>4.0.0</modelVersion>

    <groupId>com.ghy.www</groupId>
    <artifactId>Web</artifactId>
    <version>4.0-RELEASE</version>
    <packaging>war</packaging>

    <name>Web Maven Webapp</name>
    <url>http://www.example.com</url>

    <properties>
        <project.build.sourceEncoding>UTF-8</project.build.sourceEncoding>
        <maven.compiler.source>1.7</maven.compiler.source>
        <maven.compiler.target>1.7</maven.compiler.target>
    </properties>
```

```
<dependencies>
    <dependency>
        <groupId>junit</groupId>
        <artifactId>junit</artifactId>
    </dependency>

    <dependency>
        <groupId>javax.servlet</groupId>
        <artifactId>javax.servlet-api</artifactId>
    </dependency>

    <dependency>
        <groupId>org.glassfish.web</groupId>
        <artifactId>jstl-impl</artifactId>
    </dependency>

    <dependency>
        <groupId>com.ghy.www</groupId>
        <artifactId>Service</artifactId>
        <version>4.0-RELEASE</version>
        <scope>compile</scope>
    </dependency>
</dependencies>
```

图 16-61

## 11. 更改Servlet代码

更改 Servlet 的代码，如下：

```
@WebServlet(name = "Test1", urlPatterns = "/Test1")
public class Test1 extends HttpServlet {
    protected void doGet(HttpServletRequest request, HttpServletResponse
response) throws ServletException, IOException {
        UserinfoService userinfoService = new UserinfoService();
        System.out.println("Test doGet 方法执行了！ = " +
userinfoService.getUsernameByUserIdService());
        System.out.println("版本为：4.0-RELEASE");
    }
}
```

## 12. 发布Web子模块

发布 Web 子模块，如图 16-62 所示。

## 13. 运行项目

启动 Tomcat 后执行路径为 Test1 的 Servlet，程序成功运行，效果如图 16-63 所示。

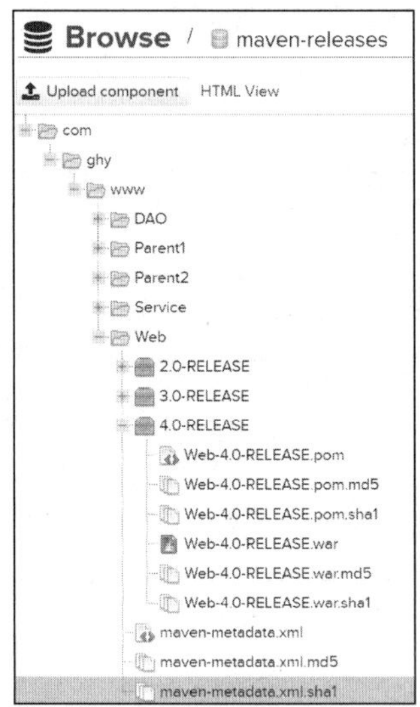


图 16-62

```
Test doGet 方法执行了！ = 返回的username值=org.springframework.context.support
 .ClassPathXmlApplicationContext@38980903, started on Thu Jan 01 08:00:00 CST 1970 org.apache.ibatis
 .session.SqlSessionFactoryBuilder@3a7ef5e9
版本为：4.0-RELEASE
```

图 16-63

### 16.2.4 配置<pluginManagement></pluginManagement>

配置<dependencyManagement></dependencyManagement>管理的是依赖，配置<pluginManagement></pluginManagement>管理的是插件，它们的作用相同，都是在子模块间继承时使用。

如果在每个子模块中使用如下配置代码将 source 源代码打包，则配置会出现冗余。

```
<build>
    <plugins>
        <plugin>
            <groupId>org.apache.maven.plugins</groupId>
            <artifactId>maven-source-plugin</artifactId>
            <version>2.2.1</version>
            <executions>
                <execution>
                    <id>attach-sources</id>
                    <goals>
```

```
                    <goal>jar</goal>
                </goals>
            </execution>
        </executions>
    </plugin>
</plugins>
</build>
```

这时，我们可以借助于<pluginManagement></pluginManagement>来减少配置的冗余。

#### 1. 编辑Parent1子模块的pom.xml配置文件

编辑 Parent1 子模块的 pom.xml 文件，增加的配置代码如下：

```
<build>
    <pluginManagement>
        <plugins>
            <plugin>
                <groupId>org.apache.maven.plugins</groupId>
                <artifactId>maven-source-plugin</artifactId>
                <version>2.2.1</version>
                <executions>
                    <execution>
                        <id>attach-sources</id>
                        <goals>
                            <goal>jar</goal>
                        </goals>
                    </execution>
                </executions>
            </plugin>
        </plugins>
    </pluginManagement>
</build>
```

#### 2. 编辑Parent2子模块的pom.xml配置文件

编辑 Parent2 子模块的 pom.xml 文件，配置代码的更改如下：

```
<parent>
    <groupId>com.ghy.www</groupId>
    <artifactId>Parent1</artifactId>
```

```
    <version>4.0-RELEASE</version>
    <relativePath>../Parent1/pom.xml</relativePath>
</parent>

    <build>
</build>
```

删除<build></build>中全部配置代码的目的是使用 Parent1 中的插件配置，否则就是在 Parent2 子模块中对 Parent1 子模块中的配置进行了重定义。

对 Parent1 和 Parent2 子模块执行发布操作。

### 3. 在Web项目中引用插件

在 Web 项目中，引用如图 16-64 所示的配置代码。

```
<build>
    <plugins>
        <plugin>
            <groupId>org.apache.maven.plugins</groupId>
            <artifactId>maven-source-plugin</artifactId>
        </plugin>
    </plugins>
</build>
```

图 16-64

### 4. 发布并成功生成源代码.jar文件

对 Web 项目执行发布操作，成功生成 source 源代码.jar 文件，如图 16-65 所示。

图 16-65